System Design through MATLAB® Control Toolbox and SIMULINK®

Springer

*London
Berlin
Heidelberg
New York
Barcelona
Hong Kong
Milan
Paris
Singapore
Tokyo*

Krishna Singh and Gayatri Agnihotri

System Design through MATLAB®, Control Toolbox and SIMULINK®

Springer

Krishna K. Singh, BE, Mtech
Electrical Engineering Department, S.V. Government Polytechnic, Shamla Hills, Bhopal (MP),
PIN 462002, India

Gayatri Agnihotri, BE, Mtech, PhD
Electrical Engineering Department, M.A. College of Technology, Bhopal (MP),
PIN 462007, India

ISBN-13: 978-1-85233-337-9 Springer-Verlag London Berlin Heidelberg

British Library Cataloguing in Publication Data
Singh, Krishna Kumari
   System Design through MATLAB control toolbox and SIMULINK
   1. MATLAB (Computer file) 2. SIMULINK (Computer file)
   3. Control toolbox (Computer file) 4. Systems design
   5. Automatic control – Computer simulation
   I. Title II. Agnihotri, Gayatri
   629.8'312'0285'5369

Library of Congress Cataloging-in-Publication Data
Singh, Krishna Kumari, 1969-
    System Design through MATLAB control toolbox and SIMULINK / Krishna Kumari
   Singh and Gayatri Agnihotri.
       p. cm.
       ISBN-13: 978-1-85233-337-9   e-ISBN-13: 978-1-4471-0697-5
       DOI:10.1007/978-1-4471-0697-5

    1. System design. 2. MATLAB. 3. SIMULINK. I. Agnihotri, Gayatri, 1947- II. Title.
   QA76.9.S88 S5673 2000
   003.'.85'011353042—dc21
                                                       00-044675

MATLAB® and SIMULINK® are registered trademarks of The MathWorks Inc., http://www.mathworks.com

Typesetting: Camera ready by authors

69/3830-543210 Printed on acid-free paper SPIN 10770940

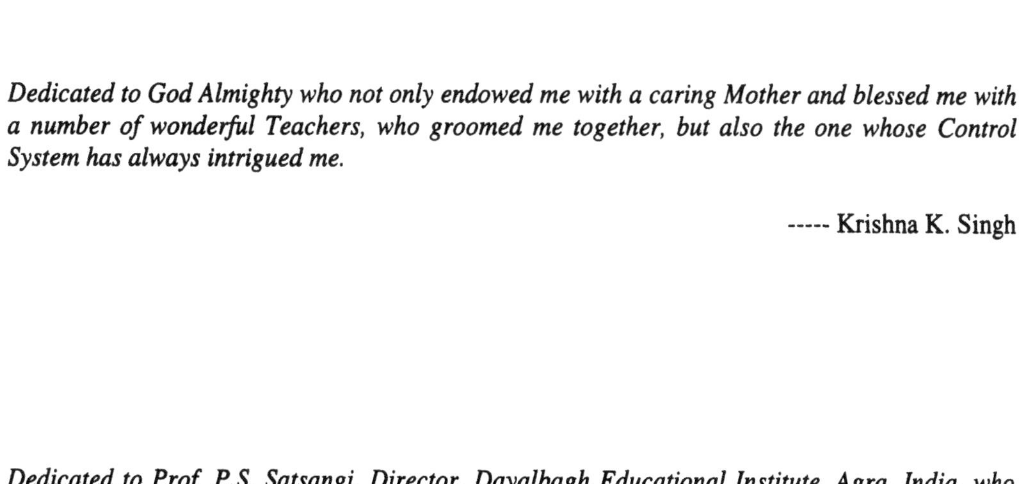

Dedicated to God Almighty who not only endowed me with a caring Mother and blessed me with a number of wonderful Teachers, who groomed me together, but also the one whose Control System has always intrigued me.

----- Krishna K. Singh

Dedicated to Prof. P.S. Satsangi, Director, Dayalbagh Educational Institute, Agra, India, who has not only been my supervisor for Ph.D. at I.I.T., Delhi, India in 1984, but also the person who instilled in me the capabilities for writing a book like this.

----- Gayatri Agnihotri

Foreword

MATLAB® has been a strong software for high-performance, numerical computations and visualisations. The fact that it provides an interactive environment with large number of built-in functions for technical computations, graphics and animations makes it an excellent tool for a very large variety of subjects starting from Linear Algebra computations, Data Analysis, Signal Processing, Optimisation, Numerical Solution of ODE's, Quadrature *etc.* and extending upto Neural Network and Fuzzy Logic. In association with SIMULINK® and Control Toolbox, it provides the scientific community with a powerful tool for the design, analysis and response study of Systems.

Going through the manuscript of this book has been a pleasant experience. The fact that in their maiden attempt, the authors have been able to introduce the intricate and subtle points of the subject in a very simple and interactive manner speaks in itself for their deep insight into the subject. Their approach towards introducing new concepts has been gradual and the inclusion of examples and exercises within the chapters will make the reader more and more confident as he/she progresses through the book. I hope that the book will serve the purpose of its readers and make them expert in using MATLAB®, Control Toolbox and SIMULINK® in design and analysis of Systems.

The book is eminently suitable for both as a textbook and reference book for the students of these novel computational tools.

Prof. P.B. Sharma,
Vice-Chancellor,
Rajiv Gandhi Proudyogiki Vishwavidyalaya,
(University of Technology of Madhya Pradesh)
Bhopal, India
19th May 2000

Preface

MATLAB®, developed by The MathWorks Inc. is fast gaining popularity in the area of simulation of systems for scientific computation. With the addition of various Toolboxes and SIMULINK®, MATLAB® proves to be a strong and indispensable tool for several specialised simulations.

The target groups of this book are:

- undergraduate students of electrical, electronics, mechanical, mechatronics, robotics and aerospace engineering undergoing their final semester of project work;
- postgraduate students for whom MATLAB® forms an integral part of the syllabus or who decide to take up simulation of Control System engineering problems as their dissertation work;
- research scholars for whom MATLAB® along with Toolboxes and SIMULINK® is an indispensable tool for simulation;
- people from research institutes and industries who wish to learn the tricks of simulating Control System problems using MATLAB® platform for their routine and/or development work.

Text presented in this book uses very simple language and explains the main and basic features of MATLAB® Control System Toolbox and SIMULINK® required for simulation of Systems under Windows environment.

The course covered by this book has been divided into nine chapters. Starting from formation of models in Chapter 1, the subject has been evolved through Chapter 9, which deals with some typical and complex applications. Illustrative examples, with gradual increase in the difficulty level on 12.5% grey background throughout the lesson along with practice problems in each chapter in deep 25% grey Practice Test box is a salient feature of this book. Problem given at the end of each chapter as an exercise not only intend to reinforce learning but also aim at helping the student evolve concepts required for simulating real life Control System problems. The text presented in this book uses a language that gives the reader a feel of personal touch close to the class room situation.

The illustrations/diagrams presented in the book have been converted to greyscale. On the waveforms legends have been provided depicting colors (*e.g.*, Y for yellow, C for cyan, G for green, M for magenta and so on.....) to make things more clear. And so, the reader does not actually ever miss the colors.

In addition, examples covered in the book that includes live simulations of various systems, are also provided at the web-site of Springer-Verlag at `ftp://ftp.springer.co.uk` with an aim to help reader not only save time but also learn with ease.

The authors feel that this book will not only serve the purpose of consolidating basic concepts of System Engineering into the mind of the readers but also help them in getting started towards using MATLAB®, Control Systems Toolbox and SIMULINK® with minimum effort and maximum amount of ease.

Constructive suggestions for improvement are what the authors seek from the readers of this book and expect them to flow in freely.

So, Happy Reading and Simulations...!

Acknowledgements

First of all we wish to express our thanks to The MathWorks Inc. for all the information and support they provided during the writing of this book. Without the help of 'The Book Program' available on their web-site, at `http://www.mathworks.com`, made especially for the writers aspiring to write on MATLAB® related topics, this book would have been a dream for us of distant future.

We also express our gratitude towards our Alma Mater, M. A. College of Technology, Bhopal, for providing all possible facilities that has gone into the making of this book. We acknowledge the efforts of Prof. R.P. Singh and Prof. A.K. Tiwari, Heads of Electronics and Computer Engineering Departments respectively, from this very Regional College of Engineering, and thank them from the bottom our hearts.

Our special thanks also goes to Dr. S.C.P. Singh, Educational Psychologist, whose consultancy we seeked time and again. His advice and direction proved to be of immense help especially while planning the difficulty level of the material incorporated in this book.

From the day that our mind conceived the idea of writing this book, Mr. Surendra Sinh, our friend has been our source of constant help and inspiration whatsoever. Despite his overwhelming responsibilities and busy schedule, he always somehow managed to find time for going through the manuscripts, correcting them and suggesting improvements. Simple words of thanks are just not sufficient to express our feelings of gratitude towards him.

In addition, we would like to thank our friends and colleagues – Prof. Vaishali Sohoni and Prof. Umesh Kumar Soni, Lecturers, Electrical Engineering, S.V. Government, Bhopal, for their constant help and constructive criticism. Our acknowledgement would be incomplete if we do not express our gratitude towards Prof. K.C. Verma, Principal, of the same Institution, for his guidance and encouragement.

Can we ever forget the contributions of Mr. Pankaj Singh and Mr. Nikesh Samaiya, our young friends who are always bubbling with energy and have wonderful head full of brilliant ideas, many of which have found place in this text.

Lastly, if it were not for the support and infinite patience of our family members, this book would have remained an uncherished dream.

We thank everyone, directly and indirectly involved in the making of this book.

Table of Contents

Introduction

Here you will find an overview of the material contained in this book

0.1 What are MATLAB®, Control System Toolbox and SIMULINK®?

MATLAB® is a software package developed by The MathWorks, Inc. for technical computations. The name stands for MATrix LABoratory. Originally, MATLAB® was developed for matrix computations in an easy, quick and interactive environment without going through the cumbersome long procedure of code writing and programming required to accomplish the same work in high level languages like FORTRAN, PASCAL or C. Later on, with addition of several Toolboxes and SIMULINK®, MATLAB® has grown into the best platform available for several specialized scientific and technical visualizations and simulations. The Control System Toolbox consists of functions specialized for System Engineering in the matrix environment of the MATLAB®. It is a collection of m-files containing algorithms to be used for modeling, analysis and design of continuous and discrete systems.

0.2 The MATLAB® Family

The family of MATLAB® is illustrated in Figure 0.1. MATLAB® in itself consists of a programming language, several built-in functions that are available to the user directly, user written functions that can be added to the MATLAB® and extra functions. It has several Toolboxes for simulating specialized problems of different areas and extensions to link up MATLAB® to other programs. SIMULINK® is a program built on top of the MATLAB® environment, which along with its specialized products enhances the power of MATLAB® for scientific computions, simulations and visualizations.

0.3 Who Should Read this Book?

This book is useful for anyone who wishes to learn and try out simulations in System Engineering using MATLAB® platform through its built-in Control System Toolbox and SIMULINK®. However, it is important to mention here, that this book does not aim at teaching the

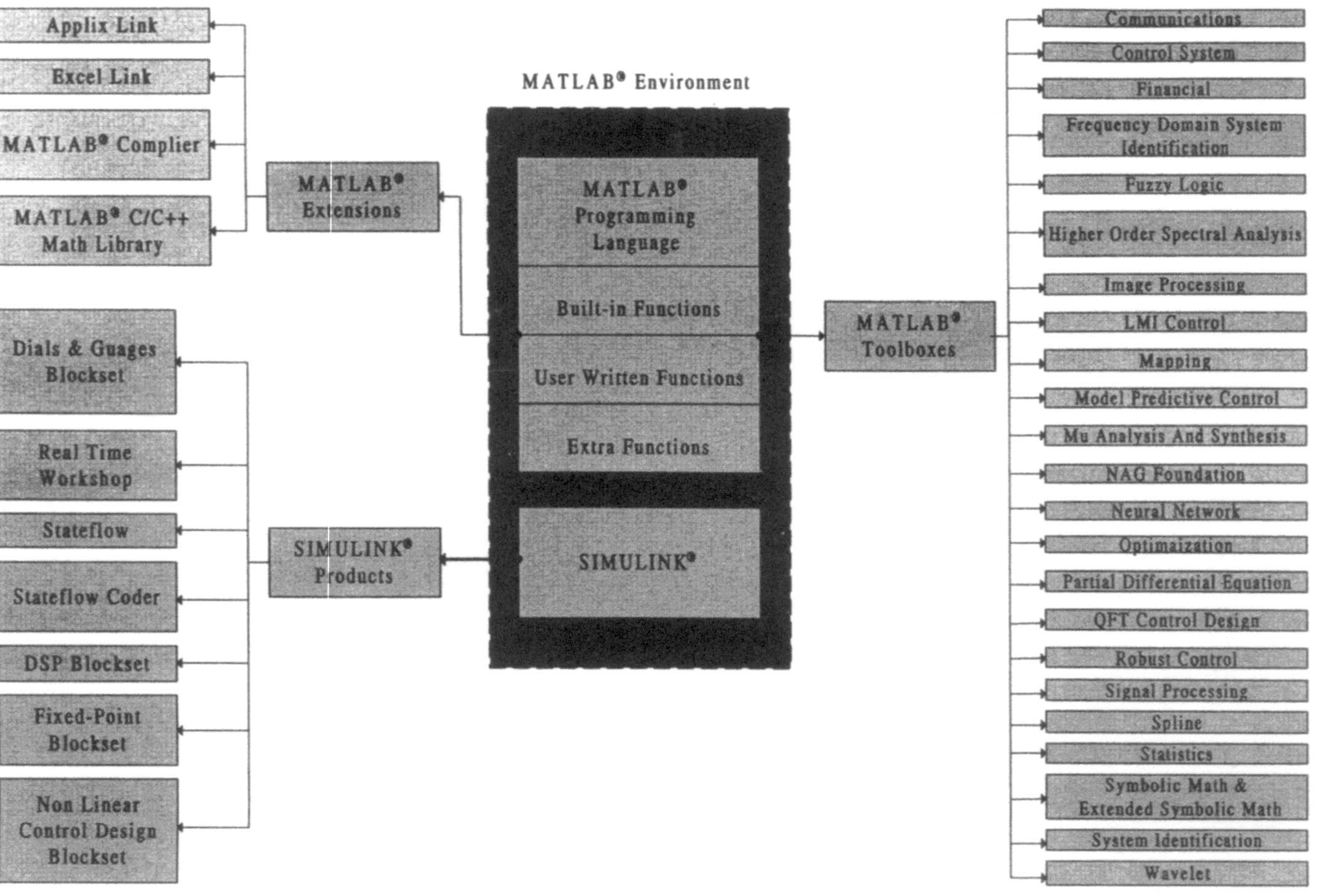

Figure 0.1. The MATLAB® family

principles of System Engineering, Control Systems, MATLAB® or even SIMULINK®. Hence, this book will disappoint you, if you wish to learn any of these. In fact, it is presumed that you already have sufficient knowledge of these. With this book, you will learn to analyze and design System Engineering problems in MATLAB® environment under Windows operating system.

0.4 What is Required to Do the Simulations

To be able to perform these simulations, the first thing you should have is a computer with monitor, keyboard, mouse or trackball and an optional printer (only in case you wish to get the printouts for the work you do). The PC should have a minimum 486 processor with math coprocessor, at least 16 MB of RAM and preloaded Windows operating system (95, 98 or NT). However, for optimum performance a Pentium, Pentium Pro or a Pentium II or III processor with 24 MB RAM or more with graphic accelerator card is recommended. You could install MATLAB® and Control System Toolbox following the instructions provided by the supplier in the Installation Manual for your platform.

In addition to the above, and as already mentioned, you should also have adequate knowledge of Windows Operating System, MATLAB® and SIMULINK® and Theoretical knowledge of Control System Engineering. This book assumes that you know how to start and run MATLAB® on your PC and invoke SIMULINK®. It also assumes that you know that for invoking Control System Toolbox functions all you have to do is to write the function with proper syntax at the command prompt of the MATLAB® command window just as you do with the MATLAB® functions. However, you can always refer to Appendix A and Appendix B for help on certain basic features of MATLAB® and SIMULINK® respectively.

0.5 Which Version is Covered?

The versions covered in this book are MATLAB® 5.3, SIMULINK® 3.0 and Control System Toolbox 4.2. But if you are using earlier versions of any of these, you may find some difference in the working environment, as a few additional functions may not be available to you. Nevertheless, the book may still be useful to you as the basic spirit of the program still remains the same i.e. help doing simulations.

0.6 What is not Covered

Apart from Control System Toolbox, MATLAB® provides other useful specialized features for Advanced System Design and Control as follows:

- LMI Control Toolbox;
- QFT Control Design Toolbox;
- Model Predictive Control Toolbox;
- Non-linear Control Design Blockset;
- Mu Analysis and Synthesis Toolbox;
- Robust Control Toolbox;

- Stateflow.

However, this book is restricted to developing only the concepts of System Engineering through the Control System Toolbox and does not explain features of the above-mentioned Advanced Control System titles.

0.7 How the Book is Organized

This book has been divided into nine chapters as follows:

Chapter 1 teaches you the concepts of system representation or modeling;
Chapter 2 teaches you model manipulation techniques;
Chapter 3 teaches you how to extract information from the models;
Chapter 4 teaches you how to perform model analysis of control systems;
Chapter 5 teaches you the use of two GUIs, the LTI Viewer, and the Root Locus Design GUI;
Chapter 6 teaches you the use of SIMULINK® for analysis and design of systems;
Chapter 7 teaches you how to design compensators for systems;
Chapter 8 teaches you to apply the skills acquired so far to some simple real life situations;
Chapter 9 teaches you to apply the skills acquired to some complex systems.

In addition to the above, following appendices have also been provided for quick reference:

Appendix A explains certain basic features of MATLAB® regarding matrix manipulations essential for system analysis and design;
Appendix B explains certain basic features of SIMULINK®;
Appendix C provides a reference to the functions of Control System Toolbox along with several relevant information pertaining to System Engineering;
Appendix D provides you with information about the contents and use of the material provided on the web-site of Springer-Verlag at `ftp://ftp.springer.co.uk`;
Appendix E provides you with umpteen number of tips useful for analysis and design of systems in question and answer form.

The concepts are elucidated with several illustrative examples in each chapter along with Practice Test boxes for you to check what you have learnt. In addition, end of chapter exercises is also given to check how much you have grasped and also to reinforce concepts developed during the chapters.

0.8 Accessory Material Available via the Internet

The material accompanying this book put on the web-site of Springer-Verlag at `ftp://ftp.springer.co.uk` has been specially prepared for you, keeping in mind that you are in the learning process. It contains:

- workspace saved as MATLAB® file with **.MAT** extension, containing resulting models generated in various chapters to cross check your results and save your time, which you may load directly before you start your learning session;

- commands used for generating various examples given in different chapters saved as MSWord 97 file with **.doc** extension, or MATLAB® m-file with **.m** extension, which you may use to troubleshoot a command that you have tried, or, simply to save your typing time spent in entering the commands at the MATLAB® command prompt using the copy and paste technique, which you as window user are quite familiar with;
- the SIMULINK® models generated in the various chapters of this book are saved as the SIMULINK® model files with **.mdl extension,** for quick and easy access to check what parameters have been set for the various blocks or even for simulations.

Complete details of the material are provided in Appendix D.

0.9 How to Use this Book

It entirely depends upon the individual, how he extracts the best out of anything. However, you can use this book for any of the following purposes:

- as a Learning Guide
- as a Review Tool
- as a Quick Reference

The   solutions   provided   on   the   web-site   of   Springer-Verlag   at `ftp://ftp.springer.co.uk` are only to help you out at the time of need or crisis. It is advised that you first try out the problems yourself. Only when you fail, you should take the help of the material provided on the web-site.

0.10 Conventions Used in this Book

The book uses the following type forms as conventions:

Times New Roman, Regular, 10 points	Text portion of the book
Times New Roman, Regular, 12 points	Introductory text for the chapters
Times New Roman, Bold, 16 points	Chapter Titles
Times New Roman Bold 14 points	Level 2 Main Headings in the chapters
Times New Roman Bold 12 points	Level 3 Main Headings in the chapters
Times New Roman Bold, Italic 10 points	Level 4 Sub-Headings in the chapters
Times New Roman Bold 10 points	MATLAB®/Control System Toolbox Commands
Times New Roman Italic 10 points	User supplied values
Tahoma Regular 8 points, 0.5 character spacing	Examples

0.11 A Word of Caution

As you progress through this book, keep in mind that the examples given to help development of concepts are mostly of fictitious system and may not necessarily represent a real system unless otherwise mentioned.

0.12 Before You Start

The activities in each chapter are designed to be followed in a sequence. You will find it difficult to comprehend certain concepts if you have casually flipped through the preceding pages. Hence, it is recommended that you devote enough time to each chapter and try to comprehend the maximum in one sitting. Do not forget to try out the illustrative examples yourself and do solve the in-between practice problems in the practice boxes and the end-of-chapter exercises. In case you get stuck pondering over something you had learnt in MATLAB®, SIMULINK® or in Control System Toolbox, which you cannot recall now, feel free to refer the appendices given at the end of the book for a quick reference.

Chapter 1

System Representation and Modeling

In this chapter, you will learn to create models for various systems and assign properties to the models so generated.

Study and analysis of any system requires some kind of its equivalent representation that describes the various components of the system and the relationship between them. This equivalent representation of the system, also known as model of the system may be of several types. The Control System Toolbox of MATLAB® supports generation of models for Linear Time-Invariant (LTI) systems in the following four main forms:

- zero/pole/gain (zpk) model of the form $\dfrac{k(s - z_1)(s - z_2)\ldots\ldots(s - z_n)}{(s - p_1)(s - p_2)\ldots\ldots(s - p_n)}$

- transfer function (tf) model of the form $\dfrac{a_n s^n + \ldots\ldots + a_2 s^2 + a_1 s + a}{b_m s^m + \ldots\ldots + b_2 s^2 + b_1 s + b}$

- state space (ss) model of the form $\begin{aligned} \dot{x} &= ax + bu \\ y &= cx + du \end{aligned}$

- frequency response data (frd) model in which response of a system at different frequencies are involved

These models can be created for a continuous-time system or a discrete-time system that may be either single-input/single-output (SISO) type or multiple-input/multiple-output (MIMO) type. Storing several systems with same properties under a single variable name as an array is also possible.

Before we learn how to create a model in MATLAB® environment using Control System Toolbox, let us first study the properties of these models.

1.1 Properties of the Models

Properties of the models represent the information attached with different fields of the models. MATLAB® refers to these properties as LTI Properties, which are subdivided into two categories:

1) Generic Properties
2) Model-specific Properties

Table 1.1. Generic properties of the LTI systems

Property Name	Description	Property Value	Purpose	Remarks
IoDelay Matrix	Transport delays between input and output	• Ny by Nu matrix where IoDelay(i, j) specifies time delay between input j and output i • Scalar for same delay for all input-output pairs	To specify time delay in both input and output channel pair	• Default value is zero • Expressed in seconds for continuous-time models • Expressed as integer multiples of sample period for discrete-time models (to obtain the delay times in seconds, multiply time delay, td by the sample time Ts of the model *i.e.*, time delay for discrete-time models = td*Ts)
InputDelay	Transport delays in the input channels	• Nu by 1 vector of time delays for each input • Scalar for same delay for each input	To specify time delay in input channel	• Default value is zero • Specify time delays in seconds for continuous-time models • Specify time delays as integers for discrete-time models (integer value V means a delay of V sampling periods *i.e.*, InputDelay for discrete-time models = V*Ts)
InputGroup	Groups of input channels	M by 2 Cell Array for M input groups	To create different groups of input channels and assign a single name to each group	• Default value 0 by 2 empty cell • Value {i, 1} is the vector of channel indices for Group 1 • Value {i, 2} is the vector of channel indices for Group 2
InputName	Names of input channels	Nu by 1 cell array of strings	To give names to the individual input channels	• Default name is empty string for all channels *i.e.*, ' ' and also for channels having no name • String for single-input models • Cell vector of strings for multiple-input (*i.e.*, MIMO and MISO) models • The number of cells is the same as the number of inputs for multiple-output (*i.e.*, MIMO and SIMO) models

Table 1.1. (continued)

Property Name	Description	Property Value	Purpose	Remarks
Notes	User notes	Cell array of strings	To store additional information	Default value is an empty string
OutputDelay	Transport delays in the output channels	• Ny by 1 vector of time delays for each output • Scalar for same delay for each output	To specify time delay in output channel	Same as for InputDelay
OutputGroup	Groups of output channels	P by 2 cell array for P output groups	To give names to the individual output channels	Same as for InputGroup
OutputName	Names of output channels	Ny by 1 cell array of strings	To create different groups of output channels and assign a single name to the group	Same as for InputName
Ts	Sample time in seconds	Positive scalar	To keep track of the sample time (in seconds) of discrete-time	Default value is zero for continuous-time systems and −1 or [ ] for discrete-time systems with sample time unspecified
User Data	Additional information of data	Arbitrary, any MATLAB® data type	To accommodate arbitrary user-supplied data	Default value is an empty string

Note:
- Nu = Number of input
- Ny = Number of outputs
- M = Number of input groups
- P = Number of output groups

1.1.1 Generic Properties

Generic properties are the properties shared by all four types of models mentioned above and supported by the Control System Toolbox. The details of these are given in Table 1.1.

1.1.2 Model-specific Properties

Other LTI properties are specific to the four types of models supported by Control System Toolbox of MATLAB®. The details of these properties are given in Table 1.2, 1.3, 1.4 and 1.5 for zpk, tf, ss and frd models respectively.

Table 1.2.   zpk-specific properties of the LTI systems

Property Name	Description	Property Value	Remarks
k	Gains for each channel	• Scalar for SISO zpk models • Ny by Nu real valued matrix for MIMO models • Array of size [Ny Nu S1.......Sp] for S1 by ...Sp arrays of zpk models	-
p	Poles for each channel	• Row vector for SISO zpk models • Ny by Nu cell array of row vectors for MIMO zpk model • Cell array of size [Ny Nu S1.......Sp] for S1 by ...Sp arrays of zpk models	• poles may be real or complex conjugate
variable	Transfer Function Variable	• String 's' and 'p' for continuous time systems • String 'z', 'q' and 'z^{-1}' for discrete time systems	• default value are 's' and 'z' for continuous and discrete systems respectively • 'p' is equivalent to 's' • ' z^{-1} ' variable name allows for DSP (Digital Signal Processing) convention for display of discrete transfer functions
z	Zeros for each channel	same as for poles	same as for poles

Note:
- Nu = Numbér of inputs
- Ny = Number of outputs

Table 1.3. tf-specific properties of the LTI systems

Property Name	Description	Property Value	Remarks
den	Denominator data	• Row vector for SISO tf models • Ny by Nu cell array of row vectors for MIMO tf models • Cell array of size [Ny Nu S1.......Sp] for S1 by ...Sp arrays of tf models	-
num	Numerator data	same as den	same as den
variable	Transfer function variable name (string)	• String 's' and 'p' for continuous time systems • String 'z', 'q' and 'z^{-1}' for discrete time systems	• default value are 's' and 'z' for continuous and discrete systems respectively • 'p' is equivalent to 's' • ' z^{-1} ' variable name allows for DSP (Digital Signal Processing) convention for display of discrete transfer functions

Note:
- Nu = Number of inputs
- Ny = Number of outputs

Table 1.4. ss-specific properties of LTI systems

Property Name	Description	Property Value	Remarks
a	State to state matrix A	• Nx by Nx real valued matrix for MIMO ss model • array of size [Nx Nx S1....Sp] for S1 by Sp arrays of ss models	system matrix
b	Input to state matrix B	• Nx by Nu real valued matrix for MIMO ss model • array of size [Nx Nu S1....Sp] for S1 by Sp arrays of ss models	input matrix
c	State to output matrix C	• Ny by Nx real valued matrix for MIMO ss model • array of size [Ny Nx S1....Sp] for S1 by Sp arrays of ss models	output matrix
d	Input to output matrix D	• Ny by Nu real valued matrix for MIMO ss model • array of size [Ny Nu S1....Sp] for S1 by Sp arrays of ss models • D=0 or [ ] meaning zero matrix of adequate dimensions.	transmission or feedthrough matrix
e	State to state descriptor state matrix, E	• Nx by Nx real valued matrix for MIMO ss model • array of size [Nx Nx S1....Sp] for S1 by Sp arrays of ss models • [ ] means identity matrix of size Nx	• Default value is 1 or [ ] for standard state space models • E must be a non-singular matrix
StateName	Name of States	• String for first-order models • Cell vector of strings for models with two or more states • Cell array of Strings of size Nx by 1	default value is empty string *i.e.,* ' '

Note:
- Nu = Number of inputs
- Ny = Number of outputs
- Nx = Number of states

Table 1.5. frd-specific properties of LTI systems

Property Name	Description	Property Value	Remarks
Frequency	Frequency data points	Nf by 1 real-valued vector	vector of real numbers of size Nf
Response data	Frequency response	Complex-valued multidimensional array	• array of size [Ny Nu Nf] for a MIMO frd model • array of size [Ny Nu Nf S1....Sp] for S1 by Sp arrays of frd model • the number of response data should be the same as the number of frequency values
Units	Units for frequency	String 'rad/s' or 'Hz'	default value is rad/sec

Note:
- Nu = Number of inputs
- Ny = Number of outputs
- Nf = Number of frequency data points

Before we learn to manipulate the above-mentioned properties of a model, let us learn how to create a model in the first place. We will learn more about the LTI properties in Chapter 3.

1.2 Creating a Model

You already know that the backbone of MATLAB® is matrices. Hence, it should be easy for you to envisage that the basis of formulation of models must also lie in matrices. Individual parameters required by different functions provided in the Control System Toolbox are entered in the form of matrices, arrays or cell arrays depending upon the dimension of the model. In the lines to follow, you will learn how to create the zero/pole/gain model, transfer function model, state space model and frequency response data model supported by Control System Toolbox of MATLAB®.

1.2.1 Zero/Pole/Gain (zpk) Model

There are several methods by which the zero/pole/gain model of a system can be generated. Consider the MIMO model indicated below. It has as many inputs as the number of columns and as many outputs as the number of rows.

$$\text{ModelResult} = \begin{bmatrix} \dfrac{k_{11}(s-z_{11a})(s-z_{11b})\cdots}{(s-p_{11a})(s-p_{11b})\cdots} & \dfrac{k_{12}(s-z_{12a})(s-z_{12b})\cdots}{(s-p_{12a})(s-p_{12b})\cdots} & \cdots \\ \dfrac{k_{21}(s-z_{21a})(s-z_{21b})\cdots}{(s-p_{21a})(s-p_{21b})\cdots} & \dfrac{k_{22}(s-z_{22a})(s-z_{22b})\cdots}{(s-p_{22a})(s-p_{22b})\cdots} & \cdots \\ \vdots & \vdots & \\ \vdots & \vdots & \cdots \end{bmatrix}$$

The syntax for obtaining this zpk model is:

- ModelResult = **zpk** ({[z_{11a} z_{11b}...] [z_{12a} z_{12b}...]...;[z_{21a} z_{22b}...] [z_{22a} z_{22b}...]...}, {[p_{11a} p_{11b}...] [p_{12a} p_{12b}...]...;[p_{21a} p_{22b}...] [p_{22a} p_{22b}...]...}, [k_{11} k_{12}...; k_{21} k_{22}...], *tsys*, 'property1', value1, 'property2', value2,..., refmodel*)

- ModelResult = **zpk** (*N*)

where,

ModelResult...................... is the user-specified name of the zpk model so generated

z_{11a}, z_{11b}............................ specifies the real or complex conjugate zeros of the model

p_{11a}, p_{11b}............................ specifies the real or complex conjugate poles of the model

k_1, k_2................................. specifies the real valued gains of the model

tsys .... specifies the sample time for discrete system (optional) (if undetermined, put tsys=-1 or [ ])

property1, *property2*........ specifies the names of the properties which may be generic or zpk model-specific (optional)

value1, *value2* .... specifies the values of the properties mentioned in *property1*, *property2*...

refmodel............................ is the name of the model whose properties are to be inherited by the new zpk model ModelResult generated (optional)

N.. is a scalar or matrix such that ModelResult is a static gain *N* zpk model

Note that for a SISO system, zeros and poles are vectors while gain is scalar. While for a MIMO system, with Nu inputs and Ny outputs, zeros and poles are Ny by Nu cell arrays in which there are as many rows as outputs and as many columns as inputs, such that $z(i, j)$ and $p(i, j)$ represent respectively the vectors of zeros and poles of the zpk model from input j to output i and $k(i, j)$ is then the scalar gain of zpk model from input j to output i.

Examples:

Let the zero, pole, gain and sample time parameters of systemA1 and systemA2 be specified as given below:

```
zsysA1=[-1];
zsysA2=[2 0.4+i  0.4-i];
psysA1=[3];
psysA2=[6  -8+0.5i  -8-0.5i];
ksysA1=5;
ksysA2=-20;
tsys=0.1;
```

With the data given above, we can obtain a number of models with different combinations of zeros, poles, gains and sample time of the two systems as follows:

```
modelA1 = zpk (zsysA1, psysA1, ksysA1)
returns
```

$$\text{modelA1} = \left[\frac{5(s+1)}{(s-3)} \right]$$

Zero/pole/gain:

```
5 (s+1)
-------
 (s-3)
```

```
modelA2 = zpk( zsysA1, psysA1, ksysA1, tsys)
returns

Zero/pole/gain:
5 (z+1)
-------
 (z-3)

Sampling time: 0.1
```

$$\text{modelA2} = \left[\frac{5(z+1)}{(z-3)} \right]$$

Sampling Time:0.1

```
modelA3 = zpk (zsysA1, psysA2, ksysA1, tsys)
returns

Zero/pole/gain:
        5 (z+1)
-------------------------
(z-6) (z^2 + 16z + 64.25)

Sampling time: 0.1
```

$$\text{modelA3} = \frac{5(z+1)}{(z-6)(z^2+16z+64.25)}$$

Sampling time: 0.1

```
modelA4 = zpk ({[zsysA1] [zsysA2]}, {[psysA1] [psysA2]}, [ksysA1 ksysA2])
returns

Zero/pole/gain from input 1 to output:
5 (s+1)
-------
 (s-3)

Zero/pole/gain from input 2 to output:
-20 (s-2) (s^2 - 0.8s + 1.16)
-----------------------------
 (s-6) (s^2 + 16s + 64.25)
```

$$\text{modelA4} = \left[\frac{5(s+1)}{(s-3)} \quad \frac{-20(s-2)(s^2-0.8s+1.16)}{(s-6)(s^2+16s+64.25)} \right]$$

```
modelA5 = zpk ({[zsysA1]; [zsysA2]}, {[psysA1]; [psysA2]}, [ksysA1; ksysA2], tsys)
returns

Zero/pole/gain from input to output...

       5 (z+1)
#1:  -------
        (z-3)

     -20 (z-2) (z^2 - 0.8z + 1.16)
#2:  -----------------------------
      (z-6) (z^2 + 16z + 64.25)

Sampling time: 0.1
```

$$\text{modelA5} = \left[\begin{array}{c} \dfrac{5(z+1)}{(z-3)} \\[2mm] \dfrac{-20(z-2)(z^2-0.8z+1.16)}{(z-6)(z^2+16z+64.25)} \end{array} \right]$$

Sampling Time: 0.1

Instead of first specifying the values of the zero, pole and gain matrices and then subsequently proceeding ahead to obtain a zpk model, you can obtain the zpk model using direct values of the zeros, poles and the gains and yet obtain the same result. This is illustrated in the forthcoming examples.

```
modelA6 = zpk (-1, 3, 5, 0.1)
returns

Zero/pole/gain:
5 (z+1)
-------
 (z-3)

Sampling time: 0.1
```

modelA6 =
same as modelA2

```
modelA7 = zpk([-1], [6  -8+0.5i  -8-0.5i], 5, 0.1)
returns

Zero/pole/gain:

        5 (z+1)
---------------------------
(z-6) (z^2 + 16z + 64.25)

Sampling time: 0.1
```

modelA7 =
same as modelA3

```
modelA8 = zpk ({[-1]; [2 0.4+i 0.4-i]}, {[3]; [6 -8+0.5i -8-0.5i]}, [5; -20], 0.1)
returns

Zero/pole/gain from input to output...
        5 (z+1)
#1:  --------
       (z-3)

      -20 (z-2) (z^2 - 0.8z + 1.16)
#2:  ---------------------------------
      (z-6) (z^2 + 16z + 64.25)

Sampling time: 0.1
```

modelA8 =
same as modelA5

```
modelA9 = zpk (5)
returns

Zero/pole/gain:
5
```

modelA9 = $\begin{bmatrix} 5 \end{bmatrix}$

a static gain matrix zpk model

In the examples above, the properties of the models were left unaltered from their default values. In the examples to follow, you will learn how to assign and manipulate the properties of the zpk models created above. Remember that zpk model has all of the ten generic properties and the four model-specific properties as listed in Table 1.1 and 1.2 respectively. Any of the properties mentioned in these tables can be changed.

Examples:

Let us first start with modelA5 and change the variable to q, the default InputName to 'Parameter1', OutputName to 'Result1 and Result2' respectively.

```
modelA10 = zpk ({[zsysA1]; [zsysA2]}, {[psysA1]; [psysA2]}, [ksysA1; ksysA2], tsys,
InputName', 'Parameter1', 'OutputName', {'Result1'; 'Result2'}, 'Variable', 'q')
returns
```

Zero/pole/gain from input "Parameter1" to output...

```
            5 (1 + q)
  Result1:  ---------
            (1 - 3q)


          -20 (1 - 2q) (1 - 0.8q + 1.16q^2)
  Result2: ------------------------------------
           (1 - 6q) (1 + 16q + 64.25q^2)
```

Sampling time: 0.1

> modelA10 =
> compare with modelA5

Properties can also be inherited from other models of similar type as indicated below.

```
modelA11 = zpk ({[-5]; [-4 0.3+7i 0.3-7i]}, {[2]; [9 1+0.5i 1-0.5i]}, [-10; 0.01], modelA10)
returns
```

Zero/pole/gain from input "Parameter1" to output...

```
            -10 (z+5)
  Result1:  ---------
            (z-2)


          0.01 (z+4) (z^2 - 0.6z + 49.09)
  Result2: -------------------------------
           (z-9) (z^2 - 2z + 1.25)
```

Sampling time: 0.1

$$\mathrm{modelA11} = \begin{bmatrix} \dfrac{-10(z+5)}{(z-2)} \\ \dfrac{0.01(z+4)(z^2 - 0.6z + 49.09)}{(z-9)(z^2 - 2z + 1.25)} \end{bmatrix}$$

Sampling Time: 0.1

ModelA11 has inherited all of the properties from modelA10 except for the variable, which is not inherited and remains unchanged unless otherwise specified

Let us create another model with one input and five outputs and try changing the InputName to 'Fuel', OutputName to 'a, b, c, d and e', OutputGroup of first 2 channels to 'Heat', the following two channels to 'Pressure', the last one to 'Illumination', the variable to 'p', Notes to 'A fictitious system', UserData to '100 tons of coal per day'. Consider InputDelay as '0.1 seconds' and OutputDelay as 'from 0.1 to 1 seconds in increments of 0.2 seconds':

```
modelA12 = zpk ({[5 10]; [0.2 0.4+i 0.4-i]; [1.2 2.2 3.2]; [0.45 0.9 0.6]; [1.6 7.2]}, {[3]; [6 -
9+0.5i -9-0.5i]; [0.5 1.5 2.5]; [ 1 3 6 ]; [4.8 6.4]},  [5;10;20;50;100], 'InputName', 'Fuel',
'OutputName', {'a','b','c','d','e'}, 'OutputGroup', {[1 2] 'Heat'; [3 4] 'Pressure'; [5]
'Illumination'}, 'InputDelay', 0.1, 'OutputDelay', [0.1:0.2:1] , 'var', 'p', 'Notes', 'A fictitious
system', 'UserData', '100 tons of coal per day')
returns
```

Zero/pole/gain from input "Fuel" to output...

```
                  5 (p-5) (p-10)
  a:  exp(-0.2*p) * ---------------
                       (p-3)


                  10 (p-0.2) (p^2 - 0.8p + 1.16)
  b:  exp(-0.4*p) * -------------------------------
                     (p-6) (p^2 + 18p + 81.25)
```

$$\text{c:} \quad \exp(-0.6*p) * \frac{20\ (p-1.2)\ (p-2.2)\ (p-3.2)}{(p-0.5)\ (p-1.5)\ (p-2.5)}$$

$$\text{d:} \quad \exp(-0.8*p) * \frac{50\ (p-0.45)\ (p-0.6)\ (p-0.9)}{(p-1)\ (p-3)\ (p-6)}$$

$$\text{e:} \quad \exp(-1*p) * \frac{100\ (p-1.6)\ (p-7.2)}{(p-4.8)\ (p-6.4)}$$

I/O groups:

Group name	I/O	Channel(s)
Heat	O	1,2
Pressure	O	3,4
Illumination	O	5

$$\text{modelA12} = \begin{bmatrix} e^{-0.2p}\ \dfrac{5(p-5)(p-10)}{(p-3)^2} \\[2ex] e^{-0.4p}\ \dfrac{10(p-0.2)(p^2-0.8p+1.16)}{(p-6)(p^2+18p+81.25)} \\[2ex] e^{-0.6p}\ \dfrac{20(p-1.2)(p-2.2)(p-3.2)}{(p-0.5)(p-1.5)(p-2.5)} \\[2ex] e^{-0.8p}\ \dfrac{50(p-0.45)(p-0.6)(p-0.9)}{(p-1)(p-3)(p-6)} \\[2ex] e^{-1.0p}\ \dfrac{100(p-1.6)(p-7.2)}{(p-4.8)(p-6.4)} \end{bmatrix}$$

Let us again create a zpk model with five inputs and five outputs with InputName as 'a, b, c, d, e' and OutputName as 'M, N, O, P, Q,' InputDelay of '1 to 5' for each input respectively, OutputDelay of '1 to 10 in steps of 2' for each output respectively, Sampling Time as '0.5', InputGroup for channels 1 and 2 as 'Parameter12', for channel 3 and 5 as 'Parameter35' while for channel 4 as 'Parameter4'. We shall start by creating the zero, pole and gain matrix separately as follows:

z1={0.11 0.12 0.13 0.14 0.15; 0.21 0.22 0.23 0.24 0.25; 0.31 0.32 0.33 0.34 0.35; 0.41 0.42 0.43 0.44 0.45; 0.51 0.52 0.53 0.54 0.55};

p1={[1;1] [1;2] [1;3] [1;4] [1;5]; [2;1] [2;2] [2;3] [2;4] [2;5]; [3;1] [3;2] [3;3] [3;4] [3;5]; [4;1] [4;2] [4;3] [4;4] [4;5]; [5;1] [5;2] [5;3] [5;4] [5;5]};

k1=[11 12 13 14 15; 21 22 23 24 25; 31 32 33 34 35; 41 42 43 44 45; 51 52 53 54 55];

$$\text{modelA13} = \begin{bmatrix} z^{-2}\dfrac{11(z-0.11)}{(z-1)(z-1)} & z^{-3}\dfrac{12(z-0.12)}{(z-1)(z-2)} & z^{-4}\dfrac{13(z-0.13)}{(z-1)(z-3)} & z^{-5}\dfrac{14(z-0.14)}{(z-1)(z-4)} & z^{-6}\dfrac{15(z-0.15)}{(z-1)(z-5)} \\[2ex] z^{-4}\dfrac{21(z-0.21)}{(z-2)(z-1)} & z^{-5}\dfrac{22(z-0.22)}{(z-2)(z-2)} & z^{-6}\dfrac{23(z-0.23)}{(z-2)(z-3)} & z^{-7}\dfrac{24(z-0.24)}{(z-2)(z-4)} & z^{-8}\dfrac{25(z-0.25)}{(z-2)(z-5)} \\[2ex] z^{-6}\dfrac{31(z-0.31)}{(z-3)(z-1)} & z^{-7}\dfrac{32(z-0.32)}{(z-3)(z-2)} & z^{-8}\dfrac{33(z-0.33)}{(z-3)(z-3)} & z^{-9}\dfrac{34(z-0.34)}{(z-3)(z-4)} & z^{-10}\dfrac{35(z-0.35)}{(z-3)(z-5)} \\[2ex] z^{-8}\dfrac{41(z-0.41)}{(z-4)(z-1)} & z^{-9}\dfrac{42(z-0.42)}{(z-4)(z-2)} & z^{-10}\dfrac{43(z-0.43)}{(z-4)(z-3)} & z^{-11}\dfrac{44(z-0.44)}{(z-4)(z-4)} & z^{-12}\dfrac{45(z-0.45)}{(z-4)(z-5)} \\[2ex] z^{-10}\dfrac{51(z-0.51)}{(z-5)(z-1)} & z^{-11}\dfrac{52(z-0.52)}{(z-5)(z-2)} & z^{-12}\dfrac{53(z-0.53)}{(z-5)(z-3)} & z^{-13}\dfrac{45(z-0.45)}{(z-4)(z-5)} & z^{-14}\dfrac{55(z-0.55)}{(z-5)(z-5)} \end{bmatrix}$$

modelA13 = zpk (z1, p1, k1, 'InputGroup', {[1 2] 'Parameter12'; [3 5] 'Parameter35'; [4] Parameter4'}, 'InputName', {'a';'b';'c';'d';'e'}, 'OutputName', {'M'; 'N'; 'O'; 'P'; 'Q'}, 'Ts', 0.5, InputDelay', [1:5], 'OutputDelay', [1:2:10])
returns

Zero/pole/gain from input "a" to output...

$$M: \quad z^{(-2)} * \frac{11\,(z-0.11)}{(z-1)^2}$$

$$N: \quad z^{(-4)} * \frac{21\,(z-0.21)}{(z-2)\,(z-1)}$$

$$O: \quad z^{(-6)} * \frac{31\,(z-0.31)}{(z-3)\,(z-1)}$$

$$P: \quad z^{(-8)} * \frac{41\,(z-0.41)}{(z-4)\,(z-1)}$$

$$Q: \quad z^{(-10)} * \frac{51\,(z-0.51)}{(z-5)\,(z-1)}$$

Zero/pole/gain from input "b" to output...

$$M: \quad z^{(-3)} * \frac{12\,(z-0.12)}{(z-1)\,(z-2)}$$

$$N: \quad z^{(-5)} * \frac{22\,(z-0.22)}{(z-2)^2}$$

$$O: \quad z^{(-7)} * \frac{32\,(z-0.32)}{(z-3)\,(z-2)}$$

$$P: \quad z^{(-9)} * \frac{42\,(z-0.42)}{(z-4)\,(z-2)}$$

$$Q: \quad z^{(-11)} * \frac{52\,(z-0.52)}{(z-5)\,(z-2)}$$

Zero/pole/gain from input "c" to output...

$$M: \quad z^{(-4)} * \frac{13\,(z-0.13)}{(z-1)\,(z-3)}$$

$$N: \quad z^{(-6)} * \frac{23\,(z-0.23)}{(z-2)\,(z-3)}$$

$$O: \quad z^{(-8)} * \frac{33\,(z-0.33)}{(z-3)^2}$$

$$P: \quad z^{(-10)} * \frac{43\,(z-0.43)}{(z-4)\,(z-3)}$$

```
              53 (z-0.53)
Q:  z^(-12) * -----------
              (z-5) (z-3)
```

Zero/pole/gain from input "d" to output...
```
              14 (z-0.14)
M:  z^(-5) * -----------
             (z-1) (z-4)
```

```
              24 (z-0.24)
N:  z^(-7) * -----------
             (z-2) (z-4)
```

```
              34 (z-0.34)
O:  z^(-9) * -----------
             (z-3) (z-4)
```

```
               44 (z-0.44)
P:  z^(-11) * -----------
              (z-4)^2
```

```
               54 (z-0.54)
Q:  z^(-13) * -----------
              (z-5) (z-4)
```

Zero/pole/gain from input "e" to output...
```
              15 (z-0.15)
M:  z^(-6) * -----------
             (z-1) (z-5)
```

```
              25 (z-0.25)
N:  z^(-8) * -----------
             (z-2) (z-5)
```

```
               35 (z-0.35)
O:  z^(-10) * -----------
              (z-3) (z-5)
```

```
               45 (z-0.45)
P:  z^(-12) * -----------
              (z-4) (z-5)
```

```
               55 (z-0.55)
Q:  z^(-14) * -----------
              (z-5)^2
```

```
I/O groups:
    Group name     I/O   Channel(s)
    Parameter12    I       1,2
    Parameter35    I       3,5
    Parameter 4    I       4
```

Sampling time: 0.5

Is it not quite boring to specify the matrices for each of the z, p and k parameters?

"Could not there be a direct method to enter the model as a rational expression, exactly in the same way as it appears in the books or we write down in our notebooks, without bothering how to form matrices for zeros and poles and what not? After all computers are for making things easy and this is getting all too complicated!"

If this is what you feel, then here is a solution to your problem. You can definitely specify the zpk model as a rational expression in 's' or 'z' as the case may be respectively of a continuous or a discrete system. All you have to do is to specify

s = **zpk** ('s')
or,
z = **zpk** ('z', tsys)

at the command prompt of MATLAB® command window, press the return key and then go ahead entering the rational expression of the zpk model in the usual way. Examples given below should make things more clear.

Example:

First specify the variable 's'.

```
s=zpk('s');
```

Now, you can enter the model directly as a rational expression.

```
modelA14=[0.01*(s+7)/(s-0.3)      0.5*(s-0.9)/((s-0.1)*(s-0.4));      10*(s+0.2)/(s-6)
2*(s+0.5)/((s-0.1-i)*(s-0.1+i))]
returns
```

```
Zero/pole/gain from input 1 to output...
      0.01 (s+7)
#1:  -----------
       (s-0.3)

       10 (s+0.2)
#2:  -----------
       (s-6)
```

$$modelA14 = \begin{bmatrix} \dfrac{0.01(s+7)}{(s-0.3)} & \dfrac{10(s+0.2)}{(s-6)} \\ \dfrac{0.5(s-0.9)}{(s-0.1)(s-0.4)} & \dfrac{2(s+0.5)}{(s^2-0.2s+1.01)} \end{bmatrix}$$

```
Zero/pole/gain from input 2 to output...

        0.5 (s-0.9)
#1:  -----------------
      (s-0.1) (s-0.4)

        2 (s+0.5)
#2:  -------------------
      (s^2 - 0.2s + 1.01)
```

You can extend similar treatment to discrete-time models too by first specifying the variable 'z'.

```
z=zpk('z',0.5);
```

```
modelA15=[1/(z+0.3)  0.5*(z-0.2)/(z-0.1)*(z+0.4) ; 10/(z-6)  2*(z+0.5)/((z-0.1+i)*(z-0.1-i))]
```

returns

Zero/pole/gain from input 1 to output...
```
         1
 #1:  -------
       (z+0.3)
```

```
         10
 #2:  -----
       (z-6)
```

Zero/pole/gain from input 2 to output...
```
      0.5 (z-0.2) (z+0.4)
 #1:  -------------------
           (z-0.1)
```

```
          2 (z+0.5)
 #2:  -------------------
      (z^2 - 0.2z + 1.01)
```

Sampling time: 0.5

Properties can be assigned to the models so generated in the same way as was done earlier.

Zero/pole/gain LTI model arrays can also be obtained by specifying array dimensions for z, p and k variables. Let us consider a system as shown in Figure 1.1 and obtain a zpk array model for the same.

```
z={[111] [121]; [211] [221]; [311] [321]};
z(:,:,2)={[112] [122]; [212] [222]; [312] [322]};
z(:,:,3)={[113] [123]; [213] [223]; [313] [323]};
z(:,:,4)={[114] [124]; [214] [224]; [314] [324]};

p={[1.11 11.1] [1.21 12.1]; [2.11 21.1] [2.21 22.1]; [3.11 31.1] [3.21 32.1]};
p(:,:,2)={[1.12 11.2] [1.22 12.2]; [2.12 21.2] [2.22 22.2]; [3.12 31.2] [3.22 32.2]};
p(:,:,3)={[1.13 11.3] [1.23 12.3]; [2.13 21.3] [2.23 22.3]; [3.13 31.3] [3.23 32.3]};
p(:,:,4)={[1.14 11.4] [1.24 12.4]; [2.14 21.4] [2.24 22.4]; [3.14 31.4] [3.24 32.4]};

k=[111 121; 211 221; 311 321];
k(:,:,2)=[112 122; 212 222; 312 322];
k(:,:,3)=[113 123; 213 223; 313 323];
k(:,:,4)=[114 124; 214 224; 314 324];

modelA16(:,:,1)=zpk(z(:,:,1),p(:,:,1),k(:,:,1));
modelA16(:,:,2)=zpk(z(:,:,2),p(:,:,2),k(:,:,2));
modelA16(:,:,3)=zpk(z(:,:,3),p(:,:,3),k(:,:,3));
modelA16(:,:,4)=zpk(z(:,:,4),p(:,:,4),k(:,:,4));

modelA16
```
returns

```
Model modelA16(:,:,1,1)
=========================
```

   Zero/pole/gain from input 1 to output...

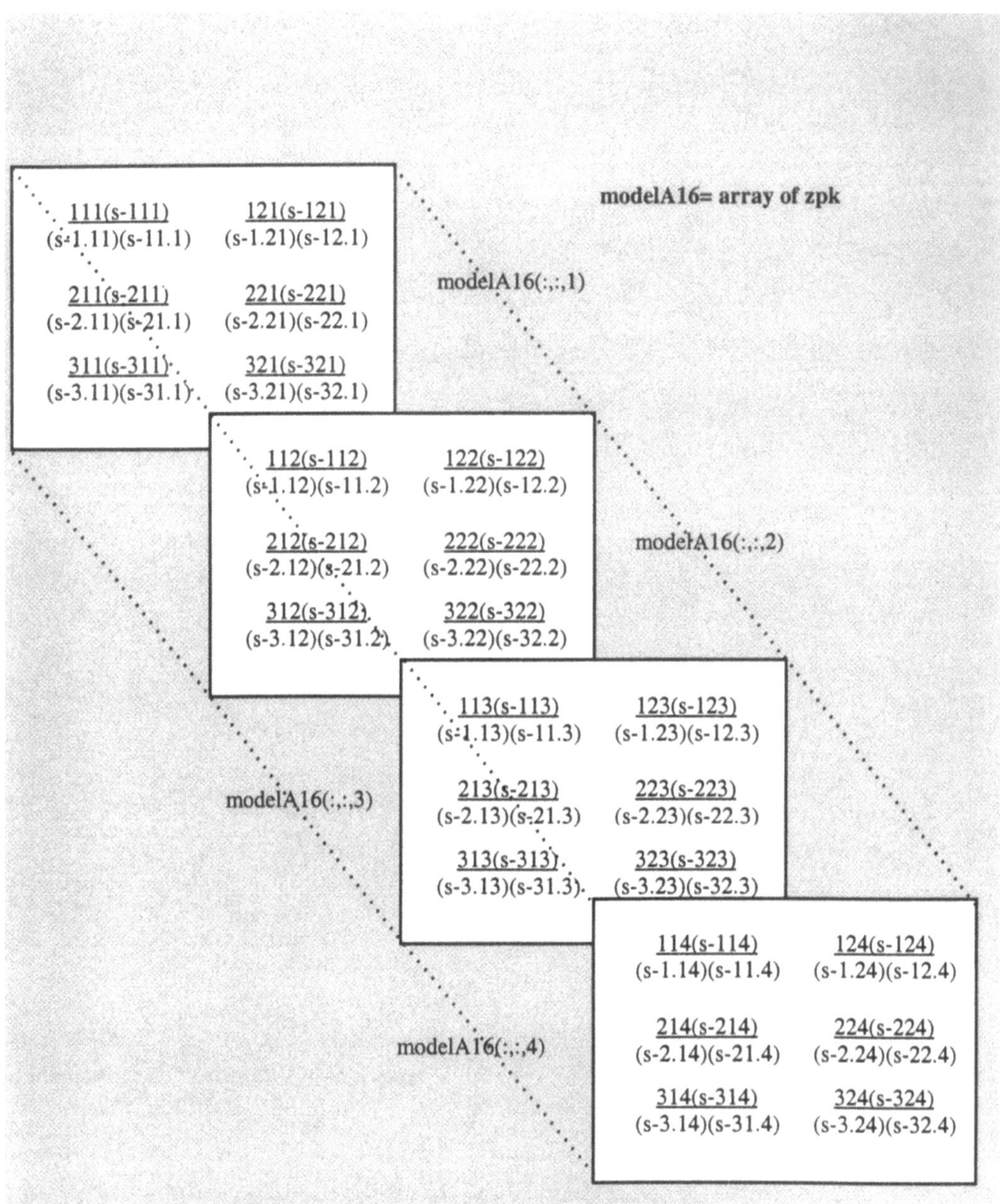

Figure 1.1. zpk array model

```
           111 (s-111)
 #1:  ------------------
       (s-1.11) (s-11.1)


           211 (s-211)
 #2:  ------------------
       (s-2.11) (s-21.1)


            311 (s-311)
 #3:  ------------------
       (s-3.11) (s-31.1)


Zero/pole/gain from input 2 to output...
           121 (s-121)
 #1:  ------------------
       (s-1.21) (s-12.1)


           221 (s-221)
 #2:  ------------------
       (s-2.21) (s-22.1)


           321 (s-321)
 #3:  ------------------
       (s-3.21) (s-32.1)


Model modelA16(:,:,2,1)
========================

Zero/pole/gain from input 1 to output...
           112 (s-112)
 #1:  ------------------
       (s-1.12) (s-11.2)


           212 (s-212)
 #2:  ------------------
       (s-2.12) (s-21.2)


           312 (s-312)
 #3:  ------------------
       (s-3.12) (s-31.2)


Zero/pole/gain from input 2 to output...
           122 (s-122)
 #1:  ------------------
       (s-1.22) (s-12.2)


           222 (s-222)
 #2:  ------------------
       (s-2.22) (s-22.2)


           322 (s-322)
 #3:  ------------------
       (s-3.22) (s-32.2)


Model modelA16(:,:,3,1)
========================
```

```
Zero/pole/gain from input 1 to output...
        113 (s-113)
 #1:  -----------------
      (s-1.13) (s-11.3)


        213 (s-213)
 #2:  -----------------
      (s-2.13) (s-21.3)


        313 (s-313)
 #3:  -----------------
      (s-3.13) (s-31.3)

Zero/pole/gain from input 2 to output
        123 (s-123)
 #1:  -----------------
      (s-1.23) (s-12.3)


        223 (s-223)
 #2:  -----------------
      (s-2.23) (s-22.3)


        323 (s-323)
 #3:  -----------------
      (s-3.23) (s-32.3)

Model modelA16(:,:,4,1)
=======================

Zero/pole/gain from input 1 to output...
        114 (s-114)
 #1:  -----------------
      (s-1.14) (s-11.4)


        214 (s-214)
 #2:  -----------------
      (s-2.14) (s-21.4)


        314 (s-314)
 #3:  -----------------
      (s-3.14) (s-31.4)

Zero/pole/gain from input 2 to output...
        124 (s-124)
 #1:  -----------------
      (s-1.24) (s-12.4)


        224 (s-224)
 #2:  -----------------
      (s-2.24) (s-22.4)


        324 (s-324)
 #3:  -----------------
      (s-3.24) (s-32.4)

4x1 array of continuous-time zero-pole-gain models.
```

With these sixteen zpk models of various systems, we are sure that you can now go ahead developing a zpk model successfully for any desired system. Then, why don't you test yourself? Take up the practice test given below in the practice box…

Practice Test 1.1.

1. Create the following zpk models by separately specifying the z, p and k matrices:

(a)

$$\text{modelAp1} = \left[\frac{10(s+1)(s-1)}{(s+3)(s-5)(s-7)} \right]$$

(b)

$$\text{modelAp2} = \left[\frac{10z(z+5)}{(z+4+2i)(z+4-2i)} \right]$$

Sampling Time=0.5

(c)

$$\text{modelAp3} = \left[\begin{array}{c} \dfrac{(z+1)(z+0.2)}{(z+5)(z+0.4i)(z-0.4i)} \\ \dfrac{10z(z-1)}{(z+2)(z+5)} \end{array} \right]$$

Sampling Time = 0.1

(d)

$$\text{modelAp4} = \left[\begin{array}{cc} \dfrac{10(s+1)}{(s+4)(s-3)} & \dfrac{20(s+1)(s-4i)(s+4i)}{s(s+2)(s-3)(s+5)} \\ \dfrac{5(s+0.1)}{(s+3)(s+7)(s+9)} & \dfrac{15(s+8)(s+7)}{s(s+2+2i)(s+2-2i)} \end{array} \right]$$

InputName = a, b

OutputName = r, s

Variable = p

InputDelay = 1

OutputDelay = 2

InputGroup = Channel Input1

OutputGroup = Channel Output2

Notes = This is a practice problem

UserData = Created by me!

2. Create the above models directly (*i.e.*, without specifying separate matrices for any of the z, p or k parameters).

3. Create the above models directly as a rational expression in the variable 's' and 'z'.

4. Change the variable for:
- modelAp1 of question 1(a) to 'p'
- modelAp2 of question 1(b) to 'q'
- modelAp3 of question 1(c) to 'z^{-1}'

5. Create a zpk array modelAp5 having dimensions 2 by 2 by 2. You can choose the values of z, p and k parameters on your own.

1.2.2 Transfer Function (tf) Model

Like the zpk model, the Control System Toolbox of MATLAB® also provides function for creation of transfer function model of a system. Consider the MIMO tf model shown below. It has as many inputs as the number of columns and as many outputs as the number of rows.

$$
\text{ModelResult} = \begin{bmatrix}
\dfrac{a_{11n}s^n + \ldots + a_{112}s^2 + a_{111}s + a_{11}}{b_{11m}s^m + \ldots + b_{112}s^2 + b_{111}s + b_{11} \cdot} & \dfrac{a_{12n}s^n + \ldots + a_{122}s^2 + a_{121}s + a_{12}}{b_{12m}s^m + \ldots + b_{122}s^2 + b_{121}s + b_{12} \cdot} & \ldots \\
\dfrac{a_{21n}s^n + \ldots + a_{212}s^2 + a_{211}s + a_{21}}{b_{21m}s^m + \ldots + b_{212}s^2 + b_{211}s + b_{21} \cdot} & \dfrac{a_{22n}s^n + \ldots + a_{222}s^2 + a_{221}s + a_{22}}{b_{22m}s^m + \ldots + b_{222}s^2 + b_{221}s + b_{22} \cdot} & \ldots \\
\vdots & \vdots & \\
\vdots & \vdots & \ldots
\end{bmatrix}
$$

The syntax for obtaining this transfer function model is:

- ModelResult = **tf**({[$a_{11n}\ldots a_{112}\ a_{111}\ a_{11}$] [$a_{12n}\ldots a_{122}\ a_{121}\ a_{12}$]...; [$a_{21n}\ldots a_{212}\ a_{211}\ a_{21}$] [$a_{22n}\ldots a_{222}\ a_{221}\ a_{22}$...]}, {[$b_{11m}\ldots b_{112}\ b_{111}\ b_{11}$] [$b_{12m}\ldots b_{122}\ b_{121}\ b_{12}$]...; [$b_{21m}\ldots b_{212}\ b_{211}\ b_{21}$] [$b_{22m}\ldots b_{222}\ b_{221}\ b_{22}$...]}, *tsys*, *'property1'*, *value1*, *'property2'*, *value2*,..., *refmodel*)

- ModelResult = **tf** (*N*)

The **tf** function used above gives the result in descending powers of 's' unless specified otherwise. However, the DSP oriented specification of discrete transfer function requires transfer function to be expressed in ascending power of 'z^{-1}'. This can be obtained directly using the **filt** function as follows:

- ModelResult = **filt** ({[$a_{11n}\ldots a_{112}\ a_{111}\ a_{11}$] [$a_{12n}\ldots a_{122}\ a_{121}\ a_{12}$]...; [$a_{21n}\ldots a_{212}\ a_{211}\ a_{21}$] [$a_{22n}\ldots a_{222}\ a_{221}\ a_{22}$...]}, {[$b_{11m}\ldots b_{112}\ b_{111}\ b_{11}$] [$b_{12m}\ldots b_{122}\ b_{121}\ b_{12}$]...; [$b_{21m}\ldots b_{212}\ b_{211}\ b_{21}$] [$b_{22m}\ldots b_{222}\ b_{221}\ b_{22}$...]}, *tsys*, *'property1'*, *value1*, *'property2'*, *value2*,..., *refmodel*)

- ModelResult = **filt** (*N*)

where,

ModelResult......................	is the user-specified name of the tf model so generated
$a_{11n} \ldots a_{112}\, a_{111}\, a_{11}$	specifies the numerator coefficients of the model ordered in the descending powers of 's' and 'p' or the ascending powers of 'q' and ' z^{-1} '
$b_{11m} \ldots b_{112}\, b_{111}\, b_{11}$............	specifies the denominator coefficients of the model ordered in the descending powers of 's' and 'p' or the ascending powers of 'q' and ' z^{-1} '
tsys	specifies the sample time for discrete system (optional) (if undetermined, put tsys=-1 or [ ])
property1, property2........	specifies the names of the properties which may be generic or zpk model-specific (optional)
value1, value2	specifies the values of the properties mentioned in *property1, property2...*
refmodel	is the name of model whose properties are to be inherited by the new tf model ModelResult so generated (optional)
N...	is a scalar or matrix such that ModelResult is a static gain N tf model

Note that for a SISO system, numerator and denominator coefficients are row vectors in descending powers of s and z or ascending powers of q and z^{-1} as specified for continuous- and discrete-time systems respectively. While for a MIMO system, with Nu inputs and Ny outputs, numerator and denominator coefficients are Ny by Nu cell arrays of row vectors in which there are as many rows as outputs and as many columns as inputs such that *Num(i,j)* and *Den(i,j)* respectively would represent the vectors of coefficients of numerator and denominator of the transfer function model from input *j* to output *i*.

Examples:

Let there be two systems, whose numerator and denominator parameters are specified as given below:

```
nsysA1 = [1 -3];
nsysA2 = [-2 4];
dsysA1 = [5 7 -9];
dsysA2 = [0 -6 8];
tsys = 0.2;
```

With the above data, we can obtain a number of models with different combinations of the numerator, denominator and the sample time of the two systems as follows:

```
modelA17 = tf (nsysA1, dsysA1)
returns
```

$$modelA17 = \left[\frac{(s-3)}{(5s^2 + 7s - 9)} \right]$$

```
Transfer function:
  s - 3
-----------------
5 s^2 + 7 s - 9
```

```
modelA18 = tf (nsysA2, dsysA2)
returns
```

```
Transfer function:
2 s - 4
-------
6 s - 8
```

$$\text{modelA18} = \frac{2s - 4}{6s - 8}$$

```
modelA19 = tf (nsysA2, dsysA1, tsys)
returns

Transfer function:

  -2 z + 4
---------------
5 z^2 + 7 z - 9

Sampling time: 0.2
```

$$\text{modelA19} = \frac{-2z + 4}{5z^2 + 7z - 9}$$

Sampling Time: 0.2

```
modelA20 = tf ({[nsysA1]; [nsysA2]}, {[dsysA1]; [dsysA2]}, tsys)
returns

Transfer function from input to output...
        z - 3
 #1: ----------------
     5 z^2 + 7 z - 9

     2 z - 4
 #2: -------
     6 z - 8

Sampling time: 0.2
```

$$\text{modelA20} = \begin{bmatrix} \dfrac{z - 3}{5z^2 + 7z - 9} \\ \dfrac{2z - 4}{6z - 8} \end{bmatrix}$$

Sampling Time: 0.2

```
modelA21 = tf ({[nsysA1]; [nsysA2]}, {[dsysA1]; [dsysA2]})
returns

Transfer function from input to output...
        s - 3
 #1: ----------------
     5 s^2 + 7 s - 9

     2 s - 4
 #2: --------
     6 s - 8
```

$$\text{modelA21} = \begin{bmatrix} \dfrac{s - 3}{5s^2 + 7s - 9} \\ \dfrac{2s - 4}{6s - 8} \end{bmatrix}$$

Instead of specifying the numerator and denominator matrices first and then proceeding with the generation of tf model, you can obtain the tf model by using direct values of numerator and denominator coefficients as is illustrated in the forthcoming examples.

```
modelA22 = tf ([1 -3], [5 7 9])
returns

Transfer function:
   s - 3
----------------
5 s^2 + 7 s + 9
```

modelA22 = same as modelA17

```
modelA23 = tf ([-2 4], [5 7 -9], 0.2)
```

```
returns

Transfer function:
  -2 z + 4
----------------
5 z^2 + 7 z - 9

Sampling time: 0.2

modelA24 = tf ({[1 -3]; [-2 4]}, {[5 7 -9]; [0 -6 8]}, 0.2)
returns

Transfer function from input to output...
          z - 3
 #1:  ----------------
      5 z^2 + 7 z - 9

        2 z - 4
 #2:  ---------
        6 z - 8

Sampling time: 0.2

modelA25 = tf ({[1 -3]; [-2 4]}, {[5 7 -9]; [0 -6 8]})
returns

Transfer function from input to output...
          s - 3
 #1:  ----------------
      5 s^2 + 7 s - 9

        2 s - 4
 #2:  ---------
        6 s - 8

modelA26 = tf (2)
returns

Transfer function:
2
```

modelA23 =
same as modelA19

modelA24 =
same as modelA20

modelA25=
same as modelA21

modelA26 = [2]
a static gain matrix tf model

Let us now try to change the properties of some of the tf models generated above. The method followed is the same as for zpk models. Remember that tf model has all of the ten generic properties along with the three model-specific properties as listed in Table 1.1 and 1.3.

Let us begin with modelA20 and change its InputName to 'a', OutputName to 'x' and 'y', OutputDelay to '0.1 and 0.5', variable to 'p' and Notes to 'This is a tf model'.

```
modelA27 = tf ({[nsysA1]; [nsysA2]}, {[dsysA1]; [dsysA2]}, 'InputName', {'a'}, 'OutputName',
{'x', 'y'}, 'OutputDelay', [0.1;  0.5], 'var', 'p', 'Notes', 'This is a tf model')
returns

Transfer function from input "a" to output...
```

```
         p - 3
x:  exp(-0.1*p) * ----------------
              5 p^2 + 7 p - 9

         2 p - 4
y:  exp(-0.5*p) * --------
              6 p - 8
```

$$modelA27 = \left[\begin{array}{c} e^{-0.1p}\, \dfrac{(p-3)}{(5p^2+7p-9)} \\[2ex] e^{-0.5p}\, \dfrac{(2p-4)}{(6p-8)} \end{array} \right]$$

As we learnt earlier while working with zpk models, the properties of tf models can also be inherited from a similar model as indicated below.

```
modelA28 = tf ({[2 -4]; [-3 7]}, {[0.3 0.7 -0.9]; [5 -0.6 0.8]}, modelA27)
returns
```

```
Transfer function from input "a" to output...
          2 s - 4
x:  exp(-0.1*s) * ----------------------
             0.3 s^2 + 0.7 s - 0.9

          -3 s + 7
y:  exp(-0.5*s) * --------------------
             5 s^2 - 0.6 s + 0.8
```

modelA28 =

inherits the property of modelA27 except for the variable property which remains the same as default unless otherwise specified

Let us now generate a model in DSP format using the **filt** function:

```
modelA29=filt({[nsysA1]; [nsysA2]}, {[dsysA1]; [dsysA2]}, tsys)
returns
```

```
Transfer function from input to output...
        1 - 3 z^-1
#1:  --------------------
      5 + 7 z^-1 - 9 z^-2

      2 - 4 z^-1
#2:  ---------------
      6 z^-1 - 8 z^-2
```

Sampling time: 0.2

$$modelA29 = \left[\begin{array}{c} \dfrac{1 - 3z^{-1}}{5 + 7z^{-1} - 9z^{-2}} \\[2ex] \dfrac{2 - 4z^{-1}}{6z^{-1} - 8z^{-2}} \end{array} \right]$$

Sampling Time: 0.2

Next, we generate a tf model with 5 inputs and 6 outputs. Let the inputs have names 'a, b, c, d, e' and the outputs have names 'm, n, o, p, q, r'. Let the InputGroup for channel 1 and 2 be 'Electric Power', while for 3, 4 and 5 be 'Firewood'. Let the OutputGroup for channel 1, 4 and 5 be 'Light' while for 2, 3 and 6 be 'Air'. Also, let the variable be 'q' and Sample Time be '1' second. The InputDelay should vary from '1 to 5 seconds' respectively for each input, while, the OutputDelay should vary from '1 to 12 seconds with increments of 2 seconds' respectively for each output. Let the parameter Notes be 'This is an example model' and the parameter UserData be ' Fictitious'.

modelA30 =

$$
\begin{bmatrix}
q^2\,\dfrac{1+q+q^2}{1+q+q^2} & q^3\,\dfrac{1+q+2q^2}{1+q+2q^2} & q^4\,\dfrac{1+q+3q^2}{1+q+3q^2} & q^5\,\dfrac{1+q+4q^2}{1+q+4q^2} & q^6\,\dfrac{1+q+5q^2}{1+q+5q^2} \\[2ex]
q^4\,\dfrac{2+2q+q^2}{2+2q+6q^2} & q^5\,\dfrac{2+2q+2q^2}{2+2q+7q^2} & q^6\,\dfrac{2+2q+3q^2}{2+2q+8q^2} & q^7\,\dfrac{2+2q+4q^2}{2+2q+9q^2} & q^8\,\dfrac{2+2q+5q^2}{2+2q+10q^2} \\[2ex]
q^6\,\dfrac{3+3q+q^2}{3+3q+11q^2} & q^7\,\dfrac{3+3q+2q^2}{3+3q+12q^2} & q^8\,\dfrac{3+3q+3q^2}{3+3q+13q^2} & q^9\,\dfrac{3+3q+4q^2}{3+3q+14q^2} & q^{10}\,\dfrac{3+3q+5q^2}{3+3q+15q^2} \\[2ex]
q^8\,\dfrac{4+4q+q^2}{4+4q+16q^2} & q^9\,\dfrac{4+4q+2q^2}{4+4q+17q^2} & q^{10}\,\dfrac{4+4q+3q^2}{4+4q+18q^2} & q^{11}\,\dfrac{4+4q+4q^2}{4+4q+19q^2} & q^{12}\,\dfrac{4+4q+5q^2}{4+4q+20q^2} \\[2ex]
q^{10}\,\dfrac{5+5q+q^2}{5+5q+21q^2} & q^{11}\,\dfrac{5+5q+2q^2}{5+5q+22q^2} & q^{12}\,\dfrac{5+5q+3q^2}{5+5q+23q^2} & q^{13}\,\dfrac{5+5q+4q^2}{5+5q+24q^2} & q^{14}\,\dfrac{5+5q+5q^2}{5+5q+25q^2} \\[2ex]
q^{12}\,\dfrac{6+6q+q^2}{6+6q+26q^2} & q^{13}\,\dfrac{6+6q+2q^2}{6+6q+27q^2} & q^{14}\,\dfrac{6+6q+3q^2}{6+6q+28q^2} & q^{15}\,\dfrac{6+6q+4q^2}{6+6q+29q^2} & q^{16}\,\dfrac{6+6q+5q^2}{6+6q+30q^2}
\end{bmatrix}
$$

```
n1={[1 1 1] [1 1 2] [1 1 3] [1 1 4] [1 1 5];  [2 2 1] [2 2 2] [2 2 3] [2 2 4] [2 2 5]; [3 3 1] [3 3
2] [3 3 3] [3 3 4] [3 3 5];  [4 4 1] [4 4 2] [4 4 3] [4 4 4] [4 4 5]; [5 5 1] [5 5 2] [5 5 3] [5 5 4]
[5 5 5];  [6 6 1] [6 6 2] [6 6 3] [6 6 4] [6 6 5]};

d1={[1 1 1] [1 1 2] [1 1 3] [1 1 4] [1 1 5]; [2 2 6] [2 2 7] [2 2 8] [2 2 9] [2 2 10];  [3 3 11] [3
3 12] [3 3 13] [3 3 14] [3 3 15];  [4 4 16] [4 4 17] [4 4 18] [4 4 19] [4 4 20];  [5 5 21] [5 5
22] [5 5 23] [5 5 24] [5 5 25];  [6 6 26] [6 6 27] [6 6 28] [6 6 29] [6 6 30]};

modelA30 = tf (n1,d1, 'InputName', {'a','b','c','d','e'},  'OutputName', {'m','n','o','p','q','r'},
'InputGroup', {[1 2] 'Electric Power'; [3:5] 'Firewood'}, 'OutputGroup', {[1 4 5] 'Light'; [2 3 6],
'Air'}, 'Var', 'q', 'Ts', 1, 'InputDelay', [1:5], 'OutputDelay', [1:2:12], 'Notes', 'This is an example
model', 'UserData', 'Fictitious')
returns

Transfer function from input "a" to output...
          1 + q + q^2
m:  q^2 * -----------
          1 + q + q^2


        2 + 2 q + q^2
n:  q^4 * ----------------
        2 + 2 q + 6 q^2


        3 + 3 q + q^2
o:  q^6 * ----------------
        3 + 3 q + 11 q^2


        4 + 4 q + q^2
p:  q^8 * ----------------
        4 + 4 q + 16 q^2


         5 + 5 q + q^2
q:  q^10 * ----------------
         5 + 5 q + 21 q^2
```

```
            6 + 6 q + q^2
r:  q^12 * ------------------
            6 + 6 q + 26 q^2
```

Transfer function from input "b" to output...

```
            1 + q + 2 q^2
m:  q^3 * --------------
            1 + q + 2 q^2


            2 + 2 q + 2 q^2
n:  q^5 * ----------------
            2 + 2 q + 7 q^2


            3 + 3 q + 2 q^2
o:  q^7 * ----------------
            3 + 3 q + 12 q^2


            4 + 4 q + 2 q^2
p:  q^9 * ----------------
            4 + 4 q + 17 q^2


            5 + 5 q + 2 q^2
q:  q^11 * ----------------
            5 + 5 q + 22 q^2


            6 + 6 q + 2 q^2
r:  q^13 * ----------------
            6 + 6 q + 27 q^2
```

Transfer function from input "c" to output...

```
            1 + q + 3 q^2
m:  q^4 * --------------
            1 + q + 3 q^2


            2 + 2 q + 3 q^2
n:  q^6 * ----------------
            2 + 2 q + 8 q^2


            3 + 3 q + 3 q^2
o:  q^8 * ----------------
            3 + 3 q + 13 q^2


            4 + 4 q + 3 q^2
p:  q^10 * ----------------
            4 + 4 q + 18 q^2


            5 + 5 q + 3 q^2
q:  q^12 * ----------------
            5 + 5 q + 23 q^2


            6 + 6 q + 3 q^2
r:  q^14 * ----------------
            6 + 6 q + 28 q^2
```

Transfer function from input "d" to output...

```
            1 + q + 4 q^2
m:  q^5 * --------------
            1 + q + 4 q^2
```

```
         2 + 2 q + 4 q^2
n:  q^7 * ----------------
         2 + 2 q + 9 q^2


         3 + 3 q + 4 q^2
o:  q^9 * -----------------
         3 + 3 q + 14 q^2


          4 + 4 q + 4 q^2
p:  q^11 * -----------------
          4 + 4 q + 19 q^2


          5 + 5 q + 4 q^2
q:  q^13 * -----------------
          5 + 5 q + 24 q^2


          6 + 6 q + 4 q^2
r:  q^15 * -----------------
          6 + 6 q + 29 q^2
```

Transfer function from input "e" to output...

```
         1 + q + 5 q^2
m:  q^6 * ---------------
         1 + q + 5 q^2


         2 + 2 q + 5 q^2
n:  q^8 * -----------------
         2 + 2 q + 10 q^2


          3 + 3 q + 5 q^2
o:  q^10 * -----------------
          3 + 3 q + 15 q^2


          4 + 4 q + 5 q^2
p:  q^12 * -----------------
          4 + 4 q + 20 q^2


          5 + 5 q + 5 q^2
q:  q^14 * -----------------
          5 + 5 q + 25 q^2


          6 + 6 q + 5 q^2
r:  q^16 * -----------------
          6 + 6 q + 30 q^2
```

I/O groups:

Group name	I/O	Channel(s)
Electric Power	I	1,2
Firewood	I	3,4,5
Light	O	1,4,5
Air	O	2,3,6

Sampling time: 1

Similar to the zpk modelA14 andmodelA15, you can enter the tf model as a rational expression too. The steps involved remain the same. Let us do this for a continuous transfer function model first.

```
s = tf('s');

modelA31 = [ (s^2+2*s-3)/(4*s^3-3*s^2+s-1)   s/(5*s^4+6*s-3);  (s-7)/(9*s^3-3*s^2+7*s-8)  (s^3+2*s^2-3*s+4)/(4*s^3-3*s^2+s-1)]
returns
```

```
Transfer function from input 1 to output...
        s^2 + 2 s - 3
 #1: ---------------------
       4 s^3 - 3 s^2 + s - 1
           s - 7
 #2: -------------------------
       9 s^3 - 3 s^2 + 7 s - 8

Transfer function from input 2 to output...
           s
 #1: ----------------
       5 s^4 + 6 s - 3

       s^3 + 2 s^2 - 3 s + 4
 #2: ----------------------
       4 s^3 - 3 s^2 + s - 1
```

$$modelA31=
\begin{bmatrix}
\dfrac{s^2+2s-3}{4s^3-3s^2+s-1} & \dfrac{s}{5s^4+6s-3} \\
\dfrac{s-7}{9s^3-3s^2+7s-8} & \dfrac{s^3+2s^2-3s+4}{4s^3-3s^2+s-1}
\end{bmatrix}$$

Let us next extend the same treatment to a discrete transfer function model.

```
z = tf('z', 0.5);

modelA32 = [ (4*z^2+3*z-3)/(4*z^3-3*z^2+7*z-1)   (5*z-1)/(2*z^4-3*z);  (4*z-7)/(9*z^3-3*z+7*z-8)  (z^3+4*z^2-3*z+2)/(6*z^5-9*z^2+z^3-z+1)]
returns
```

```
Transfer function from input 1 to output...
        4 z^2 + 3 z - 3
 #1: ------------------------
       4 z^3 - 3 z^2 + 7 z - 1

          4 z - 7
 #2: ----------------
       9 z^3 + 4 z - 8

Transfer function from input 2 to output...
        5 z - 1
 #1: ------------
       2 z^4 - 3 z

        z^3 + 4 z^2 - 3 z + 2
 #2: --------------------------
       6 z^5 + z^3 - 9 z^2 - z + 1

Sampling time: 0.5
```

$$modelA32=
\begin{bmatrix}
\dfrac{4z^2+3z-3}{4z^3-3z^2+7z-1} & \dfrac{5z-1}{2z^4-3z} \\
\dfrac{4z-1}{9z^3+4z-8} & \dfrac{z^3+4z^2-3z+2}{6z^5+z^3-9z^2-z+1}
\end{bmatrix}$$

Sampling Time: 0.5

Transfer function LTI model arrays too can be obtained by specifying array dimensions for numerator and denominator coefficients. Let us consider the system shown in Figure 1.2 and obtain a tf array model for the same:

```
n={[111 1] [121 1]; [211 1] [221 1]; [311 1] [321 1]};
n(:,:,2)={[112 2] [122 2]; [212 2] [222 2]; [312 2] [322 2]};
n(:,:,3)={[113 3] [123 3]; [213 3] [223 3]; [313 3] [323 3]};
n(:,:,4)={[114 4] [124 4]; [214 4] [224 4]; [314 4] [324 4]};

d={[1.11 11.1 1] [1.21 12.1 1]; [2.11 21.1 1] [2.21 22.1 1]; [3.11 31.1 1] [3.21 32.1 1]};
d(:,:,2)={[1.12 11.2 2] [1.22 12.2 2]; [2.12 21.2 2] [2.22 22.2 2]; [3.12 31.2 2] [3.22 32.2 2]};
d(:,:,3)={[1.13 11.3 3] [1.23 12.3 3]; [2.13 21.3 3] [2.23 22.3 3]; [3.13 31.3 3] [3.23 32.3 3]};
d(:,:,4)={[1.14 11.4 4] [1.24 12.4 4]; [2.14 21.4 4] [2.24 22.4 4]; [3.14 31.4 4] [3.24 32.4 4]};

modelA33(:,:,1)=tf(n(:,:,1),d(:,:,1));
modelA33(:,:,2)=tf(n(:,:,2),d(:,:,2));
modelA33(:,:,3)=tf(n(:,:,3),d(:,:,3));
modelA33(:,:,4)=tf(n(:,:,4),d(:,:,4));

modelA33
returns

Model modelA33(:,:,1,1)
=========================

 Transfer function from input 1 to output...
          111 s + 1
  #1:  ----------------------
      1.11 s^2 + 11.1 s + 1


          211 s + 1
  #2:  ----------------------
      2.11 s^2 + 21.1 s + 1


          311 s + 1
  #3:  ----------------------
      3.11 s^2 + 31.1 s + 1

 Transfer function from input 2 to output...
          121 s + 1
  #1:  ----------------------
      1.21 s^2 + 12.1 s + 1


          221 s + 1
  #2:  ----------------------
      2.21 s^2 + 22.1 s + 1


          321 s + 1
  #3:  ----------------------
      3.21 s^2 + 32.1 s + 1


Model modelA33(:,:,2,1)
=========================

 Transfer function from input 1 to output...
          112 s + 2
  #1:  ----------------------
      1.12 s^2 + 11.2 s + 2
```

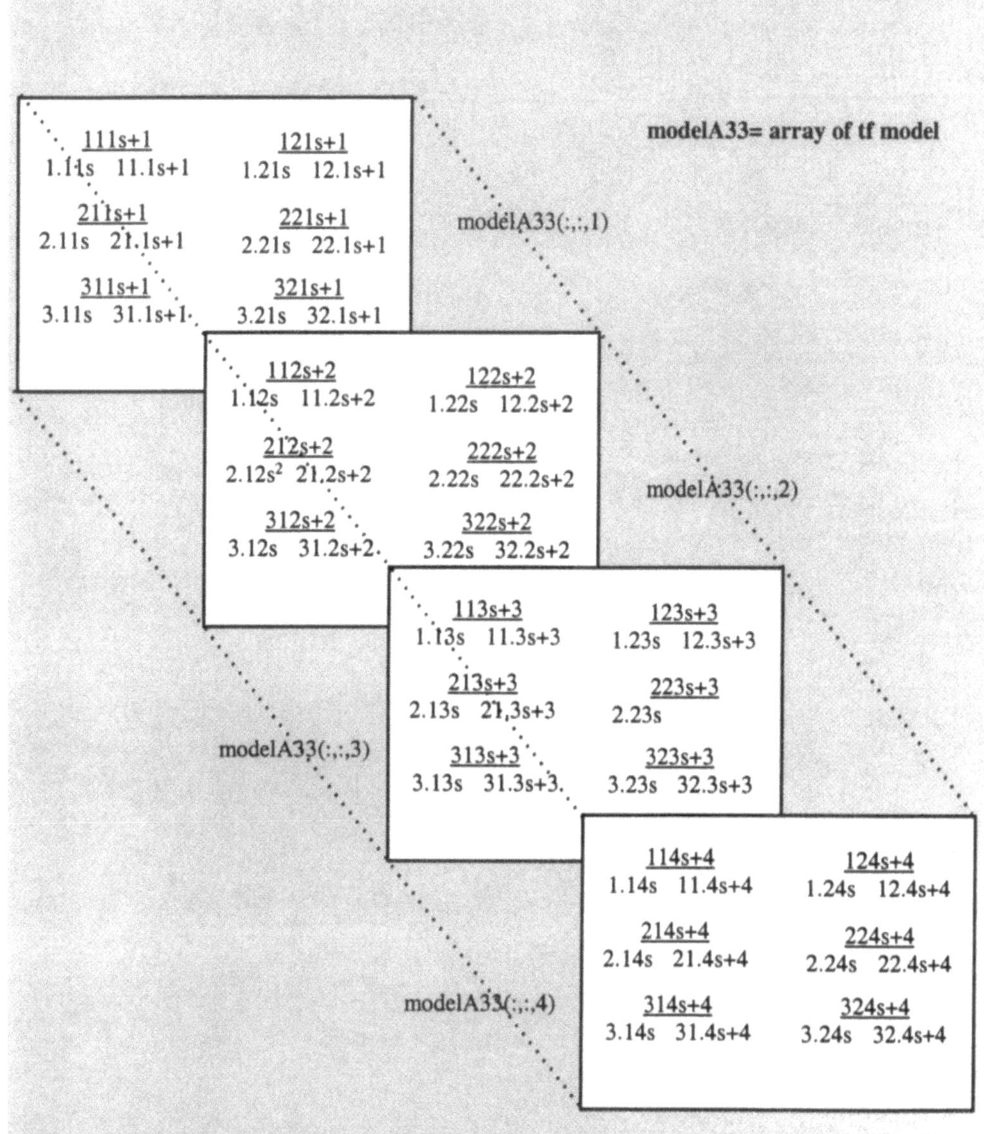

Figure 1.2. tf array model

```
             212 s + 2
  #2:  ----------------------
       2.12 s^2 + 21.2 s + 2


             312 s + 2
  #3:  ----------------------
       3.12 s^2 + 31.2 s + 2


Transfer function from input 2 to output...
             122 s + 2
  #1:  ----------------------
       1.22 s^2 + 12.2 s + 2


             222 s + 2
  #2:  ----------------------
       2.22 s^2 + 22.2 s + 2


             322 s + 2
  #3:  ----------------------
       3.22 s^2 + 32.2 s + 2


Model modelA33(:,:,3,1)
========================

Transfer function from input 1 to output...
             113 s + 3
  #1:  ----------------------
       1.13 s^2 + 11.3 s + 3


             213 s + 3
  #2:  ----------------------
       2.13 s^2 + 21.3 s + 3


             313 s + 3
  #3:  ----------------------
       3.13 s^2 + 31.3 s + 3


Transfer function from input 2 to output...
             123 s + 3
  #1:  ----------------------
       1.23 s^2 + 12.3 s + 3


             223 s + 3
  #2:  ----------------------
       2.23 s^2 + 22.3 s + 3


             323 s + 3
  #3:  ----------------------
       3.23 s^2 + 32.3 s + 3


Model modelA33(:,:,4,1)
========================

Transfer function from input 1 to output...
```

```
         114 s + 4
#1:  ----------------------
      1.14 s^2 + 11.4 s + 4


         214 s + 4
#2:  ----------------------
      2.14 s^2 + 21.4 s + 4


         314 s + 4
#3:  ----------------------
      3.14 s^2 + 31.4 s + 4

Transfer function from input 2 to output...


         124 s + 4
#1:  ----------------------
      1.24 s^2 + 12.4 s + 4


         224 s + 4
#2:  ----------------------
      2.24 s^2 + 22.4 s + 4


         324 s + 4
#3:  ----------------------
      3.24 s^2 + 32.4 s + 4


4x1 array of continuous-time transfer functions.
```

After creating these seventeen transfer function models, we are sure that you are ready to take up a practice test. Then, what are you waiting for? Go ahead…

Practice Test 1.2.

1. Create the following tf models by separately specifying the numerator and denominator matrices:

(a)

$$modelAp6 = \left[\frac{s^2 - 2s + 1}{s^3 - 3s^2 + 4s - 3} \right]$$

(b)

$$modelAp7 = \left[\frac{z^2 - 3z + 7}{z^3 + 2z - 3} \right]$$

$$Sampling\ Time = 0.1$$

(c)

$$\text{modelAp8} = \begin{bmatrix} \dfrac{z}{z+1} & \dfrac{2z}{z^2-3z+4} \\[2ex] \dfrac{2z}{z^2-2z+1} & \dfrac{1}{z-1} \end{bmatrix}$$

Sampling Time : 0.5

d)

$$\text{modelAp9} = \begin{bmatrix} \dfrac{2s+3}{s^4-4s+2-1} & \dfrac{s^2+1}{3s^2-4s+8} & \dfrac{1}{s+1} \\[2ex] \dfrac{s}{s^2-3s+2} & \dfrac{3s^4-2s}{5s^5+6s^4-2s+5} & \dfrac{s^2-1}{s^3-2s+1} \\[2ex] \dfrac{3s^2-4s+7}{s^5-4s^3+3s-6} & \dfrac{s+3}{s^2-2s+1} & \dfrac{s}{s+3} \end{bmatrix}$$

InputName = a, b, c

OutputName = m, n, o

InputGroup = for channel 1,2 - micro, for channel 3 - pico

OutputGroup = measurements

InputDelay = 1 to 3

OutputDelay = 5 to 8

Variable = p

Notes = I can create tf models.

UserData = Matlab is easy!

2. Create the above tf models from a rational expression in the variable 's' and 'z'.

3. Change the variable for:
- modelAp6 of question 1(a) to 'p'
- modelAp7 of question 1(b) to 'q'
- modelAp8 of question 1© to 'z^{-1}'

4. Create a tf array modelAp10 having dimensions 2 by 2 by 2. You can choose the values of numerator and denominator coefficients on your own.

1.2.3 State Space (ss) Model

Control System Toolbox of MATLAB® supports two types of state space models:

- Standard State Space Model
- Descriptor State Space Model

1.2.3.1 Standard State Space (ss) Model

For a continuous-time system represented as:

$$dx/dt = ax+bu$$
$$y = cx+du \, ,$$

or, a discrete-time system represented as:

$$x_{(n+1)} = ax_n+bu_n$$
$$y_n = cx_n+du_n \, ,$$

the syntax for obtaining the standard state space model is:

- ModelResult = **ss** ([*a*], [*b*], [*c*], [*d*], *tsys*, '*property1*', '*value1*', '*property2*', '*value2*',…, *refmodel*)

- ModelResult = **ss** (*N*)
  to specify static gain matrix ss model

where,

ModelResult………………	is the user-specified name of the ss model so generated
a ………………………	is 2D real valued system matrix
b ………………………	is 2D real valued input matrix
c…………………………	is 2D real valued output matrix
d ………………………	is 2D real valued transmission or feedthrough matrix
tsys ……………………	specifies the sample time for discrete system (if undetermined, put tsys=-1 or [ ])
property1, *property2*………	specifies the names of the properties that may be generic or ss model-specific (optional)
value1, *value2* ………………	specifies the values of the properties mentioned in *property1*, *property2*…
refmodel …………………	is the name of model whose properties are to be inherited by the new model ss model ModelResult so generated (optional)
N…………………………	is a scalar or matrix such that ModelResult is a static gain *N* ss model (the resulting model is nothing but the *d* matrix which gives the relationship between the input and the output)

It is obvious that matrix *b* and matrix *d* have the same number of columns, while matrix *c* and matrix *d* have the same number of rows. If *d* is equated to zero, it is assumed that *d* is a zero matrix of adequate dimensions.

Examples:

Let us specify

```
a  = [0 3; -2 -5];
b = [1 1; 1 1];
c = [2 1; 1 0];
d = [0];
tsys = 0.5;
```

With the above data, we can obtain the model of the system as follows:

```
modelA34=ss(a,b,c,d)
returns
```

a =

	x1	x2
x1	0	3
x2	-2	-5

b =

	u1	u2
x1	1	1
x2	1	1

c =

	x1	x2
y1	2	1
y2	1	0

d =

	u1	u2
y1	0	0
y2	0	0

Continuous-time model.

```
modelA35=ss(a,b,c,d,tsys)
returns
```

a =

	x1	x2
x1	0	3
x2	-2	-5

b =

	u1	u2
x1	1	1
x2	1	1

c =

	x1	x2
y1	2	1
y2	1	0

d =

	u1	u2
y1	0	0
y2	0	0

Sampling time: 0.5
Discrete-time model.

modelA34 =

$$\dot{x} = \begin{bmatrix} 0 & 3 \\ -2 & -5 \end{bmatrix} x + \begin{bmatrix} 1 & 1 \\ 1 & 1 \end{bmatrix} u$$

$$y = \begin{bmatrix} 2 & 1 \\ 1 & 0 \end{bmatrix} x + \begin{bmatrix} 0 & 0 \\ 0 & 0 \end{bmatrix} u$$

modelA35 =

$$\dot{x} = \begin{bmatrix} 0 & 3 \\ -2 & -5 \end{bmatrix} x + \begin{bmatrix} 1 & 1 \\ 1 & 1 \end{bmatrix} u$$

$$y = \begin{bmatrix} 2 & 1 \\ 1 & 0 \end{bmatrix} x + \begin{bmatrix} 0 & 0 \\ 0 & 0 \end{bmatrix} u$$

Sampling time: 0.5

Instead of first specifying a, b, c and d matrices separately and then going ahead to generate ss model using these, you can generate the same ss model by using direct values of a, b, c and d parameter matrices. Forthcoming illustrations will clarify the concept.

```
modelA36=ss([0 3 −7; -2 −5 6; -2 −5 −9], [1 1 ; 1 1 ; 1 1], [2 5 1; 1 0 8], [1 5; 4 6])
returns
```

```
a =
                x1          x2          x3
        x1       0           3          -7
        x2      -2          -5           6
        x3      -2          -5          -9

b =
                u1          u2
        x1       1           1
        x2       1           1
        x3       1           1

c =
                x1          x2          x3
        y1       2           5           1
        y2       1           0           8

d =
                u1          u2
        y1       1           5
        y2       4           6
```

$$
\text{modelA36}=
$$
$$
\dot{x} = \begin{bmatrix} 0 & 3 & -7 \\ -2 & -5 & 6 \\ -2 & -5 & -9 \end{bmatrix} x + \begin{bmatrix} 1 & 1 \\ 1 & 1 \\ 1 & 1 \end{bmatrix} u
$$
$$
y = \begin{bmatrix} 2 & 5 & 1 \\ 1 & 0 & 8 \end{bmatrix} x + \begin{bmatrix} 1 & 5 \\ 4 & 6 \end{bmatrix} u
$$

Continuous-time model.

```
modelA37 = ss (10)
returns

d =
                u1
        y1      10
```

$$
\text{modelA37} =
$$
$$
y = [10]\, u
$$
Static gain matrix ss model

Static gain.

We can manipulate the properties of ss models too. Let us do this for modelA34:

```
modelA38 = ss ([0 3 −7; -2 −5 6; -2 −5 −9], [1 1 ; 1 1 ; 1 1], [2 5 1; 1 0 8], [1 5 ; 4 6 ],
'InputName', {'Voltage', 'Current'}, 'OutputName',  {'Noise', 'Bandwidth'}, 'InputGroup', {[1 2]
'Frequency'}, 'OutputGroup', {[1 2] 'Resonance'}, 'InputDelay', 3, 'OutputDelay', 5, 'Ts', 1,
'Statename', {'Resistance', 'Inductance', 'Capacitance'}, 'Notes', 'This is a fictitious example
model', 'UserData', 'Will this example crash on 1/1/2000')
returns
```

```
a =
                Resistance  Inductance  Capacitance
    Resistance       0           3          -7
    Inductance      -2          -5           6
   Capacitance      -2          -5          -9

b =
                Voltage     Current
    Resistance       1           1
    Inductance       1           1
   Capacitance       1           1
```

$$
\text{modelA38}=
$$
$$
\dot{x} = \begin{bmatrix} 0 & 3 & -7 \\ -2 & -5 & 6 \\ -2 & -5 & -9 \end{bmatrix} x + \begin{bmatrix} 1 & 1 \\ 1 & 1 \\ 1 & 1 \end{bmatrix} u
$$
$$
y = \begin{bmatrix} 2 & 5 & 1 \\ 1 & 0 & 8 \end{bmatrix} x + \begin{bmatrix} 1 & 5 \\ 4 & 6 \end{bmatrix} u
$$

```
c =
                Resistance       Inductance       Capacitance
        Noise        2                5                1
    Bandwidth        1                0                8

d =
                Voltage      Current
        Noise        1           5
    Bandwidth        4           6

Input delays (listed by channel): 3  3
Output delays (listed by channel): 5  5

I/O groups:
    Group name    I/O    Channel(s)
    Frequency      I        1,2
    Resonance      O        1,2

Sampling time: 1
Discrete-time model.
```

Since ss models are not expressed as rational functions, there is neither a direct method nor a need for generating state space models directly in a way similar to zpk modelA14 and modelA15 or tf modelA31 and modelA32. However, state space LTI model arrays can be obtained by specifying array dimensions of a, b, c and d matrices. Let us consider the system shown in Figure 1.3 and obtain ss array model for the same.

```
a=[111 121; 211 221];
a(:,:,2)=[112 122; 212 222];
a(:,:,3)=[113 123; 213 223];
a(:,:,4)=[114 124; 214 224];

b=[11; 21];
b(:,:,2)=[32; 42];
b(:,:,3)=[53; 63];
b(:,:,4)=[74; 84];

c=[1110 1210];
c(:,:,2)=[1120 1220];
c(:,:,3)=[1130 1230];
c(:,:,4)=[1140 1240];

d=0;

modelA39(:,:,1)=ss(a(:,:,1),b(:,:,1),c(:,:,1),d);
modelA39(:,:,2)=ss(a(:,:,2),b(:,:,2),c(:,:,2),d);
modelA39(:,:,3)=ss(a(:,:,3),b(:,:,3),c(:,:,3),d);
modelA39(:,:,4)=ss(a(:,:,4),b(:,:,4),c(:,:,4),d);

modelA39
returns

Model modelA39(:,:,1,1)
========================
```

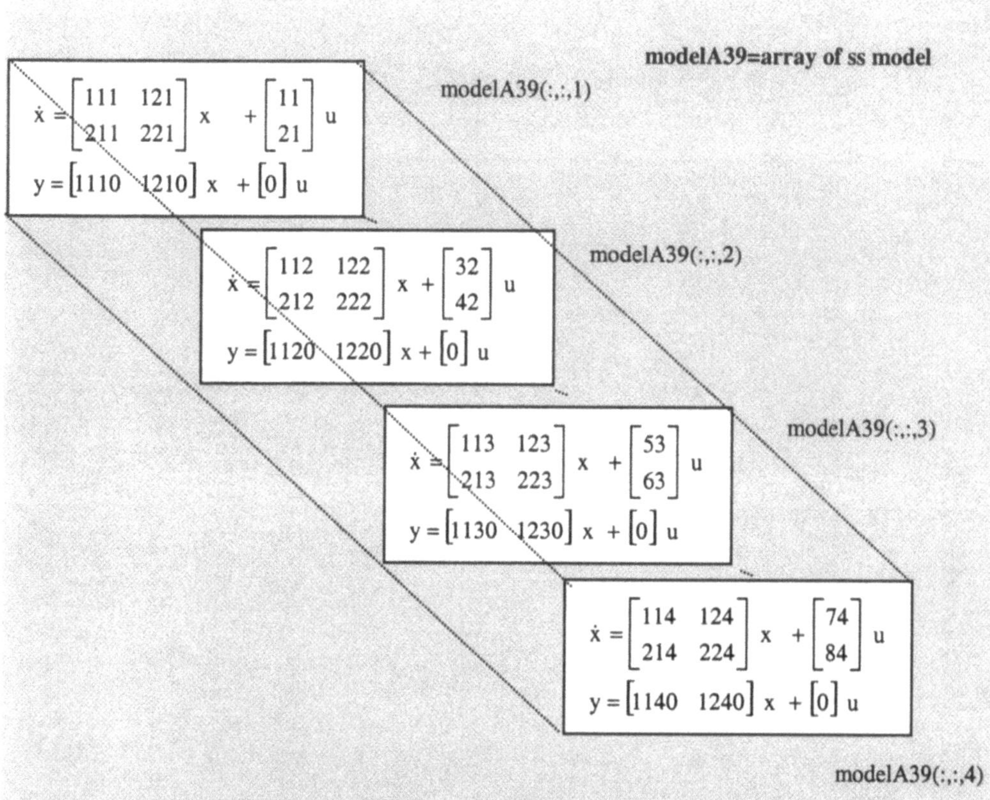

$$\dot{x} = \begin{bmatrix} 111 & 121 \\ 211 & 221 \end{bmatrix} x + \begin{bmatrix} 11 \\ 21 \end{bmatrix} u$$

$$y = \begin{bmatrix} 1110 & 1210 \end{bmatrix} x + \begin{bmatrix} 0 \end{bmatrix} u$$

$$\dot{x} = \begin{bmatrix} 112 & 122 \\ 212 & 222 \end{bmatrix} x + \begin{bmatrix} 32 \\ 42 \end{bmatrix} u$$

$$y = \begin{bmatrix} 1120 & 1220 \end{bmatrix} x + \begin{bmatrix} 0 \end{bmatrix} u$$

$$\dot{x} = \begin{bmatrix} 113 & 123 \\ 213 & 223 \end{bmatrix} x + \begin{bmatrix} 53 \\ 63 \end{bmatrix} u$$

$$y = \begin{bmatrix} 1130 & 1230 \end{bmatrix} x + \begin{bmatrix} 0 \end{bmatrix} u$$

$$\dot{x} = \begin{bmatrix} 114 & 124 \\ 214 & 224 \end{bmatrix} x + \begin{bmatrix} 74 \\ 84 \end{bmatrix} u$$

$$y = \begin{bmatrix} 1140 & 1240 \end{bmatrix} x + \begin{bmatrix} 0 \end{bmatrix} u$$

Figure 1.3. ss array model

```
a =
                x1        x2
        x1      111       121
        x2      211       221

b =
                u1
        x1      11
        x2      21

c =
                x1        x2
        y1      1110      1210

d =
                u1
        y1      0


Model modelA39(:,:,2,1)
========================
```

```
a =

             x1        x2
     x1      112       122
     x2      212       222

b =

             u1
     x1      32
     x2      42

c =

             x1        x2
     y1      1120      1220

d =

             u1
     y1      0
```

Model modelA39(:,:,3,1)
========================

```
a =

             x1        x2
     x1      113       123
     x2      213       223

b =

             u1
     x1      53
     x2      63

c =

             x1        x2
     y1      1130      1230

d =

             u1
     y1      0
```

Model modelA39(:,:,4,1)
========================

```
a =

             x1        x2
     x1      114       124
     x2      214       224

b =

             u1
     x1      74
     x2      84

c =

             x1        x2
     y1      1140      1240
```

```
d =

                  u1
       y1          0
```

4x1 array of continuous-time state-space models.

1.2.3.2 Descriptor State Space Model
For a system represented as:

$$e.dx/dt = ax+bu$$
$$y = cx+du \ ,$$

or, a discrete-time system represented as:

$$e.x_{(n+1)} = ax_n+bu_n$$
$$y_n = cx_n+du_n \ ,$$

descripter state space model can be obtained using the syntax given below:

ModelResult = **dss**([*a*], [*b*], [*c*], [*d*], [*e*], *tsys*, '*property1*', *value1*, '*property2*', *value2*,...,
       refmodel)

where,
ModelResult..................... is the user-specified name of the dss model so generated
e.. is a Nx by Nx real valued matrix
the rest of the parameters are same as in case of standard state space models.

Example:

Let us now try to create a dss model shown in the box given below:

```
modelA40 =

  ⎡  2    3    4  - 3 ⎤        ⎡ 0    3  - 7    5 ⎤      ⎡ 1  1  1 ⎤
  ⎢ - 2   5    6    7 ⎥  ẋ  =  ⎢ 1  - 2  - 5    6 ⎥  x + ⎢ 1  1  1 ⎥ u
  ⎢  3    2    7  - 2 ⎥        ⎢ 8  - 2  - 5  - 9 ⎥      ⎢ 1  1  1 ⎥
  ⎣  2  - 7    3    2 ⎦        ⎣ 3    7    6    5 ⎦      ⎣ 1  1  1 ⎦

                             ⎡ 2    5  - 6   1 ⎤      ⎡ 1  5  4 ⎤
                        y =  ⎢ 1  - 5    0   8 ⎥  x + ⎢ 4  6  9 ⎥ u
                             ⎣ 3    4  - 6   8 ⎦      ⎣ 2  1  6 ⎦
```

modelA40=dss([0 3 -7 5; 1 -2 -5 6; 8 -2 -5 -9; 3 7 6 5], [1 1 1; 1 1 1 ; 1 1 1; 1 1 1], [2 5 -6 1;
1 -5 0 8; 3 4 -6 8], [1 5 4 ; 4 6 9; 2 1 6], [2 3 4 -3; -2 5 6 7; 3 2 7 -2; 2 -7 3 2])
returns

```
a =

                  x1        x2        x3        x4
          x1       0         3        -7         5
          x2       1        -2        -5         6
          x3       8        -2        -5        -9
          x4       3         7         6         5


b =

                  u1        u2        u3
          x1       1         1         1
          x2       1         1         1
          x3       1         1         1
          x4       1         1         1


c =

                  x1        x2        x3        x4
          y1       2         5        -6         1
          y2       1        -5         0         8
          y3       3         4        -6         8


d =

                  u1        u2        u3
          y1       1         5         4
          y2       4         6         9
          y3       2         1         6


e =

                  x1        x2        x3        x4
          x1       2         3         4        -3
          x2      -2         5         6         7
          x3       3         2         7        -2
          x4       2        -7         3         2

Continuous-time model.
```

You may try different permutations and combinations on ss and dss models and when you feel confident enough, take up the practice test given below…

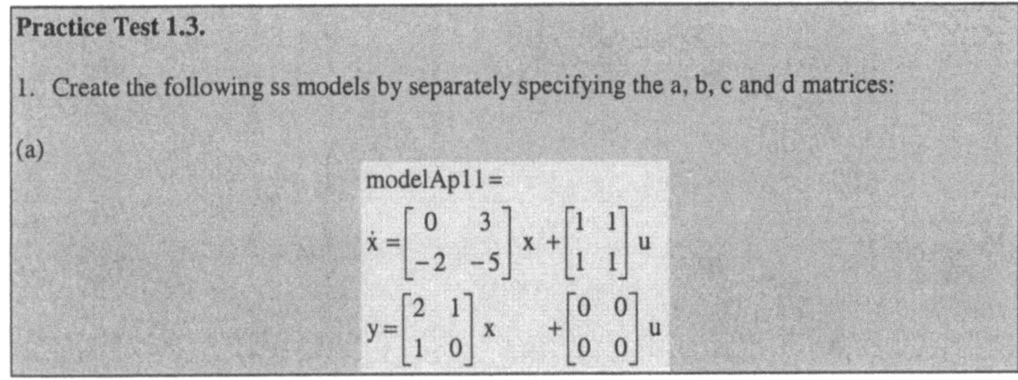

Practice Test 1.3.

1. Create the following ss models by separately specifying the a, b, c and d matrices:

(a)

$$\text{modelAp11} =$$

$$\dot{x} = \begin{bmatrix} 0 & 3 \\ -2 & -5 \end{bmatrix} x + \begin{bmatrix} 1 & 1 \\ 1 & 1 \end{bmatrix} u$$

$$y = \begin{bmatrix} 2 & 1 \\ 1 & 0 \end{bmatrix} x + \begin{bmatrix} 0 & 0 \\ 0 & 0 \end{bmatrix} u$$

(b)

$$modelAp12 =$$

$$\dot{x} = \begin{bmatrix} -1 & 1 \\ 0 & -1 \end{bmatrix} x + \begin{bmatrix} 0 \\ -2 \end{bmatrix} u$$

$$y = \begin{bmatrix} -5 & 0 \end{bmatrix} x + \begin{bmatrix} 0 \end{bmatrix} u$$

$$Sampling\ Time : 0.1$$

(c)

$$modelAp13 =$$

$$\begin{bmatrix} \dot{x}_1 \\ \dot{x}_2 \end{bmatrix} = \begin{bmatrix} -1 & 0 & 0 \\ 0 & -2 & 0 \\ 0 & 0 & -3 \end{bmatrix} \begin{bmatrix} x_1 \\ x_2 \end{bmatrix} + \begin{bmatrix} 0 & 1 \\ -1 & 0 \\ 0 & 2 \end{bmatrix} \begin{bmatrix} u_1 \\ u_2 \end{bmatrix}$$

$$\begin{bmatrix} y_1 \\ y_2 \end{bmatrix} = \begin{bmatrix} 1 & 2 & -1 \\ 1 & 0 & 0 \\ 0 & 0 & -1 \end{bmatrix} \begin{bmatrix} x_1 \\ x_2 \end{bmatrix} + \begin{bmatrix} -1 & 0 \\ 0 & -1 \\ 1 & 1 \end{bmatrix} \begin{bmatrix} u_1 \\ u_2 \end{bmatrix}$$

(d)

$$modelAp14 =$$

$$[\dot{x}] = \begin{bmatrix} -1 & 0 & -1 & 0 \\ 0 & -2 & 0 & 2 \\ 3 & 0 & 0 & 3 \\ 4 & -1 & 1 & 4 \end{bmatrix} [x] + \begin{bmatrix} 0 & 0 & 1 \\ 1 & 0 & 0 \\ 0 & 1 & 0 \\ 1 & 1 & 1 \end{bmatrix} [u]$$

$$[y] = \begin{bmatrix} 0 & 2 & 0 & 0 \\ 0 & 0 & 2 & 1 \\ 5 & 0 & 1 & 2 \end{bmatrix} [x] + \begin{bmatrix} 0 \end{bmatrix} [u]$$

InputName = InputP, InputQ, InputR

OutputName = OutputM, OutputN, OutputO

StateName = State1, State2, State3, State4

InputDelay = 0.1 to 0.3

OutputDelay = 0.01 to 0.06 in steps of 0.02

InputGroup = Channel 1 and 2 - - Macro;   Channel 3 - - Micro

OutputGroup = Channel 1 and 3 - - Nano

Notes = Creating ss model is a bit tricky!

UserData = I can create ss model with confidence!

2. Create modelAp15 to modelAp18 by converting the ss models of Problem 1 above to dss model for the value of parameter e given as:

(a) $\begin{bmatrix} 0 & 3 \\ -2 & -5 \end{bmatrix}$ (b) $\begin{bmatrix} -5 & -2 \\ 1 & 3 \end{bmatrix}$

(c) $\begin{bmatrix} 0 & 0 & 1 \\ 1 & 0 & 0 \\ 1 & -1 & -2 \end{bmatrix}$ (d) $\begin{bmatrix} 1 & 2 & 3 & 4 \\ 4 & 3 & 2 & 1 \\ -3 & 2 & -4 & 1 \\ 0 & 0 & 2 & -3 \end{bmatrix}$

3. Create ss LTI array modelAp19 having 4 array dimensions. You can choose the values of the a, b, c and d parameters on your own.

4. Create ss LTI array modelAp20 having 3 array dimensions. You can choose the values of the a, b, c and d parameters on your own.

1.2.4 Frequency Response Data (frd) Model

As the name itself suggests, frequency response data model is slightly different from the models discussed so far, in the sense that it is a model derived from the experimental or simulated response of a system to various input frequencies. The syntax for creating a frd model is:

ModelResult = **frd** ($[r]$, $[f]$, *tsys* ,'*property1*', '*value1*', '*property2*', '*value2*',…, *refmodel*, '*Units*', *units*)

where,

ModelResult...................... is the user-specified name of the frd model so generated

r.. specifies complex frequency response vector of the system

f.. specifies the frequency point vector at which response of the system is observed

tsys .... specifies the sample time for discrete system (if undetermined, put tsys=-1 or [ ])

property1, *property2*........ specifies the names of the properties that may be generic or frd model-specific (optional)

value1, *value*.................... specifies the values of the properties mentioned in *property1*, *property2*…

refmodel .... is the name of model whose properties are to be inherited by the new frd model ModelResult so generated (optional)

'*Units*'.............................. is the property name to specify the unit of frequency as different from the default *i.e.*, rad/sec to Hz

The default unit for frequency is 'rad/sec' that can be changed to 'Hz' by using the **ChangeUnit** function whose syntax is as follows:

Chgunits (*ModelName*, '*unit string*')

Note that for a SISO system, r is the response vector and f is the frequency point real vector of the same length as r. While for a MIMO system with Ny outputs, Nu inputs and Nf frequency points, the response is a Ny by Nu by Nf array whose *Response(i,j,k)* specifies the frequency response from input j to output i at frequency k.

Example:

Let us consider a system whose response is as follows:

Frequency	5	10	15	20	25	30	35	40	45	50
Response	0.2	0.4	1.3	1.5	1.7	1.4	1.3	0.9	0.8	0.4

```
r = [0.2 0.4 1.3 1.5 1.7 1.4 1.3 0.9 0.8 0.4];
f = [5:5:50];
tsys = 0.25;

modelA41 = frd (r, f)
returns
```

From input 1 to:

```
Frequency(rad/s)        output 1
-----------------       ---------
        5                  0.2
       10                  0.4
       15                  1.3
       20                  1.5
       25                  1.7
       30                  1.4
       35                  1.3
       40                  0.9
       45                  0.8
       50                  0.4
```

Continuous-time frequency response data model.

```
modelA42 = frd (r, f, tsys, 'Unit', 'Hz')
returns
```

From input 1 to:

```
Frequency(Hz)   output 1
-------------   --------
       5          0.2
      10          0.4
      15          1.3
      20          1.5
      25          1.7
      30          1.4
      35          1.3
      40          0.9
      45          0.8
      50          0.4
```

modelA42=
Note that the values of frequency is being expressed in Hz instead of the default unit rad/sec

Sampling time: 0.25
Discrete-time frequency response data model.

Unit of frequency can also be changed to 'Hz' using the **ChangeUnit** function as follows:

modelA43 = Chgunits (modelA41, 'Hz')
returns

From input 1 to:

Frequency(Hz)	output 1
0.795775	0.2
1.591549	0.4
2.387324	1.3
3.183099	1.5
3.978874	1.7
4.774648	1.4
5.570423	1.3
6.366198	0.9
7.161972	0.8
7.957747	0.4

> modelA43=
> Note that the value of frequency has changed to its equivalent value in Hz as per the following relationship:
> $1 Hz = 2\pi$ rad/sec

Continuous-time frequency response data model.

modelA44 = Chgunits (modelA42, 'rad/s')
returns

From input 1 to:

Frequency(rad/s)	output 1
31.415927	0.2
62.831853	0.4
94.247780	1.3
125.663706	1.5
157.079633	1.7
188.495559	1.4
219.911486	1.3
251.327412	0.9
282.743339	0.8
314.159265	0.4

Sampling time: 0.25
Discrete-time frequency response data model.

The frd model can also be created by using the direct values of frequency and Response parameter, instead of specifying these first and then creating the model using these values. It is also possible to manipulate the property of such a model.

modelA45 = frd (0.2:2:20, 10:10:100, 'InputName', 'Signal Generator', 'OutputName', 'Frequency Meter', 'Ts', 0.1, 'Units', 'Hz', 'IoDelay', 1,'Notes', 'ac bridges')
returns

From input 'Signal Generator' to:

```
Frequency(Hz)    Frequency Meter
--------------   ----------------
    10                0.2
    20                2.2
    30                4.2
    40                6.2
    50                8.2
    60               10.2
    70               12.2
    80               14.2
    90               16.2
   100               18.2
```

I/O delay time (for all I/O pairs): 1

Sampling time: 0.1
Discrete-time frequency response data model.

Frequency response data model for MIMO systems can also be generated at different frequency points in the form of an array. Let us take up an example with 4 frequency points, 2 inputs and 3 outputs and develop frd model for the same. The array formation can be understood with the help of Figure 1.4.

Let us start with the formulation of f vector and r array of 3-by-2-by-4 (*i.e.*, Ny-by-Nu-by-Nf) dimension as follows:

```
f=[15:15:60];

r=[111 121;  211  221; 311  321];
r(:,:,2)=[112 122;  212  222; 312  322];
r(:,:,3)=[113 123;  213  223; 313  323];
r(:,:,4)=[114 124;  214  224; 314  324];

modelA46=frd(r,f)
returns
```

From input 1 to:

```
Frequency(rad/s)   output 1   output 2   output 3
-----------------  --------   --------   --------
     15              111        211        311
     30              112        212        312
     45              113        213        313
     60              114        214        314
```

From input 2 to:

```
Frequency(rad/s)   output 1   output 2   output 3
-----------------  --------   --------   --------
     15              121        221        321
     30              122        222        322
     45              123        223        323
     60              124        224        324
```

Continuous-time frequency response data model.

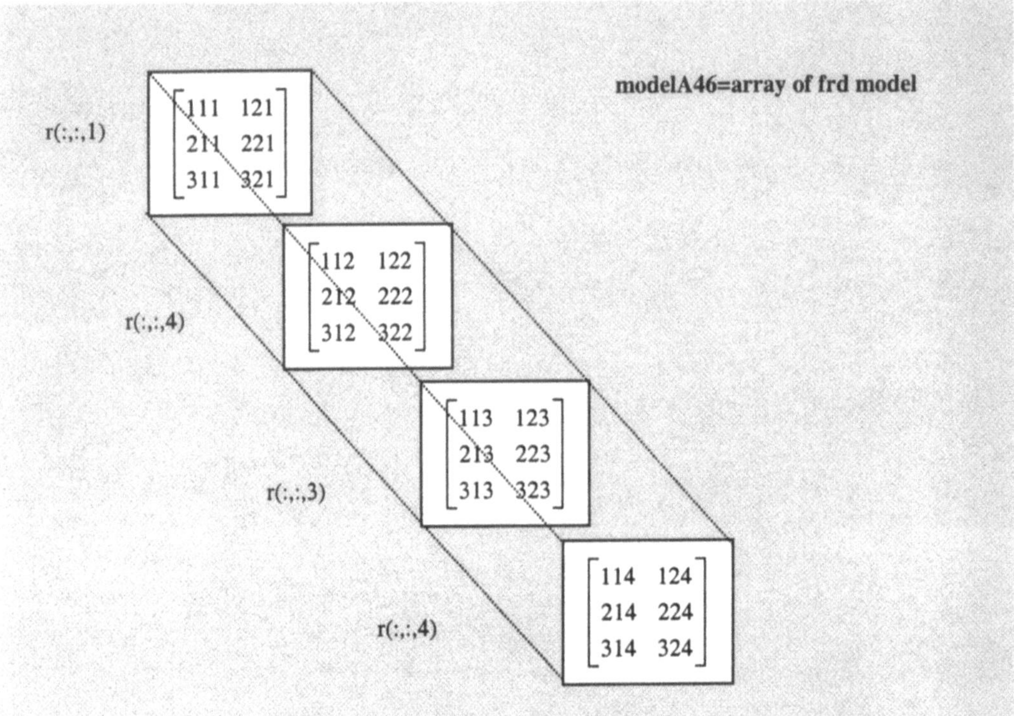

Figure 1.4. frd array model

Was all this too complicated? We don't think so. We are sure you will fare well in this practice test also...

Practice Test 1.4.

1. Create an frd model – modelAp21, for frequency variation of 1 to 100 in steps of 2 and response variation of 1.05 to 1.55 with a variation of 0.05, with sampling time of 0.01 sec and InputName as 'Wein's Bridge' and OutputName as 'Transducer'.

2. Create an frd modelAp22 with frequency points as 0 to $\pi/2$ at an interval of $\pi/6$ with 1 input resulting into 2 responses which are **sin** function of frequency and **cos** function of frequency.

1.2.5 Generation of Random Stable Models

In the previous sections, we specified the values of the parameters and the Control System Toolbox generated a model. In addition to this, the Control System Toolbox also provides functions for the generation of random stable models for continuous as well as discrete systems. These models may be created using any of the following functions:

- **rss** function to obtain a random state space stable model for continuous-time system, with syntax as:

ModelResult = **rss** (*n*, *Ny*, *Nu*, *S1*… *Sp*)

- **drss** function to obtain a random state space stable model for discrete-time system, with syntax as:

ModelResult = **drss** (*n*, *Ny*, *Nu*, *S1*…*Sp*)

- **rmodel** function to obtain a random state space stable model for a continuous-time system which can be used to obtain the **num** and **den** parameters of tf model or the **a**, **b**, **c** and **d** parameters of ss model, with syntax as:

[num,den] = **rmodel** (*n*, *Ny*, *Nu*)
[a,b,c,d] = **rmodel** (*n*, *Ny*, *Nu*)

- **drmodel** function to obtain a random state space stable model for a discrete-time system which can be used to obtain the **num** and **den** parameters of tf model or the **a**, **b**, **c** and **d** parameters of dss model, with syntax as:

[num,den] = **drmodel** (*n*, *Ny*, *Nu*)

[a,b,c,d]   = **drmodel** (*n*, *Ny*, *Nu*)

where,
n .... specifies the order of the model
Ny .... specifies the number of output of the model (default value is 1)
Nu .... specifies the number of input of the model (default value is 1)
S1, S2… Sp .... for *S1*-by-*S2*-by-… *Sp* array of model

Note that all the discrete models thus obtained have unspecified sampling time. Examples considered below will further elucidate this concept.

Examples:

```
modelA47=rss(3,4,5,1,2)
returns

Model modelA47(:,:,1,1)
==========================

a =
              x1        x2        x3
      x1   -0.36837   0.20275   0.14925
      x2   -0.23638  -0.64783   0.51501
      x3    0.086654 -0.52916  -0.59924
```

modelA47=
a stable continuous state space model of 1-by-2 array of 3rd order (*i.e.*, 3 no. of states), with 4 outputs, 5 inputs.

```
b =
              u1        u2        u3        u4        u5
    x1      1.254     0.57115   0.81562    0.6686    -0.01979
    x2      0         0         0          1.1908    -0.15672
    x3      0         0.69      1.2902    -1.2025    -1.6041

c =
              x1        x2        x3
    y1      0.38034   4.3192e-005  0.42818
    y2     -1.0091    -0.31786    0.89564
    y3     -0.019511   1.095      0.73096
    y4     -0.048221  -1.874      0.57786

d =
              u1        u2        u3        u4        u5
    y1      0         0        1.0823     0         0
    y2      0         0.23788   0         0         0.56896
    y3      0.94089   0         0         0        -0.82171
    y4      0         0         0        -0.9499    0


Model modelA47(:,:,1,2)
=========================

a =
              x1        x2        x3
    x1     -1.5197    0.28462    0.82205
    x2      0.28462  -1.0451    -0.22072
    x3      0.82205  -0.22072   -2.3343

b =
              u1        u2        u3        u4        u5
    x1      0.2573     0         0        -1.0106    1.6924
    x2     -1.0565    0.52874   -2.1707    0         0.59128
    x3      1.4151    0.21932   -0.059188  0.50774   0

c =
              x1        x2        x3
    y1      0         0         0.11844
    y2      0         0         0.31481
    y3      0.5689    -1.4751    1.4435
    y4     -0.25565    0        -0.35097

d =
              u1        u2        u3        u4        u5
    y1      0         0.32737   -0.94715   1.4725    -1.1283
    y2     -2.2023    0.23406   -0.37443   0         0
    y3      0.98634    0         0        -1.2173    0
    y4      0        -1.0039     0        -0.041227  0


1x2 array of continuous-time state-space models.
```

```
modelA48=drss(2,2,2)
returns

a =
                x1           x2
      x1      0.58343    -0.047257
      x2     -0.047257     0.14064


b =
                u1           u2
      x1       0.12864     -1.1678
      x2          0            0


c =
                x1           x2
      y1          0            0
      y2          0         0.93122


d =
                u1           u2
      y1       0.011245        0
      y2      -0.64515      0.23163


Sampling time: unspecified
Discrete-time model.

[a,b,c,d]=rmodel(3,4,5)
returns

a =
   -0.1342    0.5449    -0.0355
    0.5449   -0.2014    -0.2424
   -0.0355   -0.2424     0.5391

b =
    0.0129    0.0928    0.6085        0         0
    0.3840    0.0353    0.0158    0.5869    0.6315
    0.6831    0.6124       0      0.0576    0.7176

c =
        0     0.3533     0.7275
    0.0841       0       0.4784
        0     0.6756     0.5548
    0.4418       0       0.1210

d =
    0.4508    0.2548    0.9084    0.0784       0
    0.7159       0      0.2319    0.6408       0
    0.8928    0.2324       0         0      0.9943
        0     0.8049    0.0498       0      0.4398
```

modelA48=
a stable discrete state space model of 2nd order (*i.e.*, 2 number of states), with 2 inputs and unspecified sampling time.

value of a, b, c and d matrices are obtained for a discrete-time stable state space model of 3rd order with 4 outputs and 5 inputs

Similarly, you can obtain numerator and denominator parameters with **rmodel** function.

```
[num,den] = rmodel ( 4, 3, 2)
returns

num =
      0      0      0      0    -1.2919
      0      0      0   -0.0729    0.1079
      0      0  -0.3306  0.1501    0.2574
```

> coefficients of numerator and denominator with descending power of s is obtained for a continuous-time system of 4th order (highest power of s) with 2 inputs and 3 outputs.

```
den =
   1.0000   5.5249   7.0145   1.8952      0
```

Similarly, you can obtain numerator and denominator parameters with **drmodel** function for discrete system.

Ready to check yourself?

Practice Test 1.5.

1. Using **rss** and **drss** functions obtain the following:

- modelAp23 and modelAp24 of 3-by-2 array 4th-order, with 1 input and 5 outputs.
- modelAp25 and modelAp26 of 1-by-2 array of 2nd order, with 2 inputs and 3 outputs.

2. Using **rmodel and drmodel** functions obtain the following:

- numerator and denominator of 5th-order model, with 3 input and 2 outputs.
- a,b,c,d matrices of 3rd-order model, with 4 inputs and 2 outputs.

Exercise for Chapter 1:

1. Create the following zpk modelAe1 with properties as indicated:

$$modelAe1 = \begin{bmatrix} \dfrac{10(s+1)}{(s+4)(s-3)} & 0 \\[2em] \dfrac{15(s+8)(s+7)}{s(s+2+2i)(s+2-2i)} & \dfrac{20(s+1)(s-4i)(s+4i)}{s(s+2)(s-3)(s+5)} \\[2em] -1 & \dfrac{5(s+0.1)}{(s+3)(s+7)(s+9)} \end{bmatrix}$$

$$\text{InputName} = \text{a, b}$$

$$\text{OutputName} = \text{r, s, t}$$

$$\text{Variable} = \text{p}$$

$$\text{InputDelay} = 0.01$$

$$\text{OutputDelay} = 0.02 \text{ to } 0.18 \text{ in steps of } 0.06$$

$$\text{InputGroup} = \text{Channel Input1}$$

$$\text{OutputGroup} = \text{Channel Output2}$$

$$\text{Notes} = \text{This is a simple problem}$$

$$\text{UserData} = \text{I can create models and assign properties to them.}$$

2. Create the following tf modelAe2 which inherits the properties from zpk modelAe1 created above:

$$\text{modelAe2} = \begin{bmatrix} \dfrac{3s^4 - 2s}{5s^5 + 6s^4 - 2s + 5} & \dfrac{s^2 + 1}{3s^2 - 4s + 8} \\ -1 & \dfrac{1}{s+1} \\ \dfrac{s+3}{s^2 - 2s + 1} & 0 \end{bmatrix}$$

3. Create the following ss modelAe3 which inherits the properties from zpk modelAe1 created above:

$$\text{modelAe3} =$$

$$[\dot{x}] = \begin{bmatrix} -1 & -4 & -6 \\ 3 & -2 & 5 \\ 7 & 2 & -3 \end{bmatrix} [x] + \begin{bmatrix} -1 & 1 \\ -1 & 3 \\ 0 & 2 \end{bmatrix} [u]$$

$$[y] = \begin{bmatrix} 1 & 2 & -1 \\ 1 & 0 & -3 \\ 0 & -2 & -1 \end{bmatrix} [x] + \begin{bmatrix} -1 & 0 \\ 0 & -2 \\ -1 & 1 \end{bmatrix} [u]$$

4. Create the following dss modelAe4 which inherits the properties from zpk modelAe1 created above:

$$\text{modelAe4} =$$

$$\begin{bmatrix} -1 & 0 & 0 \\ 0 & -2 & 0 \\ 0 & 0 & -3 \end{bmatrix} [\dot{x}] = \begin{bmatrix} -1 & -4 & -6 \\ 3 & -2 & 5 \\ 7 & 2 & -3 \end{bmatrix} [x] + \begin{bmatrix} -1 & 1 \\ -1 & 3 \\ 0 & 2 \end{bmatrix} [u]$$

$$[y] = \begin{bmatrix} 1 & 2 & -1 \\ 1 & 0 & -3 \\ 0 & -2 & -1 \end{bmatrix} [x] + \begin{bmatrix} -1 & 0 \\ 0 & -2 \\ -1 & 1 \end{bmatrix} [u]$$

5. Create frd modelAe5 with data given below so as to inherit the properties from zpk modelAe1 created above:

Frequency:		2 KHz	20 KHz	200 KHz	2000 KHz
Response:					
Input1	Output1	0.2	0.4	0.8	1.6
	Output2	2	4	8	16
	Output3	20	40	80	160
Input2	Output1	0.2×10^{-3}	0.4×10^{-3}	0.8×10^{-3}	1.6×10^{-3}
	Output2	2×10^{-3}	4×10^{-3}	8×10^{-3}	16×10^{-3}
	Output3	200×10^{-3}	400×10^{-3}	800×10^{-3}	160×10^{-3}

6. Create the following tf modelAe6 with the properties as indicated:

$$\begin{bmatrix} q^2 \dfrac{1+q+q^2}{1+q+q^2} & q^3 \dfrac{1+q+2q^2}{1+q+2q^2} \\ q^4 \dfrac{2+2q+q^2}{2+2q+6q^2} & q^5 \dfrac{2+2q+2q^2}{2+2q+7q^2} \end{bmatrix}$$

InputName = alpha, beta

OutputName = gamma, theta

Sampling Time = 1

InputDelay = 1 to 2

OutputDelay = 1 to 4 in steps of 2

Notes = This is an exercise model

UserData = This is a fictitious system!

7. Create the following zpk modelAe7 which inherits the properties from tf modelAe6 created above:

$$\begin{bmatrix} e^{-0.2p}\dfrac{5(p-5)(p-10)}{(p-3)} & e^{-0.6p}\dfrac{20(p-1.2)(p-2.2)(p-3.2)}{(p-0.5)(p-1.5)(p-2.5)} \\ e^{-0.4p}\dfrac{10(p-0.2)(p^2-0.8p+1.16)}{(p-6)(p^2+18p+81.25)} & e^{-0.8p}\dfrac{50(p-0.45)(p-0.6)(p-0.9)}{(p-1)(p-3)(p-6)} \end{bmatrix}$$

8. Create the following ss modelAe8 which inherits the properties from tf modelAe6 created above:

$$\dot{x} = \begin{bmatrix} -1 & -3 \\ 2 & 5 \end{bmatrix} x + \begin{bmatrix} 0 & 1 \\ 1 & 0 \end{bmatrix} u$$

$$y = \begin{bmatrix} 2 & -1 \\ 1 & 0 \end{bmatrix} x + \begin{bmatrix} 1 & 0 \\ 0 & 2 \end{bmatrix} u$$

Sampling Time : 0.5

9. Create the following dss modelAe9 which inherits the properties from tf modelAe6 created above:

$$\begin{bmatrix} 1 & 3 \\ -2 & 1 \end{bmatrix}\dot{x} = \begin{bmatrix} 0 & 3 \\ -2 & -5 \end{bmatrix} x + \begin{bmatrix} 1 & -2 \\ -5 & 1 \end{bmatrix} u$$

$$y = \begin{bmatrix} -2 & -1 \\ 1 & 0 \end{bmatrix} x + \begin{bmatrix} -1 & 2 \\ -2 & 3 \end{bmatrix} u$$

Sampling Time : 0.5

10. Create frd modelAe10 with data given below so as to inherit the properties from tf modelAe6 created above:

Frequency:		10 Hz	100 Hz	1000 Hz	10000 Hz	100000 Hz
Response:						
Input1	Output1	1.2	0	9.4	0.012	0.55
	Output2	2.4	0.1	4.9	0.043	0.62
Input2	Output1	4.8	1.7	6.5	0.056	0.42
	Output2	9.6	7.2	5.6	0.098	0.76

11. Create a stable ss modelAe11 with dimension 5 by 4 by 3 by 2. Choosing property values on your own, assign properties to the model so generated.

12. Create a discrete stable tf modelAe12 of order 3 having 2 inputs and 4 outputs.

Chapter 2

Model Manipulation

In this chapter, you will learn to manipulate the models of systems/subsystems to obtain a larger system by performing arithmetic operations on them and/or interconnecting them.

The usual procedure for complete analysis of a large system with several small and distinct units, is to develop models for various parts of the system, and then manipulate these models to get complete representation of the system. These manipulations usually involve 'Arithmetic Operations' and 'Interconnections'.

MATLAB® supports several arithmetic operations and equivalent interconnections on the system models. In this chapter, you will learn how to achieve this. However, before we start, let us first study the 'Precedence Rule' and the 'Law of Property Inheritance' for the Linear Time Invariant (LTI) models which is followed during each manipulation on the model.

2.1 The Precedence Rule and the Law of Property Inheritance

Arithmetic operations can be done on LTI models of all types. In a similar way, interconnection of LTI models is also possible for all types of models. The type of resulting LTI model from these arithmetic operations and interconnections is not arbitrary, but is decided by the 'Precedence Rule' and the 'Law of Property Inheritance', which are explained below.

2.1.1 The Precedence Rule

According to the 'Precedence Rule', the models have implicit order of priority. The zpk model takes precedence over tf model, ss model takes precedence over zpk model and thus automatically over tf model, and frd model takes precedence over all the other three. This can be summarised as:

First Precedence:	frd model
Second Precedence:	ss model
Third Precedence:	zpk model
Fourth and Last Precedence:	tf model

Table 2.1. The Law of Property Inheritance

Property Name	Remarks
InputName and OutputName	• If the models involved in the operations have the same input or output names, then the resulting model will also have the same name. • If the models involved have different names, then the resulting model does not inherit either of the names and these properties remain unspecified. • If for several models involved, these properties have been specified for only one model, then the resulting model will also have the same name.
InputGroup and OutputGroup	Same as InputName and OutputName
Sampling Time	• If the models involved in the operations have the same Sampling Time, then the resulting model will also have the same Sampling Time as the original ones involved. • If the models involved in the operations are such that Sampling Time for one is specified while for others it is not, then the resulting model will have the same Sampling Time as that of the model for which Sampling Time was specified. • Operations cannot be carried out on models for which different Sampling Time have been specified.
InputDelay and OutputDelay	Models involved must have identical delay times.
Variable	This property is inherited from the operands of the tf or zpk model on which operations were carried out. However, in case of conflict the property value of the resulting model is decided as follows: For continuous-time models • 'p' has precedence over 's'. For discrete-time models • 'z^{-1}' has precedence over 'q' and 'z'. • 'q' has precedence over 'z' .
Notes	Most of the operations ignore this property.
UserData	It is usually ignored during most of the operations.

Table 2.2. Arithmetic operations on the LTI models

Operation	Description	Effect on Models/Remarks
+	Addition of LTI models	equivalent to connecting the models in parallel
-	Subtraction of LTI models	
*	Multiplication of LTI models	equivalent to connecting the models in series
inv	Inversion of LTI models	• equivalent to inverting the input-output relationship • valid only on square systems with same number of inputs and outputs • for a system with transfer function $y=G(s)u$, **inv** operation is equivalent to producing a system with transfer function as $H(s)=G(s)^{-1}$ *i.e.*, $u=H(s)y$
/	Right Divide of LTI models	equivalent to the operation **inv**(model1) * (model2)
\	Left Divide of LTI models	equivalent to the operation (model1) * **inv**(model2)
.'	Transposition of LTI models	• for tf models, cell arrays num and den are transposed • for zpk models, cell arrays z, p, k are transposed • for ss models, a, b, c, d, e matrices are transposed • for frd models, the matrix of frequency response data at each frequency is transposed
'	Pertransposition of LTI models	for a system with transfer function as $H(s)$ this is equivalent to a system with transfer function $[H(-s)]^{T}$
^	Power of LTI model	• LTI model powers • same as multiplication *i.e.*, connecting the same system in series
stack	Creating array of LTI models	stacks LTI models/arrays along specified array dimension

Thus, in any operation involving different types of models, the result would be:

- an frd model .... if there is one frd model involved in the operations
- an ss model .... if there is no frd model and at least one ss model involved in the operations
- a zpk model .... if there are no frd or ss models but at least one zpk model involved in the operations
- a tf model................... if there are no other types of models but only tf models involved in the operations

While doing operations on different types of models, MATLAB® converts all the models into one type of model according to the 'Precedence Rule', and then the operations are carried out. Therefore, the result after completion of the operations is naturally the type of model to which all the models were first converted by MATLAB® before carrying out the operations. To get the resulting model of a particular desired type overruling the 'Precedence Rules', you should either first convert all the models into the type desired and then proceed with the manipulations or you should convert the result obtained to the desired type. The method of conversion of one type of model to another is explained in Chapter 3.

2.1.2 The Law of Property Inheritance

If, operations are being carried out on systems with different properties, it is again a point to ponder as to what properties would be inherited by the resulting model.

The 'Law of Property Inheritance' gives a solution to this problem. This is summarised property-wise in Table 2.1.

2.2 Arithmetic Operation on Models

Keeping in mind that all the models developed earlier in this book have their genesis in a matrix, it is possible to have all types of arithmetic operations, which could be performed, on a matrix, conducted on these models too. The various arithmetic operations and their effects are summarised in the Table 2.2.

These arithmetic operations on the models are explained in more details in the lines to follow.

2.2.1 Addition and Subtraction

As mentioned in Table 2.2, addition and/or subtraction of two or more LTI models is actually equivalent to connecting them in parallel, which means that the inputs of all the models are connected together and similarly the outputs of all the models are connected together. The only restriction being, all the systems should be either continuous or discrete with the same sampling time. There is no limitation as far as the type of the models is concerned. The parallel connection or the addition and subtraction of the models can be achieved using any of the following syntax:

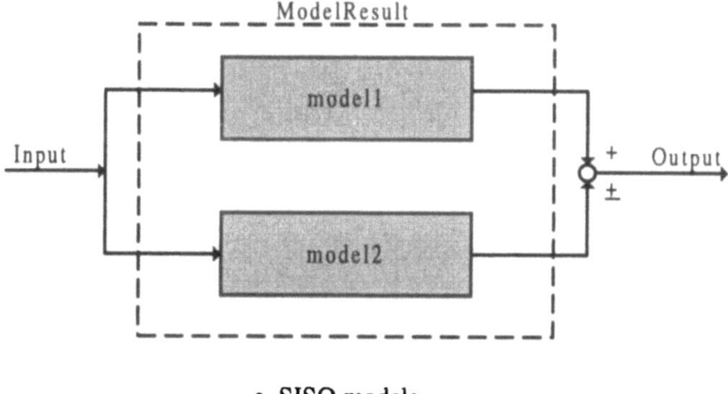

a. SISO models

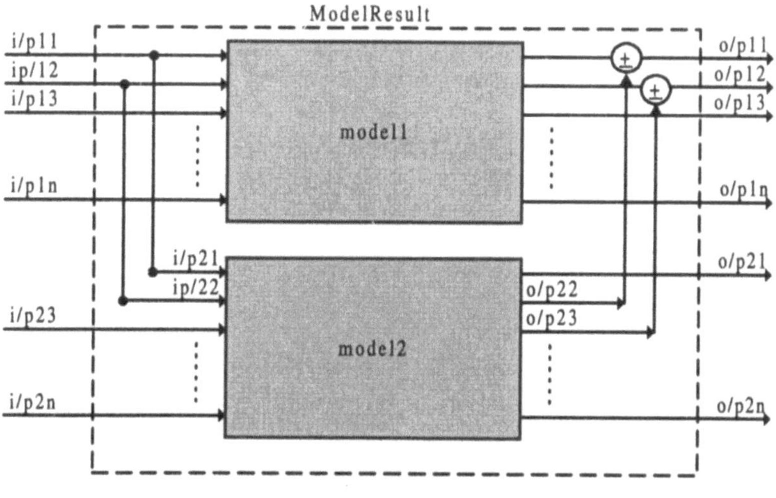

b. MIMO models

Figure 2.1. Addition/subtraction operation or parallel connection of LTI models

- ModelResult = *model1* ± *model2*...

- ModelResult = **parallel** (*model1*, *model2*, *i/p1*,*i/p2*,*o/p1*,*o/p2*)

where,

ModelResult......................	is user-specified name of the resulting model after addition/parallel operation
model1, model2................	are the names of the models involved
i/p1, *i/p2*...........................	are vectors of inputs of the two models whose indices indicate the inputs of *model1* to be connected to the to the inputs of *model2*
o/p1, *o/p2*	are vectors of outputs of the two models whose indices indicate the outputs of *model1* to be added to/subtracted from the outputs of *model2*

It is to be noted that the addition and subtraction operation can be performed on any number of models, but Control System Toolbox supports parallel operation only on two models. Also, note that while doing parallel connection on a pair of MIMO model, connection between various input and output pairs is possible, but, it is not so with addition/subtraction operation wherein complete models are involved. For a SISO model, obviously there in no need to specify the input and the output pair.

In case of state space models with parameters as a1, b1, c1, d1 for the first model and a2, b2, c2, d2 for the second model, the result of parallel connection (or addition/subtraction operation) is:

$$\begin{bmatrix} a1 & 0 \\ 0 & b1 \end{bmatrix} \ , \ \begin{bmatrix} b1 \\ b2 \end{bmatrix} \ , \ \begin{bmatrix} c1 & c2 \end{bmatrix} \ , \ d1+d2$$

Addition or the parallel connection for SISO models and MIMO models can be better understood by the respective block diagrams given in Figure 2.1.

Let us consider some examples for better understanding.

Example:

Let us understand the concepts of addition/subtraction operation or parallel connection with the help of models generated in Chapter 1. Keep an open eye as to how the 'Precedence Rule' and the 'Law of Property Inheritance' are followed during the operation.

modelB1= modelA1 + modelA17
returns

Zero/pole/gain:

5 (s-0.776) (s+0.7536) (s+2.462)

  (s-3) (s+2.213) (s-0.8133)

- The same result can be obtained using parallel function as follows:
  modelB1 = **parallel** (modelA1, modelA17)
- modelA1 was zpk model and modelA17 was tf model but the resultant modelB1 is a zpk model as zpk model has precedence over tf model.

modelB2 = modelA10 - modelA26 + modelA37
returns

a =

	x1	x2	x3	x4
x1	3	0	0	0
x2	0	6	-5.159	-1.3032
x3	0	0	-16	-4.0156
x4	0	0	16	0

- Parallel function will not operate as there are more than two models involved.
- modelA37 is ss model. Hence the resulting modelB7 is also a ss model as per the precedence rule.
- Properties of modelA10 are inherited by resulting modelB7.

```
b =
            Parameter1
     x1        4.4721
     x2        6.5928
     x3        17.395
     x4            0

c =
              x1          x2          x3          x4
   Result1    4.4721       0           0           0
   Result2       0      -6.5928     17.215      4.3488

d =
            Parameter1
   Result1        13
   Result2       -12

Sampling time: 0.1
Discrete-time model.
```

modelB3=modelA9-modelA45
returns

From input 'Signal Generator' to:

Frequency(Hz)	Frequency Meter
10	4.8 - 4.898587e-017i
20	2.8 - 1.077689e-015i
30	0.8 - 3.086110e-015i
40	- 1.2 - 6.074248e-015i
50	- 3.2 + 1.909015e-014i
60	- 5.2 - 1.498968e-014i
70	- 7.2 - 2.091697e-014i
80	- 9.2 - 2.782398e-014i
90	-11.2 + 7.939722e-014i
100	-13.2 + 8.474163e-014i

Sampling time: 0.1
Discrete-time frequency response data model.

modelB4 = modelA40 - modelA41
returns

From input 1 to:

Frequency(rad/s)	output 1	output 2	output 3
5	-0.355748+0.162751i	3.198686+0.039843i	-0.068164+0.232388i
10	-0.542889+0.234113i	3.001492+0.102423i	-0.251015+0.365711i
15	-1.403102+0.314925i	2.121351+0.152040i	-1.087445+0.501572i
20	-1.549696+0.388874i	1.949021+0.194327i	-1.201211+0.623757i
25	-1.687670+0.451541i	1.781424+0.229278i	-1.300824+0.726702i
30	-1.320975+0.501715i	2.116364+0.256978i	-0.892793+0.808933i
35	-1.152841+0.539722i	2.252099+0.277899i	-0.682395+0.871180i
40	-0.685737+0.566713i	2.687315+0.292791i	-0.173646+0.915409i
45	-0.521391+0.584223i	2.821094+0.302544i	0.030641+0.944164i
50	-0.060894+0.593885i	3.252859+0.308062i	0.528696+0.960120i

modelB3 = **parallel** (modelA9,-modelA45)
gives the same result

modelB4 = **parallel** (model40,-model41)
gives the same result

From input 2 to:

Frequency(rad/s)	output 1	output 2	output 3
5	3.644252+0.162751i	5.198686+0.039843i	-1.068164+0.232388i
10	3.457111+0.234113i	5.001492+0.102423i	-1.251015+0.365711i
15	2.596898+0.314925i	4.121351+0.152040i	-2.087445+0.501572i
20	2.450304+0.388874i	3.949021+0.194327i	-2.201211+0.623757i
25	2.312330+0.451541i	3.781424+0.229278i	-2.300824+0.726702i
30	2.679025+0.501715i	4.116364+0.256978i	-1.892793+0.808933i
35	2.847159+0.539722i	4.252099+0.277899i	-1.682395+0.871180i
40	3.314263+0.566713i	4.687315+0.292791i	-1.173646+0.915409i
45	3.478609+0.584223i	4.821094+0.302544i	-0.969359+0.944164i
50	3.939106+0.593885i	5.252859+0.308062i	-0.471304+0.960120i

From input 3 to:

Frequency(rad/s)	output 1	output 2	output 3
5	2.644252+0.162751i	8.198686+0.039843i	3.931836+0.232388i
10	2.457111+0.234113i	8.001492+0.102423i	3.748985+0.365711i
15	1.596898+0.314925i	7.121351+0.152040i	2.912555+0.501572i
20	1.450304+0.388874i	6.949021+0.194327i	2.798789+0.623757i
25	1.312330+0.451541i	6.781424+0.229278i	2.699176+0.726702i
30	1.679025+0.501715i	7.116364+0.256978i	3.107207+0.808933i
35	1.847159+0.539722i	7.252099+0.277899i	3.317605+0.871180i
40	2.314263+0.566713i	7.687315+0.292791i	3.826354+0.915409i
45	2.478609+0.584223i	7.821094+0.302544i	4.030641+0.944164i
50	2.939106+0.593885i	8.252859+0.308062i	4.528696+0.960120i

Continuous-time frequency response data model.

modelB5 = parallel (modelA13,-modelA32,[2 3],[1 2],[3,4],[2 1])
returns

Zero/pole/gain from input "A" to output...

$$M: \quad \frac{11\ (z-0.11)}{z^2\ (z-1)^2}$$

$$N: \quad \frac{21\ (z-0.21)}{z^4\ (z-2)\ (z-1)}$$

modelB5=

modelA32 is connected in parallel to modelA13 such that

- input1 of modelA32 and input2 of modelA13 are connected together
- input2 of modelA32 and input3 of modelA13 are connected together
- output2 of modelA32 is subtracted from output3 of modelA13
- output1 of modelA32 is subtracted from output4 modelA13

Properties of modelA13 is inherited by resultant modelB5 which is zpk model according to the 'Precedence Rule'

$$Q: \quad \frac{3.1401e\text{-}014\ (z+0.07757)\ (z-0.001114)\ (z+23.15)\ (z-24.06)\ (z+3.535e\text{-}006)\ (z^2 + 0.9211z + 557.7)\ (z^2 - 0.4009z + 1.061e013)}{z^{10}\ (z-5)\ (z-1)}$$

$$O: \quad \frac{31\ (z-0.31)}{z^6\ (z-3)\ (z-1)}$$

$$P: \frac{3.4116e\text{-}014\ (z\text{-}6.146)\ (z+5.753)\ (z\text{-}1.008)\ (z^2 + 1.013z + 1.057)\ (z^2 + 0.01563z + 1.1e012)}{z^8\ (z\text{-}4)\ (z\text{-}1)}$$

Zero/pole/gain from input "D" to output...

$$M: \frac{1.3873e\text{-}014\ (z\text{-}0.1392)\ (z^2 + 0.1652z + 1.009e015)}{z^5\ (z\text{-}1)\ (z\text{-}4)}$$

$$N: \frac{6.3507e\text{-}014\ (z+3486)\ (z\text{-}3443)\ (z\text{-}0.206)\ (z^2 - 43.12z + 8.753e011)}{z^7\ (z\text{-}2)\ (z\text{-}4)}$$

$$Q: \frac{-3.4174e\text{-}013\ (z+8261)\ (z\text{-}8261)\ (z+10.3)\ (z+0.3876)\ (z\text{-}0.2256)\ (z^2 - 18.67z + 108.6)\ (z^2 + 12.82z + 111.4)\ (z^2 - 4.621z + 114.6)}{z^{13}\ (z\text{-}5)\ (z\text{-}4)}$$

$$O: \frac{5.3712e\text{-}014\ (z+9.537e005)\ (z\text{-}9.53e005)\ (z\text{-}648.1)\ (z\text{-}26.46)}{z^9\ (z\text{-}3)\ (z\text{-}4)}$$

$$P: \frac{2.5198e\text{-}013\ (z\text{-}203.6)\ (z\text{-}0.6206)\ (z^2 + 5849z + 1.148e007)\ (z^2 - 5645z + 1.109e007)}{z^{11}\ (z\text{-}4)^2}$$

Zero/pole/gain from input "E" to output...

$$M: \frac{15\ (z\text{-}0.15)}{z^6\ (z\text{-}1)\ (z\text{-}5)}$$

$$N: \frac{25\ (z\text{-}0.25)}{z^8\ (z\text{-}2)\ (z\text{-}5)}$$

$$Q: \frac{4.2796e\text{-}013\ (z\text{-}1.831)\ (z\text{-}0.99)\ (z\text{-}0.5522)\ (z+1.15)\ (z+1.489)\ (z^2 + 1.662z + 1.303)\ (z^2 - 1.395z + 1.123)\ (z^2 + 0.1245z + 1.24)\ (z^2 + 8.053e008)}{z^{14}\ (z\text{-}5)^2}$$

$$O: \frac{-7.1403e\text{-}014\ (z\text{-}0.06121)\ (z+0.267)\ (z\text{-}0.003675)\ (z^2 - 5.825z + 16.91)\ (z^2 + 5.829z + 17.05)\ (z^2 + 1.088z + 1.732e010)}{z^{10}\ (z\text{-}3)\ (z\text{-}5)}$$

$$-1.9946\text{e-}013 \ (z\text{-}3.123) \ (z\text{+}2.119) \ (z\text{-}0.4477) \ (z^2 + 2.134z + 1.484)$$

$$(z^2 - 1.828z + 1.195) \ (z^2 + 0.142z + 1.38) \ (z^2 - 0.09763z + 2.325\text{e}009)$$

$$P: \ \text{---}$$

$$z^{12} \ (z\text{-}4) \ (z\text{-}5)$$

Zero/pole/gain from input "B" to output...

$$12 \ (z\text{-}0.12)$$

$$M: \ \text{-----------------}$$

$$z^3 \ (z\text{-}1) \ (z\text{-}2)$$

$$22 \ (z\text{-}0.22)$$

$$N: \ \text{-----------}$$

$$z^5 \ (z\text{-}2)^2$$

$$-1.0645\text{e-}013 \ (z\text{-}4.035\text{e}004) \ (z\text{+}4.03\text{e}004) \ (z\text{+}1.112\text{e}004) \ (z\text{-}1.112\text{e}004)$$

$$(z\text{+}43.52) \ (z\text{-}0.8057)$$

$$Q: \ \text{---}$$

$$z^{11} \ (z\text{-}5) \ (z\text{-}2)$$

$$-0.44444 \ (z\text{-}3.542) \ (z\text{+}1.416) \ (z\text{-}0.8025) \ (z\text{-}0.32) \ (z^2 + 0.9625z + 1.04)$$

$$(z^2 - 4.527z + 7.146) \ (z^2 + 0.06261z + 2.141)$$

$$O: \ \text{---}$$

$$z^7 \ (z\text{-}3) \ (z\text{-}2) \ (z\text{-}0.8089) \ (z^2 + 0.8089z + 1.099)$$

$$- \ (z\text{-}4.001) \ (z\text{-}0.42) \ (z\text{-}0.1506) \ (z^2 - 3.477z + 3.027) \ (z^2 + 2.915z + 2.266)$$

$$(z^2 - 1.406z + 1.389) \ (z^2 + 1.455z + 1.601) \ (z^2 - 0.166z + 1.142)$$

$$P: \ \text{---}$$

$$z^9 \ (z\text{-}4) \ (z\text{-}2) \ (z\text{-}0.1506) \ (z^2 - 0.5994z + 1.66)$$

Zero/pole/gain from input "C" to output...

$$13 \ (z\text{-}0.13)$$

$$M: \ \text{-----------------}$$

$$z^4 \ (z\text{-}1) \ (z\text{-}3)$$

$$-1.8283\text{e-}014 \ (z\text{-}1.583\text{e}006) \ (z\text{+}1.582\text{e}006) \ (z^2 + 499.3z + 1.317\text{e}007)$$

$$N: \ \text{---}$$

$$z^6 \ (z\text{-}2) \ (z\text{-}3)$$

$$2.7383\text{e-}013 \ (z\text{+}0.41) \ (z\text{-}0.5206) \ (z\text{-}0.07105) \ (z\text{+}1.103\text{e}005) \ (z\text{-}1.103\text{e}005)$$

$$(z^2 + 2.113z + 1.446) \ (z^2 - 1.746z + 1.132) \ (z^2 + 0.154z + 1.326)$$

$$Q: \ \text{---}$$

$$z^{12} \ (z\text{-}5) \ (z\text{-}3)$$

```
   -0.16667 (z+4.718) (z-3.432) (z-1.174) (z+1.366) (z+0.3769) (z-0.3298)

          (z-0.2883) (z^2 - 4.171z + 4.842) (z^2 + 1.076z + 1.139)

                              (z^2 - 0.1416z + 2.12)
O: ------------------------------------------------------------------
      z^8 (z-3)^2 (z-1.1) (z-0.2882) (z+0.3768) (z^2 + 1.011z + 1.395)

   -2.5 z (z-4.004) (z-2.979) (z+1.12) (z-0.43) (z^2 - 2.146z + 1.239)


         (z^2 + 1.522z + 1.062) (z^2 + 0.6502z + 1.017) (z^2 - 0.9337z + 1.442)
P: ----------------------------------------------------------------------------
            z^11 (z-4) (z-3) (z-1.145) (z^2 + 1.145z + 1.31)


I/O groups:
    Group name      I/O    Channel(s)
    Parameter12      I       1,4
    Parameter35      I       3,5
    Parameter 4      I       2
    Result24         O       2,5
    Result 13        O       1,4


Sampling time: 0.5
```

Got the concepts? Better check it up! Take a practice test…

Practice Test 2.1.

1. Create the following models:

$$
\text{modelBp1} =
\begin{bmatrix}
\dfrac{0.1(s+1)}{s(s+3)(s-6)} & \dfrac{0.2(s-4)}{s(s+3)(s-6)} & \dfrac{0.4(s+2)}{s(s+3)(s-6)} \\[2mm]
\dfrac{0.5(s-6)}{s(s-2)(s-3)} & \dfrac{0.6(s-7)}{s(s-2)(s-3)} & \dfrac{0.7(s-5)}{s(s-2)(s-3)} \\[2mm]
\dfrac{0.8(s-3)}{s(s+2)(s-5)} & \dfrac{0.9(s-1)}{s(s+2)(s-5)} & \dfrac{1(s-3)}{s(s+2)(s-5)}
\end{bmatrix}
$$

$$
\text{modelBp2} =
\begin{bmatrix}
\dfrac{s^2-2s+1}{2s^5-3s^2+4s-1} & \dfrac{s^2+4s+3}{2s^5-3s^2+4s-1} & \dfrac{3s+1}{2s^5-3s^2+4s-1} & \dfrac{s^2+1}{2s^5-3s^2+4s-1} \\[2mm]
\dfrac{3s^3-s^2-2s+1}{2s^5-3s^2+4s-1} & \dfrac{s+1}{2s^5-3s^2+4s-1} & \dfrac{5s^2+1}{2s^5-3s^2+4s-1} & \dfrac{s}{2s^5-3s^2+4s-1} \\[2mm]
\dfrac{4s}{2s^5-3s^2+4s-1} & \dfrac{4s^4-3s^3+5}{2s^5-3s^2+4s-1} & \dfrac{s^2+2s-7}{2s^5-3s^2+4s-1} & \dfrac{s^2+2s}{2s^5-3s^2+4s-1} \\[2mm]
\dfrac{5s-2}{2s^5-3s^2+4s-1} & \dfrac{2s^2+3}{2s^5-3s^2+4s-1} & \dfrac{3s+5s-2}{2s^5-3s^2+4s-1} & \dfrac{s^2-1}{2s^5-3s^2+4s-1}
\end{bmatrix}
$$

a) Add the two models just created and name the resulting model as modelBp3.

b) Connect the two models just created in parallel to obtain a new model -- modelBp4 such that input1 and input2 of both the models are connected together while output3 and output4 of both the models are added respectively.

c) Connect the two models just created in parallel to obtain a new modelBp5 such that input2 and input3 of modelBp1 is connected to input1 and input2 of modelB2 while output1 and output3 of modelBp1 are added to output2 and 4 of modelBp2 respectively.

d) Connect the two models just created in parallel to obtain a new modelBp6 such that input1 and input4 of modelBp1 is connected to input2 and input3 of modelBp2 while output3 and output1 of modelBp1 is subtracted from output2 and output4 of modeBp2 respectively.

2.2.2 Multiplication

As mentioned in the Table 2.2, multiplication of two or more LTI models is infact equivalent to connecting them in series. This means that the output of first model is connected to the input of the second model and so on. The restriction for the choice of these models is again the same as for parallel connection *i.e.,* all the systems should be either all continuous or all discrete. In the latter case, all the models must have the same sampling time. There is no limitation as far as the types of the model are concerned. The series connection or the multiplication of the models can be achieved by any of the following ways:

- ModelResult = *model1 * model1…*

- ModelResult = **series** (*model1, model2, o/p1, i/p2*)

where,

ModelResult……………….. is user-specified name of the resulting model after multiplication/series operation

model1, model2……………. are the names of the models involved

o/p1, i/p2…………………. are the vector of output-input pair of *model1* and *model2* respectively whose indices indicate which output of *model1* is to be connected to the input of *model2*

It is to be noted that the multiplication operation can be performed on any number of models, but Control System Toolbox supports the series operation only on two models. Also, while doing series connection on a pair of MIMO model, connection between various output-input pairs is possible, but, it is not so with multiplication operation. For a SISO model, obviously there in no need to specify the input and the output pair.

In case of state space models with parameters as a1, b1, c1, d1 for the first model and a2, b2, c2, d2 for the second model, the result of series connection or multiplication operation is:

$$\begin{bmatrix} a1 & b1c2 \\ 0 & a2 \end{bmatrix}, \begin{bmatrix} b1d2 \\ b2 \end{bmatrix}, \begin{bmatrix} c1 & d1c2 \end{bmatrix}, d1d2$$

The multiplication or the series connection for SISO models and MIMO models can be better understood by the block diagrams given in Figure 2.2.

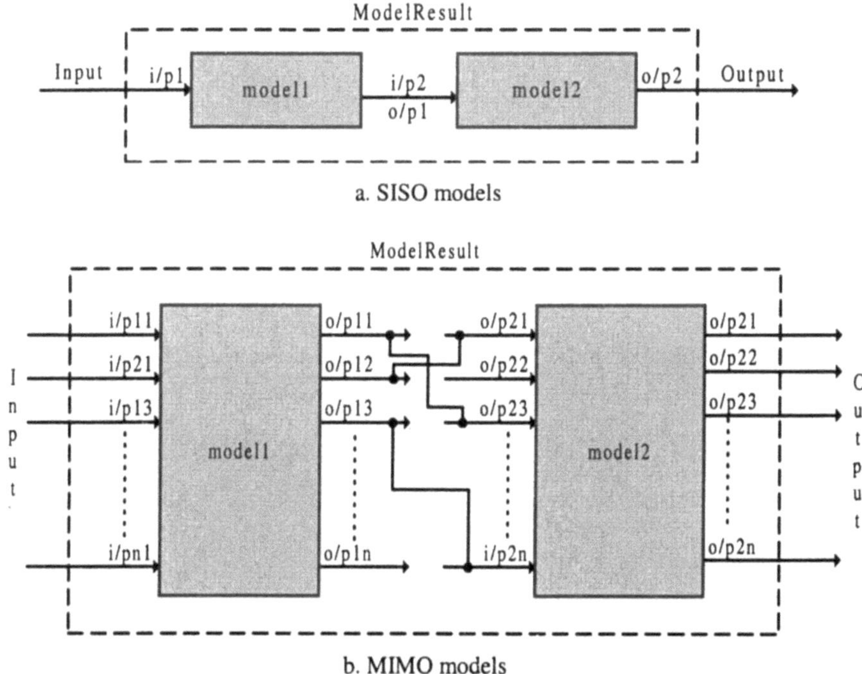

Figure 2.2. Multiplication Operation or Series Connection on LTI Models

Example:

Let us do the multiplication operation on some of the models created in Chapter 1 of this book to get a better understanding of the concept. Watch as to how the 'Precedence Rule' and the 'Law of Property Inheritance' are followed during operations.

```
modelB6 = modelA1 * modelA17
returns

Zero/pole/gain:
     (s+1) (s-3)
--------------------------
(s-3) (s+2.213) (s-0.8133)
```

modelB6 = series(modelA1, modelA17)
gives the same result

```
modelB7 = modelA10 *(- modelA26) * modelA37
```

returns

a =

	x1	x2	x3	x4
x1	3	0	0	0
x2	0	6	-4.2302	-0.99287
x3	0	0	-16	-4.0156
x4	0	0	16	0

b =

	u1
x1	63.246
x2	63.246
x3	251.18
x4	0

c =

	x1	x2	x3	x4
Result1	-6.3246	0	0	0
Result2	0	25.298	-26.754	-6.2794

d =

	u1
Result1	-100
Result2	400

Sampling time: 0.1
Discrete-time model.

modelB8=modelA9*(-modelA45)
returns

From input 'Signal Generator' to:

```
Frequency(Hz)  output 1
-------------  --------
      10         - 1
      20         -11
      30         -21
      40         -31
      50         -41
      60         -51
      70         -61
      80         -71
      90         -81
     100         -91
```

I/O delay time (for all I/O pairs): 1

Sampling time: 0.1
Discrete-time frequency response data model.

modelB9 = modelA40*(- modelA41)
returns

From input 1 to:

> - Series function will not operate as there are more than two models involved.
> - modelA37 is ss model. Hence the resulting modelB7 is also a ss model as per the precedence rule.
> - Properties of modelA10 are inherited by resulting modelB7.

> modelB8 = **series**(modelA9, -modelA45)
> gives the same result

> modelB9 = **series** (modelA40,-modelA41)
> gives the same result

Frequency(rad/s)	output 1	output 2	output 3
5	0.031150-0.032550i	-0.679737-0.007969i	-0.026367-0.046478i
10	0.057155-0.093645i	-1.360597-0.040969i	-0.059594-0.146284i
15	0.134032-0.409402i	-4.447756-0.197653i	-0.276322-0.652044i
20	0.074543-0.583311i	-5.173531-0.291490i	-0.448183-0.935636i
25	-0.020961-0.767620i	-5.918421-0.389772i	-0.678600-1.235394i
30	-0.110635-0.702401i	-4.922909-0.359769i	-0.710090-1.132507i
35	-0.191307-0.701638i	-4.617729-0.361268i	-0.802887-1.132535i
40	-0.192837-0.510041i	-3.228584-0.263512i	-0.653718-0.823868i
45	-0.222887-0.467379i	-2.896875-0.242035i	-0.664512-0.755331i
50	-0.135643-0.237554i	-1.461144-0.123225i	-0.371478-0.384048i

From input 2 to:

Frequency(rad/s)	output 1	output 2	output 3
5	-0.768850-0.032550i	-1.079737-0.007969i	0.173633-0.046478i
10	-1.542845-0.093645i	-2.160597-0.040969i	0.340406-0.146284i
15	-5.065968-0.409402i	-7.047756-0.197653i	1.023678-0.652044i
20	-5.925457-0.583311i	-8.173531-0.291490i	1.051817-0.935636i
25	-6.820961-0.767620i	-9.318421-0.389772i	1.021400-1.235394i
30	-5.710635-0.702401i	-7.722909-0.359769i	0.689910-1.132507i
35	-5.391307-0.701638i	-7.217729-0.361268i	0.497113-1.132535i
40	-3.792837-0.510041i	-5.028584-0.263512i	0.246282-0.823868i
45	-3.422887-0.467379i	-4.496875-0.242035i	0.135488-0.755331i
50	-1.735643-0.237554i	-2.261144-0.123225i	0.028522-0.384048i

From input 3 to:

Frequency(rad/s)	output 1	output 2	output 3
5	-0.568850-0.032550i	- 1.679737-0.007969i	-0.826367-0.046478i
10	-1.142845-0.093645i	- 3.360597-0.040969i	-1.659594-0.146284i
15	-3.765968-0.409402i	-10.947756-0.197653i	-5.476322-0.652044i
20	-4.425457-0.583311i	-12.673531-0.291490i	-6.448183-0.935636i
25	-5.120961-0.767620i	-14.418421-0.389772i	-7.478600-1.235394i
30	-4.310635-0.702401i	-11.922909-0.359769i	-6.310090-1.132507i
35	-4.091307-0.701638i	-11.117729-0.361268i	-6.002887-1.132535i
40	-2.892837-0.510041i	- 7.728584-0.263512i	-4.253718-0.823868i
45	-2.622887-0.467379i	- 6.896875-0.242035i	-3.864512-0.755331i
50	-1.335643-0.237554i	- 3.461144-0.123225i	-1.971478-0.384048i

Continuous-time frequency response data model.

modelB10 = series (modelA13,-modelA32,[3 4],[2 1])
returns

modelB10=

modelA32 is connected in series with modelA13 such that

- output3 and output4 of modelA13 are connected to negative input2 and negative input1 of modelA32 respectively.

Properties of modelA13 is inherited by resulting modelB10

Zero/pole/gain from input "A" to output...

$$-118.5\ z\ (z-3.604)\ (z-1)\ (z-0.3947)\ (z+0.5494)\ (z^2 - 1.153z + 0.4036)$$

$$(z^2 + 0.2412z + 1.518)$$

#1: z^(-6) * --

$$z^3\ (z-1.145)\ (z-1)^2\ (z-3)\ (z-4)\ (z-0.1506)\ (z^2 + 1.145z + 1.31)$$

$$(z^2 - 0.5994z + 1.66)$$

$$-5.1667\ (z-3.94)\ (z+3.718)\ (z+1.078)\ (z+0.5298)\ (z-1)\ (z-0.456)$$

$$(z-0.2921)\ (z^2 - 1.707z + 0.8126)\ (z^2 + 0.7587z + 1.397)$$

#2: z^(-6) * --

$$z^2\ (z-1)^2\ (z-1.1)\ (z-0.8089)\ (z-3)\ (z-4)\ (z-0.2882)\ (z+0.3768)$$

$$(z^2 + 0.8089z + 1.099)\ (z^2 + 1.011z + 1.395)$$

Zero/pole/gain from input "B" to output...

$$-122\ z\ (z-3.606)\ (z-2)\ (z-0.4027)\ (z+0.5477)\ (z^2 - 1.152z + 0.4043)$$

$$(z^2 + 0.2378z + 1.518)$$

#1: z^(-7) * --

$$z^3\ (z-1.145)\ (z-2)^2\ (z-3)\ (z-4)\ (z-0.1506)\ (z^2 + 1.145z + 1.31)$$

$$(z^2 - 0.5994z + 1.66)$$

$$-5.3333\ (z-3.94)\ (z-2)\ (z-0.4652)\ (z-0.2937)\ (z+1.064)$$

$$(z+0.5333)\ (z+3.729)\ (z^2 - 1.704z + 0.8099)\ (z^2 + 0.7571z + 1.395)$$

#2: z^(-7) * --

$$z^2\ (z-1.1)\ (z-0.8089)\ (z-2)^2\ (z-3)\ (z-4)\ (z-0.2882)\ (z+0.3768)$$

$$(z^2 + 0.8089z + 1.099)\ (z^2 + 1.011z + 1.395)$$

Zero/pole/gain from input "C" to output...

$$-125.5\ z\ (z-3.607)\ (z-3)\ (z-0.4105)\ (z+0.5461)\ (z^2 - 1.152z + 0.405)$$

$$(z^2 + 0.2346z + 1.518)$$

#1: z^(-8) * --

$$z^3\ (z-1.145)\ (z-3)^3\ (z-4)\ (z-0.1506)\ (z^2 + 1.145z + 1.31)$$

$$(z^2 - 0.5994z + 1.66)$$

$$-5.5\ (z-3.94)\ (z-3)\ (z+3.739)\ (z-0.4745)\ (z-0.2951)\ (z+1.049)$$

$$(z+0.5367)\ (z^2 - 1.701z + 0.8074)\ (z^2 + 0.7556z + 1.393)$$

#2: z^(-8) * --

$$z^2\ (z-1.1)\ (z-0.8089)\ (z-3)^3\ (z-4)\ (z-0.2882)\ (z+0.3768)$$

$$(z^2 + 0.8089z + 1.099)\ (z^2 + 1.011z + 1.395)$$

Zero/pole/gain from input "D" to output...

```
         -129 z (z-4) (z-3.609) (z-0.4182) (z+0.5446) (z^2 - 1.152z + 0.4059)

                                    (z^2 + 0.2316z + 1.518)
 #1:  z^(-9) * ----------------------------------------------------------------
           z^3 (z-1.145) (z-3) (z-4)^3 (z-0.1506) (z^2 + 1.145z + 1.31)

                                    (z^2 - 0.5994z + 1.66)
```

```
        -5.6667 (z-4) (z-3.941) (z+3.748) (z+1.036) (z+0.5402)

            (z-0.4839) (z-0.2964) (z^2 - 1.698z + 0.805) (z^2 + 0.7542z + 1.392)
 #2:  z^(-9) * ----------------------------------------------------------------
           z^2 (z-1.1) (z-0.8089) (z-3) (z-4)^3 (z-0.2882) (z+0.3768)

              (z^2 + 0.8089z + 1.099) (z^2 + 1.011z + 1.395)
```

Zero/pole/gain from input "E" to output...

```
         -132.5 z (z-5) (z-3.611) (z-0.4258) (z+0.5431) (z^2 - 1.153z + 0.4069)

                                    (z^2 + 0.2288z + 1.518)
 #1:  z^(-10) * ----------------------------------------------------------------
           z^3 (z-1.145) (z-3) (z-4) (z-5)^2 (z-0.1506) (z^2 + 1.145z + 1.31)

                                    (z^2 - 0.5994z + 1.66)
```

```
        -5.8333 (z-5) (z-3.941) (z+3.757) (z+1.023) (z+0.5437)

            (z-0.4934) (z-0.2976) (z^2 - 1.695z + 0.8027)

                                    (z^2 + 0.7529z + 1.39)
 #2:  z^(-10) * ----------------------------------------------------------------
           z^2 (z-1.1) (z-0.8089) (z-3) (z-4) (z-5)^2 (z-0.2882)

            (z+0.3768) (z^2 + 0.8089z + 1.099) (z^2 + 1.011z + 1.395)
```

I/O groups:

Group name	I/O	Channel(s)
Parameter12	I	1,2
Parameter35	I	3,5
Parameter 4	I	4

Sampling time: 0.5

Learnt the secrets of multiplication and series connection of LTI models? If you need a little more practice, take up the practice test given in the dark grey box...

Practice Test 2.2.

1. Multiply the models -- modelBp1 and modelBp2 (created in Practice Test 2.1 Question 1) to obtain a new model -- modelBp7.

2. Connect modelBp1 and modelBp2 in series such that the output1 and output3 of modelBp1 are connected to input3 and input2 of modelBp2 respectively. Name the resulting model as modelBp8.

2.2.3 Inversion

As mentioned in the Table 2.2, inversion of a LTI model actually means inverting the input-output relationship. It is valid only on square systems with the same number of inputs and outputs, having no delay in the input or output channels. The system can be continuous or discrete.

For a system with transfer function:

$y = G(s)u$ for a continuous model
or,
$y = G(z)u$ for a discrete model

the inverse operation is equivalent to producing a system with transfer function:

$H(s) = G(s)^{-1}$ or, $H(z) = G(z)^{-1}$
i.e.,
$u = H(s)y$ or, $u = H(z)y$

for continuous-time and discrete-time models respectively.

Syntax for performing inverse operation on a model is similar to one followed for matrix inversion. It can be written down as:

ModelResult = **inv** (*model1*)

where,
ModelResult...................... is user-specified name of the resulting model after inversion operation
model1............................. is the name of the square model involved

For state space model with parameters, a, b, c, d, the result of inversion is:

$$a - bd^{-1}c , \quad bd^{-1} , \quad -d^{-1}c , \quad d^{-1}$$

where matrix d should be non-singular.

This operation can be understood more clearly with the help of block diagram given in Figure 2.3.

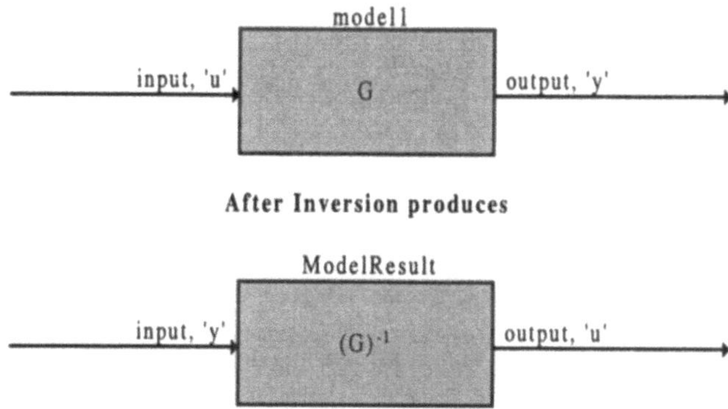

Figure 2.3. Inversion Operation on LTI Models

Example:

```
modelB11=inv(modelA1)
returns

Zero/pole/gain:
0.2 (s-3)
----------
 (s+1)

modelB12=inv(modelA36)
returns

a =
             x1        x2        x3
    x1        0      3.7143    -9.1429
    x2       -2     -4.2857     3.8571
    x3       -2     -4.2857    -11.143

b =
             u1        u2
    x1    -0.14286   0.28571
    x2    -0.14286   0.28571
    x3    -0.14286   0.28571

c =
             x1        x2        x3
    y1       0.5     2.1429    -2.4286
    y2      -0.5    -1.4286     0.28571

d =
             u1        u2
    y1    -0.42857   0.35714
    y2     0.28571  -0.071429

Continuous-time model.
```

To get a clear picture as to how the input and output behave on inversion, let us create a square tf model and observe.

```
modelB13=tf({[1 5] [1];[1] [0]}, {[1 3] [1]; [1] [1 3]},'InputName',{'a' 'b'}, 'OutputName',{'m' 'n'})
```

returns

```
Transfer function from input "a" to output...
     s + 5
 m:  -----
     s + 3

 n:  1

Transfer function from input "b" to output...
 m:  1

 n:  0
```

$$modelB13 = \begin{bmatrix} \dfrac{s+5}{s+3} & 1 \\ 1 & 0 \end{bmatrix}$$

inputname : a,b

outputname: m,,n

Now, let us invert the model created above...

```
modelB14=inv(modelB13)
returns

Transfer function from input "m" to output...
 a:  0

 b:  1

Transfer function from input "n" to output...
 a:  1

     -s - 5
 b:  ------
     s + 3
```

$$modelB14 = \begin{bmatrix} 0 & 1 \\ 1 & \dfrac{-s-5}{s+3} \end{bmatrix}$$

inputname : m,n

outputname: a,b

Getting the feel of the inversion operation? Try out a test and check yourself...

Practice Test 2.3.

1. Create the following models and invert them (leave their names as such).

a)

$$modelBp9 = \begin{bmatrix} 0 & 1 \\ \dfrac{s+5}{2s-1} & 1 \end{bmatrix}$$

$$InputName \quad = Alpha, \quad Beta$$
$$OutputName \quad = Rho, \quad Delta$$

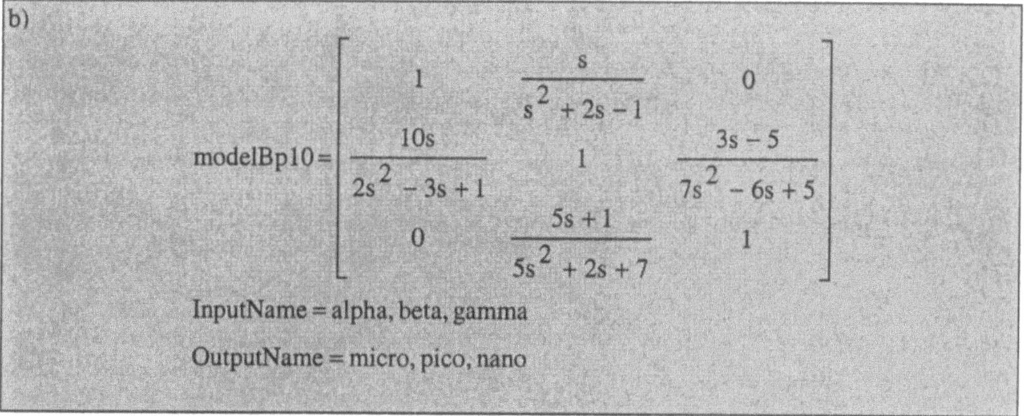

$$
\text{modelBp10} = \begin{bmatrix} 1 & \dfrac{s}{s^2 + 2s - 1} & 0 \\[2ex] \dfrac{10s}{2s^2 - 3s + 1} & 1 & \dfrac{3s - 5}{7s^2 - 6s + 5} \\[2ex] 0 & \dfrac{5s + 1}{5s^2 + 2s + 7} & 1 \end{bmatrix}
$$

2.2.4 Right Divide

As mentioned in the Table 2.2, right divide of two LTI models is actually equivalent to the operation **inv**(model1)*(model2). This can be achieved either by writing the equivalent statement or directly as follows:

ModelResult = *model1* \ *model2*

where,

ModelResult .... is user-specified name of the resulting model after right divide operation

model1, model2 .... are the names of the models involved

The operation can be better understood by the block diagram given in Figure 2.4.

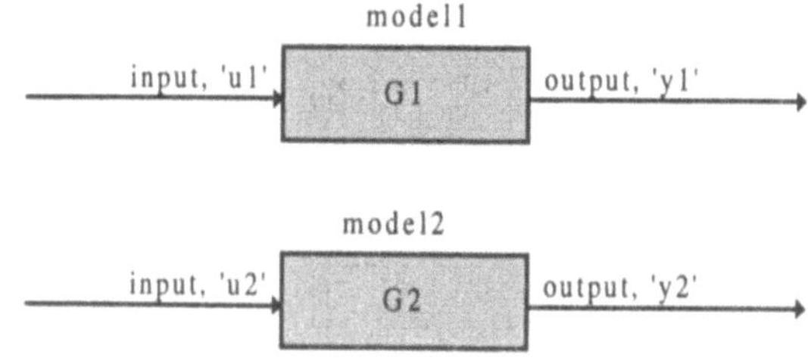

After the Right Divide operation
model1\model 2 results into

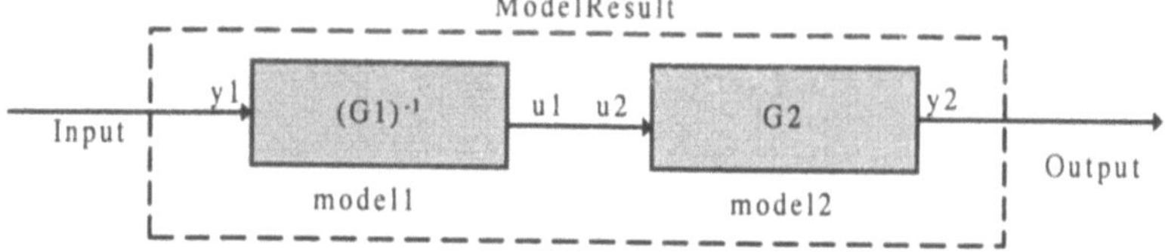

Figure 2.4. Right Divide operation on LTI models

Example:

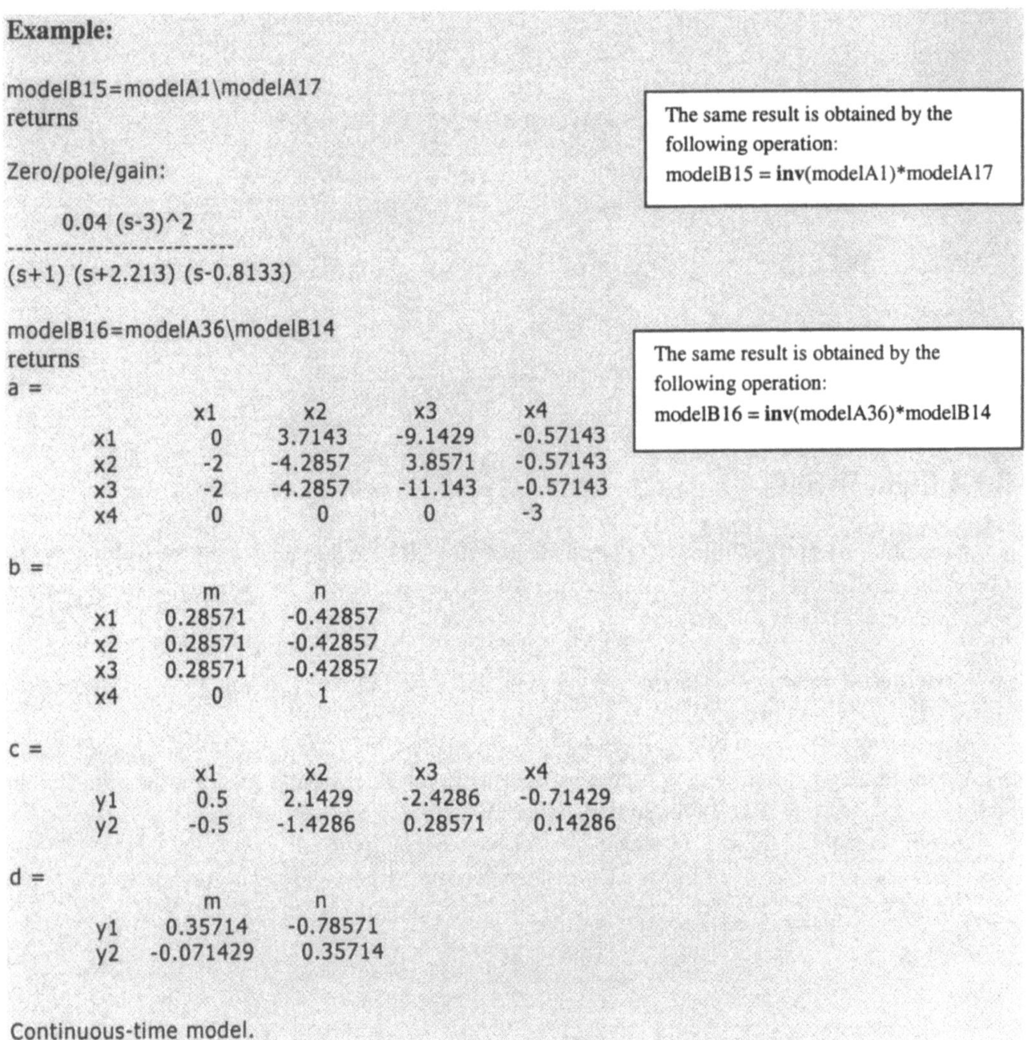

```
modelB15=modelA1\modelA17
returns

Zero/pole/gain:

    0.04 (s-3)^2
---------------------------
(s+1) (s+2.213) (s-0.8133)

modelB16=modelA36\modelB14
returns
a =
                  x1        x2        x3        x4
          x1        0      3.7143   -9.1429   -0.57143
          x2       -2     -4.2857    3.8571   -0.57143
          x3       -2     -4.2857   -11.143   -0.57143
          x4        0        0         0        -3

b =
                  m         n
          x1    0.28571   -0.42857
          x2    0.28571   -0.42857
          x3    0.28571   -0.42857
          x4       0         1


c =
                  x1        x2        x3        x4
          y1     0.5      2.1429   -2.4286   -0.71429
          y2    -0.5     -1.4286    0.28571   0.14286

d =
                  m          n
          y1    0.35714   -0.78571
          y2   -0.071429   0.35714

Continuous-time model.
```

The same result is obtained by the following operation:
modelB15 = **inv**(modelA1)*modelA17

The same result is obtained by the following operation:
modelB16 = **inv**(modelA36)*modelB14

Ready to take up a practice test now?

Practice Test 2.4.

1. Create modelBp11 by right dividing modelB4 by modelB5. Check the result by the standard formula for right division of matrices.

2.2.5 Left Divide

As mentioned in the Table 2.2, left divide of two LTI models is actually equivalent to the operation (model1)***inv**(model2). This can be achieved either by writing the equivalent statement or directly as follows:

ModelResult = *model1 / model2*

where,

ModelResult...................... is user-specified name of the resulting model after left divide operation

model1, model2............... are the names of the models involved

This operation can be better understood by the block diagram given in Figure2.5.

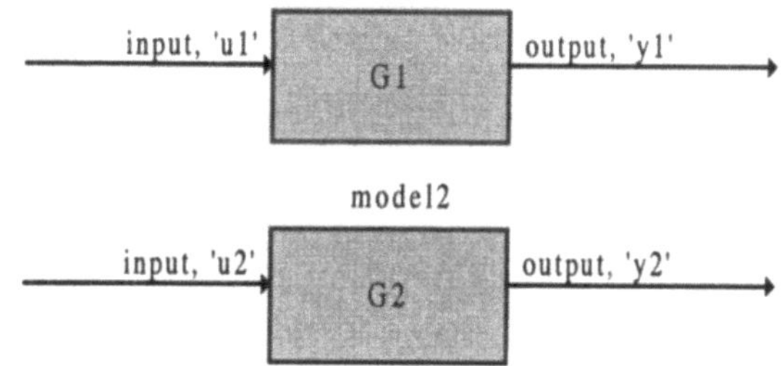

**After the Left Divide operation
model1/model 2 results into**

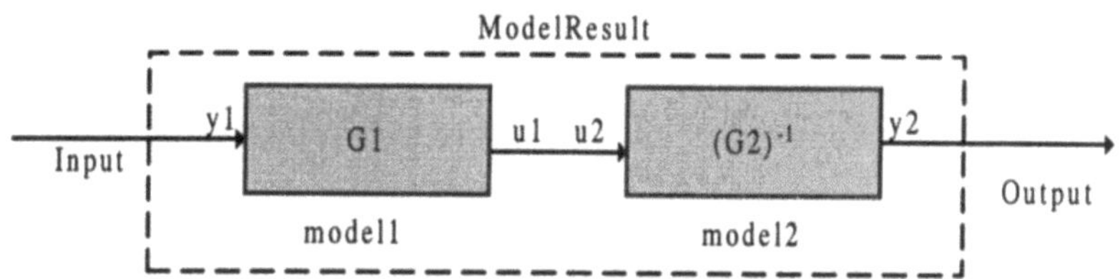

Figure 2.5. Left Divide operation on LTI models

Example:

```
modelB17=modelA1/modelA17
returns

Zero/pole/gain:
25 (s+1) (s+2.213) (s-0.8133)
------------------------------
          (s-3)^2

modelB18=modelA36/modelB14
returns

a =
              x1        x2        x3        x4
        x1     0         3        -7         2
        x2    -2        -5         6         2
        x3    -2        -5        -9         2
        x4     0         0         0        -3
```

The same result is obtained by the following operation:

modelB17 = modelA1*inv(modelA17)

```
b =

                    a          b
        x1          2          1
        x2          2          1
        x3          2          1
        x4          1          0

c =

                    x1         x2         x3         x4
        y1          2          5          1          2
        y2          1          0          8          8

d =

                    a          b
        y1          6          1
        y2          10         4

Continuous-time model.
```

Ready for a practice test?

2.2.6 Transposition

As mentioned in the Table 2.2, transposition operation has different action on different types of LTI models and results into the following:

- for tf models, cell arrays *num* and *den* are transposed
- for zpk models, cell arrays *z*, *p* and matrices *k* are transposed
- for ss models, transposition produces a^T, c^T, b^T, d^T, e^T matrices
- for frd models, the matrix of frequency response data at each frequency is transposed

Syntax for performing transposition operation on a model is the same as the one followed for matrix transposition:

ModelResult = (*model1*).'

where,
ModelResult.............. is user-specified name of the resulting model after transposition operation
model1....................... is the name of the model involved

Example:

```
modelB19=(modelA14).'
```

returns

Zero/pole/gain from input 1 to output...

```
     0.01 (s+7)
#1:  ----------
      (s-0.3)

      0.5 (s-0.9)
#2:  ----------------
     (s-0.1) (s-0.4)
```

Zero/pole/gain from input 2 to output...

```
     10 (s+0.2)
#1:  ----------
       (s-6)

         2 (s+0.5)
#2:  --------------------
     (s^2 - 0.2s + 1.01)
```

modelB20=(modelA32).'
returns

Transfer function from input 1 to output...

```
      4 z^2 + 3 z - 3
#1:  -------------------------
     4 z^3 - 3 z^2 + 7 z - 1

       5 z - 1
#2:  -----------
     2 z^4 - 3 z
```

Transfer function from input 2 to output...

```
        4 z - 7
#1:  ----------------
     9 z^3 + 4 z - 8

     z^3 + 4 z^2 - 3 z + 2
#2:  ---------------------------
     6 z^5 + z^3 - 9 z^2 - z + 1
```

Sampling time: 0.5

modelB21=(modelA40).'
returns

$$modelB19 =
\begin{bmatrix}
\dfrac{0.01(s + 7)}{(s - 0.3)} & \dfrac{10(s + 0.2)}{(s - 6)} \\[2ex]
\dfrac{0.5(s - 0.9)}{(s - 0.1)(s - 0.4)} & \dfrac{2(s + 0.5)}{s^2 - 0.2s + 1.01}
\end{bmatrix}$$

$$modelB20 =
\begin{bmatrix}
\dfrac{4z^2 + 3z - 3}{4z^3 - 3z^2 + 7z - 1} & \dfrac{4z - 7}{9z^3 + 4z - 8} \\[2ex]
\dfrac{5z - 1}{2z^4 - 3z} & \dfrac{z3 + 4z2 - 3z + 2}{6z^5 + z^3 - 9z^2 - z + 1}
\end{bmatrix}$$

Sampling time: 0.5

$$modelB21 =$$

$$\begin{bmatrix}
2 & -2 & 3 & 2 \\
3 & 5 & 2 & -7 \\
4 & 6 & 7 & 3 \\
-3 & 7 & -2 & 2
\end{bmatrix} \dot{x} =
\begin{bmatrix}
0 & 1 & 8 & 3 \\
3 & -2 & -2 & 7 \\
-7 & -5 & -5 & 6 \\
5 & 6 & -9 & 5
\end{bmatrix} x +
\begin{bmatrix}
2 & 1 & 3 \\
5 & -5 & 4 \\
-6 & 0 & -6 \\
1 & 8 & 8
\end{bmatrix} u$$

$$y =
\begin{bmatrix}
1 & 1 & 1 & 1 \\
1 & 1 & 1 & 1 \\
1 & 1 & 1 & 1
\end{bmatrix} x +
\begin{bmatrix}
1 & 4 & 2 \\
5 & 6 & 1 \\
4 & 9 & 6
\end{bmatrix} u$$

```
a =

            x1          x2          x3          x4
    x1       0           1           8           3
    x2       3          -2          -2           7
    x3      -7          -5          -5           6
    x4       5           6          -9           5

b =

            u1          u2          u3
    x1       2           1           3
    x2       5          -5           4
    x3      -6           0          -6
    x4       1           8           8

c =

            x1          x2          x3          x4
    y1       1           1           1           1
    y2       1           1           1           1
    y3       1           1           1           1

d =

            u1          u2          u3
    y1       1           4           2
    y2       5           6           1
    y3       4           9           6

e =

            x1          x2          x3          x4
    x1       2          -2           3           2
    x2       3           5           2          -7
    x3       4           6           7           3
    x4      -3           7          -2           2

Continuous-time model.
```

Check yourself by taking the following test...

Practice Test 2.6.

1. Create modelBp13, modelBp14 and modelBp15 by transposing modelA15, modelA29, and modelA35. Observe the resulting models carefully.

2.2.7 Pertransposition

As mentioned in the Table 2.2, for a continuous LTI model with transfer function as H(s), pertransposition operation is equivalent to a model with transfer function $[H(-s)]^T$. Similarly for a discrete LTI model, pertransposition on model H(z) returns a model $[H(z^{-1})]^T$. Hence, we observe that pertransposition is actually equivalent to obtaining conjugate transpose of a model.

Syntax to pertranspose an LTI model is similar to the one followed while working with matrices and is:

$$\text{ModelResult} = (model1)'$$

where,

ModelResult .... is user-specified name of the resulting model after pertransposition operation

model1 .... is the name of the model involved

Example:

```
modelB22 = (modelA14)'
returns

Zero/pole/gain from input 1 to output...
      0.01 (s-7)
 #1:  ----------
      (s+0.3)

      -0.5 (s+0.9)
 #2:  ----------------
      (s+0.1) (s+0.4)

Zero/pole/gain from input 2 to output...
      10 (s-0.2)
 #1:  ----------
      (s+6)

       -2 (s-0.5)
 #2:  --------------------
      (s^2 + 0.2s + 1.01)
```

$$
\text{modelB22}=
\begin{bmatrix}
\dfrac{0.01(s-7)}{(s+0.3)} & \dfrac{10(s-0.2)}{(s+6)} \\[2ex]
\dfrac{-0.5(s+0.9)}{(s+0.1)(s+0.4)} & \dfrac{-2(s-0.5)}{(s^2+0.2s+1.01)}
\end{bmatrix}
$$

```
modelB23 =(modelA32)'
returns

Transfer function from input 1 to output...
      3 z^3 - 3 z^2 - 4 z
 #1:  ----------------------
      z^3 - 7 z^2 + 3 z - 4

      z^4 - 5 z^3
 #2:  -----------
      3 z^3 - 2

Transfer function from input 2 to output...
      7 z^3 - 4 z^2
 #1:  -----------------
      8 z^3 - 4 z^2 - 9

      2 z^5 - 3 z^4 + 4 z^3 + z^2
 #2:  ------------------------------
      z^5 - z^4 - 9 z^3 + z^2 + 6

Sampling time: 0.5
```

$$
\text{modelB23} =
\begin{bmatrix}
\dfrac{3z^3 - 2z^2 - 4z}{z^3 - 7z^2 + 3z - 4} & \dfrac{7z^3 - 4z^2}{8z^3 - 4z^2 - 9} \\[2ex]
\dfrac{z^4 - 5z^3}{3z^3 - 2} & \dfrac{2z^5 - 3z^4 + 4z^3 + z^2}{z^5 - z^4 - 9z^3 + z^2 + 6}
\end{bmatrix}
$$

Sampling Time: 0.5

```
modelB24=(modelA40)'
returns
```

$$
\text{modelB24} =
$$

$$
\begin{bmatrix} 2 & -2 & 3 & 2 \\ 3 & 5 & 2 & -7 \\ 4 & 6 & 7 & 3 \\ -3 & 7 & -2 & 2 \end{bmatrix} \dot{x} = \begin{bmatrix} 0 & -1 & -8 & -3 \\ -3 & 2 & 2 & -7 \\ 7 & 5 & 5 & -6 \\ -5 & -6 & 9 & -5 \end{bmatrix} x + \begin{bmatrix} -2 & -1 & -3 \\ -5 & 5 & -4 \\ 6 & 0 & 6 \\ -1 & -8 & -8 \end{bmatrix} u
$$

$$
y = \begin{bmatrix} 1 & 1 & 1 & 1 \\ 1 & 1 & 1 & 1 \\ 1 & 1 & 1 & 1 \end{bmatrix} x + \begin{bmatrix} 1 & 4 & 2 \\ 5 & 6 & 1 \\ 4 & 9 & 6 \end{bmatrix} u
$$

a =

	x1	x2	x3	x4
x1	0	-1	-8	-3
x2	-3	2	2	-7
x3	7	5	5	-6
x4	-5	-6	9	-5

b =

	u1	u2	u3
x1	-2	-1	-3
x2	-5	5	-4
x3	6	0	6
x4	-1	-8	-8

c =

	x1	x2	x3	x4
y1	1	1	1	1
y2	1	1	1	1
y3	1	1	1	1

d =

	u1	u2	u3
y1	1	4	2
y2	5	6	1
y3	4	9	6

e =

	x1	x2	x3	x4
x1	2	-2	3	2
x2	3	5	2	-7
x3	4	6	7	3
x4	-3	7	-2	2

Continuous-time model.

Feeling confident with the operations on the models? Better check…

Practice Test 2.7.

1. Create modelBp16, modelBp17 and modelBp18 by transposing modelA15, modelA29, and modelA35. Observe the resulting models carefully.

2.2.8 Power of Model

As mentioned in the Table 2.2, power of a model is equivalent to:

- raising the power of the model by the factor specified
- multiplying the model by itself as many times as specified
- connecting the model to itself in series as many times as specified

The only limitation is that the model should be square *i.e.,* it should have the same number of inputs as the number of outputs.

Syntax for raising the power of an LTI model is same as the one followed while dealing with matrices and is:

ModelResult = (*model1*)^n

where,

ModelResult...................... is user-specified name of the resulting model after raising the model to power *n*

model1............................. is the name of the model

n.. *is an integer indicating the power to which the model is to be raised*

Example:

```
modelB25=(modelA14)^3
returns
```

> same result can be obtained as follows:
> modelB25 = modelA14*modelA14*modelA14

```
Zero/pole/gain from input 1 to output...

      1e-006 (s+9.991e004) (s+106.7) (s-6) (s-0.9004) (s-0.4)^2 (s-0.3128)

                (s-0.3) (s-0.2838) (s-0.198) (s-0.1)^2 (s+0.1982) (s+0.3815)
 #1:  ------------------------------------------------------------------------
            (s-0.3)^4 (s-0.4)^3 (s-0.1)^3 (s-6)^2 (s^2 - 0.2s + 1.01)


      0.001 (s+5.021e004) (s-6)^2 (s-2.17) (s-0.3249) (s-0.2817) (s+0.2413)

            (s+0.2) (s^2 - 0.579s + 0.1736) (s^2 - 0.2s + 1.01) (s^2 + 2.22s + 2.401)
 #2:  ------------------------------------------------------------------------------
            (s-6)^4 (s-0.4) (s-0.3)^2 (s-0.1) (s^2 - 0.2s + 1.01)^3

Zero/pole/gain from input 2 to output...
      5e-005 (s+5.021e004) (s-2.17) (s-0.9) (s-0.4)^2 (s-0.3249) (s-0.3)

            (s-0.2817) (s+0.2413) (s^2 - 0.2s + 0.01) (s^2 - 0.579s + 0.1736)

                              (s^2 + 2.22s + 2.401)
 #1:  ----------------------------------------------------------------------
            (s-0.3)^3 (s-0.4)^4 (s-0.1)^4 (s-6) (s^2 - 0.2s + 1.01)^2
```

```
       0.05 (s+406.2) (s-6)^2 (s-1.615) (s-0.4) (s-0.2732) (s-0.1) (s+0.4719)

           (s+0.2311) (s^2 - 0.6964s + 0.2877) (s^2 - 0.2s + 1.01)

                                        (s^2 + 1.375s + 1.984)
  #2:  ----------------------------------------------------------------
           (s-6)^3 (s-0.4)^2 (s-0.3) (s-0.1)^2 (s^2 - 0.2s + 1.01)^4
```

```
modelB26=(modelA40)^2
returns
```

a =

	x1	x2	x3	x4	x5
x1	0	3	-7	5	6
x2	1	-2	-5	6	6
x3	8	-2	-5	-9	6
x4	3	7	6	5	6
x5	0	0	0	0	0
x6	0	0	0	0	1
x7	0	0	0	0	8
x8	0	0	0	0	3

	x6	x7	x8
x1	4	-12	17
x2	4	-12	17
x3	4	-12	17
x4	4	-12	17
x5	3	-7	5
x6	-2	-5	6
x7	-2	-5	-9
x8	7	6	5

same result can be obtained as follows:
- modelB26 = **series**(modelA40, modelA40)

or,

- modelB26 = modelA40*modelA40

b =

	u1	u2	u3
x1	7	12	19
x2	7	12	19
x3	7	12	19
x4	7	12	19
x5	1	1	1
x6	1	1	1
x7	1	1	1
x8	1	1	1

c =

	x1	x2	x3	x4	x5
y1	2	5	-6	1	19
y2	1	-5	0	8	41
y3	3	4	-6	8	23

	x6	x7	x8
y1	-4	-30	73
y2	26	-78	124
y3	29	-48	58

```
d =
                u1        u2        u3
        y1      29        39        73
        y2      46        65        124
        y3      18        22        53


e =
                x1        x2        x3        x4        x5
        x1      2         3         4         -3        0
        x2      -2        5         6         7         0
        x3      3         2         7         -2        0
        x4      2         -7        3         2         0
        x5      0         0         0         0         2
        x6      0         0         0         0         -2
        x7      0         0         0         0         3
        x8      0         0         0         0         2

                x6        x7        x8
        x1      0         0         0
        x2      0         0         0
        x3      0         0         0
        x4      0         0         0
        x5      3         4         -3
        x6      5         6         7
        x7      2         7         -2
        x8      -7        3         2

Continuous-time model.

modelB27=(modelA45)^4
returns

From input 'Signal Generator' to:

   Frequency(Hz)    Frequency Meter
   --------------   ----------------
        10          1.600000e-003
        20          2.342560e+001
        30          3.111696e+002
        40          1.477634e+003
        50          4.521218e+003
        60          1.082432e+004
        70          2.215335e+004
        80          4.065869e+004
        90          6.887475e+004
       100          1.097199e+005

I/O delay time (for all I/O pairs): 4

Sampling time: 0.1
Discrete-time frequency response data model.
```

That was simple! Check if it really was...

Practice Test 2.8.

1. Create modelBp19 and modelBp20 by raising modelA1 and modelA37 to the power of 2 and 4 respectively.

2.2.9 Stacking

As mentioned in the Table 2.2, **stack** function is used to stack LTI models/arrays along some array dimension. Syntax to achieve this is:

ModelResult = **stack** (*arraydimension, model1, model2...*)

where,

ModelResult .... is user-specified name of the resulting model after stacking operation

arraydimension .... is the dimension of the array along which models are to be stacked. (Note that the array dimension is different from the Input-Output dimensions of the concerned models)

model1, model2 .... are the names of the models involved having same Input-Output dimensions

Example:

```
modelB28=stack(1,modelA15,modelA32)
returns

Model modelB28(:,:,1,1)
=========================

  Zero/pole/gain from input 1 to output...
         1
  #1:  -------
       (z+0.3)


         10
  #2:  -----
       (z-6)

  Zero/pole/gain from input 2 to output...
       0.5 (z-0.2) (z+0.4)
  #1:  -------------------
            (z-0.1)


          2 (z+0.5)
   #2:  -------------------
        (z^2 - 0.2z + 1.01)

Model modelB28(:,:,2,1)
=========================

  Zero/pole/gain from input 1 to output...
```

```
           (z+1.319) (z-0.5687)
  #1:  ------------------------------------
      (z-0.1506) (z^2 - 0.5994z + 1.66)


             0.44444 (z-1.75)
  #2:  ------------------------------------
      (z-0.8089) (z^2 + 0.8089z + 1.099)

Zero/pole/gain from input 2 to output...
               2.5 (z-0.2)
  #1:  ------------------------------------
      z (z-1.145) (z^2 + 1.145z + 1.31)


         0.16667 (z+4.725) (z^2 - 0.7246z + 0.4233)
  #2:  -------------------------------------------------
      (z-1.1) (z-0.2882) (z+0.3768) (z^2 + 1.011z + 1.395)

Sampling time: 0.5

2x1 array of discrete-time zero-pole-gain models.
```

Take up a small practice test now...

2.3 Interconnection of Models

The models created may also need to be interconnected sometimes. Control System Toolbox supports several interconnection functions to facilitate this. These are summarised in Table 2.3.These interconnection functions are explained in more detail in the lines to follow.

Table 2.3.   Interconnection of the LTI models

Interconnection Operator	Description
[,]	concatenates horizontally
[;]	concatenates vertically
append	appends models in block diagram configuration
augstate	augments the output by appending states
connect	forms ss model from a block diagonal LTI object for an arbitrary interconnection matrix. It works with append to produce an arbitrary interconnection scheme to a set of LTI models
feedback	forms the feedback interconnection of two models
lft	produces the Redheffer Star product of two models
parallel	forms parallel connection of two models
series	forms series connection of two models

2.3.1 Concatenation

As indicated in Table 2.3, the models under consideration can be concatenated horizontally, vertically or along any desired dimension. The concatenation of LTI models is similar to the matrix concatenation and can be achieved as:

- Horizontal Concatenation:

  Similar to matrix concatenation as:
  ModelResult = **horzcat** (*model1, model2,...*)

  or, alternatively as:

  ModelResult = [*model1   model2...*] or [*model1, model2,...*]

- Vertical Concatenation:

  Similar to matrix concatenation as:
  ModelResult = **vertcat** (*model1, model2,...*)

  or, alternatively as:

  Model Result = [*model1; model2;...*]

- Concatenation along a dimension:

This is similar to the stack operation and can be done either using the **stack** function explained in Article 2.2.9 or following the method of matrix concatenation as:

ModelResult = **cat** (*dimension, model1, model2,...*)

where,
ModelResult...................... is user-specified name of the resulting model after concatenation
model1, model2................ are the names of the models to be concatenated
dimension......................... is the dimension along which the models are to be concatenated

Example:

```
modelB29=cat(2,modelA15,modelA32)
returns

Zero/pole/gain from input 1 to output...

        1
#1:  -------
     (z+0.3)

        10
#2:  -----
     (z-6)
```

compare modelB28 with modelB27

```
Zero/pole/gain from input 2 to output...
      0.5 (z-0.2) (z+0.4)
 #1:  --------------------
           (z-0.1)

        2 (z+0.5)
 #2:  -------------------
      (z^2 - 0.2z + 1.01)

Zero/pole/gain from input 3 to output...
          (z+1.319) (z-0.5687)
 #1:  -----------------------------------
      (z-0.1506) (z^2 - 0.5994z + 1.66)

             0.44444 (z-1.75)
 #2:  -----------------------------------
      (z-0.8089) (z^2 + 0.8089z + 1.099)

Zero/pole/gain from input 4 to output...
                2.5 (z-0.2)
 #1:  -----------------------------------
      z (z-1.145) (z^2 + 1.145z + 1.31)

          0.16667 (z+4.725) (z^2 - 0.7246z + 0.4233)
 #2:  ----------------------------------------------------------------
      (z-1.1) (z-0.2882) (z+0.3768) (z^2 + 1.011z + 1.395)

Sampling time: 0.5
```

Ready for a practice test?

Practice Test 2.10.

1. Create modelBp23 and modelBp24 by horizontal and vertical concatenation of modelA15 and modelA31. Can you explain the results?

2.3.2 Append

As indicated in Table 2.3, **append** function appends the LTI models involved to result into a model with unconnected inputs and outputs. The syntax for achieving this is:

ModelResult = **append** (*model1*, *model2*…)

where,
ModelResult..................... is user-specified name of the resulting model after appending
model1, *model2*................ are the names of the models to be appended

The result of the **append** function produces a model which has block diagram as shown in Figure 2.6.

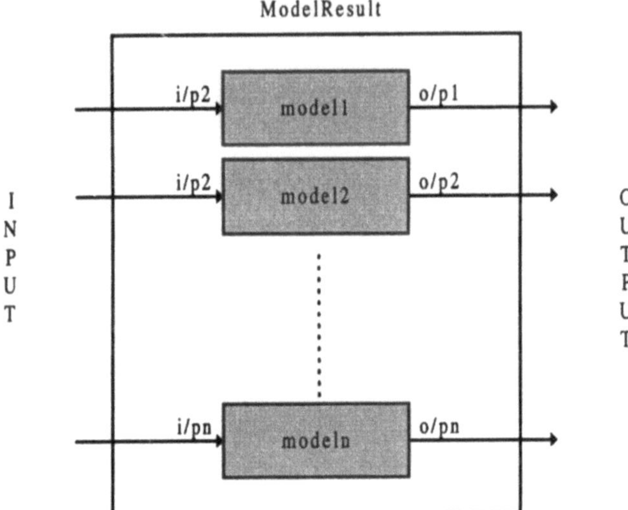

Figure 2.6. Result of **append** function

This is equivalent to model that has block diagonal transfer function as indicated below:

$$\begin{bmatrix} \text{model1} & 0 & 0 & \ldots\ldots & 0 \\ 0 & \text{model2} & 0 & \ldots\ldots & 0 \\ 0 & 0 & \text{model3} & \ldots\ldots & 0 \\ 0 & 0 & 0 & \ldots\ldots & \text{modeln} \end{bmatrix}$$

For state space models with two systems having parameters as a1, b1, c1, d1 and as a2, b2, c2, d2, the **append** function appends all of the four respective matrices.

Example:

```
modelB30 = append(modelA1,modelA4,modelA14)
returns

Zero/pole/gain from input 1 to output...
      5 (s+1)
 #1:  -------
       (s-3)

 #2:  0

 #3:  0

 #4:  0

Zero/pole/gain from input 2 to output...
 #1:  0
```

```
      5 (s+1)
#2:   -------
      (s-3)

#3:  0

#4:  0

Zero/pole/gain from input 3 to output...
 #1:  0

     -20 (s-2) (s^2 - 0.8s + 1.16)
#2:  -------------------------------
     (s-6) (s^2 + 16s + 64.25)

#3:  0

#4:  0

Zero/pole/gain from input 4 to output...
 #1:  0

#2:  0

     0.01 (s+7)
#3:  ----------
     (s-0.3)

     10 (s+0.2)
#4:  ----------
     (s-6)

Zero/pole/gain from input 5 to output...
 #1:  0

#2:  0

     0.5 (s-0.9)
#3:  ---------------
     (s-0.1) (s-0.4)

         2 (s+0.5)
#4:  -------------------
     (s^2 - 0.2s + 1.01)
```

Check yourself by taking a practice test…

Practice Test 2.11.

1. Append the following sets of models to create modelBp25, modelBp26 and modelBp27 respectively and explain the result:

a) modelA17, modelA21 and modelA30

b) modelA32 and modelA35

c) modelA1, modelA17, modelA32 and modelA35.

2.3.3 Augstate

As indicated in Table 2.3, **augstate** function appends the state vector of a state space LTI model to the output vector. The syntax for achieving this is:

ModelResult = **augstate** (*ssmodel*)

where,
ModelResult...................... is user-specified name of the resulting model after appending
ssmodel............................ is the name of the state space LTI model

Consider a state space model with its usual four parameters a, b, c and d. **Augstate** function appends the states x to the output y and the resulting model is:

dx/dt =ax+bu

$$\begin{bmatrix} y \\ x \end{bmatrix} = \begin{bmatrix} c \\ I \end{bmatrix} x + \begin{bmatrix} d \\ 0 \end{bmatrix} u$$

Note that **augstate** function is obviously valid for state space models only.

Example:

```
modelB31= augstate(modelA36)
returns

a =
              x1        x2        x3
      x1       0         3        -7
      x2      -2        -5         6
      x3      -2        -5        -9

b =
              u1        u2
      x1       1         1
      x2       1         1
      x3       1         1

c =
              x1        x2        x3
      y1       2         5         1
      y2       1         0         8
      y3       1         0         0
      y4       0         1         0
      y5       0         0         1

d =
              u1        u2
      y1       1         5
      y2       4         6
      y3       0         0
      y4       0         0
      y5       0         0
```

modelB31=

$$\dot{x} = \begin{bmatrix} 0 & 3 & -7 \\ -2 & -5 & 6 \\ -2 & -5 & -9 \end{bmatrix} x + \begin{bmatrix} 1 & 1 \\ 1 & 1 \\ 1 & 1 \end{bmatrix}$$

$$y = \begin{bmatrix} 2 & 5 & 1 \\ 1 & 0 & 8 \\ 1 & 0 & 0 \\ 0 & 1 & 0 \\ 0 & 0 & 1 \end{bmatrix} x + \begin{bmatrix} 1 & 5 \\ 4 & 6 \\ 0 & 0 \\ 0 & 0 \\ 0 & 0 \end{bmatrix} u$$

Example:

Let us use the **connect** function to obtain the diagram shown below. We will directly use modelB30 as it gives the result of the **append** operation on modelA1, modelA4 and modelA14.

Try a practice test now…

> **Practice Test 2.12.**
>
> 1. Create modelBp28 by augstating modelA38.

2.3.4 Connect

As indicated in the Table 2.3, **connect** is a general type of function through which any type of connection of the LTI models can be accomplished to obtain a state space model as the result.

The usual steps to be followed to accomplish this are:

- first, append all the given models to get a resultant model with unconnected inputs and outputs using the **append** function explained in Article 2.3.2
- next, use the **connect** function to connect the models in the desired way

Syntax for achieving this is:

ModelResult = **connect** (*modelappend, N, in, out*)

where,

ModelResult…………………	is user-specified name of the resulting model after connection
modelappend…………………	is the resultant model under consideration which in turn has been obtained after appending different models
N………………………………	is the interconnection matrix with as many rows as the number of inputs that get their signals from different outputs. The first element of each row is the input and the following elements are the outputs with are to be connected to the input. The negative sign of the output elements specifies negative input to the summing point due to that output.
in ………………………………	is the input matrix, the elements of which specify the external inputs of the system after the connect operation
out ……………………………	is the output matrix, the elements of which specify the external outputs of the system after the connect operation

Connect function can be better understood with the help of forthcoming example.

```
I/O groups:
   Group name   I/O   Channel(s)
   states        O    3,4,5

Continuous-time model.
```

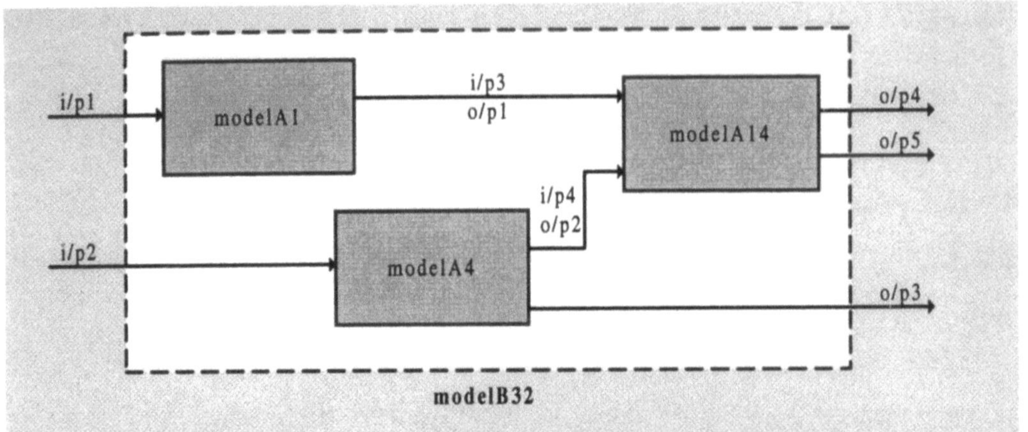

- [3 1; 4 2] is the interconnection matrix where input3 gets signal from output1, while, input4 gets signal from output2.
- [1 2] matrix specifies the external input of the system *i.e.*, of modelB32.
- [3 4 5] matrix specifies the external output of the system *i.e.*, of modelB32.
- the resultant matrix modelB32 is a state space matrix.

modelB32=connect(modelB30, [3 1; 4 2], [1 2], [3 4 5])
returns

a =

	x1	x2	x3	x4	x5
x1	3	0	0	0	0
x2	0	3	0	0	0
x3	0	0	6	-4.2302	-0.99287
x4	0	0	0	-16	-4.0156
x5	0	0	0	16	0
x6	1.2083	0	0	0	0
x7	35.214	0	0	0	0
x8	0	3.1623	0	0	0
x9	0	0	0	0	0
x10	0	6.3246	0	0	0
x11	0	0	0	0	0

	x6	x7	x8	x9	x10
x1	0	0	0	0	0
x2	0	0	0	0	0
x3	0	0	0	0	0
x4	0	0	0	0	0
x5	0	0	0	0	0
x6	0.3	0	0	0	0
x7	0	6	0	0	0
x8	0	0	0.5	-0.04	0
x9	0	0	1	0	0
x10	0	0	0	0	0.2
x11	0	0	0	0	1

	x11
x1	0
x2	0
x3	0
x4	0
x5	0
x6	0
x7	0
x8	0
x9	0
x10	-1.01
x11	0

b =

	u1	u2
x1	4.4721	0
x2	0	4.4721
x3	0	4.4721
x4	0	17.761
x5	0	0
x6	1.3509	0
x7	39.37	0
x8	0	3.5355
x9	0	0
x10	0	7.0711
x11	0	0

c =

	x1	x2	x3	x4	x5
y1	0	0	-17.889	18.918	4.4402
y2	0.044721	0	0	0	0
y3	44.721	0	0	0	0

	x6	x7	x8	x9	x10
y1	0	0	0	0	0
y2	0.27019	0	0.70711	-0.6364	0
y3	0	7.874	0	0	1.4142

	x11
y1	0
y2	0
y3	0.70711

d =

	u1	u2
y1	0	-20
y2	0.05	0
y3	50	0

Continuous-time model.

Ready for a practice test?

Practice Test 2.13.

1. Use the **connect** function to obtain modelBp29 as shown in the forthcoming figure.

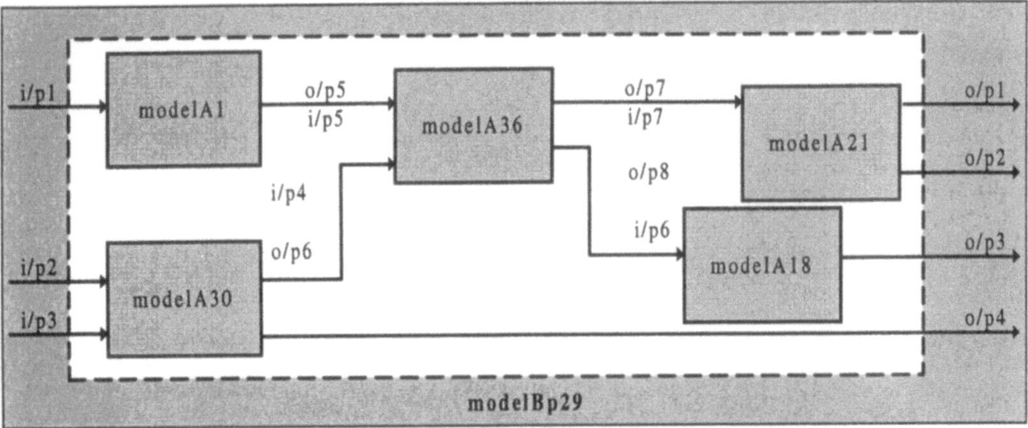

2.3.5 Feedback

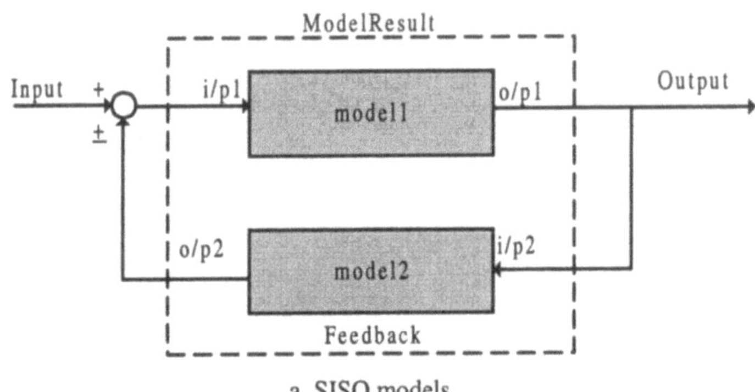

a. SISO models

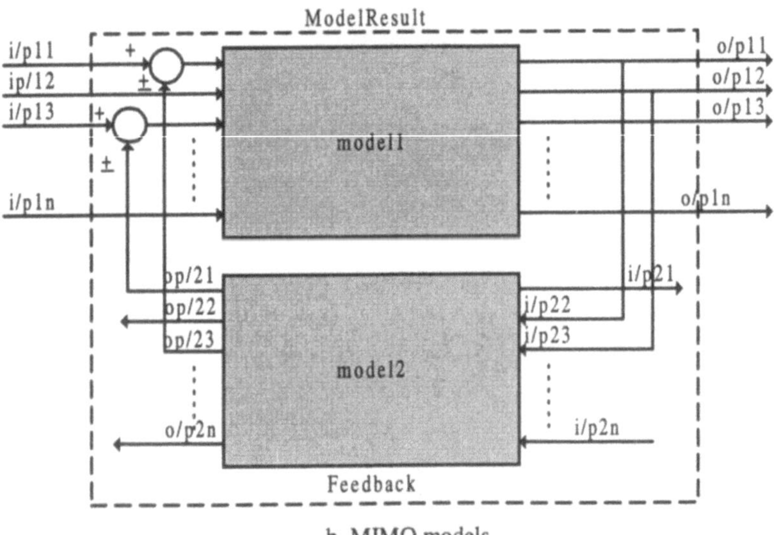

b. MIMO models

Figure 2.7. Feedback connection of LTI models

As indicated in Table 2.3, **feedback** function connects two models in feedback configuration. Its syntax is:

ModelResult = **feedback** (*model1*, *model2*, *fin*, *fout*, *sign*)

where,

ModelResult......................	is user-specified name of the resulting model
model1.............................	is the model in the forward path
model2.............................	is the model in the feedback path
fin.......................................	specifies the vector whose indicies indicate the input of forward path model1 to receive signals from the output of the feedback path model2
fout....................................	specifies the output vector of forward path model1 involved in the feedback and to be connected to the input of feedback path model2
sign....................................	specifies the sign of feedback (by default it is negative). It is specified as +1 or -1(optional)

Feedback connection can be better understood by the block diagrams given in Figure 2.7 for SISO and MIMO models.

Example:

```
modelB33=feedback(modelA22, modelA1, +1)
returns

Zero/pole/gain:
     0.2 (s-3)^2
----------------------------
(s+1.885) (s-1.485) (s-3)
```

- modelA22 is the model in forward path
- modelA1 is the model in feedback path
- +1 specifies positive feedback

```
modelB34=feedback(modelA22, modelA9)
returns

Zero/pole/gain:
     0.2 (s-3)
-------------------
(s+3.698) (s-1.298)
```

- modelA22 is the model in forward path
- modelA9 is a static gain model in feedback path
- no sign indicates negative feedback

```
modelB35=feedback(modelA22,1)
returns

Transfer function:
   s - 3
----------------
5 s^2 + 8 s - 12
```

- indicates unity feedback loop

```
modelB36=feedback(1,modelA22)
returns
```

- indicates unity forward path gain

```
Transfer function:
5 s^2 + 7 s - 9
-----------------
5 s^2 + 8 s - 12
```

```
modelB37=feedback(modelA34, modelA21, [2 1], [2] )
returns
```

a =

	x1	x2	x3	x4	x5
x1	-0.33333	3	-0.2	0.3	0.44444
x2	-2.3333	-5	-0.2	0.3	0.44444
x3	1	0	-1.4	0.9	0
x4	0	0	2	0	0
x5	0.5	0	0	0	1.3333

b =

	u1	u2
x1	1	1
x2	1	1
x3	0	0
x4	0	0
x5	0	0

c =

	x1	x2	x3	x4	x5
y1	2	1	0	0	0
y2	1	0	0	0	0

d =

	u1	u2
y1	0	0
y2	0	0

Continuous-time model.

- input2 and input 1of modelA34 in the forward path receive signal from outpput of modelA21 in the feedback path
- output2 of modelA34 in the forward path is involved in the feedback so as to form input of modelA21 in the feedback path
- no sign specifies positive feedback

Ready for a practice test?

Practice Test 2.14.

1. Create a modelBp30 connecting

- input2 and input1 of modelA30 in the forward path to receive signal from the outputs of modelA21 in the feedback path.
- output2 of modelA30 to input of modelA21.

2. Create a modelBp31 using modelA13 in the forward path and modelA31 in the feedback path, connecting

- the outputs of modelA30 to the input5 and input2 of modelA13 and
- connecting the output4 and output2 of modelA13 to the inputs of modelA31.

2.3.6 Linear Fractional Transformation (lft) Connection

As indicated in the Table 2.3, Linear Fractional Transformation (**lft**) connection produces a Redheffer Star Product of two models. The syntax for this function is:

ModelResult = **lft** (*model1*, *model2*, *n1*, *n2*)

where,

ModelResult...................... is user-specified name of the resulting model

model1, model2.................. are the names of the models to be connected

n1 .... specifies the first n1 outputs of *model2* to be connected to the last n1 inputs of *model1*

n2 .... specifies the last n2 outputs of *model1* to be connected to the first n2 inputs of *model2*

The **lft** connection can be better understood by the block diagrams given in Figure 2.8.

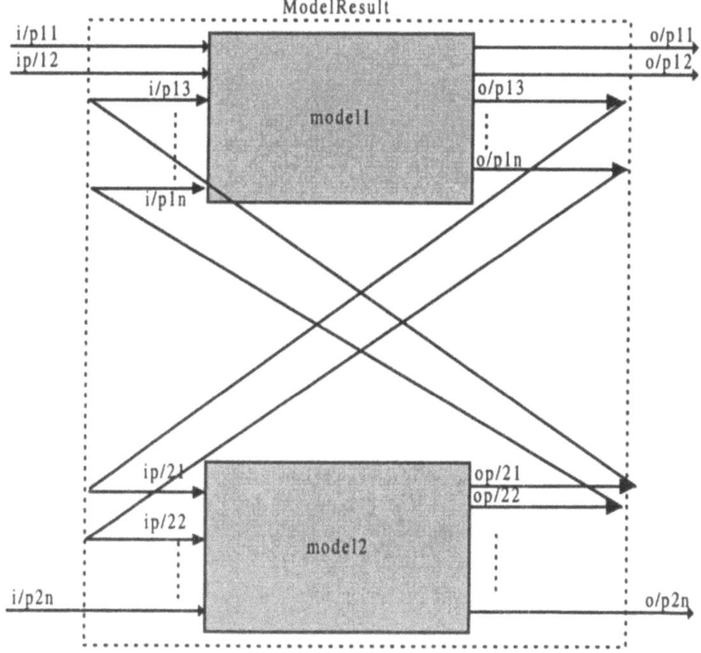

Figure 2.8. Linear Fractional Transformation connection of LTI models

Example:

```
modelB38=lft(modelA14,modelA31,1 ,2)
returns
```

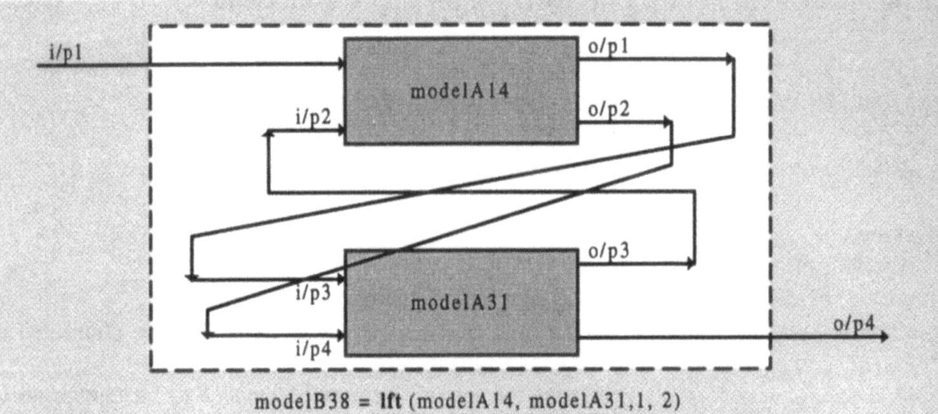

modelB38 = lft (modelA14, modelA31,1, 2)

Zero/pole/gain:

$$2.5 \; (s+3.285) \; (s+1.201) \; (s-0.7859) \; (s-0.8184) \; (s-0.4538) \; (s-0.2916) \; (s+0.2109)$$

$$(s^2 - 1.309s + 0.5843) \; (s^2 + 0.8727s + 0.7217) \; (s^2 - 1.283s + 1.246)$$

$$(s^2 - 0.7303s + 1.047) \; (s^2 - 0.2592s + 1.049) \; (s^2 + 0.4782s + 1.161)$$

$$\rule{12cm}{0.4pt}$$

$$(s-0.785) \; (s-0.818) \; (s-0.8384) \; (s-0.4665) \; (s-0.3) \; (s+1.213) \; (s-6) \; (s^2 - 1.37s + 0.6097)$$

$$(s^2 + 0.068s + 0.3056) \; (s^2 + 0.8743s + 0.6717) \; (s^2 - 0.3998s + 0.7774)$$

$$(s^2 + 0.4517s + 1.132) \; (s^2 - 0.463s + 1.394)$$

modelB39=lft(modelA14,modelA31,2 ,2)
returns

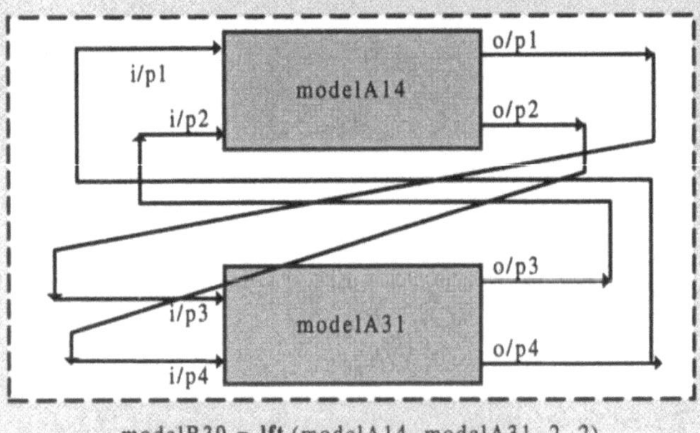

modelB39 = lft (modelA14, modelA31, 2, 2)

Empty zero-pole-gain model.

modelB40=lft(modelA14, modelA31, 1, 1)
returns

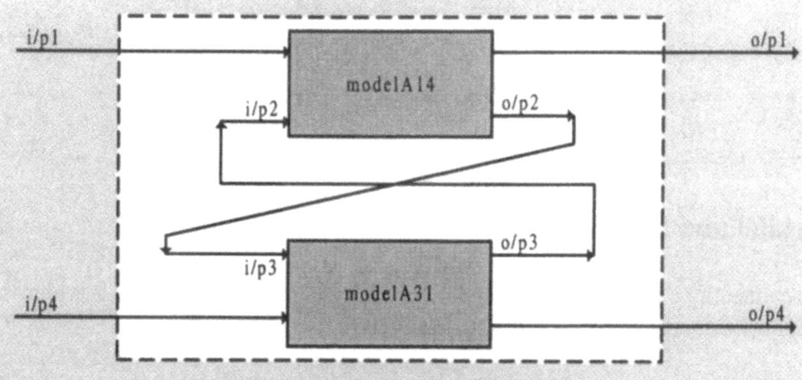

Zero/pole/gain from input 1 to output...

$$
\#1: \quad \frac{0.01\,(s+3.822)\,(s-1.079)\,(s-0.8194)\,(s-0.2934)\,(s+0.2038)\,(s^2 - 0.3165s + 1.145)\,(s^2 - 1.968s + 92)}{(s-0.4)\,(s-0.3)\,(s-0.1)\,(s+0.2532)\,(s-6)\,(s^2 - 2.023s + 1.124)\,(s^2 + 0.8194s + 1.748)}
$$

$$
\#2: \quad \frac{1.1111\,(s-7)\,(s-0.818)\,(s+0.2)\,(s^2 + 0.068s + 0.3056)\,(s^2 - 0.2s + 1.01)}{(s-0.785)\,(s+0.2532)\,(s-6)\,(s^2 - 2.023s + 1.124)\,(s^2 + 0.4517s + 1.132)\,(s^2 + 0.8194s + 1.748)}
$$

Zero/pole/gain from input 2 to output...

$$
\#1: \quad \frac{0.1\,s\,(s-0.9)\,(s-0.818)\,(s^2 + 0.068s + 0.3056)\,(s^2 - 0.2s + 1.01)}{(s-0.4)\,(s-0.462)\,(s+1.194)\,(s+0.2532)\,(s-0.1)\,(s^2 - 2.023s + 1.124)\,(s^2 - 0.732s + 1.088)\,(s^2 + 0.8194s + 1.748)}
$$

$$
\#2: \quad \frac{0.25\,(s+3.283)\,(s+1.201)\,(s-0.7855)\,(s-0.4507)\,(s+0.2541)\,(s^2 - 2.04s + 1.128)\,(s^2 - 0.711s + 0.9847)\,(s^2 - 1.318s + 1.343)\,(s^2 + 0.5094s + 1.125)\,(s^2 + 0.7733s + 1.784)}{(s-0.818)\,(s-0.785)\,(s-0.462)\,(s+1.194)\,(s+0.2532)\,(s^2 - 2.023s + 1.124)\,(s^2 + 0.068s + 0.3056)\,(s^2 - 0.732s + 1.088)\,(s^2 + 0.4517s + 1.132)\,(s^2 + 0.8194s + 1.748)}
$$

Ready to test yourself? Go ahead...

> **Practice Test 2.15.**
>
> 1. Create a modelBp32 connecting modelA38 and modelB24 such that
>
> - the first 3 outputs of modelB24 is connected to last 3 inputs of modelA38
> - the last 2 outputs of modelB24 is connected to first 2 inputs of modelA38.

2.3.7 Parallel and Series Connection

These interconnections have been explained in Article 2.2.1 and Article 2.2.2 under the heading Addition/Subtraction and Multiplication respectively.

2.4 The Linear Time Invariant (LTI) Subsystem

Till now we have learnt to create models and operate upon them to get larger models. In the forthcoming paragraphs, we will learn the *vice versa, i.e.,* to get models of smaller subsystems from the larger ones.

Obtaining subsystems which are subsets giving the relationship between the inputs and outputs of larger systems is quite easy. If we perceive the model of any system in terms of a large matrix, it is quite convenient to see that sub-matrix of the system forms the subsets of the LTI models. By the usual method of indexing, we can obtain the subsystem of a larger model. In addition, by the same method of referencing indices, we can alter the properties of a subsystem.

In the lines to follow, you will learn to deal with subsystems through examples.

2.4.1 Obtaining a Subsystem

As mentioned above the simplest way to obtain a subsystem is by referring it by indices. The example considered below will make things more clear.

Example:

Let us consider a large system, say modelA13 and obtain its subsystems:

```
modelB41=modelA13(4,5)
returns

Zero/pole/gain from input "E" to output "P":

          45 (z-0.45)
z^(-12) * -------------
          (z-4) (z-5)

I/O groups:
    Group name    I/O    Channel(s)
    Parameter35    I         1
    Result24       O         1
```

> modelB41= subsystem of modelA13 with
> - output 4 (*i.e.,* p)
> - input 5 (*i.e.,* e)

```
Sampling time: 0.5

modelB42=modelA13(:,4)
returns

Zero/pole/gain from input "D" to output...
              14 (z-0.14)
  M:  z^(-5) * -----------
              (z-1) (z-4)

              24 (z-0.24)
  N:  z^(-7) * -----------
              (z-2) (z-4)

              34 (z-0.34)
  O:  z^(-9) * -----------
              (z-3) (z-4)

               44 (z-0.44)
  P:  z^(-11) * -----------
                (z-4)^2

               54 (z-0.54)
  Q:  z^(-13) * -----------
                (z-5) (z-4)

I/O groups:
   Group name      I/O   Channel(s)
   Parameter 4      I     1
   Result24         O     2,4
   Result 13        O     1,3

Sampling time: 0.5

modelB43=modelA13([2 4],[1 3])
returns

Zero/pole/gain from input "A" to output...
              21 (z-0.21)
  N:  z^(-4) * -----------
              (z-2) (z-1)

              41 (z-0.41)
  P:  z^(-8) * -----------
              (z-4) (z-1)

Zero/pole/gain from input "C" to output...
              23 (z-0.23)
  N:  z^(-6) * -----------
              (z-2) (z-3)

               43 (z-0.43)
  P:  z^(-10) * -----------
               (z-4) (z-3)

I/O groups:
```

modeB42=subsystem of model13 with
- output all *i.e.*, 1 to 6 (*i.e.*, m to q)
- input 4 (*i.e.*, d)

```
   Group name    I/O   Channel(s)
   Parameter12    I        1
   Parameter35    I        2
    Result24      O       1,2
```

Sampling time: 0.5

```
modelB44=modelA13(1:2:5,1:2:6)
returns
```

Zero/pole/gain from input "A" to output...

$$M: z^{\wedge}(-2) * \frac{11\ (z-0.11)}{(z-1)^{\wedge}2}$$

$$O: z^{\wedge}(-6) * \frac{31\ (z-0.31)}{(z-3)\ (z-1)}$$

$$Q: z^{\wedge}(-10) * \frac{51\ (z-0.51)}{(z-5)\ (z-1)}$$

Zero/pole/gain from input "C" to output...

$$M: z^{\wedge}(-4) * \frac{13\ (z-0.13)}{(z-1)\ (z-3)}$$

$$O: z^{\wedge}(-8) * \frac{33\ (z-0.33)}{(z-3)^{\wedge}2}$$

$$Q: z^{\wedge}(-12) * \frac{53\ (z-0.53)}{(z-5)\ (z-3)}$$

Zero/pole/gain from input "E" to output...

$$M: z^{\wedge}(-6) * \frac{15\ (z-0.15)}{(z-1)\ (z-5)}$$

$$O: z^{\wedge}(-10) * \frac{35\ (z-0.35)}{(z-3)\ (z-5)}$$

$$Q: z^{\wedge}(-14) * \frac{55\ (z-0.55)}{(z-5)^{\wedge}2}$$

I/O groups:
```
   Group name    I/O   Channel(s)
   Parameter12    I        1
   Parameter35    I       2,3
    Result 13     O       1,2
```

Sampling time: 0.5

2.4.2 Modifying a System/Subsystem

Just as in the case of matrices, a system (which also has its genesis in matrices) can be modified by altering its subsystems. The subsystem may be referred to by its indices as usual, and equating this expression to a new value changes the values in the subsystem, eventually resulting into change in the original system. The examples to follow will make the procedure more clear.

Example:

Consider modelA21, which is a model having one input and two outputs. Let us convert it to a model with two inputs and two outputs using modelA17 and modelA9 as follows:

```
modelB45=[modelA21 [modelA9; modelA17]]
returns

Zero/pole/gain from input 1 to output...
           0.2 (s-3)
 #1:  --------------------
       (s+2.213) (s-0.8133)

       0.33333 (s-2)
 #2:  --------------
       (s-1.333)

Zero/pole/gain from input 2 to output...
 #1:  5

           0.2 (s-3)
 #2:  --------------------
       (s+2.213) (s-0.8133)

modelB45(1,2)=modelA26
returns

Zero/pole/gain from input 1 to output...
        5 (s+1)
 #1:  --------
        (s-3)

       -20 (s-2) (s^2 - 0.8s + 1.16)
 #2:  --------------------------------
        (s-6) (s^2 + 16s + 64.25)

Zero/pole/gain from input 2 to output...
 #1:  2

           0.2 (s-3)
 #2:  --------------------
       (s+2.213) (s-0.8133)
```

> the response of the system from input 2 to output 1 changes from value of modelA9 (*i.e.*, value 5) to corresponding value of modelA26 (*i.e.*, value 2)

Check what you have learnt by taking up the following practice test…

Practice Test 2.16.

1. Extract the following values from modelA30 and change them to 10 and 50 respectively:

- input4 to output3
- input6 to output5

2.4.3 Subsystem Manipulations

All the methods of model manipulations, which are possible on a LTI model (as mentioned in Article 2.1) are possible on LTI subsystems also. Both, the 'Arithmetic Operations' and the 'Interconnections' can be done on LTI subsystems. Thus, like LTI systems, LTI subsystems can also be added, subtracted, multiplied, inverted, transposed, stacked, appended, connected as desired and what not! The 'Precedence Rule' and the 'Law of Property Inheritance' are followed in the same way as in case of LTI systems. The syntax for operations and interconnection also remain same.

Forthcoming illustrations will further elucidate the concept.

Example:

```
modelB46=series(modelB45,modelA14)
returns

Zero/pole/gain from input 1 to output...

       0.05 (s-174.3) (s-0.4355) (s-0.2044) (s^2 + 0.5516s + 1.378) (s^2 - 8.067s + 22.62)
 #1: ------------------------------------------------------------------------------------
               (s-0.3) (s-0.4) (s-0.1) (s-3) (s-6) (s^2 + 16s + 64.25)

       50 (s-6) (s+0.8705) (s+0.1559) (s^2 - 0.2136s + 1.044) (s^2 + 15.39s + 71.99)
 #2: ------------------------------------------------------------------------------------
               (s-6)^2 (s-3) (s^2 + 16s + 64.25) (s^2 - 0.2s + 1.01)

Zero/pole/gain from input 2 to output...

       0.02 (s-0.275) (s^2 - 1.571s + 0.6429) (s^2 + 9.746s + 25.76)
 #1: ----------------------------------------------------------------
               (s-0.3) (s-0.4) (s-0.8133) (s-0.1) (s+2.213)

       20 (s+2.174) (s-0.7375) (s+0.1318) (s^2 - 0.1686s + 0.8688)
 #2: ---------------------------------------------------------------
               (s-6) (s+2.213) (s-0.8133) (s^2 - 0.2s + 1.01)

modelB47=(modelA44).'
returns

From input 1 to:

  Frequency(rad/s)  output 1
  ----------------  --------
```

```
  31.415927     0.2
  62.831853     0.4
  94.247780     1.3
 125.663706     1.5
 157.079633     1.7
 188.495559     1.4
 219.911486     1.3
 251.327412     0.9
 282.743339     0.8
 314.159265     0.4
```

Sampling time: 0.25
Discrete-time frequency response data model.

modelB48=feedback(modelB45, modelA31)
returns

Zero/pole/gain from input 1 to output...

```
     15 (s-5.438) (s+1.194) (s-0.462) (s-0.785) (s-0.818) (s-1.106) (s-0.1216)

        (s^2 + 4.064s + 4.16) (s^2 - 0.4979s + 0.2481) (s^2 + 0.068s + 0.3056)

        (s^2 - 0.732s + 1.088) (s^2 + 0.4517s + 1.132) (s^2 + 5.4s + 18.71)
 #1: -----------------------------------------------------------------------
   (s+11.81) (s+6.297) (s+2.163) (s+1.15) (s+0.594) (s-0.7856) (s-0.4486)

        (s-5.71) (s+0.02056) (s^2 - 2.076s + 1.09) (s^2 - 0.5347s + 0.3451)

        (s^2 - 0.7528s + 1.096) (s^2 + 0.4505s + 1.614) (s^2 - 1.813s + 4.464)


    -20 (s-3) (s-0.818)^2 (s-0.462) (s+1.194) (s+2.223) (s+0.1146) (s^2 - 3.834s + 3.793)

        (s^2 + 0.068s + 0.3056)^2 (s^2 - 0.732s + 1.088) (s^2 - 1.302s + 2.138)

                              (s^2 + 1.065s + 2.315)
 #2: -----------------------------------------------------------------------
     (s+11.81) (s+6.297) (s+2.163) (s+1.15) (s+0.594) (s-0.7856) (s-0.4486)

         (s-5.71) (s+0.02056) (s^2 - 2.076s + 1.09) (s^2 - 0.5347s + 0.3451)

         (s^2 - 0.7528s + 1.096) (s^2 + 0.4505s + 1.614) (s^2 - 1.813s + 4.464)
```

Zero/pole/gain from input 2 to output...

```
      2 (s-6.012) (s-3) (s+2.217) (s+1.164) (s-0.8663) (s-0.818)^2 (s-0.785)

         (s-0.4278) (s^2 + 0.068s + 0.3056)^2 (s^2 + 16.04s + 65)

                    (s^2 - 0.7197s + 1.114) (s^2 + 0.4517s + 1.132)
 #1: ---------------------------------------------------------------------
    (s+11.81) (s+6.297) (s+2.163) (s+1.15) (s+0.594) (s-0.7856) (s-0.4486)

         (s-5.71) (s+0.02056) (s^2 - 2.076s + 1.09) (s^2 - 0.5347s + 0.3451)
```

$$
\text{10.2} \frac{(s^2 - 0.7528s + 1.096)\ (s^2 + 0.4505s + 1.614)\ (s^2 - 1.813s + 4.464)}{(s+1.194)\ (s-1.066)\ (s-0.818)\ (s-0.785)\ (s-0.462)\ (s-2.599)}
$$

$$
(s-3)\ (s+0.1612)\ (s^2 + 0.068s + 0.3056)\ (s^2 + 5.732s + 9.093)
$$

$$
\#2: \frac{(s^2 - 0.732s + 1.088)\ (s^2 + 0.4517s + 1.132)\ (s^2 - 1.494s + 2.561)}{(s+11.81)\ (s+6.297)\ (s+2.163)\ (s+1.15)\ (s+0.594)\ (s-0.7856)\ (s-0.4486)}
$$

$$
(s-5.71)\ (s+0.02056)\ (s^2 - 2.076s + 1.09)\ (s^2 - 0.5347s + 0.3451)
$$

$$
(s^2 - 0.7528s + 1.096)\ (s^2 + 0.4505s + 1.614)\ (s^2 - 1.813s + 4.464)
$$

Ready for a practice test?

> **Practice Test 2.17.**
>
> 1. Create as many subsystems as possible from modelB24 and modelB26 and try out all the arithmetic operations and interconnections mentioned in Table 2.2 and Table 2.3 on them.

Exercise for Chapter 2:

Create the following models and comment on the results:

1. modelBe1 = (modelAe1)$^{-1}$ + modelAe4 * (modelAe7).'

2. modelBe2 = (modelAe1/modelAe3)^{2} \ (modelAe4/modelAe9)'

3. modelBe3 = stack (1,modelAe1, modelAe2) - modelAe9

4. modelBe4 = (modelAe12) - (modelAe6^{-1})^{2}

5. modelBe5 = (modelAe5/modelAe10)' - modelAe11

6. modelBe6 = concatenate modelAe1&modelAe2 along dimension 2 and add to modelAe3

7. modelBe7 = append modelAe5, modelAe10 and modelAe12 and subtract from modelAe12

8. modelBe8 = augstate modelAe9 and modelAe8 respectively and add the results

9. model Be9 = **lft**(modelAe1, modelAe2) + **feedback**(modelAe4, modelAe9)

10. modelBe12 = series((modelBe10, modelBe11)

   where,
   modelBe10 = inverse of **series**(modelAe3, modelAe2,3,1)
   modelBe11 = **feedback**(modelAe6,1), [1,1],[2,1])

Also, draw the equivalent connection diagram for modelBe15.

11. modelBe15 = (model Be13)' / (modelBe14)^2

where,
modelBe13 and modelBe14 are as given below:

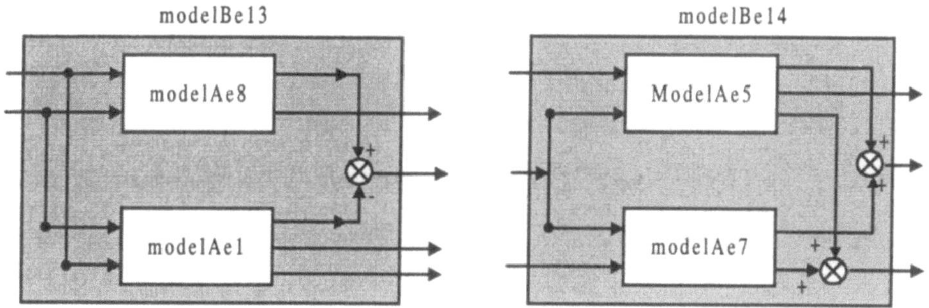

12. modelBe17=**augastate**(modelBe16)^2

where,
modelBe16 is as given below:

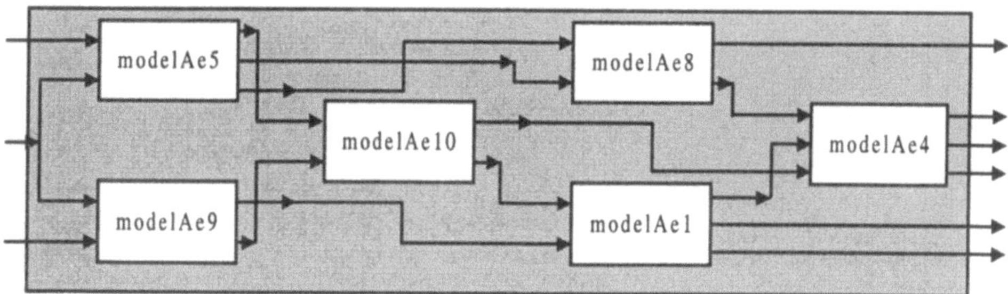

Chapter 3

Getting Information from the Models

In this chapter, you will learn to obtain information about the models generated in the previous chapters.

Till now, you were creating models, assigning values to their properties, operating upon them, interconnecting them and deriving subsystems out of larger systems. Nevertheless, so far, having given a model, you did not try to obtain information about them. The Control System Toolbox of MATLAB® provides function for data retrieval form any given system, whether it is information about the model's generic properties or the model's specific properties.

In this Chapter, you will learn the techniques of data extraction from all the four types of models about which we discussed in Chapter 1.

3.1 Model-specific Information

Control System Toolbox provides special functions for extraction of model-specific information for each of the four kinds of models. Use of these functions for the purpose is explained in the forthcoming paragraphs.

3.1.1 Zero/Pole/Gain Model

Model-specific information attached with a zpk model are the values of zeros, poles, gains and sampling time. These can be extracted using the **zpkdata** function for which the syntax is:

[z, p, k, tsys] = **zpkdata** (*ModelName*, 'v')

where, as we know already,

ModelName	specifies the name of the zpk model under consideration
z	returns the zeros of the zpk model under consideration *i.e., ModelName*
p	returns the poles of the zpk model under consideration *i.e., ModelName*
k	returns the gains of the zpk model under consideration *i.e., ModelName*
tsys	returns the sampling time of the zpk model under consideration *i.e., ModelName*

v .... is optional which when incorporated in the syntax returns the values of the elements of z, p and k as vectors instead of vector dimension (valid only for SISO models)

Examples considered below will make things more clear.

Example:

To begin with, let us extract data from some of the models developed in the previous chapters.

```
[z,p,k,tsys]=zpkdata(modelA1)
returns
```

	compare the results with modelA1

```
z =
   [-1]
p =
   [3]
k =
   5
tsys =
   0
```

```
[z,p,k,tsys]=zpkdata(modelA2)
returns
```

	compare the reults with modelA2

```
z =
   [-1]
p =
   [3]
k =
   5
tsys =
   0.1000
```

```
[z,p,k,tsys]=zpkdata(modelA3)
returns
```

	result obtained for pole indicates the dimension of the pole vector

```
z =
   [-1]
p =
   [3x1 double]
k =
   5
tsys =
   0.1000
```

For a SISO model, like modelA3 above, the elements of pole vector P can be found out by the use of optional 'v' as follows:

```
[z,p,k,tsys]=zpkdata(modelA3,'v')
returns
```

	using optional 'v', gives the value of the elements of the pole vector

```
z =
```

```
     -1
p =
   6.0000
  -8.0000 + 0.5000i
  -8.0000 - 0.5000i
k =
     5
tsys =
   0.1000


[z,p,k,tsys]=zpkdata(modelA12,'v')
returns


z =
   [2x1 double]
   [3x1 double]
   [3x1 double]
   [3x1 double]
   [2x1 double]
p =
   [        3]
   [3x1 double]
   [3x1 double]
   [3x1 double]
   [2x1 double]
k =
     5
    10
    20
    50
   100
tsys =
     0
```

to get the values of zeros and poles from input1 ('Fuel') to all the outputs, type as follows:

```
z{:,1}
returns


ans =
     5
    10
ans =
   0.2000
   0.4000 + 1.0000i
   0.4000 - 1.0000i
ans =
   1.2000
   2.2000
   3.2000
ans =
   0.4500
   0.9000
   0.6000
ans =
   1.6000
```

```
  7.2000
```

Similarly, to get the values of poles from input1 to output 2, output3, oputput4 and output5 type as follows:

```
p{2:5,1}
returns

ans =
  6.0000
 -9.0000 + 0.5000i
 -9.0000 - 0.5000i
ans =
  0.5000
  1.5000
  2.5000
ans =
  1
  3
  6
ans =
  4.8000
  6.4000
```

3.1.2 Transfer Function Model

Similar to zpk model, data extraction from tf model *i.e.*, the values of numerator, denominator and sample time can be obtained using the **tfdata** function for which the syntax is:

   [num, den, tsys] = **tfdata** (*ModelName*, 'v')

   where, as we know already,
   ModelName...................... specifies the name of the tf model under consideration
   num .... returns the numerator of the tf model under consideration *i.e.*, *ModelName*
   den................................... returns the denominator of the tf model under consideration *i.e.*, *ModelName*
   tsys .... returns the sampling time of the tf model under consideration *i.e.*, *ModelName*
   v .... is optional which when incorporated in the syntax, returns the values of the elements of num and den vectors instead of vector dimension (valid only for SISO models)

Example:

Data can be extracted from tf models in the same way as from the zpk models. Let us extract information from modelA27.

```
[n,d,tsys]=tfdata(modelA27)
returns
```

```
n =
   [1x3 double]
   [1x2 double]

d =
   [1x3 double]
   [1x2 double]
tsys =
   0
```

The values of n and d denote that they are cell arrays. Therefore, their values can be obtained by the usual cell indices referencing method.

```
n{:,1}
returns
```

```
ans =
   0   1   -3
ans =
   -2   4
```

```
d{:,1}
returns
```

```
ans =
   5   7   -9
ans =
   -6   8
```

Similarly,
```
[n,d,tsys]=tfdata(modelA32)
returns
```

```
n =
   [1x4 double]   [1x5 double]
   [1x4 double]   [1x6 double]
d =
   [1x4 double]   [1x5 double]
   [1x4 double]   [1x6 double]
tsys =
   0.5000
```

Values of numerator and denominator can be obtained as follows:

```
n{1:2,1:2}
returns
```

```
ans =
   0   4   3   -3
ans =
   0   0   4   -7
ans =
   0   0   0   5   -1
```

```
ans =
   0   0   1   4   -3   2

d{1:2,1:2}
returns
ans =
   4  -3   7  -1
ans =
   9   0   4  -8
ans =
   2   0   0  -3   0
ans =
   6   0   1  -9  -1   1
```

3.1.3 State Space Model

As for zpk and tf models, data extraction from ss model and dss model, *i.e.,* the values of matrices a, b, c, d, e and sample time, can also be obtained using the **ssdata and dssdata** function as indicated below:

- from state space model – **ssdata** for which the syntax is:

  [a, b, c, d, tsys] = **ssdata** (*ModelName*)

- from descriptor state space model – **dssdata** for which the syntax is:

  [a, b, c, d, e, tsys] = **dssdata** (*ModelName*)

where, as we know already,

ModelName	specifies the name of the ss or dss model under consideration
a	returns the system matrix
b	returns the input matrix
c	returns the output matrix
d	returns the transmission matrix
e	returns the descriptor state matrix
tsys	returns the sampling time of the ss or dss model under consideration *i.e., ModelName*
v	is optional which when incorporated in the syntax, returns the values of the elements of a, b, c, d and e vectors instead of vector dimension (valid only for SISO models)

Example:

For ss and dss models too, the procedure for data extraction is exactly the same as for zpk or tf models. The forthcoming examples will make things more clear:

```
[a,b,c,d,tsys]=ssdata(modelA36)
returns

a =
```

```
      0    3   -7
     -2   -5    6
     -2   -5   -9
b =
      1    1
      1    1
      1    1
c =
      2    5    1
      1    0    8
d =
      1    5
      4    6
tsys =
      0


[a,b,c,d,tsys]=dssdata(modelA40)
returns

a =
      0    3   -7    5
      1   -2   -5    6
      8   -2   -5   -9
      3    7    6    5
b =
      1    1    1
      1    1    1
      1    1    1
      1    1    1
c =
      2    5   -6    1
      1   -5    0    8
      3    4   -6    8
d =
      1    5    4
      4    6    9
      2    1    6
tsys =
      2    3    4   -3
     -2    5    6    7
      3    2    7   -2
      2   -7    3    2

[a,b,c,d,tsys]=ssdata(modelA39)
returns

a(:,:,1) =
    111  121
    211  221
a(:,:,2) =
    112  122
    212  222
a(:,:,3) =
    113  123
    213  223
a(:,:,4) =
    114  124
```

```
    214   224
b(:,:,1) =
   11
   21
b(:,:,2) =
   32
   42
b(:,:,3) =
   53
   63
b(:,:,4) =
   74
   84
c(:,:,1) =
        1110        1210
c(:,:,2) =
        1120        1220
c(:,:,3) =
        1130        1230
c(:,:,4) =
        1140        1240
d(:,:,1) =
     0
d(:,:,2) =
     0
d(:,:,3) =
     0
d(:,:,4) =
     0
tsys =
     0
```

3.1.4 Frequency Response Data Model

Similar to zpk,, tf, ss and dss models discussed above, we can extract data *i.e.,* the values of response, frequency and sample time from frd models too, using the **frdata** function for which the syntax is:

[r, f, tsys, td] = **frdata**(*ModelName*)

where, as we know already,

ModelName .... specifies the name of the frd model under consideration

r .... returns the complex frequency response of the frd model under consideration *i.e., ModelName*

f .... returns the frequency point vector at which response of the frd model under consideration *i.e., ModelName* is observed

tsys .... is the sampling time of the frd model under consideration *i.e., ModelName*

Example:

```
[r,f,tsys]=frdata(modelA41)
```

```
  returns

  r(:,:,1) =
      0.2000
  r(:,:,2) =
      0.4000
  r(:,:,3) =
      1.3000
  r(:,:,4) =
      1.5000
  r(:,:,5) =
      1.7000
  r(:,:,6) =
      1.4000
  r(:,:,7) =
      1.3000
  r(:,:,8) =
      0.9000
  r(:,:,9) =
      0.8000
  r(:,:,10) =
      0.4000
  f =
       5
      10
      15
      20
      25
      30
      35
      40
      45
      50
  tsys =
       0

[r,f,tsys]=frdata(modelA46)
returns

r(:,:,1) =
   111   121
   211   221
   311   321
r(:,:,2) =
   112   122
   212   222
   312   322
r(:,:,3) =
   113   123
   213   223
   313   323
r(:,:,4) =
   114   124
   214   224
   314   324
f =
   15
```

```
    30
    45
    60
tsys =
     0
```

Can you extract model-specific data from a given model? Better check...

3.2 Direct Property Referencing Method of Getting Information

Apart from the specialised functions for the four types of models mentioned above, Control System Toolbox provides yet another more direct method for data extraction from the models that is as explained below:

*   To get information about all the properties associated with a model and the values that can be assigned to them

    [AllProps,AsgnValues] = **pnames**(*ModelName*)

*   To get information about a particular property

    ModelName.PropertyName

where,
ModelName .... specifies the name of the model under consideration
PropertyName .... is the name of the property associated with the model that is desired to be determined (it can be generic or model-specific)

Example:

To get the names of all the properties associated with modelA1 and the values that can be assigned to them, enter the following at the command prompt in MATLAB® command window:

```
[AllProps,AsgnValues] = pnames(modelA1)
```
returns

```
AllProps =
```

```
'z'
'p'
'k'
'Variable'
'Ts'
'InputDelay'
'OutputDelay'
'ioDelayMatrix'
'InputName'
'OutputName'
'InputGroup'
'OutputGroup'
'Notes'
'UserData'
AsgnValues =
'Ny-by-Nu cell of vectors (Nu = no. of inputs)'
'Ny-by-Nu cell of vectors (Ny = no. of outputs)'
'Ny-by-Nu array of double'
'[ 's' | 'p' | 'z' | 'z^-1' | 'q' ]'
'scalar'
'Nu-by-1 vector'
'Ny-by-1 vector'
'Ny-by-Nu array (I/O delays)'
'Nu-by-1 cell array of strings'
'Ny-by-1 cell array of strings'
'M-by-2 cell array if M input groups'
'P-by-2 cell array if P output groups'
'array or cell array of strings'
'arbitrary'
```

You can get the value of any one property be it generic or specific, associated with a model as follows:

```
modelB1.k
returns
```

returns value if specific property k of modelB1

```
ans =
   5
```

```
modelB20.num
returns
```

returns value of tf model-specific property numerator for modelB20

```
ans =
   [1x4 double]    [1x4 double]
   [1x5 double]    [1x6 double]
```

```
modelB26.c
returns
```

returns value of matrix c of modelB26

```
ans =
   2    5   -6    1   19   -4   -30    73
   1   -5    0    8   41   26   -78   124
   3    4   -6    8   23   29   -48    58
```

```
modelB8.r
returns
```

returns value of r for modelB8

```
ans(:,:,1) =
   -1
ans(:,:,2) =
  -11
ans(:,:,3) =
  -21
ans(:,:,4) =
  -31
ans(:,:,5) =
  -41
ans(:,:,6) =
  -51
ans(:,:,7) =
  -61
ans(:,:,8) =
  -71
ans(:,:,9) =
  -81
ans(:,:,10) =
  -91

modelA12.outputdelay
returns

ans =
    0.1000
    0.3000
    0.5000
    0.7000
    0.9000

modelA30.inputname
returns

ans =
    'a'
    'b'
    'c'
    'd'
    'e'

modelA48.iodelay
returns

ans =
     0     0
     0     0
```

	returns value of generic property OutputDelay for modelA12
	returns value of generic property InputName of ss modelA30
	returns value of IoDelays associated with modelA48. Since no such value is associated with the model, the result is a null matrix.

3.3 The 'get' Function

The Control System Toolbox of MATLAB® provides a very useful **get** function, which may be used to gather either all the information attached with a system, or information about a particular property. Its syntax is:

get(*ModelName,'Property'*)

where,

ModelName .... specifies the name of the model under consideration

Property .... specifies the name of the property associated with the model whose value is desired to be found out (optional - if omitted, result is a list of all properties)

Example:

```
get(modelA29)
returns

            num           : {2x1 cell}
            den           : {2x1 cell}
            Variable      : 'z^-1'
            Ts            : 0.2
            InputDelay    : 0
            OutputDelay   : [2x1 double]
            ioDelayMatrix : [2x1 double]
            InputName     : {''}
            OutputName    : {2x1 cell}
            InputGroup    : {0x2 cell}
            OutputGroup   : {0x2 cell}
            Notes         : {}
            UserData      : []

get(modelA38)
returns

            a             : [3x3 double]
            b             : [3x2 double]
            c             : [2x3 double]
            d             : [2x2 double]
            e             : []
            StateName     : {3x1 cell}
            Ts            : 1
            InputDelay    : [2x1 double]
            OutputDelay   : [2x1 double]
            ioDelayMatrix : [2x2 double]
            InputName     : {2x1 cell}
            OutputName    : {2x1 cell}
            InputGroup    : {1x2 cell}
            OutputGroup   : {1x2 cell}
            Notes         : {'This is a fictitious example model'}
            UserData      : 'Could this example crash on 1/1/2000'

get(modelA38,'InputName')
returns

ans =
    'Voltage'
    'Current'
```

Do you need a practice test for such a simple concept too? O.K. go ahead…

3.4 Information about Model Dimensions and Characteristics

To get information about the model dimensions and its characteristics, Control System Toolbox provides some specialised functions. These functions along with their syntax and purpose are summarised in Table 3.1.

Table 3.1. Functions for information extraction pertaining to dimensions and characteristics of LTI models

Function	Syntax	Purpose
class	**class** (*object*)	provides information about the object type (zpk, tf, ss, frd, vector, cell, char, double, sparse *etc.*)
isa	**isa** (*object*,'*classname*')	provides information whether LTI model is of a given type or not
size	**size** (*object*)	provides information about model sizes and order
ndims	**ndims** (*object*)	provides information about number of dimensions
isempty	**isempty** (*ModelName*)	provides information whether LTI model is empty or not (returns 1 for true and 0 for false)
isct	**isct** (*ModelName*)	provides information whether LTI model is continuous-time or not (returns 1 for true and 0 for false)
isdt	**isdt** (*ModelName*)	provides information whether LTI model is discrete-time or not (returns 1 for true and 0 for false)
isproper	**isproper** (*ModelName*)	provides information whether LTI model is proper (relative degree less than or equal to 0) or not (returns 1 for true and 0 for false)
issiso	**issiso** (*ModelName*)	provides information whether LTI model is SISO or not (returns 1 for true and 0 for false)
reshape	**reshape** (*ModelName, m, n, p…*)	reshapes array of LTI models

Each of the functions given in Table 3.1 are illustrated below to give you a better understanding.

Example:

```
class(modelA48)
returns

ans =
ss
```

confirms class of modelA48 as state space

```
class(modelA15.p)
returns

ans =
cell
```

confirms class of poles of modelA15 as cell

```
class(modelA10.var)
returns

ans =
char
```

confirms class of variable of modelA10 as character

```
class(modelA10.p{2,1})
returns

ans =
double
```

confirms class of poles with indicies {2,1} of modelA10 as double

```
isa(modelA10.p{2,1},'char')
returns

ans =
   0
```

confirms class of poles with indicies {2,1} of modelA10 is not character

```
isa(modelA5,'zpk')
returns

ans =
   1
```

confirms class of modelA45 is zpk

```
size(modelA37)
returns

State-space model with 1 output, 1 input, and 0 states.
```

returns size of modelA37

```
[Alpha,Beta]=size(modelA13)
returns

Alpha =
   5
Beta =
   5
```

returns number of outputs and inputs (rows and columns) of modelA13

```
ndims(modelA33)
returns

ans =
   4
```

returns dimension of modelA33

```
ndims(modelA39.b)
returns

ans =
   3
```

returns dimension of parameter b of modelA33

```
isempty(modelB39)
returns

ans =
   1
```

confirms modelA39 as empty

```
isct(modelA5)
returns

ans =
   0
```

confirms modelA45 is not a continuous-time model

```
isdt(modelA5)
returns

ans =
   1
```

confirms modelA45 is a discrete-time model

```
isproper(modelA10)
returns

ans =
   1
```

confirms modelA10 is a proper model whose relative degree is $\leq=0$

```
issiso(modelA29)
returns

ans =
   0
```

confirms modelA29 is not a Single-Input Single-Output model

```
size(modelB28)
returns

2x1 array of zero-pole-gain models

Each model has 2 outputs and 2 inputs.
```

returns size of modelB28

```
modelC1=reshape(modelB28,1,2)
returns
```

modelC1=
the 2-by-1 array of modelB28
is reshaped to a 1-by-2 array

```
Model modelC1(:,:,1,1)
======================

  Zero/pole/gain from input 1 to output...
          1
  #1:  -------
       (z+0.3)


         10
  #2:  -----
       (z-6)

  Zero/pole/gain from input 2 to output...
```

```
          0.5 (z-0.2) (z+0.4)
  #1:  --------------------
             (z-0.1)


          2 (z+0.5)
  #2:  --------------------
       (z^2 - 0.2z + 1.01)
```

Model modelC1(:,:,1,2)
=======================

```
  Zero/pole/gain from input 1 to output...
          (z+1.319) (z-0.5687)
   #1:  ------------------------------------
        (z-0.1506) (z^2 - 0.5994z + 1.66)


              0.44444 (z-1.75)
   #2:  ------------------------------------
        (z-0.8089) (z^2 + 0.8089z + 1.099)


  Zero/pole/gain from input 2 to output...
                2.5 (z-0.2)
   #1:  ------------------------------------
        z (z-1.145) (z^2 + 1.145z + 1.31)


           0.16667 (z+4.725) (z^2 - 0.7246z + 0.4233)
   #2:  ----------------------------------------------------
        (z-1.1) (z-0.2882) (z+0.3768) (z^2 + 1.011z + 1.395)
```

Sampling time: 0.5
1x2 array of discrete-time zero-pole-gain models.

modelC2=reshape(modelB28,1,1,2)
returns

Model modelC2(:,:,1,1,1)
========================

```
  Zero/pole/gain from input 1 to output...
            1
  #1:  --------
       (z+0.3)


          10
  #2:  -----
       (z-6)


  Zero/pole/gain from input 2 to output...
       0.5 (z-0.2) (z+0.4)
  #1:  --------------------
            (z-0.1)


          2 (z+0.5)
  #2:  --------------------
       (z^2 - 0.2z + 1.01)
```

modelC2=
the 2-by-1 array of modelB28 is
reshaped to a 1-by-1-by-2 array

```
Model modelC2(:,:,1,1,2)
==========================
 Zero/pole/gain from input 1 to output...
           (z+1.319) (z-0.5687)
  #1:  ----------------------------------
       (z-0.1506) (z^2 - 0.5994z + 1.66)


             0.44444 (z-1.75)
  #2:  -----------------------------------
       (z-0.8089) (z^2 + 0.8089z + 1.099)

 Zero/pole/gain from input 2 to output...
                2.5 (z-0.2)
  #1:  ----------------------------------
        z (z-1.145) (z^2 + 1.145z + 1.31)


          0.16667 (z+4.725) (z^2 - 0.7246z + 0.4233)
  #2:  ------------------------------------------------------
        (z-1.1) (z-0.2882) (z+0.3768) (z^2 + 1.011z + 1.395)

Sampling time: 0.5
1x1x2 array of discrete-time zero-pole-gain models.
```

Ready to check yourself?

Practice Test 3.3.

1. Collect information attached with modelB7 and modelB14 using functions listed in Table 3.1.

3.5 Conversion of Models

While extracting data form the models in the article given above, did you not by any chance feel like trying out the **zpkdata** function on a transfer function model? I am sure an inquisitive learner like you, must have for sure. Then, did you really take a break and go ahead with what you felt like doing? If yes, then good. If no, then why don't you try it out now? Let's do it with ModelA1 itself.

Example:

```
[n,d,tsys]=tfdata(modelA1,'v')
returns


n =
     5   5
d =
     1   -3
tsys =
     0
```

You got an answer!

3.5.1 Automatic Conversion

Do you realise what you did above? You infact carried out a model conversion - 'automatically'! In Control System Toolbox of MATLAB®, model conversion is so simple!

What actually happened is that when you typed the function at the prompt, MATLAB® first converted the zpk model of ModelA1 to transfer function data explicitly and then gave you the values of the numerator n, the denominator d, and the sampling time tsys of the system. In a similar way, you can obtain transfer function data out of state space model of a system or *vice versa*. In fact, any type of model can be converted to a different type of model barring the exception of frd model that can not be converted to any other type.

3.5.2 Conversion by Specifying

What we discussed above was actually automatic conversion of one type of model to another. You can straightaway convert one type of model into another with a general function having the following syntax as follows:

ModelNew = *modeltype* (*ModelOld*)

where,

ModelNew....................... is the user specified name of the new model generated after conversion

ModelOld......................... is the name of the model under consideration that is to be converted into another type

modeltype........................ specifies the type of resulting ModelNew into which *ModelOld* has to be converted. It could be **zpk, tf, ss** or **dss**.

For frd model, you also have to specify the frequency points and so the syntax changes slightly.

ModelNew = **frd**(*ModelOld, frequency*)

where,

frequency.......................... specifies the frequency points at which the response is desired

Forthcoming illustrations will make concepts more clear.

Example:

```
modelC3=tf(modelA5)
returns

Transfer function from input to output...
    5 z + 5
#1: --------
    z - 3
```

```
        -20 z^3 + 56 z^2 - 55.2 z + 46.4
#2:  -------------------------------------
        z^3 + 10 z^2 - 31.75 z - 385.5
```

Sampling time: 0.1

```
modelC4=ss(modelA4)
returns
```

```
a =
                x1          x2          x3          x4
        x1       3           0           0           0
        x2       0           6        -4.2302     -0.99287
        x3       0           0         -16        -4.0156
        x4       0           0          16           0

b =
                u1          u2
        x1     4.4721        0
        x2       0         4.4721
        x3       0        17.761
        x4       0           0

c =
                x1          x2          x3          x4
        y1     4.4721     -17.889     18.918      4.4402

d =
                u1          u2
        y1       5         -20
```

Continuous-time model.

```
modelC5=frd(modelA1,50:10:100)
returns
```

From input 1 to:

```
 Frequency(rad/s)        output 1
 -------------------     ----------
         50          4.976086-0.398565i
         60          4.983375-0.332502i
         70          4.987778-0.285190i
         80          4.990638-0.249649i
         90          4.992601-0.221976i
        100          4.994005-0.199820i
```

Continuous-time frequency response data model.

In addition, you can try out all possible combinations of conversion for all the models that you can recollect.

3.5.3 Continuous/Discrete Conversions

The input and/or output signals of a physical plant are mostly analog *i.e.*, continuous-time signals. Nevertheless, the digital control elements like digital computer can accept and generate only sequences of signals *i.e.*, discrete-time signals. These discrete-time signals are generated by developing two types of interfacing units:

- a sampler which transforms the analog signal to a discrete one
- a hold unit that transforms a discrete signal to an analog one

Again, the end equipment/instruments may be analog. Thus, in a system, there is definite need to convert digital signals to analog signals and/or *vice versa*. The Control System Toolbox of MATLAB® supports functions to convert:

- discrete-time functions to continuous-time functions
- continuous-time functions to discrete-time functions
- discrete-time functions to discrete-time functions with different sampling time

The discretization/interpolation method supported by the Control System Toolbox along with the limitations of each method is mentioned in the Table3.2. Similarly, Table 3.3 summarises functions to achieve discrete to continuous conversion and *vice versa*.

Table 3.2. The discretization/interpolation methods along with their limitations

Discretization/ Interpolation Method	Comments/Limitations
Zero-order hold	on performing **c2d** conversion and **d2c** reconversion on a model, it is observed • **d2c** cannot operate on models with poles at z=0 • negative real poles in the z domain are mapped to pairs of complex poles in the s domain. Hence, the **d2c** conversion produces continuous system with higher order.
First-order hold	• more accurate compared to ZOH for systems driven by smooth inputs • only **c2d** conversion is available with FOH due to causality constraints
Tutsin approximation with or without frequency prewarping and matched poles and zeros	• matched pole zero method applies only in SISO models

Table 3.3. Continuous/discrete conversion functions

Conversion	Syntax	Remarks
Continuous to Discrete conversion	ModelDiscrete = **c2d**(*ModelContinuous*, Ts, method)	where, • Ts … is the sampling time of the new discrete model
Discrete to Continuous conversion	ModelContinuous = **d2c**(*ModelDiscrete*, Ts, method)	• method is a string to select any of the following discretization methods: 'zoh' (default)
Discrete to Discrete (*i.e.*, resample discrete system)	ModelDiscreteNew = (*ModelDiscreteoOld*, Ts)	'foh' 'tutsin' 'prewarp', freq 'matched'

These conversions can be understood better through examples considered below.

Examples:

```
modelC6=d2c(modelA2)
returns

Zero/pole/gain:
5 (s+10.99)
-----------
 (s-10.99)

modelC7=c2d(modelC6,0.1)
returns

Zero/pole/gain:
5 (z+1)
-------
 (z-3)

Sampling time: 0.1

modelC8=c2d(modelC6,0.5)
returns

Zero/pole/gain:
5 (z+241)
---------
 (z-243)

Sampling time: 0.5

modelC9=d2c(modelA2,'tutsin')
returns

Zero/pole/gain:
  50
------
(s-10)

modelC10=d2c(modelA2,'prewarp',50)
returns

Zero/pole/gain:
-167.331
---------
(s+33.47)

modelC11=c2d(modelC6,0.1,'prewarp',50)
returns

Zero/pole/gain:
```

```
3.5901 (z-1.393)
--------------------
  (z-0.718)

Sampling time: 0.1

modelC12=d2c(modelA2,'matched')
returns

Zero/pole/gain:
 54.9306
----------
(s-10.99)
```

Ready for a practice test?

> **Practice Test 3.4.**
>
> 1. Create a zpk modelCp1= 1/(z+0.1) with sampling time 0.5 and convert it to continuous-time modelCp2. Reconvert the resulting modelCp2 to discrete-time model with 0.5 sampling time and check the result for pole/zero pair at z=-0.1.

3.6 A Few Words on Model Properties Again

After going through all that is given in the previous pages regarding the LTI models, as mentioned in Article 1.1 of Chapter 1, we are back to the properties of LTI models once again.

3.6.1 Overruling the Precedence Rule and the Law of Property Inheritance

In all the examples generated above, we observed that the 'Precedence Rule' and the 'Law of Property Inheritance' are being followed. And therefore, when you operate upon a group of models consisting of all types of modes *viz.*, tf, ss and zpk models, what you get is ss model as a result. However, suppose you wish to obtain the answer after operating on different types of models as the tf model, which has the last precedence. This means that you have to overrule the 'Precedence Rule'. This can be done by either of the methods given below:

- first convert all the models to tf model and then continue with the desired operation on the resulting models.

or alternatively;

- carry out the operations and then convert the resulting model to tf model.

This can be better understood with the forthcoming illustrations.

Example:

Remember modelB2, in Chapter 2, which was obtained as indicated below:

modelB2 = modelA10 - modelA26 + modelA37

where, modelA10 was zpk model, modelA26 was tf model and modelA37 was ss model. According to the precedence rule, resulting modelB2 emerged out to be ss model (which has precedence over zpk and tf models).

To get modelB2 as say, zpk model which has precedence over tf model but over which the ss model has precedence, any of the following two methods can be followed:

- You can convert each model into zpk model and then carry out the operations:

```
modelC13=zpk(modelA10)-modelA26+zpk(modelA37)
returns
```

```
Zero/pole/gain from input "Parameter1" to output...
            13 (1 - 1.462q)
  Result1:  ----------------
                (1 - 3q)

            -12 (1 + 3.365q) (1 - 14.7q + 75.23q^2)
  Result2:  ------------------------------------------
                (1 - 6q) (1 + 16q + 64.25q^2)
```

Sampling time: 0.1

or,

- you can obtain the resulting model46 which is ss model, and then convert it to zpk model as follows:

```
modelC14 = modelA10 - modelA26 + modelA37 ;
modelC14=zpk(modelC14)
returns
```

```
Zero/pole/gain from input "Parameter1" to output...
            13 (z-1.462)
  Result1:  -------------
                (z-3)

            -12 (z+3.365) (z^2 - 14.7z + 75.23)
  Result2:  -------------------------------------
                (z-6) (z^2 + 16z + 64.25)
```

Sampling time: 0.1

This is zpk model, same as obtained above *i.e.*, modelC13 except for the variable property.

Take up a practice test now and check if you have really learnt the tricks of the trade...

3.6.2 Setting/Modifying the LTI Properties

In the previous pages, we learnt to specify the properties of the model during its creation. However, the Control System Toolbox also provides method for setting or modifying a model after it has been created using the **set** function for which the syntax is:

set (*ModelName, 'Property1', Value1, 'Property2', Value2...*)

where,

ModelName .... specifies the name of the model under consideration whose properties are to be changed

Property1, Property2 .... are the names of the properties which may be generic or zpk model-specific (optional)

Value1, Value2 .... are the new values of the properties that have to be assigned to *property1, property2...*

Note that the above is not equated to *ModelName*. The **set** function only sets the properties. When you write *ModelName* at the prompt and press the enter key, MATLAB® window displays *ModelName* model with all its properties.

If the property and value parameters are omitted, and **set** function is used with following syntax:

set (*ModelName*)

the result is then, a display of all assignable properties of *ModelName* and their admissible values.

A few illustrative examples will further elucidate the concept...

Example:

```
set(modelA20,'InputName','mechanical energy', 'OutputName',{ 'ac', 'dc'}, 'Ts', 1, 'UserData',
'Generator')

modelA20
returns
```

properties of modelA20 is set as indicated

```
Transfer function from input "mechanical energy" to output...
        z - 3
ac:  ---------------
     5 z^2 + 7 z - 9
```

```
      2 z - 4
dc:  --------
      6 z - 8

Sampling time: 1

set(modelA29)
returns

        num           : Ny-by-Nu cell of row vectors (Nu = no. of inputs)
        den           : Ny-by-Nu cell of row vectors (Ny = no. of outputs)
        Variable      : [ 's' | 'p' | 'z' | 'z^-1' | 'q' ]
        Ts            : scalar
        InputDelay    : Nu-by-1 vector
        OutputDelay   : Ny-by-1 vector
        ioDelayMatrix : Ny-by-Nu array (I/O delays)
        InputName     : Nu-by-1 cell array of strings
        OutputName    : Ny-by-1 cell array of strings
        InputGroup    : M-by-2 cell array if M input groups
        OutputGroup   : P-by-2 cell array if P output groups
        Notes         : array or cell array of strings
        UserData      : arbitrary
```

Type "**ltiprops tf**" for more details.

Ready for a practice test now?

Practice Test 3.6.

1. Assign the properties of the modelAp4 and modelAp8 indicated in Question 1(d) of Practice Test 1.1 and Practice Test 1.2 of Chapter 1 respectively, using the **set** function.

3.6.3 More on Time Delays

In previous sections, we learnt how to specify the delays in input channels, output channels and I/O channels for both continuous and discrete LTI models, by specifying values for 'IoDelay', 'InputDelay' and 'OutputDelay' properties during creation of the models or using the **set** function. These delays could be same for all the channels or different for each of the input and output channels of the model. We also observed that when a tf or zpk model was converted to ss model, the time delays in the resulting ss model were redistributed (wherever possible) in such a way as to produce a net reduction in overall input-output delays, input channel delays and output channel delays. This occurred either when all or part of the IoDelay got absorbed in the input and output delay vectors, minimising the total number of input-output delays, or because the InputDelays are transferred to output delays and *vice versa*.

However, apart from these, there are few more points relating to time delays, which we shall be discussing now.

3.6.3.1 Mapping of Discrete-time Delays to Poles at the Origin

Specifying discrete-time delays to a LTI model is actually equivalent to adding poles at z=0. These are easily absorbed into the transfer function denominator or the state space equations. For example, consider a transfer function model as follows:

$$modeC15 = \frac{z^{-2}}{2z^7 + 4z^5 + 3z^3}$$

Now, specify a time delay of 0.1sampling period to it.

This model can be obtained as follows by specifying InputDelay of 2:

```
modelC15 = tf(1,[2 0 4 0 3 0 0 0],'ts',0.1,'InputDelay',[2])
returns
Transfer function:
                1
z^(-2) * ---------------------
          2 z^7 + 4 z^5 + 3 z^3

Sampling time: 0.1
```

But this model is equivalent to $modeC16 = \dfrac{1}{2z^9 + 4z^7 + 3z^5}$ that is obtained as follows:

```
z=tf('z',0.1);
modelC16=(z^-2)/(2*z^7+4*z^5+3*z^3)
returns

Transfer function:
          1
---------------------
2 z^9 + 4 z^7 + 3 z^5

Sampling time: 0.1
```

However, modelC15 being of smaller order is more efficient in terms of memory requirements and therefore computations. When you desire to map the discrete-time delays to poles at the origin, you may obtain modelC16 from modelC15 using function **delay2z** as indicated below.

Thus,

modelC16 = **delay2z** (modelC15)

results into absorption of input delay in modelC15 into the transfer function denominator to produce a 7th-order transfer function.Note that modelC16 has no input delay. Let's check...

```
modelC16 = delay2z (modelC15)
returns
```

```
Transfer function:
        1
-----------------------
2 z^9 + 4 z^7 + 3 z^5

Sampling time: 0.1

modelC16.inputdelay
returns

ans =
     0
```

3.6.3.2 Pade Approximation of Time Delays

Whenever time delays are specified in a continuous-time LTI model, the result is a non-rational model having a product term e^{-Ts}. The function **pade** computes the rational approximation of time delays in such systems. It results in a finite dimensional model, which may hide fundamental behaviour of a partial differential equation, which defined the model. Therefore, methods that rely on pole-zero cancellation may not result in effective control. In fact, **pade** approximation of a stable transfer function may not always be stable. This function can be used in different ways to obtain different results related to time delays. Use of this function along with its syntax can be summarised as follows:

- to plot the step and phase responses of nth-order **pade** approximation and compare them with the exact responses of the time delay:

  pade(T,n)

  where,
  T .... specifies the time in e^{-Ts}
  n .... specifies the order of the **pade** approximation

  Note that there is no left argument and also that **pade** approximation has unit gain at all frequencies.

- to get the nth-order **pade** approximation of the continuous-time delay e^{-Ts} in transfer function form:

  [num,den] = **pade** (T,n)

  where,
  num .... specifies the vector of numerator coefficients in descending power of s
  den..................................... specifies the vector of denominator coefficients in descending power of s
  T .... specifies the time in e^{-Ts}
  n .... specifies the order of the **pade** approximation

- to obtain a delay-free approximation ModelNew of the continuous-time delay system *ModelOld* by replacing all delays by their n^{th}-order **pade** approximation:

ModelNew = **pade** (M*odelOld*)

where,

ModelNew........................ is the user specified name of the resulting model
ModelOld........................ specifies the model under consideration on which **pade** approximation is to be performed

- to obtain a delay-free approximation ModelNew of the continuous-time-delay system *ModelOld* by replacing all delays by the nth-order **pade** approximation like above but by specifying independent approximation orders for each input, output and IoDelay:

ModelNew = **pade** (M*odelOld*, *ni*, *no*, *nio*)

where,

ModelNew........................ is the user specified name of the resulting model
ModelOld........................ specifies the model under consideration on which **pade** approximation is to be performed
ni........................ specifies the integer array for input channel such that index $ni(j)$ is the approximation order for the jth input channel
no........................ specifies the integer array for output channel such that index $no(i)$ is the approximation order for the jth output channel
nio........................ specifies the integer array for input-output channel such that index $nio(i,j)$ is the approximation order for the IoDelay from input j to output i

Note that:

- Scalar values can be used for *ni*, *no* and *nio* to specify a uniform approximation order.
- [ ] may be used where there are no InputDelay, OutputDelay or IoDelay.

Forthcoming examples given below will make things more clear.

Examples:

```
[num, den] = pade (2,3)
returns

num =
   -1    6   -15    15
den =
    1    6    15    15


pade(2,3)
```

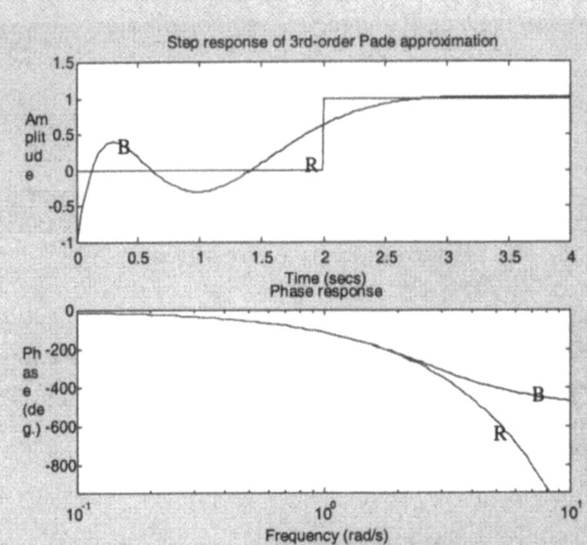


```
modelC17=pade(modelA12,2)
returns
```

Zero/pole/gain from input "Fuel" to output...

$$a: \frac{5\,(p\text{-}5)\,(p\text{-}10)\,(p^2 - 60p + 1200)^2}{(p\text{-}3)\,(p^2 + 60p + 1200)^2}$$

$$b: \frac{10\,(p\text{-}0.2)\,(p^2 - 0.8p + 1.16)\,(p^2 - 20p + 133.3)\,(p^2 - 60p + 1200)}{(p\text{-}6)\,(p^2 + 18p + 81.25)\,(p^2 + 20p + 133.3)\,(p^2 + 60p + 1200)}$$

$$c: \frac{20\,(p\text{-}1.2)\,(p\text{-}2.2)\,(p\text{-}3.2)\,(p^2 - 12p + 48)\,(p^2 - 60p + 1200)}{(p\text{-}0.5)\,(p\text{-}1.5)\,(p\text{-}2.5)\,(p^2 + 12p + 48)\,(p^2 + 60p + 1200)}$$

$$d: \frac{50\,(p\text{-}0.45)\,(p\text{-}0.6)\,(p\text{-}0.9)\,(p^2 - 8.571p + 24.49)\,(p^2 - 60p + 1200)}{(p\text{-}1)\,(p\text{-}3)\,(p\text{-}6)\,(p^2 + 8.571p + 24.49)\,(p^2 + 60p + 1200)}$$

$$e: \frac{100\,(p\text{-}1.6)\,(p\text{-}7.2)\,(p^2 - 6.667p + 14.81)\,(p^2 - 60p + 1200)}{(p\text{-}4.8)\,(p\text{-}6.4)\,(p^2 + 6.667p + 14.81)\,(p^2 + 60p + 1200)}$$

```
I/O groups:
    Group name      I/O    Channel(s)
    Heat            O       1,2
    Pressure        O       3,4
    Illumination    O       5
```

```
modelC18=pade(modelA12,[1],[2],[5])
returns
```

```
Zero/pole/gain from input "Fuel" to output...
       -5 (p-5) (p-10) (p-20) (p^2 - 60p + 1200)
  a:  -------------------------------------------
           (p-3) (p+20) (p^2 + 60p + 1200)

       -10 (p-0.2) (p-20) (p^2 - 0.8p + 1.16) (p^2 - 20p + 133.3)
  b:  -------------------------------------------------------------
         (p-6) (p+20) (p^2 + 18p + 81.25) (p^2 + 20p + 133.3)

       -20 (p-1.2) (p-2.2) (p-3.2) (p-20) (p^2 - 12p + 48)
  c:  ----------------------------------------------------
         (p-0.5) (p-1.5) (p-2.5) (p+20) (p^2 + 12p + 48)

       -50 (p-0.45) (p-0.6) (p-0.9) (p-20) (p^2 - 8.571p + 24.49)
  d:  -----------------------------------------------------------
           (p-1) (p-3) (p-6) (p+20) (p^2 + 8.571p + 24.49)

       -100 (p-1.6) (p-7.2) (p-20) (p^2 - 6.667p + 14.81)
  e:  ---------------------------------------------------
         (p-4.8) (p-6.4) (p+20) (p^2 + 6.667p + 14.81)

I/O groups:
    Group name     I/O    Channel(s)
    Heat           O      1,2
    Pressure       O      3,4
    Illumination   O      5
```

Ready to take up a practice test?

Practice Test 3.7.

1. Using **pade** 4th-order approximation, obtain a plot of step and phase response of a model with InputDelay 3.

2. Find out numerator and denominator vector in descending power of s of a 6th-order **pade** approximation of a model with InputDelay 2.

3. Obtain a delay-free approximation of model A28.

3.6.3.3 Computing Time Delays of LTI Models

In addition to the functions mentioned above, control system Toolbox of MATLAB® supports certain other functions for computing the time delays of the models. These functions along with their syntax and use are given below:

- to check if the LTI model has a time delay:

  hasdelay(*ModelName*)

where,

ModelName................. specifies the name of the model under consideration that has to be checked for presence of time delays (returns '1' if true and '0' if false)

- to get information about the total (combined delay from InputDelay, OutputDelay and IoDelay) delays between inputs and outputs of LTI model:

totaldelay (*ModelName*)

where,

ModelName...................... specifies the name of the model that has to be checked for presence of time delays.

Lets check out with some examples.

Example:

```
hasdelay(modelA45)
returns

ans =
    1

totaldelay(modelA30)
returns

ans =
    2    3    4    5    6
    4    5    6    7    8
    6    7    8    9   10
    8    9   10   11   12
   10   11   12   13   14
   12   13   14   15   16
```

Its always good to check yourself even for such simple things...

Practice Test 3.8.

1. Check all the models generated until now for time delays using **hasdelay** function.

2. Find out the value of time delays, if any, for all the models generated until now using **totaldelay** function.

Exercise for Chapter 3:

1. For all the models generated in Chapter 1 (*i.e.*, modelAe1 to modelAe12) and Chapter 2 (*i.e.*, modelBe1 to modelBe17), extract the values of all the generic and model-specific properties (as mentioned in Tables 1.1 to 1.5 of Chapter 1) associated with the models.

2. Extract information about the dimensions and characteristics of LTI models of all the models generated in Chapter 1 (*i.e.*, modelAe1 to modelAe12) and Chapter 2 (*i.e.*, modelBe1 to modelBe17) using functions mentioned in Table 3.1 of Chapter 3.1.

3. Overruling the precedence rule obtain
   - model Ae1 and modelAe2 as ss and dss models respectively
   - modelAe3 as tf model
   - modelAe5 as zpk model
   - modelAe8 as frd model (assume value of frequency points on your own)
   Comment on the result.

4. Overruling the precedence rule, obtain all the models generated in exercise of Chapter 2 (*i.e.*, modelBe1 to modelBe17) as
   - ss and dss model
   - zpk model
   - tf model
   - frd model (assume frequency response points at your own)
   Let each model have a sampling time of 0.015 seconds.

5. Use **hasdelay** and **totaldelay** functions to find out the delalys in models generated in Chapter 1 (*i.e.*, modelAe1 to modelAe12) and Chapter 2 (*i.e.*, modelBe1 to modelBe17).

6. Compute delay-free **pade** approximation for all the models generated in Chapter 1 (*i.e.*, modelAe1 to modelAe12) and Chapter 2 (*i.e.*, modelBe1 to modelBe17).

Chapter 4

Model Analysis

In this chapter, you will be analysing the models created in the previous chapters. You will learn to obtain time and frequency responses of all four types of models, which may be continuous or discrete having single-input and single-output or multiple-input and multiple-output.

After creating a number of small models for various parts of a system and then interconnecting them to obtain a complete representation of the system, the next step is to analyze these models and find out how the complete system performs under a given condition. Analysis of a model actually means study of the response of the model under different conditions and derive inference from the results.

The Control System Toolbox of MATLAB® has several built-in functions to help you assist in analysis of the models. It has functions to achieve:

- the model dynamics of a system
- the time response a system
- the frequency response of a system
- the state space analysis of a system

In this chapter you will learn to achieve all these using the models you generated and manipulated in the previous chapters.

4.1 Model Dynamics of Control System

The Control System Toolbox offers functions to determine the system poles, system zeros, DC (low-frequency) gain, pole-zero map, norms, covariance of response to white noise along with natural frequency and damping of system poles. In addition, you can also sort continuous poles by real part and discrete poles by magnitude. This is explained one by one in the forthcoming paragraphs. It should be noted however, that except for the **covariance** function, rest of the functions is not available for frd model.

4.1.1 System Poles

You may recollect Chapter 3 in which you learnt how to obtain various information about the models. It also taught you that the same data could be extracted by direct property referencing method too. For example, **zpkdata**(*ModelName*) gave you values of zeros, poles and gains of the model while entering *ModelName*.**z**, *ModelName*.**p** and *ModelName*.**k** at the command prompt of MATLAB® command window also gave you the same result. Hope you also remember that for SISO zpk or tf models, the poles are simply the denominators of the roots while for MIMO zpk or tf models, the poles are computed as the union of the poles for each SISO entry. Similarly, for state space models, the poles are eigenvalues of the 'a' matrix given by 'a-λI' for simple state space models or 'a-λe' for descriptor state space models.

However, Control System Toolbox offers another function to obtain the system poles of the models, the syntax for which is:

p=**pole**(*ModelName*)

where,
ModelName .... is the name of the model under consideration

The answer to this function is the value of poles of the model, *i.e.*, p, which is a column vector. For LTI arrays with sizes [Ny, Nu, S1... Sp], p has the same dimension as the model such that p(:,1,j1,......jp) contains the poles of the model, model(:,:,j1,... jp).

Note that:
- if columns or rows have a common denominator, the roots of this denominator are counted only once
- **pole** function is not supported by frd models

Let us try finding out poles of a few models.

Example:

```
p=pole(modelA3)
returns


p =
  6.0000
 -8.0000 + 0.5000i
 -8.0000 - 0.5000i

p=pole(modelA33(:,:,4,1))
returns


p =
 -9.6359
 -0.3641
 -9.8095
 -0.1905
 -9.8709
 -0.1291
 -9.6663
 -0.3337
 -9.8181
 -0.1819
 -9.8750
 -0.1250
```

```
p=pole(modelA38)
returns

p =
  -2.1856
  -5.9072 + 2.5067i
  -5.9072 - 2.5067i

p=pole(modelA40)
returns

p =
 -56.2300
   1.1608 + 1.0195i
   1.1608 - 1.0195i
  -1.3248
```

4.1.2 System Zeros

Finding out system zeros is similar to the procedure adopted for finding out system poles. The syntax for achieving this is:

z=**zero**(*ModelName*)

where,
ModelName .... is the name of the model under consideration

Rest is all similar to the **pole** function. Whatever applies to the **pole** function applies to the **zero** function too.

Example:

```
p=zero(modelA3)
returns

p =
   -1

p=zero(modelA33)
returns

p =
  Empty array: 0-by-1-by-4

p=zero(modelA38)
returns

p =
 -11.6680
  -1.8803 + 2.4524i
  -1.8803 - 2.4524i
```

```
p=zero(modelA40)
returns


p =
  1.0e+002 *
  -1.0526
   0.0108 + 0.0101i
   0.0108 - 0.0101i
  -0.0132
```

4.1.3 Low-frequency or DC Gain

Similar to finding out poles and zeros of a model, you can obtain the low-frequency or DC gain for the given model too. The syntax for obtaining this is:

> k=**dcgain**(*ModelName*)

> where,
> *ModelName* .... is the name of the model under consideration

Example:

```
k=dcgain(modelA3)
returns

k =
  -0.0246

k=dcgain(modelA33)
returns

k(:,:,1) =
   1   1
   1   1
   1   1
k(:,:,2) =
   1   1
   1   1
   1   1
k(:,:,3) =
   1   1
   1   1
   1   1
k(:,:,4) =
   1   1
   1   1
   1   1


k=dcgain(modelA38)
returns


k =
```

```
   2.0523   6.0523
   4.7151   6.7151

k=dcgain(modelA40)
returns

k =
  -0.0422    3.9578    2.9578
   3.2120    5.2120    8.2120
   0.1187   -0.8813    4.1187
```

4.1.4 Pole-zero Map

Till now you had been obtaining the numeric values of poles and zeros. Automatic plotting of these values on complex plane is also possible by the **pzmap** function available with Control System Toolbox of MATLAB®. Syntax for this function is:

- **pzmap**(*ModelName*)

where,
ModelName .... is the name of the model under consideration

The Figure window opens up and a map of the pole and zeros of the *ModelName* is obtained on a complex plane. The poles are plotted as 'x' and the zeros as 'o'.

- [p, z]=**pzmap**(*ModelName*)

where,
p .... returns a column vector of poles of *ModelName*
z.. returns a column vector of zeros of *ModelName*
ModelName .... is the name of the model under consideration

No plot is drawn on the screen in the Figure window.

Note that for SISO system, the transfer function poles and zeros are plotted, while, for MIMO system, system poles and transmission zeros are plotted.

Example:

```
[p,z]=pzmap(modelA3)
returns
p =
   6.0000
  -8.0000 + 0.5000i
  -8.0000 - 0.5000i
z =
  -1

[p,z]=pzmap(modelA14)
returns
```

```
p =
   0.3000
   6.0000
   0.1000
   0.4000
   0.1000 + 1.0000i
   0.1000 - 1.0000i
z =
   0.1564 + 1.0534i
   0.1564 - 1.0534i
   0.8197
  -0.2148
   0.2912
```

pzmap(modelA3)
returns

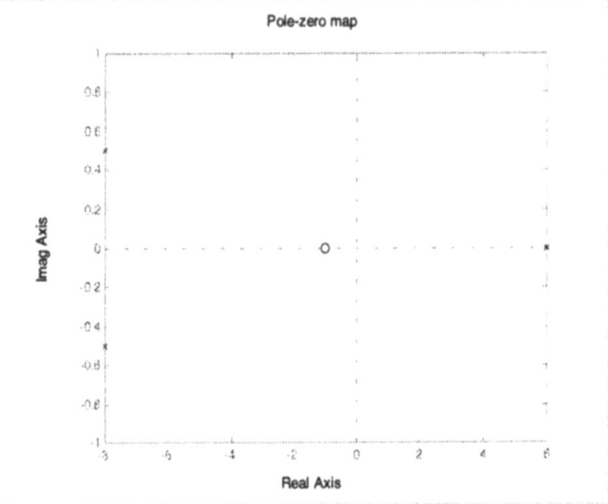

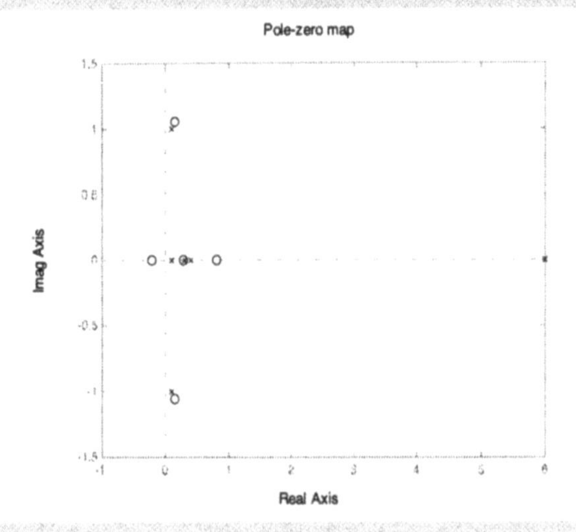

pzmap(modelA14)
returns

4.1.5 H_2 and L_∞ Norms

Control System Toolbox supports the **norm** function to find out:

- the H_2 norm of a stable continuous system, which is the root mean square of its impulse response and measures the steady state covariance or the power of the output response to unit white noise inputs.
- the infinity norm, L_∞ which is the peak gain of the frequency response of a continuous or a discrete-time system.

The syntax to obtain these norms is summarised in Table 4.1 given below.

Table 4.1. H_2 and L_∞ norms

For	Syntax	Result	Remarks
M A T R I C E S	**norm**(*ModelName*) **norm**(*ModelName*, 2)	H_2 norm of a model	• equivalent to the expression **sqrt(trace(covar(***ModelName***,1)))** • infinite for unstable models • infinite for models which are continuous and have non-zero feedthrough *i.e.,* non-zero gain at infinite frequency
	norm(*ModelName*, 1)	1- norm of a model	• the largest column sum • equivalent to the expression **max(sum(abs(***ModelName***)))**
	norm(*ModelName*, inf, tot)	infinity norm of a model	• optional 'tot' sets the desired relative accuracy on the computed infinity norm (default value is 1e-2) • the largest row sum • equivalent to the expression **max(sum(abs(***ModelName***)))**
	norm(*ModelName*, 'fro')	Frobenius norm	equivalent to the expression **sqrt(sum(diag(***ModelName*** ' * ***ModelName***)))**
V E C T O R	**norm**(*ModelName*, P)	-	equivalent to the expression **sum(abs(***ModelName***) .^P)^(1/P)**
	norm(*ModelName*)	**norm**(*ModelName*, 2)	-
	norm(*ModelName*, inf)	-	equivalent to the expression **max(abs(***ModelName***))**
	norm(*ModelName*, -inf)	-	equivalent to the expression **min(abs(***ModelName***))**

Note: *ModelName* is the name of the model under consideration

Let us try using these functions on some models…

Example:

```
norm(modelA3 )
returns
```

```
ans =
   Inf

norm(modelA48)
returns

ans =
    0.6891

norm(modelA48,2)
returns

ans =
    0.6891

norm(modelA48,inf)
returns

ans =
    0.7606

norm(modelA48,2,'fro')
returns

ans =
    0.6891
```

4.1.6 Covariance of Response to White Noise

Control System Toolbox supports **covar** function to compute the stationery covariance of the output y of an LTI model (both continuous and discrete-time), *ModelName* driven by white Gaussian noise inputs of intensity *w*. The syntax to compute the steady state output response covariance is:

[P, Q] = **covar**(*ModelName*, *w*)
(if model is tf or zpk (*i.e.,* not ss) then set Q=0)

where,
ModelName...................... specifies the name of the model under consideration
w...................................... specifies the intensity of the white Gaussian noise

When the model is an array of dimension n, P and Q are also multidimensional arrays such that P(:,:,i1......in) and Q(:,:,i1.......in) are covariance matrices for the system model (:,:,i1... in).

Note that:
- the state and output covariances are defined for stable systems only
- for continuous systems, the output response covariance P is finite only for strictly proper system

Hence, unstable systems or continuous-time models with non-zero feedthrough have infinite covariance response.

Example:

```
covar(modelB15,4)
returns

ans =
  Inf

covar(modelA48,4)
returns

ans =
   0.0005   -0.0290
  -0.0290    1.8990

covar(modelA48,9)
returns

ans =
   0.0011   -0.0653
  -0.0653    4.2728
```

4.1.7 Natural Frequency and Damping of LTI Model Poles

The damping factor and natural frequencies are two things, which play an important role in the analysis of the LTI models. Control System Toolbox has **damp** function for computing these values. The syntax for using this function is:

$[wn,z,p] = $ **damp**($ModelName$)

where,

wn..	returns the column vectors of natural frequency of poles of the model
z..	returns the column vectors of damping factor of poles of the model
p	returns the column vectors of eigenvalues or real poles of the model. It returns the same values as pole(model) upto reordering (optional)
ModelName	specifies the name of the model under consideration

Example:

```
[wn,z,p]=damp(modelA48)
returns

wn =
   []
```

```
z =
    []
p =
   0.5884
   0.1357

[wn,z,p]=damp(modelA31)
returns

wn =
   0.4620
   0.5528
   0.5528
   0.5528
   0.5528
   0.7850
   0.8180
   0.8180
   1.0429
   1.0429
   1.0641
   1.0641
   1.1941
z =
   -1.0000
    0.0615
    0.0615
    0.0615
    0.0615
   -1.0000
   -1.0000
   -1.0000
   -0.3510
   -0.3510
    0.2122
    0.2122
    1.0000
p =
    0.4620
   -0.0340 + 0.5518i
   -0.0340 - 0.5518i
   -0.0340 + 0.5518i
   -0.0340 - 0.5518i
    0.7850
    0.8180
    0.8180
    0.3660 + 0.9765i
    0.3660 - 0.9765i
   -0.2258 + 1.0399i
   -0.2258 - 1.0399i
   -1.1941

[wn,z]=damp(modelA12)
returns

wn =
    0.5000
```

```
  1.0000
  1.5000
  2.5000
  3.0000
  3.0000
  4.8000
  6.0000
  6.0000
  6.4000
  9.0139
  9.0139
z =
 -1.0000
 -1.0000
 -1.0000
 -1.0000
 -1.0000
 -1.0000
 -1.0000
 -1.0000
 -1.0000
 -1.0000
  0.9985
  0.9985
```

4.1.8 Sorting Eigenvalues

Sometimes it is required to sort out the eigenvalues of models in some order during analysis. The **esort** and **dsort** functions do just this for continuous and discrete-time models respectively. The syntax for these functions are:

- for continuous-time models
  [s,ndx] = **esort**(P)

where,

P .... specifies a vector containing the continuous-time poles

s .... returns the continuous-time poles in descending order by real part

ndx .... returns a display of indices used in sort (optional)

Unstable poles appear first in the list and then the remaining poles are listed by decreasing real parts.

- for discrete-time models
  [s,ndx] = **dsort**(P)

where,

P .... specifies a vector containing the discrete-time poles

s .... returns the discrete-time poles in descending order by magnitude

ndx .... returns a display of indices used in sort (optional)

Example:

```
p=pole(modelB45);
[s,ndx]=esort(p)
returns

s =
   6.0000
   3.0000
   0.8133
  -2.2133
  -8.0000 + 0.5000i
  -8.0000 - 0.5000i
ndx =
   2
   1
   6
   5
   3
   4
```

<table>
<tr><td colspan="2">modelB45=</td></tr>
<tr><td>

$$\begin{bmatrix} \dfrac{5(s+1)}{(s-3)} & 5 \\ \dfrac{-20(s-2)(s2-0.8s+1.16)}{(s-6)(s2+16s+64.25)} & \dfrac{0.2(s-3)}{(s+2.2.13)(s-0.8133)} \end{bmatrix}$$

</td></tr>
<tr><td colspan="2">where, poles are</td></tr>
<tr><td>p1=3</td><td>p2=6</td></tr>
<tr><td>p3=-8+0.5i</td><td>p4=-8-0.5i</td></tr>
<tr><td>p5=-2.213</td><td>p6=0.8133</td></tr>
</table>

Similarly we can sort poles for discrete models too.

```
p=pole(modelB23);
[s,ndx]=dsort(p)
returns

s =
   6.6389
   3.4695
  -2.6540
   1.2362
  -0.3681 + 0.8801i
  -0.3681 - 0.8801i
   0.9092
  -0.4368 + 0.7565i
  -0.4368 - 0.7565i
   0.8736
  -0.3624 + 0.7651i
  -0.3624 - 0.7651i
   0.1806 + 0.7549i
   0.1806 - 0.7549i
ndx =
   1
  10
  11
   7
   8
   9
  12
   4
   5
   6
  13
  14
   2
   3
```

Ready for a practice test?

4.2 Time Response Analysis of Control Systems

Performance of Control Systems is mostly studied with respect to time. The time response can be subdivided into:

- transient response *i.e.*, the manner in which response goes from initial to final state
and
- steady state response *i.e.*, the manner in which the response goes as time approaches infinity

The Control System Toolbox provides facility to analyse the behaviour of zpk, tf and ss models to some arbitrary input over the time. In the paragraphs to follow, you will learn to obtain these responses for some standard inputs like impulse and step signals.

4.2.1 Response of a Model to Standard Signals

There are two standard signals supported by Control System Toolbox to which the response of a model can be obtained. These are:

- Impulse signal
- Step signal

The syntax to obtain the response of a model to these standard signals is:

signaltype(ModelName1, 'style1',ModelName2,'style2', ..., tfinal or T)

where,

signaltype .... specifies the name of the standard signal which could be either ***impulse*** or ***step***

ModelName, ModelName2 specifies the names of the zpk, tf or ss models under consideration whose response to the signals have to be obtained

style1, style2 .... specifies the string for specifying the *'linestyle_marker_color'* for the plots of *ModelName1, ModelName2...* respectively (optional), see Table 4.2 for more information

tfinal .... specifies the final time of the signal such that simulation is obtained from t=0 to t=*tfinal* (optional)

T .... specifies the time vector for simulation (optional)

Note that:

- plot of response of *ModelName1, ModelName2...* is obtained on a single plot
- for discrete-time models with unspecified sampling time, *tfinal* is interpreted as the number of samples

Table 4.2. Plot style

Line Color			Value Marks		Line Type	
y	1 1 0	yellow	.	point	-	solid
m	1 0 1	magenta	o	circle	:	dotted
c	0 1 1	cyan	x	cross or x-mark	-.	dashdot
r	1 0 0	red	+	plus mark	--	dashed
g	0 1 0	green	*	star		
b	0 0 1	blue	s	square		
w	1 1 1	white	d	diamond		
k	0 0 0	black	v	triangle(down)		
			^	triangel(up)		
			<	triangle(left)		
			>	triangle(right)		
			p	pentagram		
			h	hexagram		

Note: if no style is specified, the plots follow the default style as listed in the Table above in cyclical order *i.e.,* for *ModelName1* - yellow color, point marks and solid line type, for *ModelName2*- magenta color, circle marks and dotted line type and so on. For plots more than the styles listed above, the default goes through one cycle and then repeats. That is, for *modelName5*, the line type will be solid again. For monochrome, systems, color has no significance and the cycling takes place for the value marks and the line types only.

Let us consider some examples to make thing more clear.

Example:

Let us start with time response plots for a period 0.5 to 1.5 seconds in steps of 0.5 seconds and draw it for models as indicated below:

- modelA31 discriminated on the graph by solid magenta line with star value marks
- modelA35 discriminated on the graph by dotted cyan line with dot value marks
- modeA14 discriminated on the graph by dot dash yellow line with right triangle value marks

```
impulse(modelA31, 'm*-', modelA35, 'c.:', modelA14,'y>-.',[0.5:0.5:1.5])
returns
```

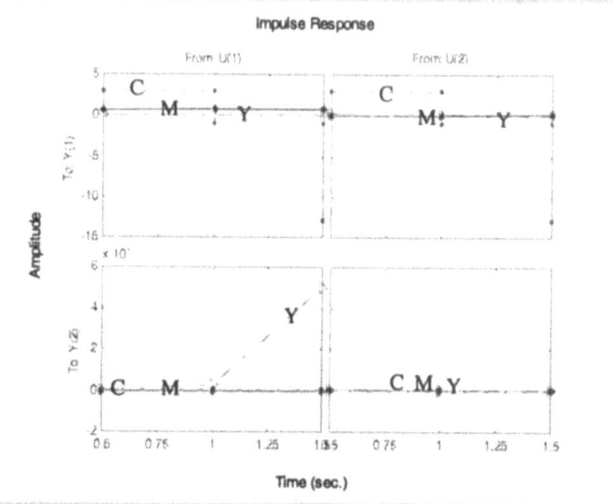

Similarly, step response can be obtained for the models but this time, instead of specifying period of time, we will specify the final time as 1 second for which the response is desired.

```
step(modelA31, 'm*-', modelA35, 'c.:', modelA14,'y>-.',1)
returns
```

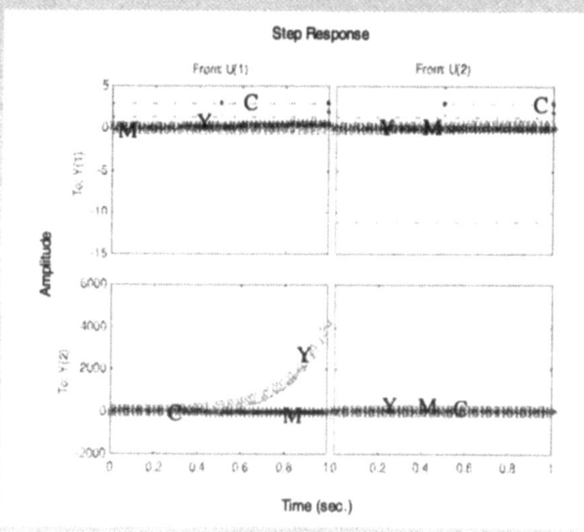

4.2.2 Response of a Model to Arbitrary Periodic Signals

The Control System Toolbox provides two more functions for obtaining the time response of the models. Using these you can obtain the response of a model to signals defined by you. The

gensig function produces a periodic signal to be used with **lsim** function, which simulates the time response of LTI models to arbitrary inputs. The syntax for these are:

- [u,t]=**gensig**(*type,period,tfinal,ts*)

where,

type	specifies the type of the signal which could be either *'sin'*, *'square'* or *'pulse'* for sine wave, square wave or periodic pulse of unit amplitude respectively
period	specifies the period of the signal
tfinal	specifies the final time of the signal such that signal is generated from t=0 to t=*tfinal* (optional)
ts	specifies the interval of the time samples in *tfinal*

Note that the signal is generated and stored in u as a vector over vector t time samples. You can see the signal you generated by using the **plot** function as **plot**(t,u).

- **lsim**(*ModelName1,'style1'*, *ModelName2*, *'style2'*.......,*u,t,x0*)

- [Y,T,X]=**lsim**(*ModelName1, u, t, x0*)

where,

ModelName1, ModelName2	specifies the names of the models under consideration
style1, style2	specifies the string for specifying the *'linestyle_marker_color'* for the plots of *ModelName1, ModelName2...* respectively (optional), as indicated in Table 4.2
u	is a matrix describing the input signal obtained from **gensig** or otherwise, having as many columns as number of inputs
t	is the regularly spaced time vector describing the signal obtained from **gensig** or otherwise
x0	specifies the non-zero initial state for ss models of the signal such that signal is generated from t=0 to t=*tfinal* (optional)
Y	returns the output used for simulation, without drawing plot on the screen. It has **length**(T), and as many columns as outputs in *ModelName*
T	returns the time vector used for simulation, without drawing plot on the screen
X	returns the state trajectory of ss model used for simulation, without drawing plot on the screen. It is a matrix of **length**(T) rows and as many columns as states

Note that for discrete-time systems:
- the time interval of vector t should match with the sampling time
- matrix u should be sampled at the same rate as the model, then, the time vector *t* could be omitted or set to an empty matrix [ ]

Example:

```
[u,t]=gensig('sin',2*pi,60);
lsim(modelA37,'b',modelA26,'r',u,t)
returns
```

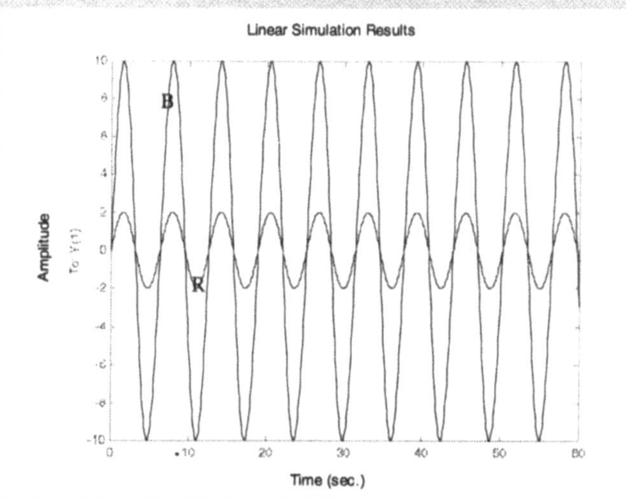

```
[u,t]=gensig('pulse',2*pi,60);
lsim(modelA27,u,t)
returns
```

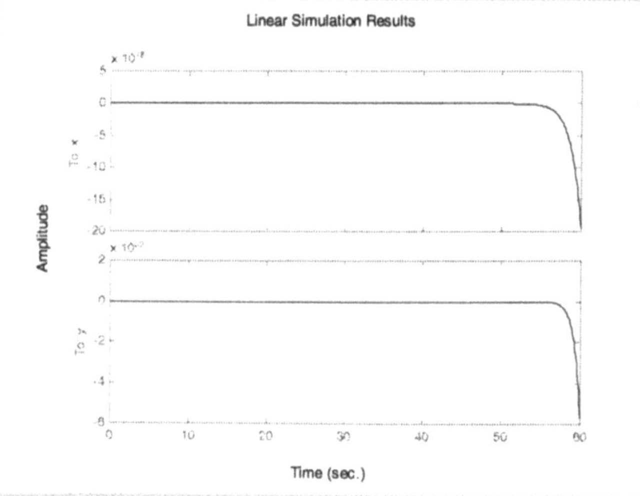

The time response values can also be obtained as follows without actually plotting the response:

```
t=0:1:5;
u=sin(t);
[Y,T]=lsim(modelB15, u,t)
returns
```

```
Y =
         0
         0
   -0.0092
    0.0462
    0.2696
    0.7569
T =
         0
         1
         2
         3
         4
         5
```

You should keep in mind that **gensig** is not the only function that can be used in conjunction with **lsim** function to generate the input signal. The input signal could be generated by any other method and used for obtaining time response.

```
t=0:0.01:5;
u=[sin(t)];
lsim(modelA27, u, t)
returns
```

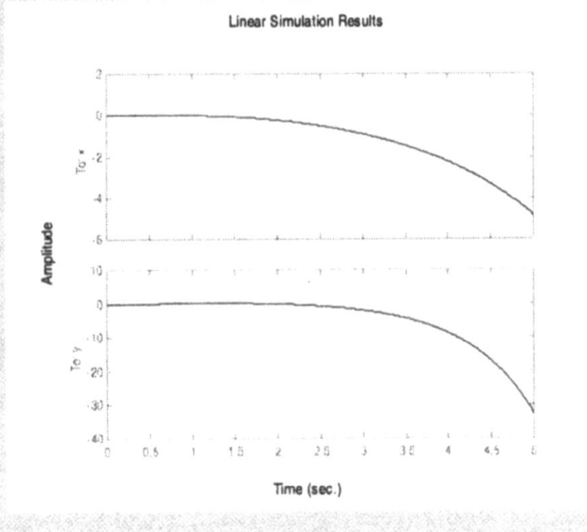

4.2.3 The Unit Step Function

You can use the **stepfun** function to obtain a unit step function as follows:

stepfun(*t*,*t0*)

where,
t .... is a vector that increases monotonically
t0 .... specifies time value

Note that a vector of **length**(t) which has values zero for t<t0 and one for t ≥ t0 is returned by this function.

Example:

```
stepfun([1 3  5 7 9 11 13 15],4)
returns

ans =
    0    0    1    1    1    1    1    1

t=0:0.1:1;
stepfun([sin(t)],0.5)
returns

ans =
    0    0    0    0    0    0    1    1    1    1    1
```

Ready to check yourself?

Practice Test 4. 2.

1. Obtain the time response of modelA23 and modelA36 for an arbitrary sinusoidal function on a plot or otherwise.

4.3 Frequency Response Analysis of Control Systems

We are all familiar that frequency response of the system plays an important role in studying the stability of a system. There are several methods by which the magnitude and phase response of an LTI model to sinusoidal input can be obtained. The Control System Toolbox of MATLAB® supports following such methods:

- Bode plot
- Singular Value plot
- Nyquist plot
- Nichols plot

You shall now learn to plot the above and use these plots to:

- evaluate the frequency response at a given frequency
- evaluate the Gain and Phase margins
- judge the stability of the system and check it by other methods

4.3.1 Obtaining Frequency Response Plots

The syntax applicable for obtaining any of the above mentioned frequency response plots is:

plotname(ModelName1, 'style1', ModelName2,'style2',…, w, {wmin, wmax})

where,

plotname.......................... specifies the type of the plot and is either **bode, sigma, nyquist** or **nichols** for Bode, singular value, Nyquist, or Nichols plot respectively

ModelName1, ModelName2 specifies the names of the models under consideration

style1, style2 .... specifies the string for specifying the *'linestyle_marker_color'* for the plots of *ModelName1, ModelName2…* respectively (optional), as indicated in Table 4.2

w..................................... specifies the user supplied vector of frequencies in rad/sec at which the response of the model is desired to be evaluated (optional)

wmin, wmax...................... specifies the user supplied minimum and maximum value of the range of frequencies in rad/sec over which the plot is desired

Apart from this, there are plot specific syntax for obtaining the frequency response of the models, which are summarized in Table 4.3.

Now we will generate these plots for some of the models we created in earlier chapters so as to make the concepts more clear…

Example:

Let us start with Bode plot and draw it for SISO models as indicated below:

- modelA26 discriminated on the graph by dashdot green line with star value marks
- modelB34 discriminated on the graph by dotted blue line with diamond value marks
- modelA1 discriminated on the graph by dashed red line with upper triangle value marks

```
bode(modelA26,'-.*g',modelB34,':db',modelA1,'--^r')
returns
```

Table 4.3. Plot-specific syntax for frequency response

Plot Type	Syntax	Remarks
Bode	[mag, phase] =**bode**(*ModelName*,w) or, [mag, phase,w] =**bode**(*ModelName*)	• returns the response magnitudes and phases (in degrees) along with the frequency vector w if specified • no plot is drawn on the screen • if model has ny outputs and nu inputs, magnitude and phase are arrays of size [ny nu **length**(w)] where mag(:,:,k) and phase(:,:,k) determine the response at the frequency w(k) • to get the magnitude in dB, type magdb=20*log10(mag)
Sigma	**sigma**(*ModelName*, *type*)	• draws the modified singular value plot of the model depending on the value of the type as follows: type=1➔ plot of inv(model) type=2➔ plot of I+ model type=3➔ plot of I+ inv(model) and, so on • *ModelName* should be a square system while using this syntax
	v=**sigma**(*ModelName*,w) or, [v,w]= **sigma**(*ModelName*)	• returns the singular values of the frequency response along with the frequency vector if unspecified • no plot is drawn on the screen • matrix v is a vector of **length**(w) columns such that v(:,k) gives the singular values in descending order at the frequency w(k)
Nyquist	[re, im] =**nyquist**(*ModelName*,w) or, [re, im,w] =**nyquist**(*ModelName*)	• returns the real parts re and imaginary part im of the frequency response along with the frequency vector if unspecified • no plot is drawn on the screen • if model has ny outputs and nu inputs, re and im are arrays of size [ny nu **length**(w)] where re(:,:,k) and im(:,:,k) determine the response at the frequency w(k) and the output is displayed as re(:,:,k)+im(:,:,k)
Nichols	[mag, phase] =**nichols**(*ModelName*,w) or, [mag, phase,w] =**nichols**(*ModelName*)	• returns the response magnitudes and phases in degrees along with the frequency vector w if specified • no plot is drawn on the screen • if model has ny outputs and nu inputs, mag and phase are arrays of size [ny nu **length**(w)] where mag(:,:,k) and phase(:,:,k) determine the response at the frequency w(k)

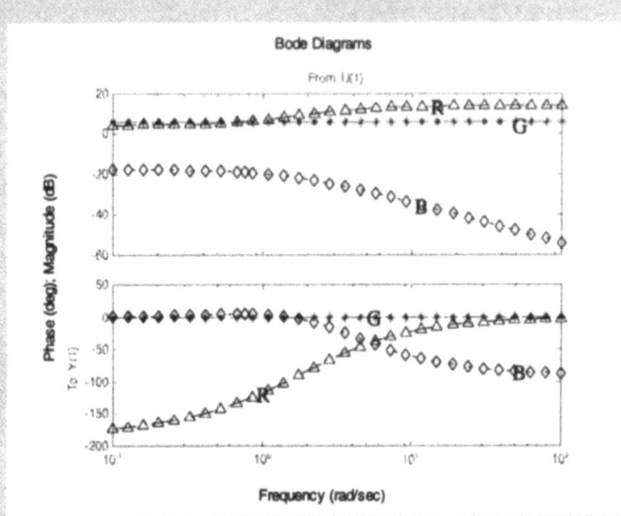

For frd modelA41 also you can obtain the Bode plot.

```
bode(modelA41)
returns
```

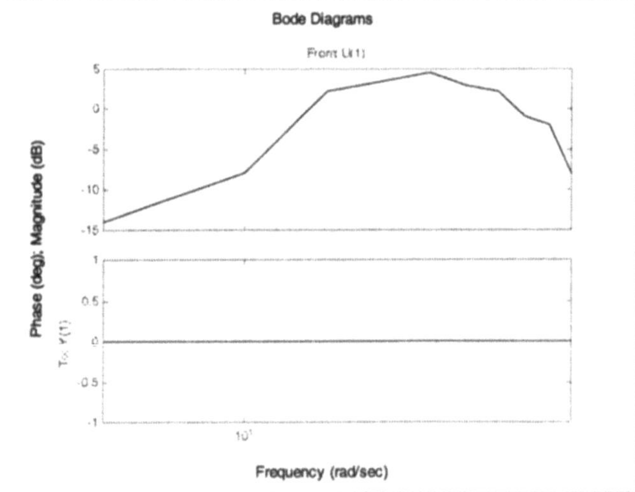

Bode plot can also be obtained for MIMO models. Observe that the plot follows the same old matrix arrangement for displaying the plots for various inputs and outputs.

Let us consider modelB45 and draw the plot for frequency range 100 to 1000.

```
bode(modelB45,{10e1,10e2})
returns
```

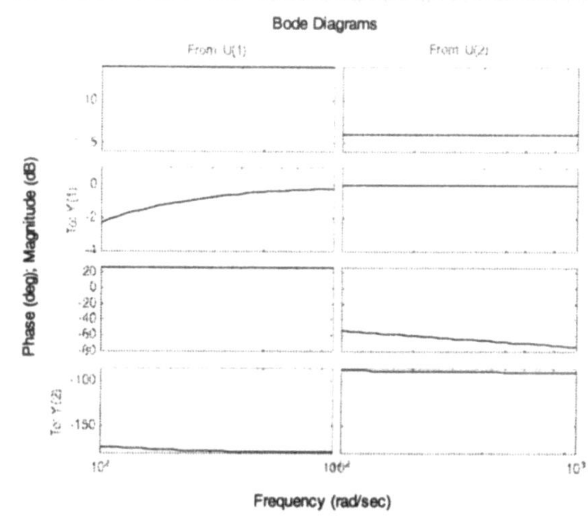

To obtain a Bode plot between a frequeny range of 10×10^1 to 10×10^4 in steps of 10×10^3 , specify the frequency range and enter the function for obtaining the Bode plot as follows:

```
bode(modelB11,[10e1:10e3:10e4])
returns
```

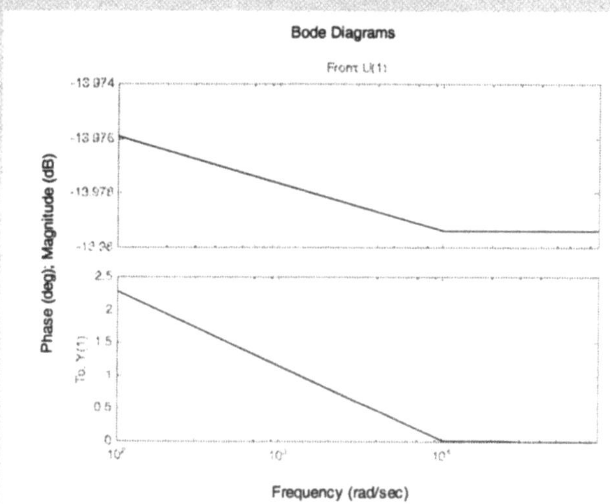

The phase and magnitude of the Bode plot can also be obtained at different frequencies without actually drawing the plot on the screen as follows:

```
[m,p,w] = bode (modelB11)
returns

m(:,:,1) =
    0.5974
```

The same values of m and p are obtained by the function
[m,p]=**bode**(modelB11,w)
the only difference is that now the list of w is not displayed.

```
m(:,:,2) =
   0.5958
m(:,:,3) =
   0.5932
m(:,:,4) =
   0.5892
m(:,:,5) =
   0.5830
m(:,:,6) =
   0.5734
m(:,:,7) =
   0.5590
m(:,:,8) =
   0.5384
m(:,:,9) =
   0.5103
m(:,:,10) =
   0.4745
m(:,:,11) =
   0.4328
m(:,:,12) =
   0.3884
m(:,:,13) =
   0.3453
m(:,:,14) =
   0.3071
m(:,:,15) =
   0.2757
m(:,:,16) =
   0.2517
m(:,:,17) =
   0.2344
m(:,:,18) =
   0.2223
m(:,:,19) =
   0.2143
m(:,:,20) =
   0.2091
m(:,:,21) =
   0.2057
m(:,:,22) =
   0.2036
m(:,:,23) =
   0.2022
m(:,:,24) =
   0.2014
m(:,:,25) =
   0.2009
m(:,:,26) =
   0.2005
m(:,:,27) =
   0.2003
m(:,:,28) =
   0.2002
m(:,:,29) =
   0.2001
m(:,:,30) =
   0.2001
```

```
p(:,:,1) =
   172.3803
p(:,:,2) =
   170.3459
p(:,:,3) =
   167.7800
p(:,:,4) =
   164.5549
p(:,:,5) =
   160.5238
p(:,:,6) =
   155.5279
p(:,:,7) =
   149.4145
p(:,:,8) =
   142.0682
p(:,:,9) =
   133.4537
p(:,:,10) =
   123.6551
p(:,:,11) =
   112.8843
p(:,:,12) =
   101.4457
p(:,:,13) =
    89.6780
p(:,:,14) =
    77.9194
p(:,:,15) =
    66.5085
p(:,:,16) =
    55.7835
p(:,:,17) =
    46.0450
p(:,:,18) =
    37.4987
p(:,:,19) =
    30.2213
p(:,:,20) =
    24.1723
p(:,:,21) =
    19.2331
p(:,:,22) =
    15.2500
p(:,:,23) =
    12.0645
p(:,:,24) =
     9.5306
p(:,:,25) =
     7.5220
p(:,:,26) =
     5.9333
p(:,:,27) =
     4.6785
p(:,:,28) =
     3.6882
p(:,:,29) =
     2.9072
```

```
p(:,:,30) =
    2.2913
w =
    0.1000
    0.1269
    0.1610
    0.2043
    0.2593
    0.3290
    0.4175
    0.5298
    0.6723
    0.8532
    1.0826
    1.3738
    1.7433
    2.2122
    2.8072
    3.5622
    4.5204
    5.7362
    7.2790
    9.2367
   11.7210
   14.8735
   18.8739
   23.9503
   30.3920
   38.5662
   48.9390
   62.1017
   78.8046
  100.0000
```

In a similar way the singular value plots and singular values can be obtained.

```
sigma(modelA15)
returns
```

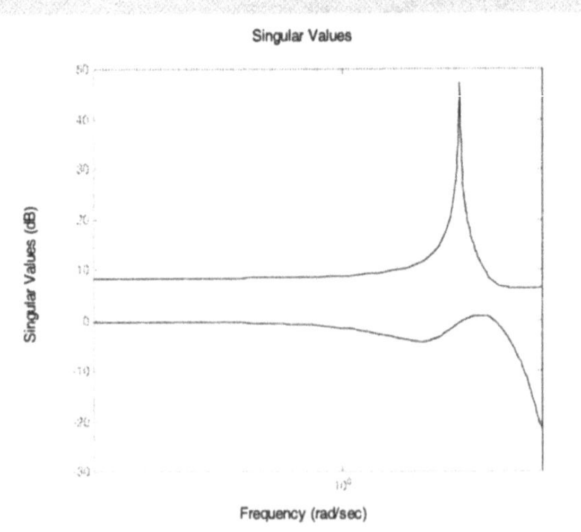

sigma(modelA27,modelA28)
returns

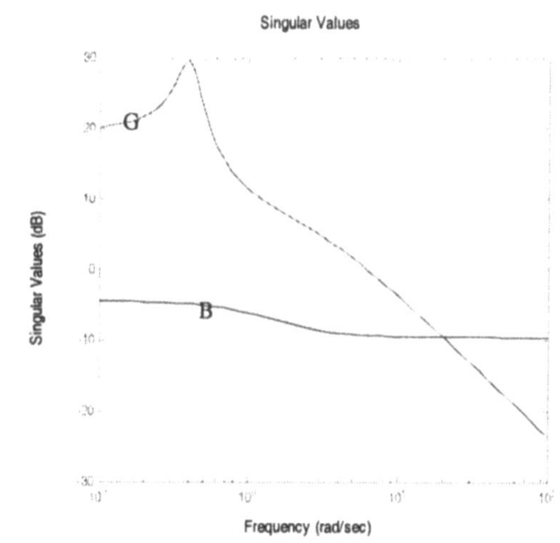

[v,w]=sigma(modelA1)
returns

v =
  Columns 1 through 7
    1.6740   1.6785   1.6857   1.6972   1.7154   1.7441   1.7889
  Columns 8 through 14
    1.8574   1.9597   2.1073   2.3105   2.5749   2.8961   3.2566
  Columns 15 through 21
    3.6266   3.9723   4.2667   4.4975   4.6662   4.7833   4.8615
  Columns 22 through 28
    4.9124   4.9449   4.9656   4.9785   4.9866   4.9917   4.9948
  Columns 29 through 30
    4.9968   4.9980

w =
  0.1000
  0.1269
  0.1610
  0.2043
  0.2593
  0.3290
  0.4175
  0.5298
  0.6723
  0.8532
  1.0826
  1.3738
  1.7433
  2.2122
  2.8072
  3.5622
  4.5204
  5.7362

```
    7.2790
    9.2367
   11.7210
   14.8735
   18.8739
   23.9503
   30.3920
   38.5662
   48.9390
   62.1017
   78.8046
  100.0000
```

The Nyquist plot can also be obtained in similar ways.

```
nyquist(modelB21)
returns
```

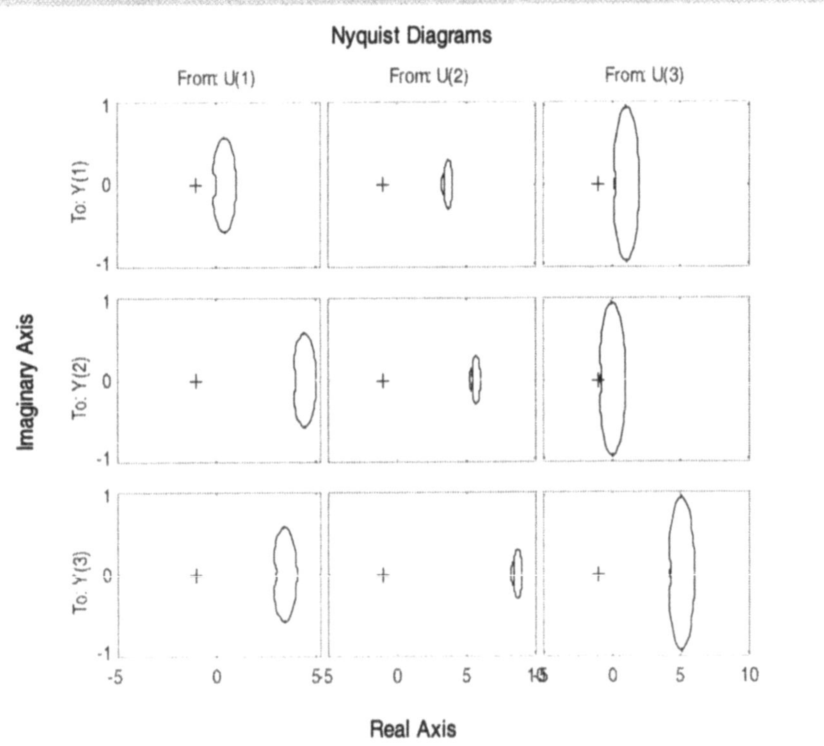

The Nichols chart can be drawn in a similar way too.

```
nichols(modelB28)
returns
```

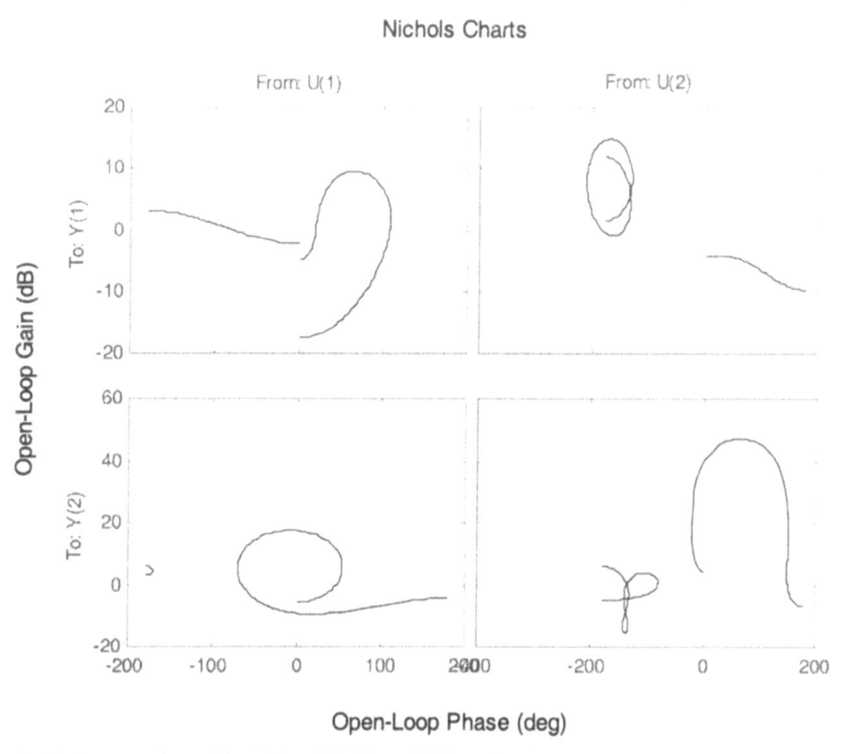

Ready for a practice test?

Practice Test 4.3

1. Obtain the real and imaginary parts of the Nyquist plot drawn above for modelB21 without obtaining the plot.

2. Obtain the magnitude and phase of the frequency response from the Nichols chart drawn for modelB28 above without actually obtaining the plot.

3. Obtain all of the frequency response plots for modelA16 in red color solid line, modelA33 in blue color dotted line, modela39 in yellow color dashed line and modelA46 in green color dashdot line on the same plot.

4.3.2 Getting Information from the Plots

During frequency analysis of a model, it is sometimes desired to evaluate frequency response of the model over a range of frequency points or at a particular frequency. The Control System

Toolbox supports two functions - **freqresp** and **evalfr** to achieve this. The potential of these functions is elucidated with illustrations in the paragraphs to follow.

4.3.2.1 Evaluating Frequency Response over a Frequency Range

As mentioned above, it is possible to evaluate the frequency response of a model over a particular frequency range using the buit-in function **freqresp** whose syntax is:

fr=**freqresp**(*ModelName,w*)

where,
fr.. returns the values of frequency response
ModelName....................... specifies the name of the models under consideration
w.. specifies the range of frequency in rad/sec over which response
is desired

If *ModelNam* is a s1-by-.............-by-sp array model having ny outputs, nu inputs and nw is the range of frequencies, then, output frequency response fr is an array of dimension [ny nu nw s1... sp] such that fr (:,:,k,:) is the response of the model at frequency w(k,:).

Note that if instead of a range, a single value of frequency is given by w, then frequency response at that particular frequency is returned.

Example:

```
fr=freqresp(modelB11,0.1:0.1:1)
returns

fr(:,:,1) =
 -0.5921 + 0.0792i
fr(:,:,2) =
 -0.5692 + 0.1538i
fr(:,:,3) =
 -0.5339 + 0.2202i
fr(:,:,4) =
 -0.4897 + 0.2759i
fr(:,:,5) =
 -0.4400 + 0.3200i
fr(:,:,6) =
 -0.3882 + 0.3529i
fr(:,:,7) =
 -0.3369 + 0.3758i
fr(:,:,8) =
 -0.2878 + 0.3902i
fr(:,:,9) =
 -0.2420 + 0.3978i
fr(:,:,10) =
 -0.2000 + 0.4000i

fr=freqresp(modelB11,78.8046)
returns

fr =
 0.1999 + 0.0102i
```

4.3.2.2 Evaluating Frequency Response at a Particular Frequency

If only the real part of the frequency response at a particular frequency is desired, then **evalfr** that is a simplified version of **freqresp** can be used. It is meant for quick evaluation of the frequency response at a single point. It is used to evaluate transfer function of the continuous and discrete-time model at the complex number s or z numerically equal to x.

fresp=**evalfr**(*ModelName*,*x*)

where,

ModelName......................	specifies the name of the models under consideration
x....................................	specifies the range of frequency in rad/sec over which response is desired
fr.....................................	returns the values of frequency response

Example:

```
fr=evalfr(modelB11,78.8046)
returns

fr =
   0.1900

fr=evalfr(modelB11,0.1)
returns

fr =
  -0.5273
```

Test whether you have got the concepts right...

Practice Test 4.4.

1. Compare the results obtained above and those obtained by invoking [m,p,w]=**bode**(modelB11). Explain the difference.

4.3.2.3 Evaluating Gain Margin, Phase Margin, Crossover Frequencies and Judging Stability

Gain margin (GM), Phase margin (PM), Gain crossover frequency (GCOF) and Phase crossover frequency (PCOF) play and important role in judging the stability of a given model. We all know that:

- for stable models,
  GM and PM are both positive
  and
  GCOF < PCOF

- for unstable models,

GM and PM are both negative
and
GCOF > PCOF

- for marginally stable models,
  GCOF = PCOF

Control System Toolbox provides facility for obtaining GM, PM, GCOF and PCOF from the Bode plot for any SISO model both continuous and discrete. The syntax to achieve these is:

- [GM,PM,GCOF,PCOF]=**margin**(*ModelName*)
  or,
- [GM,PM,GCOF,PCOF]=**margin**(*ModelName,w*)

returns numeric values of all four parameters without any plot.

- **margin**(*ModelName*......,*w*)

returns values on open loop Bode plot:

where,
ModelName...................   specifies the name of the SISO model or array of SISO model under consideration
w.............................is the user supplied vector of frequencies in rad/sec at which the response of the model is desired to be evaluated (optional)
Note that:
- *ModelName* should be SISO model or an array of SISO model
- GM=1 (0dB) and PM=0 when the closed loop is unstable.

Example:

Let us try out the functions as follows to obtain the GM, PM, GCOF and PCOF parameters on Bode plot or otherwise for modelA2:

```
margin(modelA2)
returns
```

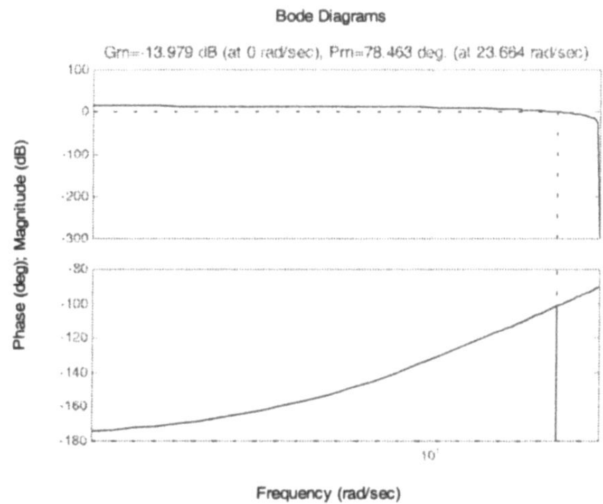


```
[GM,PM,GCOF,PCOF]=margin(modelA2)
returns

GM =
   0.2000
PM =
   78.4630
GCOF =
   0
PCOF =
   23.6640
```

to obtain GM in dB, enter

```
20*log10(GM)
returns

ans =
 -13.9794
```

Since GM is negative and GCOF < PCOF hence modelA2 is unstable.

Ready for a practice test?

Practice Test 4. 5.

1. Obtain the GM, PM, GCOF and PCOF parameters on Bode plot for modelA7 check the results obtained by obtaining the numeric values for the same. Comment on the stability of the models.

4.4 State Space Analysis of Control Systems

State space realisation forms the basis of all modern control system analysis techniques. You are already familiar with the built in functions provided by Control System Toolbox for the formulation, conversion and interconnection of the state space models. Now, you will learn the use of some other built-in functions for the analysis of state space models.

4.4.1 Eigenvalues and Eigenvectors

Eigenvalues and eigenvectors of ss model have very important role in the analysis of ss models. They form the basis of formulation of many standard matrices upon which the judgement regarding the model characteristics is based. Control System Toolbox provides following functions to obtain these:

- e=**eig**(a)

returns a vector containing the eigenvalues of a square system matrix a

- [v, d] = **eig**(*ModelName, 'nobalance'*)

returns a diagonal matrix d of eigenvalues and a full matrix v whose columns are the corresponding eigenvectors so that x*v=v*d. The string *'nobalance'* is optional, which when specified results into computations in which balancing is not done. This sometimes gives more accurate results.

- e=*eig*(a, b)

returns a vector containing the generalized eigenvalues of square matrices a and b.

- [v, d] = **eig**(a, b)

returns a diagonal matrix d of generalised eigenvalues and a full matrix v whose columns are the corresponding eigenvectors so that a*v = b*v*d.

Forthcoming examples will make these concepts more clear.

Example:

Consider 'a' matrix of modelA36, modelB16 and modelB21. Lets find out the eigenvalues and eigenvectors for them.

```
e=eig(modelA36.a)
returns

e =
 -2.1856
 -5.9072 + 2.5067i
 -5.9072 - 2.5067i
```

```
[v,d]=eig(modelA36.a)
returns

v =
  0.8841          0.2756 - 0.6301i      0.2756 + 0.6301i
 -0.4607         -0.2475 + 0.5468i     -0.2475 - 0.5468i
  0.0786         -0.0991 - 0.3961i     -0.0991 + 0.3961i
d =
 -2.1856              0                     0
     0          -5.9072 + 2.5067i           0
     0               0              -5.9072 - 2.5067i


e=eig(modelB16.a,modelB21.a)
returns


e =
  0.6671 - 0.0000i
 -0.1484 - 0.6344i
 -0.2213 + 0.0000i
 -0.1484 + 0.6344i


[v,d]=eig(modelB16.a,modelB21.a)
returns


v =
 -0.1449 - 0.2425i      0.5749 - 0.4812i     -0.7655 + 0.3701i      0.0862 - 0.7447i
 -0.4208 - 0.7039i     -0.4117 + 0.2538i      0.2068 - 0.1000i      0.0142 + 0.4834i
 -0.1278 - 0.2138i     -0.2097 - 0.1122i      0.1052 - 0.0509i      0.2090 + 0.1136i
  0.2210 + 0.3697i     -0.2694 + 0.2737i     -0.4131 + 0.1997i     -0.0807 + 0.3755i
d =
  0.6671 - 0.0000i          0                    0                     0
     0            -0.1484 - 0.6344i             0                     0
     0                 0            -0.2213 + 0.0000i                 0
     0                 0                    0            -0.1484 + 0.6344i
```

Ready for a check? Go ahead...

Practice Test 4.6.

1. Find out eigenvalues and eigenvectors for matrix 'a' of modelB1 and modelB18.

4.4.2 Initial Condition Response of ss Model

Initial condition response plays an important role in the study of the transient response of a model. Control System Toolbox provides the **initial** function the syntax of which is given below to obtain this.

- **initial**(*ModelName1*, 'style1', *ModelName2*, 'style2',............. x0, tfinal or T)

computes the unforced response of ss model with an initial condition on the states and returns initial response plot on the screen in the Figure window.

- [y,t,x]=**initial**(*ModelName1*, *x0*)

returns the output response y, time vector t used for simulation and the state trajectories x without drawing plot on the screen. The array y has **length**(t) number of rows with columns equal to the number of outputs. Similarly, x has **length**(t) rows with columns equal to the number of states.

where,

ModelName1, *ModelName2* are the names of the ss models whose response to the signals have to be obtained

style1, *style2*..................... are the string for specifying the *'linestyle_marker_color'* for the plots of *ModelName1*, *ModelName2*... respectively (optional), see Table 4.2 for more information

x0..................................... specifies the initial condition on the states

tfinal............................... specifies the final time upto which the simulation is desired *i.e.,* from t=0 to t=*tfinal* (optional)

T..................................... specifies the time vector for simulation (optional)

Example:

```
initial(modelA34,[1;2],[0:0.1:2])
returns
```

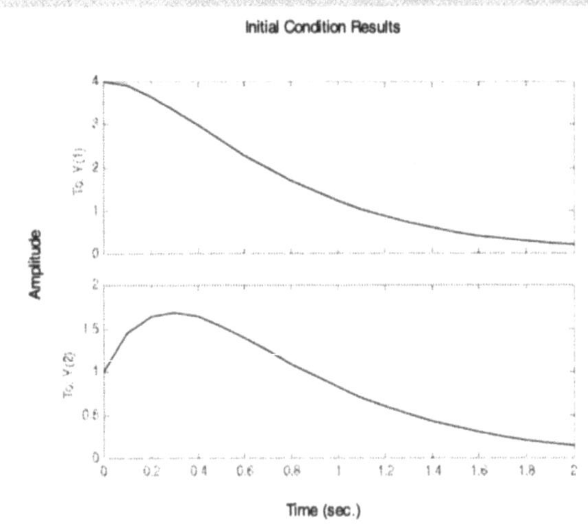


```
initial(modelA36,'r',modelB12,'g',[3;2;0])
returns
```

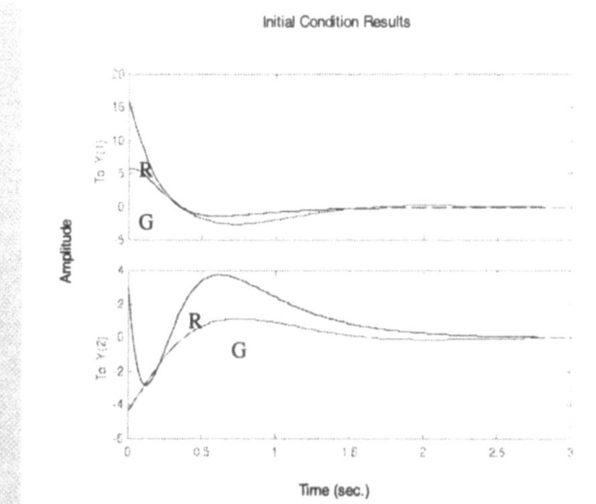

Try obtaining the output response, time vector and state trajectories for modelA34 entering [y,t,x]=**initial**(modelA34,[3,2]) at the MATLAB® command prompt.

Check if you have really understood the things…

Practice Test 4.7.

1. Obtain the initial response of modelA36 for a period ranging from 0.5 to 1.5 seconds.

4.4.3 Canonical State Space Realization

Control System Toolbox provides a built in function to produce state space realisation of the models in canonical form, which could be either of modal type or of companion type. The syntax for achieving this is:

- ModelNew=**canon**(*ModelOld*,'*type*')
  or,
- [ModelNew,T]=**canon**(*ModelOld*,'*type*')

The latter syntax returns the state transformation matrix T that relates the new state vector z to the old state vector x such that z=T*x. This syntax should however be used only when *ModelOld* is an ss model.

where,
ModelOld .... specifies the name of the model under consideration
type .... is a user supplied string specifying the class of the model which could be either '*modal*' or '*companion*'

Note that:
- the modal canonical form returns a model where the system eigenvalues appear on the diagonal.
- the companion canonical form returns a model where the characteristic polynomial appears in the right column.

A few illustrative examples will elucidate this difference.

Example:

```
modelD1=canon(modelA34,'modal')
returns
```

a =

	x1	x2
x1	-2	0
x2	0	-3

b =

	u1	u2
x1	7.2111	7.2111
x2	7.0711	7.0711

c =

	x1	x2
y1	1.1094	-0.70711
y2	0.83205	-0.70711

d =

	u1	u2
y1	0	0
y2	0	0

Continuous-time model.

```
modelD2=canon(modelA34,'companion')
returns
```

a =

	x1	x2
x1	-4.4409e-016	-6
x2	1	-5

b =

	u1	u2
x1	1	1
x2	0	0

c =

	x1	x2
y1	3	-1
y2	1	3

```
d =
                u1          u2
        y1        0           0
        y2        0           0
```
Continuous-time model.

Take a practice test now...

4.4.4 Controllability and Observability

Controllability and observability play an important role in the analysis of ss models. Control
System Toolbox offers functions to compute the controllability and observability of the models
as follows:

for controllability,
- con=**ctrb**(a,b)

returns the composite controllability matrix [b : ab : a²b : ....]

- con=**ctrb**(*ModelName*)

returns the composite controllability matrix of model with realisation (a, b, c, d), which is
equivalent to **ctrb**(*ModelName.a*, *ModelName.b*)

for observability,

- obs=**obsv**(a,c)

returns the composite observability matrix [c^T : a^Tc^T : $(a^T)^2c^T$...]

- obs=**obsv**(*ModelName*)

returns the composite observability matrix of model with realisation (a, b, c, d), which is
equivalent to **obsv**(*ModelName*.a, *ModelName*.c)

where,

ModelName .... is the name of the ss model under consideration

a,b,c .... are the system, input and output matrix of the ss model under
consideration

For, nd arrays of ss model, con and obs are an array with n+2 dimensions where con(:, :,
ji,....,jn) and obs(:, :, ji,......,jn) contains the controllability and observability matrix
respectively of the ss model(:, :, j1,......,jn).

Note that:
- a general n^{th}-order MIMO model is completely controllable if the rank of the composite matrix [b : ab : a²b : ....] is n.
- a general n^{th}-order MIMO model is completely observable if the rank of the composite matrix [c^T : $a^T c^T$: $(a^T)^2 c^T$...] is n.

Forthcoming illutrations will further elucidate the concept.

Example:

```
con=ctrb(modelA36)
returns

con =
    1    1   -4   -4  109  109
    1    1   -1   -1  -83  -83
    1    1  -16  -16  157  157
```

> a=modelA36.a;
> b=modelA36.b;
> con=**ctrb**(a,b)
> gives the same result

```
rank(con)
returns

ans =
   3
```

But, modelA36 is a third-order ss model. Hence, it is completely controllable.

```
obs=obsv(modelA36)
returns

obs =
     2    5    1
     1    0    8
   -12  -24    7
   -16  -37  -79
    34   49 -123
   232  532  601
```

> a=modelA36.a;
> c=modelA36.c;
> obs=**obsv**(a,c)
> gives the same result

```
rank(obs)
returns

ans =
   3
```

But, modelA36 is a third-order ss model. Hence, it is completely observable.

4.4.5 Controllability and Observability Gramians

The Controllability and Observability Gramians of a stable ss model can be found out respectively as follows:

wc=**gram**(*ModelName*, *'c'*)
wo=**gram**(*ModelName*, *'o'*)

where,
ModelName .... is the name of the stable ss model under consideration

Note that:
- controllability and observability gramian are positive definite only if pair (a, b) is controllable and pair (c, a) is observable respectively.
- for nd arrays of LTI model, wc and wo are arrays with n+2 dimensions such that wc(:,:,j1,... jn) and wo(:,:,j1... jn) are controllability and observability gramians of model(:,:,j1....jn).

Example:

```
wc=gram(modelA36,'c')
returns

wc =
    2.5554   -0.5342   -0.0861
   -0.5342    0.5142    0.0837
   -0.0861    0.0837    0.0837

det(wc)
returns

ans =
   0.0721
```

which is positive indicating that modelA36 is completely controllable which we had already found out earlier in the examples of last article.

```
wo=gram(modelA36,'o')
returns

wo =
    0.7783    0.9041    0.3459
    0.9041    2.9887    0.0537
    0.3459    0.0537    3.3779

det(wo)
returns

ans =
   4.7702
```

which is positive indicating that modelA36 is completely observable which we had already found out earlier in the examples of last article.

```
wc=gram(modelA48,'c')
returns
```

```
wc =
   2.0980   -0.0632
  -0.0632    0.0056

wo=gram(modelA48,'o')
returns

wo =
   0.0035   -0.0065
  -0.0065    0.8848
```

Ready to test yourself?

1. Check whether modelA34 is completely controllable and observable. Cross check the result
    by computing gramians.

4.4.6 Balancing of ss Model

Control System Toolbox supports two methods to balance a ss model.

4.4.6.1 Using Diagonal Similarity
This is achieved as follows:

- [ModelNew, T]=**ssbal**(*ModelOld*)

where,
ModelOld .... is the name of the ss model under consideration
T .... returns a diagonal similarity transformation T and a scalar α

such that $\begin{bmatrix} \mathrm{TaT}^{-1} & \mathrm{TB}/\alpha \\ \alpha\,\mathrm{cT}^{-1} & 0 \end{bmatrix}$ has approximately equal row and

column norms

ModelNew.......................... returns a model is of the form $\begin{bmatrix} \mathrm{TaT}^{-1} & \mathrm{TB}/\alpha \\ \alpha\,\mathrm{cT}^{-1} & \mathrm{d} \end{bmatrix}$ and the new state

is xn = xT

- [ModelNew, T]=**ssbal**(*ModelOld,condT*)

which in addition to above, specifies an upper bound *condT* on the condition number T so
that control is achieved over the worst case round off amplification factor that occurs due to
balancing with ill conditioned T which sometimes magnifies rounding errors.

4.4.6.2 Using Gramian-based Balancing of State Space Realisations
This is achieved as follows:

- ModelNew=**balreal**(*ModelOld*)

where,
ModelOld .... is the model under consideration
ModelNew .... returns a balanced realisation model of *ModelOld* with equal and diagonal controllability and observability gramians

- [ModelNew,G,T,Ti]=**balreal**(*ModelOld*)

where,
ModelOld .... is the model under consideration
ModelNew .... returns a balanced realisation model of *ModelOld* with equal and diagonal controllability and observability gramians as above, alongwith a vector G containing the diagonal of the balanced gramian.

Let us consider some examples for better understanding.

Example:

```
[modelD3, T] = ssbal(modelA36)
returns

a =
                x1        x2        x3
        x1       0         3       -3.5
        x2      -2        -5         3
        x3      -4       -10        -9

b =
                u1        u2
        x1       2         2
        x2       2         2
        x3       4         4

c =
                x1        x2        x3
        y1       1        2.5       0.25
        y2      0.5        0         2

d =
                u1        u2
        y1       1         5
        y2       4         6

Continuous-time model.

T =
        2     0     0
        0     2     0
        0     0     4
```

```
modelD4= balreal(modelA36)
returns

a =

                x1          x2          x3
        x1    -4.2075     -0.72808     0.24649
        x2    -3.1797     -1.0783     -3.4616
        x3     7.2935      5.9263     -8.7142


b =

                u1          u2
        x1     2.3109      2.3109
        x2     1.0274      1.0274
        x3    -2.0281     -2.0281


c =

                x1          x2          x3
        y1     2.9323      1.4461      0.12915
        y2     1.4428     -0.14121    -2.8652


d =

                u1          u2
        y1       1           5
        y2       4           6

Continuous-time model.
```

Ready to check yourself ?

> **Practice Test 4.10.**
>
> 1. Balance modelA34 by the methods you just learnt.

4.4.7 State Reduction of ss Model

It is well-known experience of control systems design engineer that the state space model of a system may not necessarily be a representation having minimum number of states. State minimisation becomes mandatory during certain analysis of the models. Control System Toolbox provides functions to achieve state reduction of the models using the **modred**, **minrreal** and **sminreal** functions whose use and syntax are described in the forthcoming articles.

4.4.7.1 The modred Function

This function is used to reduce the order of the state space models. The syntax is as follows:

ModelNew=**modred**(M*odelOld, elim, 'method*')

where,

ModelNew........................ returns the new ss model with reduced order

ModelOld .... specifies the name of the ss model under consideration

elim................................. spcifies a vector which specifies the states which are to be eliminated

method............................. is a string specifying either '*mdc*' or '*del*'

Note that:

- When the string '*mdc*' is used, the order of the ss model is reduced by eliminating the states specified in the vector *elim*. Thus, ModelNew now has *elim* states less in comparison to *ModelOld*. The derivative of the states to be eliminated is set to zero and the remaining states are solved. ModelNew now has matching steady state response.

- When the string '*del*' is used, the order of the ss model is reduced by simply deleting the states specified in the vector *elim*. ModelNew, which now has *elim* states less in comparison to *ModelOld*, may not necessarily but may produce better approximations in the frequency domain.

Illustrations discussed below will make the concepts more clear.

Example:

```
modelD5=modred(modelA36,3,'mdc')
returns

a =
              x1        x2
      x1    1.5556    6.8889
      x2   -3.3333   -8.3333

b =
              u1        u2
      x1    0.22222   0.22222
      x2    1.6667    1.6667
c =
              x1        x2
      y1    1.7778    4.4444
      y2   -0.77778  -4.4444

d =
              u1        u2
      y1    1.1111    5.1111
      y2    4.8889    6.8889

Continuous-time model.
```

Now, we shall compare the steady state response of modelA36 and modelD5 and observe that their DC gains really match

```
step(modelA36,'r-',modelD5,'g-.')
returns
```

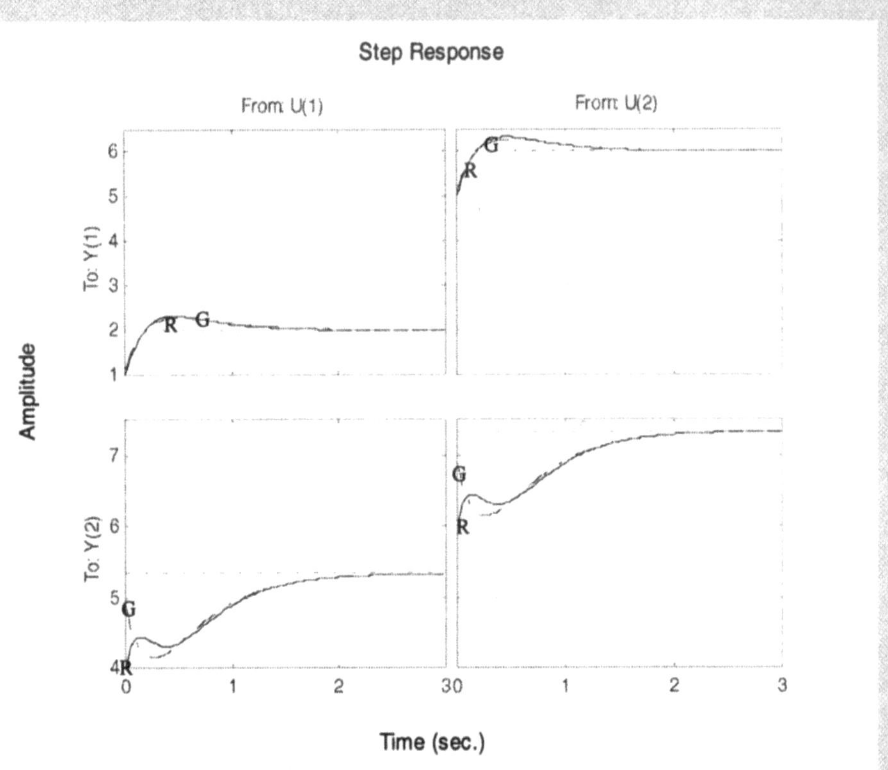


```
modelD6=modred(modelA36,3,'del')
returns

a =
              x1        x2
      x1       0         3
      x2      -2        -5

b =
              u1        u2
      x1       1         1
      x2       1         1

c =
              x1        x2
      y1       2         5
      y2       1         0

d =
              u1        u2
      y1       1         5
      y2       4         6

Continuous-time model.
```

Now, we shall compare the frequency response (Bode plot) of modelA36 and modelD6 and observe that they really match.

```
bode(modelA36,'r-',modelD6,'g-.')
returns
```

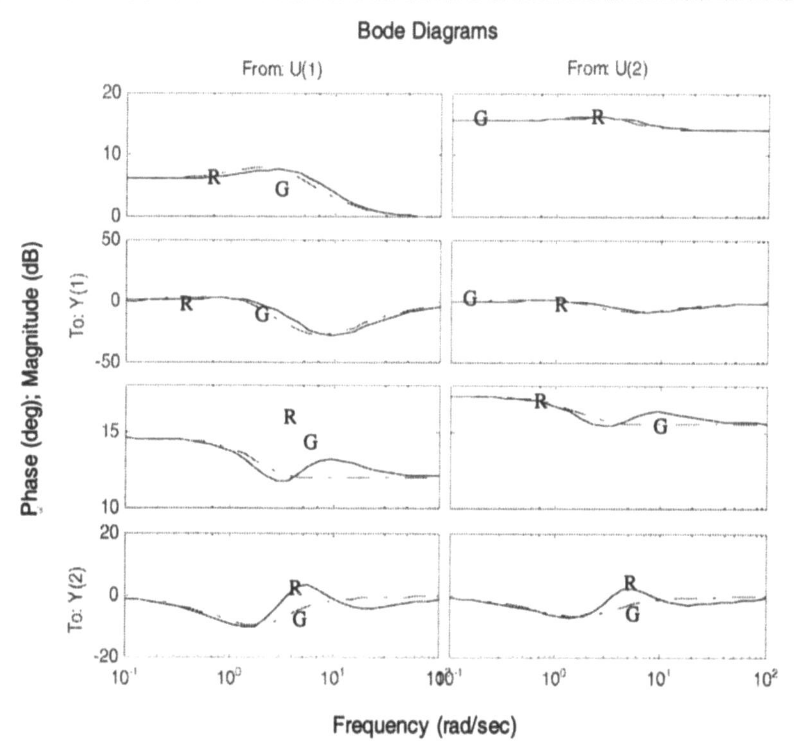

4.4.7.2 The minreal Function

The **minreal** function eliminates the uncontrollable or unobservable states in ss models and cancels pole-zero pairs in tf or zpk models. Thus, the output ModelNew has minimal order and the same response characteristics as the original ModelOld. The syntax for this is:

- ModelNew=**minreal**(*ModelOld*)

where,
ModelNew........................ returns the new ss model with reduced order
ModelOld .... is the name of the ss model under consideration
tolerance .... is the tolerance used for state elimination or pole-zero cancellation (greater tolerance forces more cancellation) (optional)

- (ModelNew, u)=**minreal**(*ModelOld*)

returns an orthogonal matrix u such that (uau^T, ub, cu^T) is Kalman decomposition of parameters (a, b, c). We will read about this in more detail while doing Kalman filters in Chapter 7.

Example:

```
modelD7=minreal(modelB33)
returns

Zero/pole/gain:
    0.2 (s-3)
-------------------
(s+1.885) (s-1.485)
```

> modelD7=
> compare with modelB33
> Observe the pole-zero
> cancellation at s=3

4.4.7.3 The sminreal Function

This function eliminates the states of the ss *ModelOld* that does not affect the input and output response *i.e.,* the unobservable states. The syntax for this is:

ModelNew=**sminreal**(*ModelOld*)

where,
ModelNew........................ returns the new ss model with reduced order
ModelOld .... is the name of the ss model under consideration

Note that the ModelNew may not necessarily be minimal and may have a higher order in comparison to the one resulting from use of **minreal** function. However, **sminreal** retains the state structure of model while **minreal** may not.

Example:

Let us consider modelA34 and modelA36 and concatenate them. Observe the states, input and output in the resulting modelD8.

```
modelD8=cat(2,modelA34,modelA36)
returns
```

a =

	x1	x2	x3	x4	x5
x1	0	3	0	0	0
x2	-2	-5	0	0	0
x3	0	0	0	3	-7
x4	0	0	-2	-5	6
x5	0	0	-2	-5	-9

b =

	u1	u2	u3	u4
x1	1	1	0	0
x2	1	1	0	0
x3	0	0	1	1
x4	0	0	1	1
x5	0	0	1	1

```
c =
                x1          x2          x3          x4          x5
        y1       2           1           2           5           1
        y2       1           0           1           0           8

d =
                u1          u2          u3          u4
        y1       0           0           1           5
        y2       0           0           4           6
```

Continuous-time model.

Now let us derive a subsystem form modelD8 as follows and observe the states in the resulting modelD9.

```
modelD9=modelD8(1,1)
returns

a =
                x1          x2          x3          x4          x5
        x1       0           3           0           0           0
        x2      -2          -5           0           0           0
        x3       0           0           0           3          -7
        x4       0           0          -2          -5           6
        x5       0           0          -2          -5          -9

b =
                u1
        x1       1
        x2       1
        x3       0
        x4       0
        x5       0

c =
                x1          x2          x3          x4          x5
        y1       2           1           2           5           1

d =
                u1
        y1       0
```

Continuous-time model.

There is certainly a difference between the states, input and output of modelD9 and modelD8. ModelD9 has retained them all. To eliminate the unobservable states from modelD9, try out the **sminreal** function on modelD9.

```
modelD10=sminreal(modelD9)
returns

a =
                x1          x2
        x1       0           3
        x2      -2          -5
```

```
b =

                 u1
        x1        1
        x2        1

c =

                 x1         x2
        y1        2          1

d =

                 u1
        y1        0

Continuous-time model.
```

Ready for a practice test?

Practice Test 4.11.

1. Concatenate model modelA34 and modelB37 and use all of the three-minimisation functions on the resulting model to obtain minimal ss model.

Exercise for Chapter 4:

For all the models generated in exercise for Chapter 1 and 2 (transform the models to relevant type where ever necessary) .

1. Obtain system poles, system zeros, DC gain and pole-zero map.
2. Obtain time response of the models to unit step input, unit impulse input.

3. Obtain H_2 and L_∞ norms using all the syntax summarised in Table 4.1 applicable to the model.

4. Obtain covariance of response to white noise assuming suitable value for the intensity of white Gaussian noise.

5. Obtain natural frequency and damping of LTI model poles.

6. Obtain sort the eigenvalues of the models in descending order.

7. Obtain time response to following signals:

    - sine signal of period 0 to 2π for time 0 to 10 seconds of unit amplitude.
    - square signal of period 0 to 2π for time 0 to 10 seconds of unit amplitude.
    - pulse signal of period 0 to 2π for time 0 to 10 seconds of unit amplitude.

8. Obtain frequency response plots using all syntax summarised in Table 4.3.

9. Obtain frequency response for a frequency range given by the vector [10:10:100] rad/sec.

10. Obtain frequency response at a frequency of 50 rad/sec.

11. Judge the stability of the models by finding out GM, PM, GCOF, PCOF.

12. Check the controllability and observability of the models by use of observability and controllability gramians or otherwise.

13. Find out the initial condition response of the models assuming the value of the initial states as 0.1, 0.2, 0.3... and so on for a time period 0 to 5 seconds.

14. Obtain canonical state space realisations of 'modal' as well as 'companion' type.

15. Balance the models using diagonal similarity and gramian based techniques.

16. Reduce the states of the models assuming suitable values of the relevant parameters using **modred**, **minreal** and **sminreal** functions.

Chapter 5

The Control System Toolbox's GUIs

In this chapter, you will learn to use the two Graphic User Interfaces provided with the Control System Toolbox -- **The LTI Viewer** and **The Root Locus Design GUI**. You will learn to obtain various response plots for LTI models and get information from them using the LTI Viewer. You will also learn to design a series (cascade) or feedback compensator using the Root Locus Design GUI for a SISO model, such that the system performs in the controlled way.

In earlier chapters, you learnt to obtain the time and frequency response of a model using the built-in functions provided with the Control System Toolbox of MATLAB®. In addition to these built-in functions, Control System Toolbox also provides two interactive Graphical User Interfaces (GUIs) called

- the LTI Viewer - for performing linear analysis like viewing, manipulating and analysing the response of LTI systems comprising of several models.

- the Root Locus Design GUI - for obtaining and analysing the root locus plot of SISO LTI models of zpk, tf or ss type and design a series (cascade) or a feedback compensator and then proceed with the analysis of the resulting model by obtaining the system responses.

In this chapter, you will learn to use these GUIs to obtain the desired response of the systems.

5.1 The LTI Viewer

As already mentioned, this interactive GUI is used for viewing, manipulating and analysing the response of LTI systems comprising of several models. It can be used for obtaining upto six response plots per LTI viewer of models each of which have the same number of inputs and outputs. The response plots that can be obtained using LTI Viewer can be any of the types as indicated in Table 5.1

It is also possible to invoke more than one LTI Viewer at the same time for say analysing the response of system with different number of inputs and outputs.

Table 5.1. Response types available with LTI Viewer

Model Type	Response Type
• ss models	• Initial plot
• zpk models • tf models • ss models	• PZmap • Impulse response plot • Step response plot • Lsim
• all models	• Bode plot • Nyquist plot • Nichols plot • Singular value plot

5.1.1 Initialising the LTI Viewer

LTI Viewer can be initialised or invoked by any of the two methods explained below:

i. Obtaining an empty LTI Viewer window:
    By simply entering **ltiview** at the command prompt of the MATLAB® command window. In this case, an empty LTI Viewer window appears on the screen. Models whose responses are desired to be studied are then loaded into the LTI Viewer workspace and the desired response types are specified thereafter.

ii. Obtaining an LTI Viewer window with plots on it:
    Invoking, LTI Viewer, loading the models and specifying the plot types that are desired to be displayed can be done simultaneously by using the same **ltiview** function with the following syntax:

• **ltiview**({'*plottype1*',............, '*plottype6*'}, *ModelName1*, ..., *ModelName6*)

• **ltiview**('*plottype*', *ModelName1*, *options*)

where,

ModelName1...ModelName6 specifies the names of the models under consideration having same number of inputs and outputs in the workspace whose responses are desired to be displayed in the LTI Viewer when it appears on the screen

'*plottype1*',... '*plottype6*' ...... specifies the name of the types of responses desired to be displayed for the corresponding models. It can be **bode, impulse, initial, lsim, nichols, nyquist, pzmap, sigma** and **step** which give you Bode plot, impulse response, initial state response (for ss models only), LTI model response to general input, Nichols chart, Nyquist plot, plots of poles and zeros, singular values of the frequency response and step response respectively.

options............................. specifies additional input arguments supported by different response types to get a plot of desired appearance (optional depending on the type of the model -- if *plottype* is '**lsim**' or '**initial**', the *input signal* or the initial conditions *x0* must be specified while other arguments may or may not be specified. It operates only when one *plottype* for the models is specified, the number of models, however, can be upto six).

The initially displayed types of plots in the LTI Viewer can be changed. Other models, which were not initially specified, can be loaded into the LTI Viewer workspace later on and their responses can be displayed in a manner similar to way as indicated in Method 1 explained above.

5.1.2 More about the ltiview Function

In addition to the above, the **ltiview** function can also have some more syntax other than those mentioned above to produce some different results as indicated below:

- **ltiview**('clear', Hviewers)

clears the plots and data from the LTI viewer with the handles provided in Hviewers.

- **ltiview**('currrent', *ModelName1, plotstr1, ……., ModelNamen, plotstrn, Hviewers*)

appends the responses of the systems in model# to the LTI viewers with the handles provided in Hviewers. Note that for this to happen, the systems must have the same dimensions as those in the LTI Viewer identified by the Hviewers. If the dimensions are different, the LTI Viewer identified by Hviewers is first cleared and only the new responses are shown.

5.1.3 The LTI Viewer Environment

Once you have invoked the LTI Viewer by any of the two methods explained above, what appears on your computer screen is the LTI Viewer window (Figure 5.1) which may or may not be displaying responses depending on the method you used to initialise it. However, one thing that remains common and appears in all LTI Viewer windows is

- the titlebar with **LTI Viewer** written on it on the left hand side and the familiar three function buttons on the right hand side for minimising, maximising and closing the LTI Viewer window as shown below:

- the top menu bar, which is displaying three words File, Tools and Help. The details of the menu of the LTI Viewer are given in Table 5.2

- the plot region that is either empty or displays the plot depending on the method that was used to initialise the LTI Viewer.

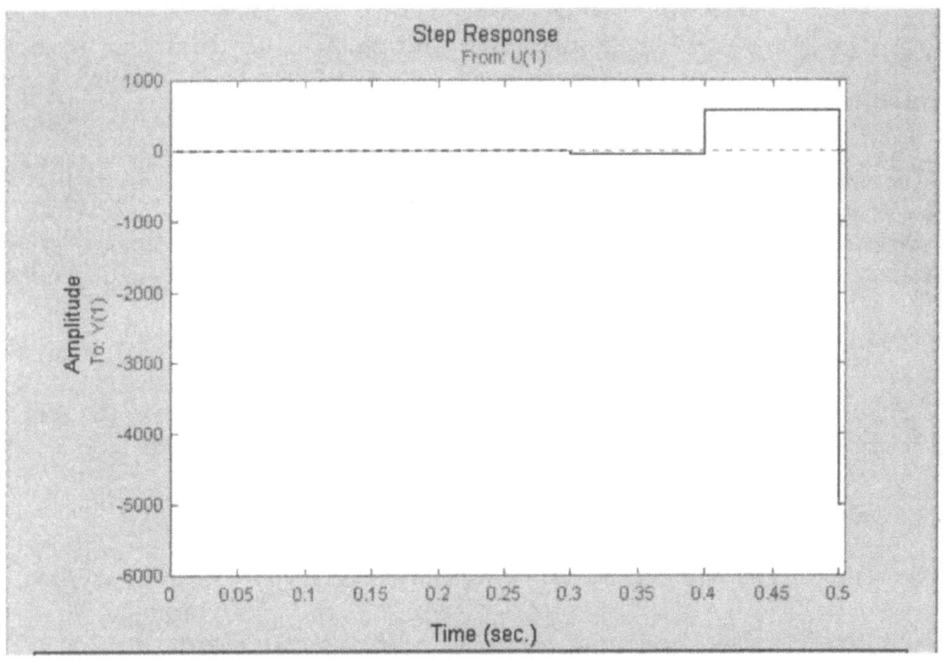

- the bottom statusbar displaying context sensitive information or help tip, which, as you proceed through your plotting and analysis may be an instruction, a hint or an error message regarding the current step or the next step you may attempt. The usual messages appearing as the help tip could be a message indicating that an unsupported function has been tried to be performed, a function has been completed or an additional information on the use of the LTI Viewer control.

In the pages to follow, you will learn to manipulate the desired response plots for various models in the LTI Viewer window and get information from these plots, using the various main-menus, their sub-menus and the right click menus.

Table 5.2. The LTI Viewer main menus

Main Menu	Sub-Menu	Purpose
File (used to manipulate data into and out of the LTI Viewer)	New Viewer (Ctrl+N)	opens a new LTI Viewer window (Figure 5.1) captioned **LTI Viewer** in the title bar.
	Import	opens the MATLAB® workspace window (Figure 5.2) captioned **LTI Browser** in the tilte bar and 'Select the systems to import' written at the top, displaying the names of all the models loaded into the MATLAB® workspace to allow you to import new models into the LTI Viewer workspace.
	Export	opens the Export window (Figure 5.3) captioned **Export LTI Models/Compensators** in the title bar displaying export list of the models, which can be exported from the LTI Viewer's workspace to the MATLAB®'s workspace or a storage device.
	Delete System	opens the **LTI Browser** window again but this time with 'Select the systems to delete' written at the top to remove one or more models from the LTI Viewer's workspace (Figure 5.4).
	Refresh System	clicking this updates the response you are analysing in the LTI Viewer incorporating any changes that you might have made.
	Print (Ctrl+P)	opens the old and much familiar **Print** window to allow printing the response being displayed in the LTI Viewer.
	Print to Figure	sends the plot on the LTI Viewer to the **Figure** window of MATLAB® such that the Figure window features are available enabling you to improve the appearance of response plot.
	Close Viewer (Ctrl+W)	closes the already open LTI Viewer (clicking the cross on the top right side of the LTI Viewer window does the same)
Tools (used to control the output/display of the LTI Viewer)	Viewer Configuration	opens the Viewer Configuration window (Figure 5.5) captioned **Available LTI Viewer Configurations** in the title bar to allow you to choose the response arrangement and response type (which need not be same for all the models) for the models to be displayed in the LTI Viewer window
	Response Preferences (Ctrl+R)	opens the Response Preferences window (Figure 5.6) captioned **Response Preferences** in the title bar which provides you with options to automatically generate (default) or manually define the parameters for • the time vector, Y-axes range and plot option for time domain analysis of the models • frequency vector and units of magnitude, phase and frequency for frequency domain analysis of the models to be displayed in the LTI Viewer window.
	Linestyle Preferences (Ctrl+L)	opens the Linestyle Preferences window (Figure 5.7) captioned **Linestyle Preferences** in the title bar to allow you to set color order, marker order and linestyle order for systems, inputs, outputs and channels for the response of the models to be displayed in the LTI Viewer window.

Table 5.2. (continued)

Main Menu	Sub-menu	Purpose
Help (used to get help on various topics)	Overview	opens the LTI Viewer Help window captioned **LTI Viewer Help** with 'Overview' written in the small text box at the top left corner of the window and the body of the window providing text explaining the LTI Viewer (Figure 5.8a).
	Response Preferences	opens the LTI Viewer Help window captioned **LTI Viewer Help** with 'Response Preferences' written in the small text box at the top left corner of the window and the body of the window providing text explaining the features of the Response Preferences window, which is a sub-menu of the tools menu (Figure 5.8b).
	Linestyle Preferences	opens the LTI Viewer Help window captioned **LTI Viewer Help** with 'Linestyle Preferences' written in the small text box at the top left corner of the window and the body of the window providing text explaining the features of the Linestyle Preferences window, which is a sub-menu of the tools menu (Figure 5.8c).

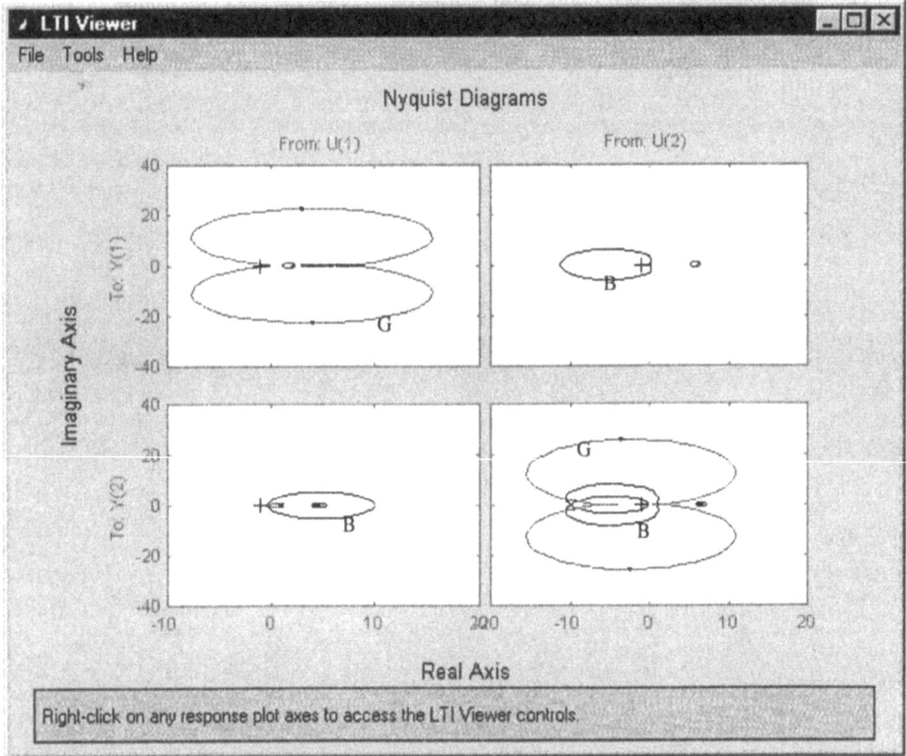

Figure 5.1. The LTI Viewer Window (obtained by entering **ltiview**('nyquist', modelA14, modelA31, modelA36) at the MATLAB® command prompt)

Figure 5.2. The Import Window

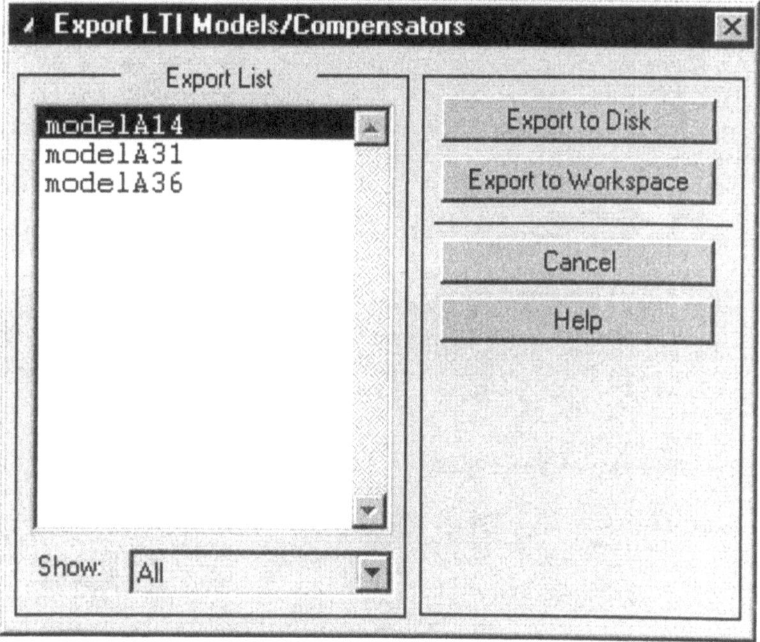

Figure 5.3. The Export Window

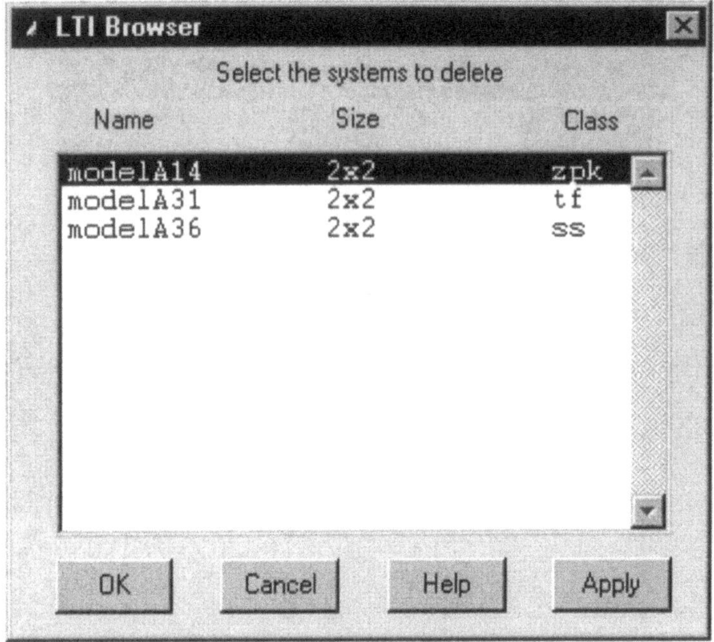

Figure 5.4. The Delete System Window

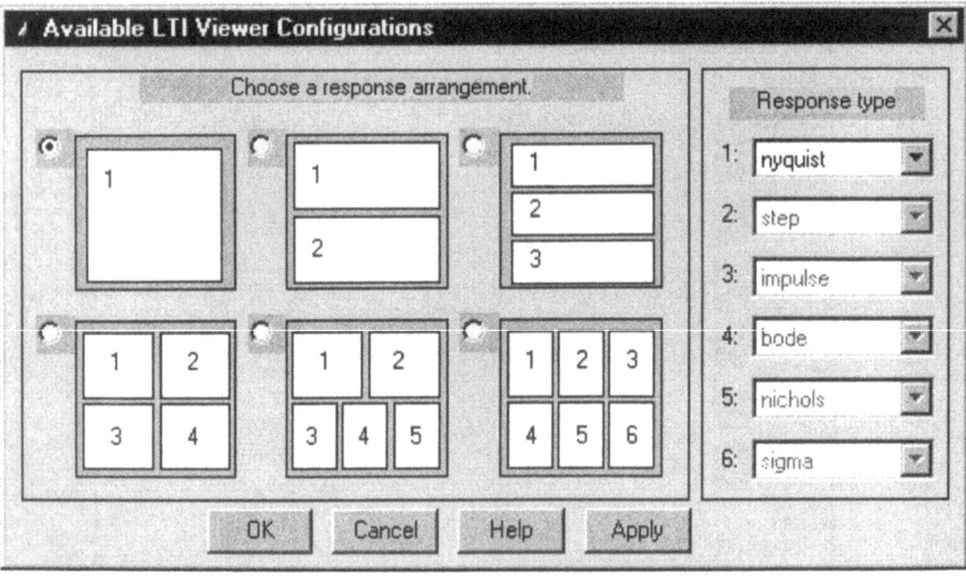

Figure 5.5. The Viewer Configuration Window

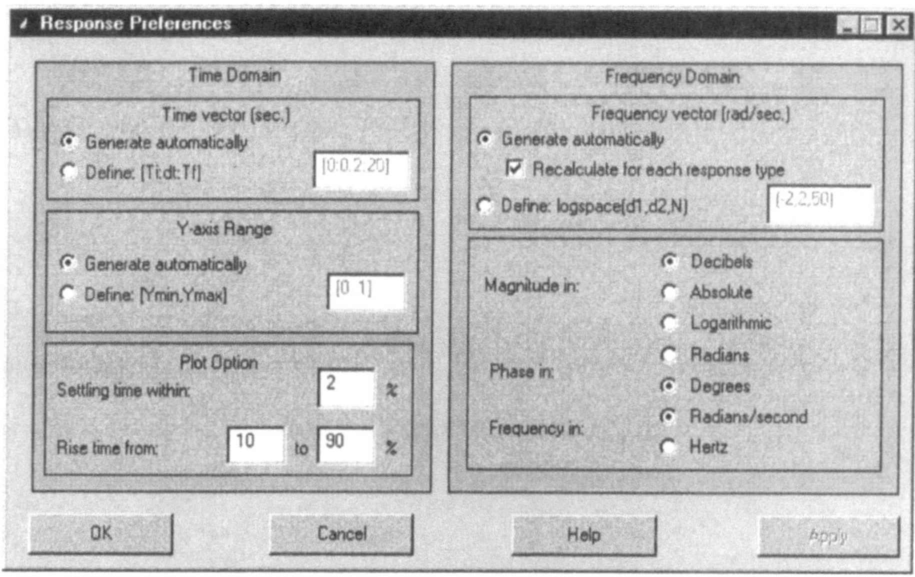

Figure 5.6. The Response Preferences Window

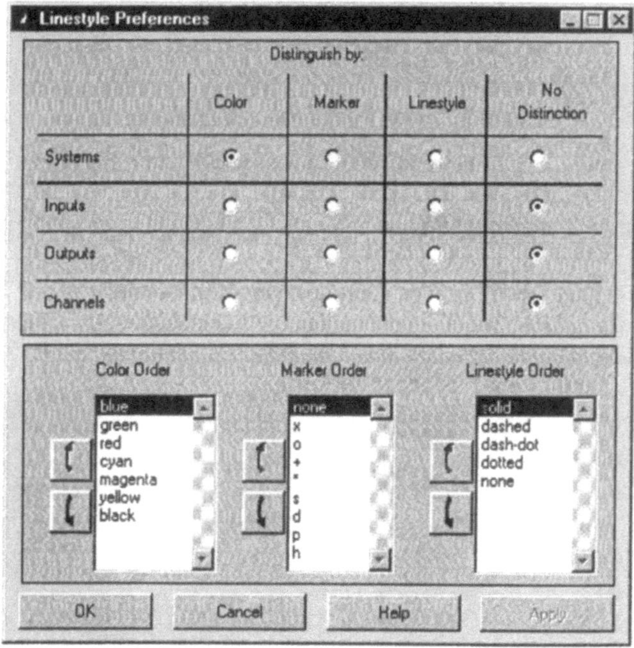

Figure 5.7. The Linestyle Preferences Window

a. LTI Overview Help Window

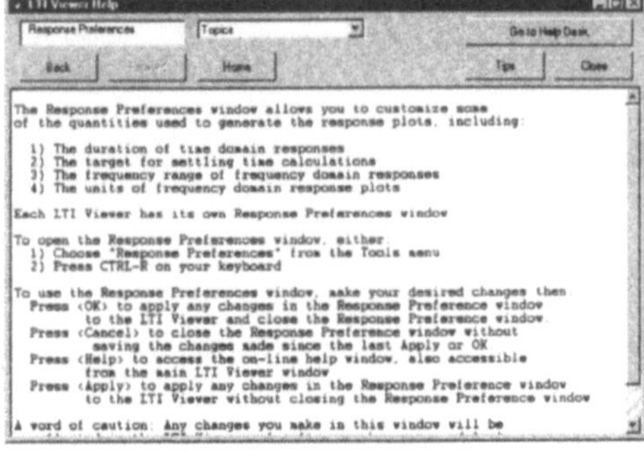

b. LTI Response Preferences Help Window

c. LTI Linestyle Preferences Help Window

Figure 5.8. LTI Viewer Help Window

5.1.4 The Bottom Command Bar

All the windows mentioned in Table 5.2 have a common bottom bar that has four command buttons with OK, Cancel, Help and Apply written on them (Figure5.9). Clicking on them has the following effect, with which you as a window user are quite familiar:

- OK button -- applies the changes made in the concerned window to the LTI Viewer and closes the window.
- Cancel -- closes the concerned window without applying any changes. The value of the parameters remain the same as earlier. Any change made is ignored when this command button is clicked.
- Help -- accesses the on-line help window
- Apply -- does the same as OK command button without closing the concerned window.

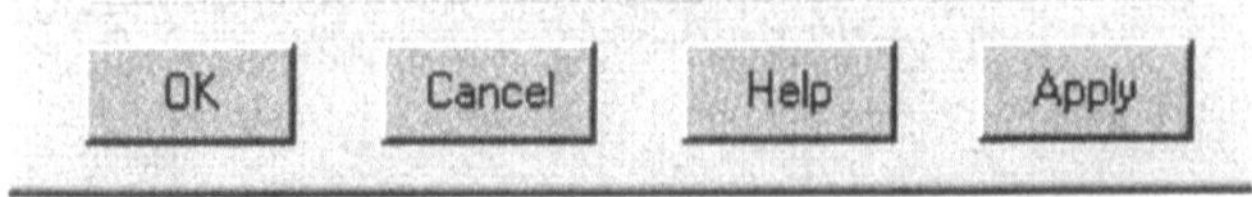

Figure 5.9. Bottom command bar of all the windows

5.1.5 The Right Click Menus

When you click the right mouse button on any portion of the plot, the right click menu is invoked displaying menus depending upon the type of the plot on which the mouse button was clicked. This is shown in Figure 5.10 for Bode plot. Therefore, the right click menu is rightly said to be plot specific or plot sensitive. The menus displayed by the right click menu are summarised in Table 5.3 for different types of the plots.

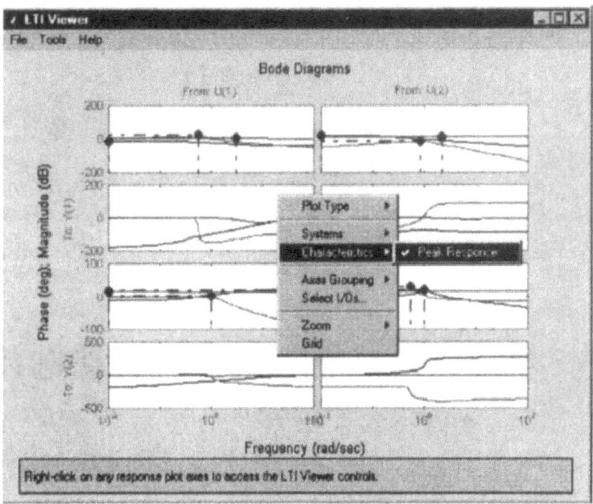

a. Obtained by entering **ltiview**('bode', modelA14, modelA31, modelA36) at the MATLAB® command prompt

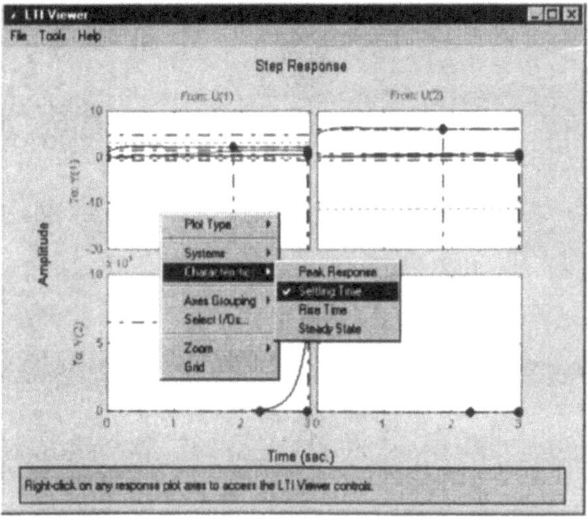

b. Obtained by entering **ltiview**('step', modelA14, modelA31, modelA36) at the MATLAB® command prompt or by choosing **Step** sub-menu from the **Plot Type** main menu of the right click menu when Figure(a) was being displayed

Figure 5.10. Right click menu of LTI Viewer

Table 5.3. Right click menus

Type of Plot ⇒	Step	Impulse	Bode	Nyquist	Sigma	PZmap	Lsim	Initial	
Right Click Menu Items ⇓	Sub-menu Items								**Remarks**
Plot Type (for all plot types)	Step	Step	Step	Step	Step	Step	Step	Step	allows you to choose a plot type or change from one plot type to another which is desired to be displayed on the LTI Viewer. The plot being displayed has a tick mark (✔) against it.
	Impulse	Impulse	Impulse	Impulse	Impulse	Impulse	Impulse	Impulse	
	Bode	Bode	Bode	Bode	Bode	Bode	Bode	Bode	
	Nyquist	Nyquist	Nyquist	Nyquist	Nyquist	Nyquist	Nyquist	Nyquist	
	Nichols	Nichols	Nichols	Nichols	Nichols	Nichols	Nichols	Nichols	
	Sigma	Sigma	Sigma	Sigma	Sigma	Sigma	Sigma	Sigma	
	PZmap	PZmap	PZmap	PZmap	PZmap	PZmap	PZmap	PZmap	
	Lsim	Lsim	Lsim	Lsim	Lsim	Lsim	Lsim	Lsim	
	Initial	Initial	Initial	Initial	Initial	Initial	Initial	Initial	
Systems (for all plot types)	model1 (linestyle1)	model1 (linestyle1)	model1 (linestyle1)	model1 (linestyle1)	model1 (linestyle1)	model1 (linestyle1)	model1 (linestyle1)	model1 (linestyle1)	displays the names of the models that are currently loaded in the LTI Viewer's workspace along with the linestyle (and array dimension for array models) of their responses. You can select of deselect the model by clicking at it which places or removes tick mark (✔) against the name or the model respectively.
	model2 (linestyle2)	model2 (linestyle2)	model2 (linestyle2)	model2 (linestyle2)	model2 (linestyle2)	model2 (linestyle2)	model2 (linestyle2)	model2 (linestyle2)	
	model3 (linestyle3)	model3 (linestyle3)	model3 (linestyle3)	model3 (linestyle3)	model3 (linestyle3)	model3 (linestyle3)	model3 (linestyle3)	model3 (linestyle3)	
	model4 (linestyle4)	model4 (linestyle4)	model4 (linestyle4)	model4 (linestyle4)	model4 (linestyle4)	model4 (linestyle4)	model4 (linestyle4)	model4 (linestyle4)	
	model5 (linestyle5)	model5 (linestyle5)	model5 (linestyle5)	model5 (linestyle5)	model5 (linestyle5)	model5 (linestyle5)	model5 (linestyle5)	model5 (linestyle5)	
	model6 (linestyle6)	model6 (linestyle6)	model6 (linestyle6)	model6 (linestyle6)	model6 (linestyle6)	model6 (linestyle6)	model6 (linestyle6)	model6 (linestyle6)	

Table 5.3. (continued)

Type of Plot ⇒	Step	Impulse	Bode	Nyquist	Sigma	PZmap	Lsim	Initial	
Right Click Menu Items ⇓	Sub-menu Items								**Remarks**
Characteristics	Peak Response	Peak Response	Peak Response	-	Peak Response	-	-	Peak Response	Marks peak response on the plot
	Settling Time	Settling Time	-	-	-	-	-	-	Marks settling time on the plot
	Rise Time	-	-	-	-	-	-	-	Marks rise time on the plot
	Steady State	-	-	-	-	-	-	-	Marks steady state on the plot
Axes Grouping (only for MIMO models)	Opens the I/O Selector window captioned **I/O Selector: Plottype** (where 'plottype' could be the name of any of the nine responses of the models) in the title bar (Figure 5.11). The InputNames are displayed in the column and the OutputNames in the rows. With the help of this window, you can hide or display the response of all the inputs amd outputs or the response of a particular input to a particular output.								
Select I/Os (only for MIMO models)	None	None	None	None	None	-	-	None	Allows grouping of input and output channels (Figure 5.12).
	All	All	All	All	All	-	-	All	
	Inputs	Inputs	Inputs	Inputs	Inputs	-	-	Inputs	
	Outputs	Outputs	Outputs	Outputs	Outputs	-	-	Outputs	
Select from LTI Array... (only for LTI array models)	Opens the '**Model Selector for LTI Arrays**' window, captioned the same. for alltypes of response plots (Figure 5.13). You can chose the arrays depending on the selection criteria chosen for the models plots of which are desired through this window.								
Zoom (for all types of plots) Sub-menu Items	In-Y	In-Y	In-Y	In-Y	In-Y	In-Y	In-Y	In-Y	Zoom in the x direction
	ln-X	ln-X	ln-X	ln-X	ln-X	ln-X	ln-X	ln-X	Zoom in the y direction
	X-Y	X-Y	X-Y	X-Y	X-Y	X-Y	X-Y	X-Y	Zoom in both x and y directions
Grid	Toggles the grid lines. When grid is on, there is a tick (✔) mark in front of the word **Grid** on the Right Click menu.								

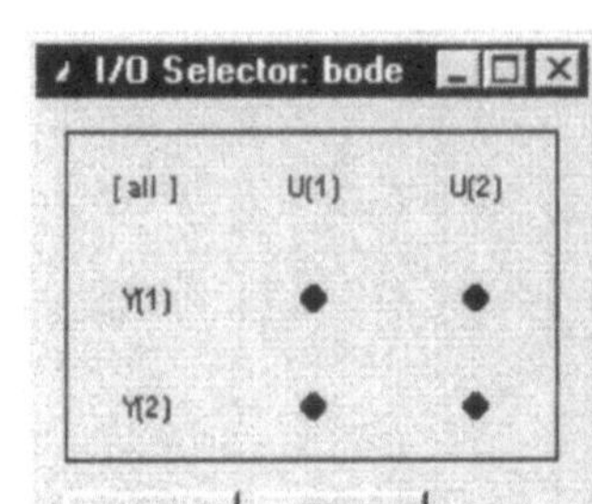

Figure 5.11. I/O Selector Window (obtained by entering **ltiview**('bode', modelA14, modelA31, modelA36) at the MATLAB® command prompt)

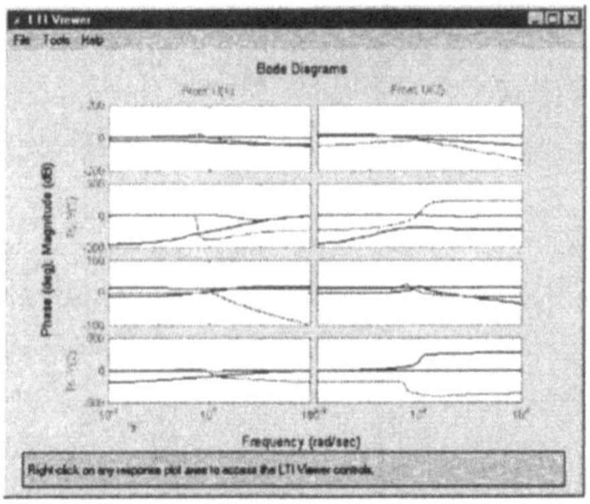

a. None

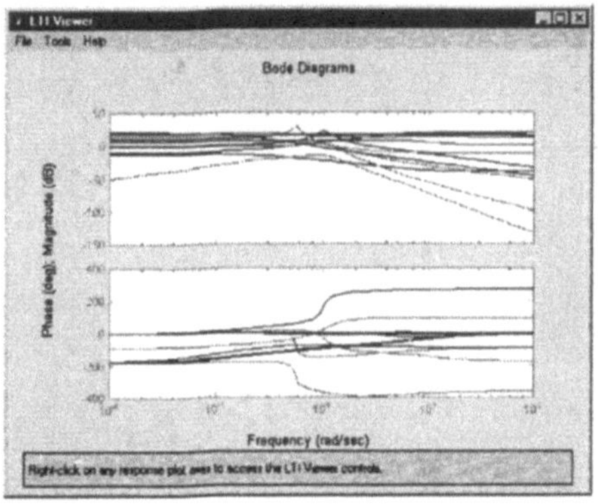

b. All

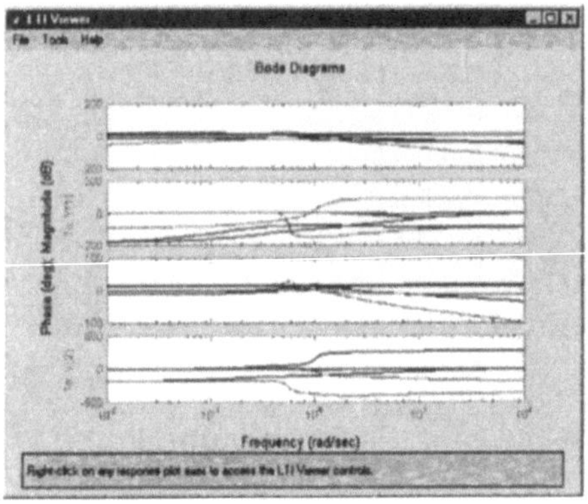

c. Inputs

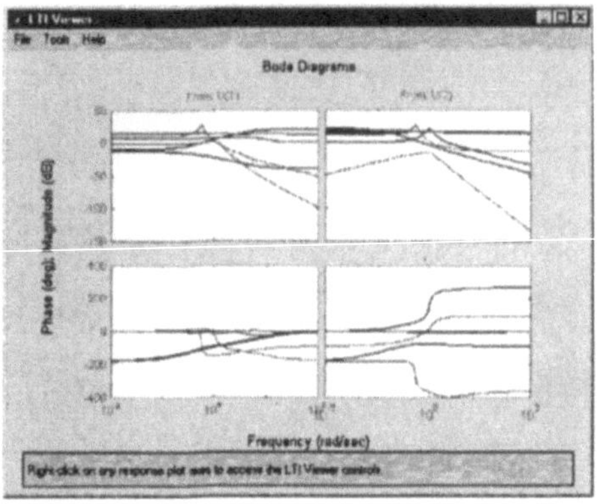

d. Outputs

Figure 5.12. Result of selecting I/O's (obtained by entering **ltiview**('bode', modelA14, modelA31, modelA36) at the MATLAB® command prompt)

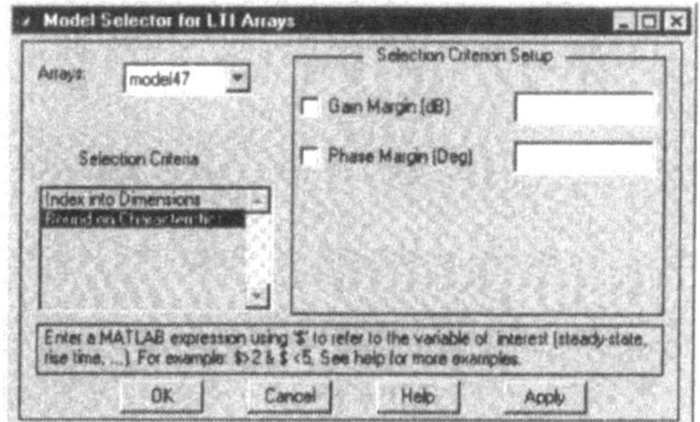

a. Selection Criteria - Index into Dimensions   b. Selection Criteria - Bound on Characteristics

Figure 5.13. Model Selector for LTI Arrays Window

5.1.6 More about Clicking on the Plots

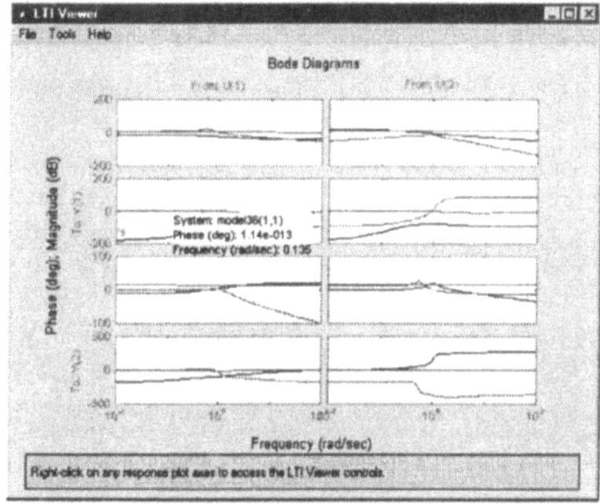

a. Left click of the mouse

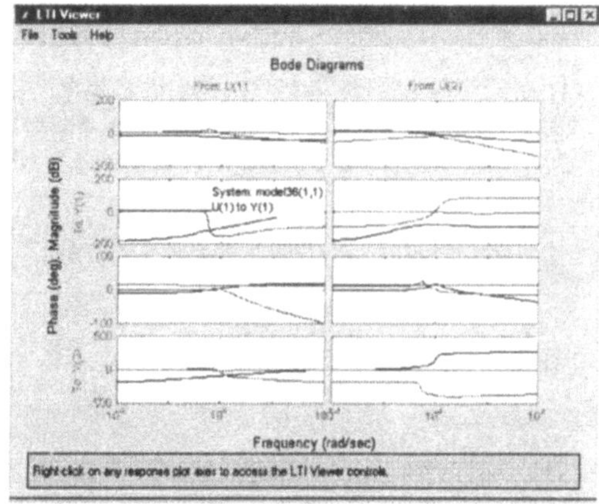

b. Right click of the mouse

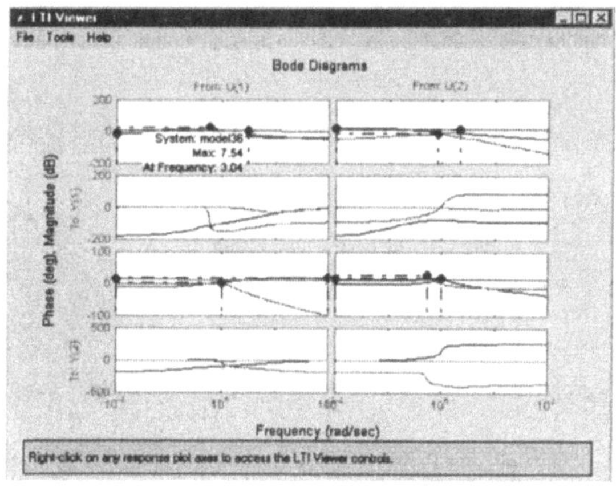

c. Right or left click of the mouse on the plot

Figure 5.14. Clicking of mouse on plots (obtained by entering **ltiview**('bode', modelA14, modelA31, modelA36) at the MATLAB® command prompt)

Clicking the right and left mouse buttons and holding it over the plots provides you with several other information regarding the plots depending upon the type of the plot. This is explained in detail as below:

- Clicking and holding the left mouse button at any point of the plot provides you with the response values of that plot at that point (Figure 5.14a).

- Clicking and holding the right mouse button at any point of the plot provides you with the I/O and model information (Figure 5.14b).

- Clicking and holding either the left or the right mouse button at the marker point of the plot provides you with the values at that point of the plot (Figure 5.14c).

5.1.7 A Few Words on the LTI Arrays Response Plots

Just like any other model, the response plot of an LTI Array model can also be obtained in the LTI Viewer. Figure 5.15 shows how the Nyquist plot of modelA47, which is a 2-by-1 array having 5 inputs and 4 outputs, is obtained by entering **ltiview**('nyquist', modelA47) at the MATLAB® command prompt. Notice that the LTI Viewer window is displaying responses for all dimensions. Figure 5.16a displays the response of the array for 1-by-1 dimension and Figure 5.16b displays the response of the array for 1-by-2 dimension.

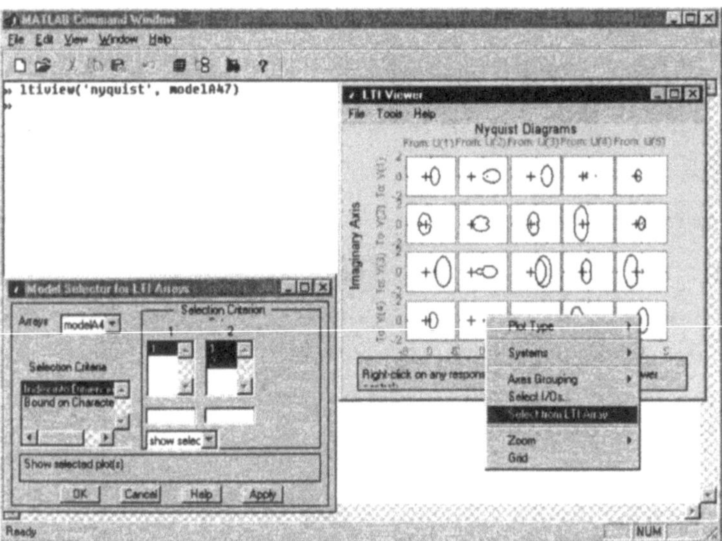

Figure 5.15. Response plot of LTI Arrays (obtained by entering **ltiview**('nyquist', modelA47) at the MATLAB® command prompt)

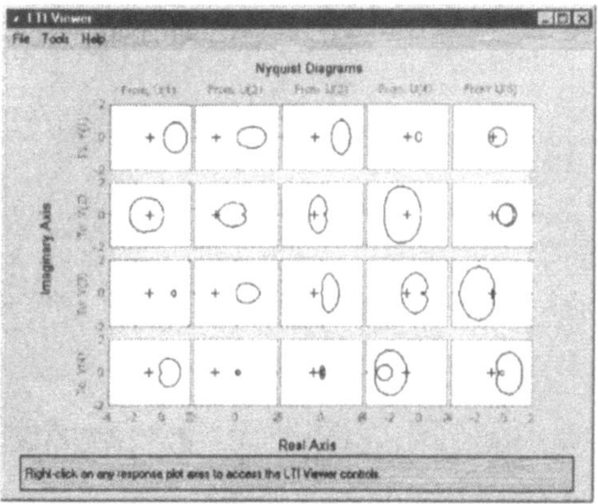

a. Response of the Array for 1-by-1 dimension

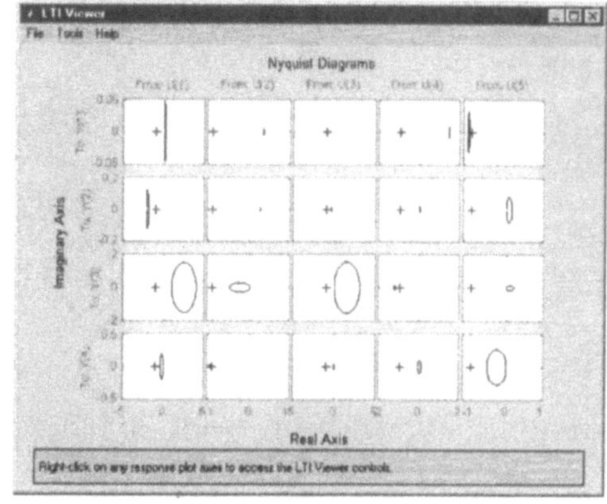

b. Response of the Array for 1-by-2 dimension

Figure 5.16. Dimensionwise Response of the Array (modelA47)

5.1.7.1 The Model Selector for LTI Arrays Window

As you have already seen in Figure 5.13, Model Selector for LTI Arrays Window consists of three parts:

- part 1-- which allows you to select the array models from the drop down list of LTI arrays which have been loaded into the LTI Viewer 's workspace.

- part 2 -- which allows you to choose from the available selection criteria:
  a) Index into Dimensions when the Window appears as shown in Figure 5.13a,
  b) Bound on Characteristics when the Window appears as shown in Figure 5.13b

- part 3 which depending upon the Selection Criteria chosen in part 2, appears as shown in Figure 5.13a and b respectively.
  a) When Index into Dimensions is chosen in part 2, this part of the window allows you to select the index of the dimensions of the model to display or hide from the drop down box. For example for modelA47, which is an array of 1-by-2 dimensions, you can display or hide the plots for all the dimensions, 1-by-1 or 1-by-2 dimensions of the model by selecting appropriate options.
  b) If 'Bound on Characteristics' is chosen, index of the model displays the list of characteristics (as displayed by the right click menu for a particular plot type), with a check box on the right hand side and text box on the left hand side of the characteristic name. For example, currently the LTI Viewer is displaying the Nyquist plot for modelA47 hence the selection criterion setup displays Gain Margin and Phase Margin when 'Bound on Characteristics' is chosen. (If step response plot were chosen to be displayed for the same model then the list of Selection Criterion setup would have displayed Peak Response, Settling Time, Rise Time and Steady State). You can enter the value of these characteristics directly when the LTI Viewer displays the plots fulfilling these criterion, or you can enter expressions using relational operators available with MATLAB® and the $ sign as the variable name for indexing to obtain desired plots.

Illustrations given below will further clarify this concept...

Example:

ltiview
returns

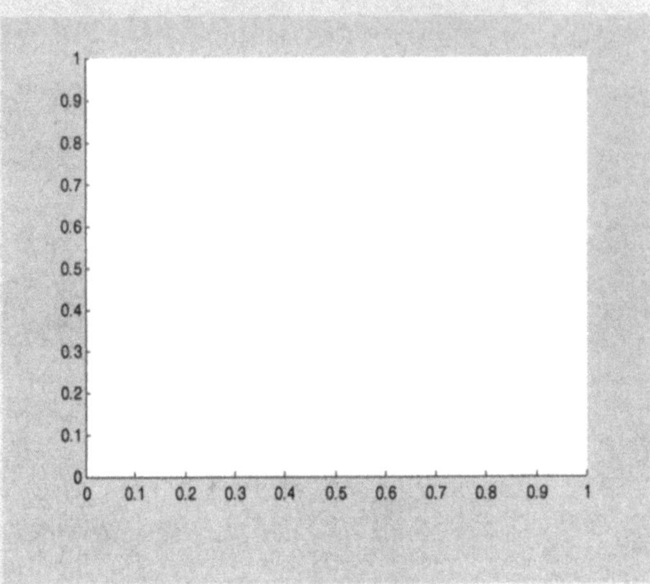

ltiview({'bode','step','sigma'},modelA14,modelA31,modelA36)
returns

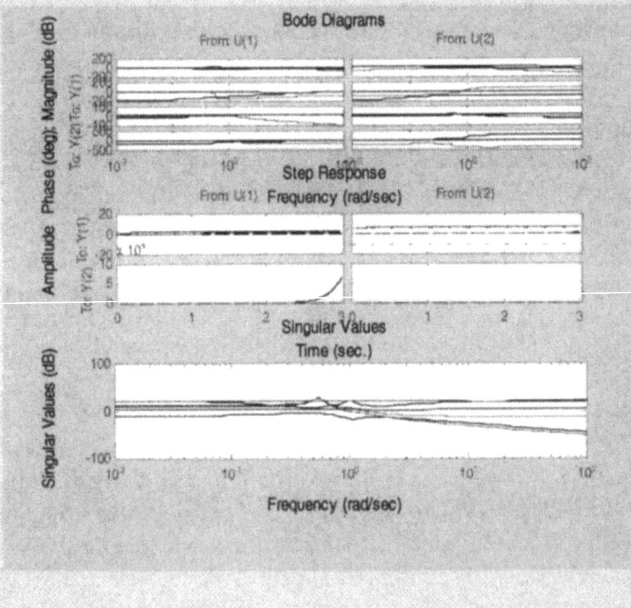

```
ltiview('bode',modelA14,[5 55 555 5555 55555],'g+--',modelA31,'bd-
',modelA36,{10e2,10e4},'r*.:')
returns
```

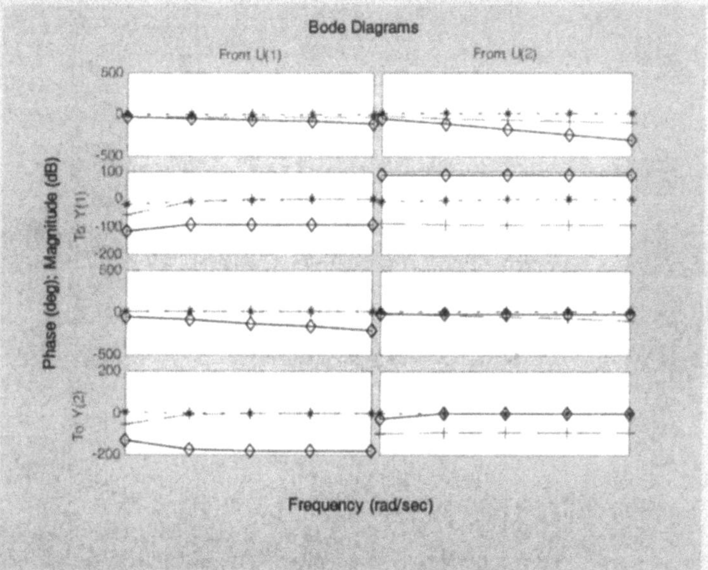

```
ltiview('initial',modelA34,[1,2])
returns
```

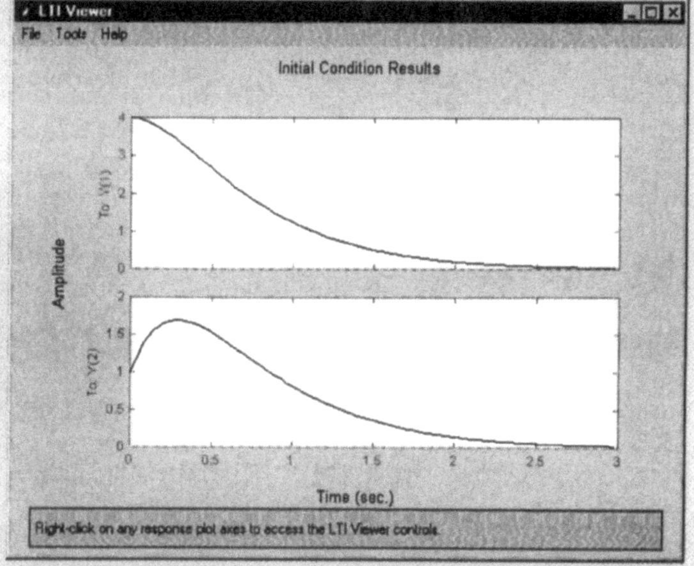

```
[u,t]=gensig('sin',2*pi,60);
ltiview('lsim',modelA37,modelA26,u,t)
returns
```

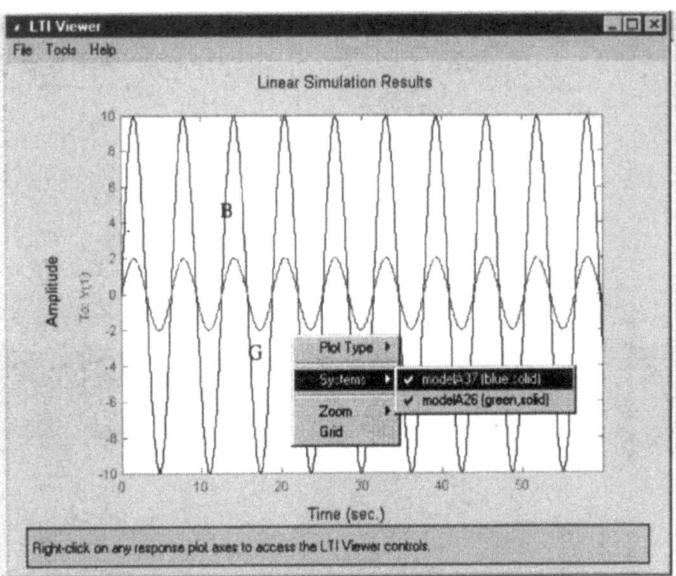

Note the right click menu items above. It has only four main menus.

ltiview({'nichols','sigma','pzmap','impulse'},modelA14, modelA31, modelA36)
returns

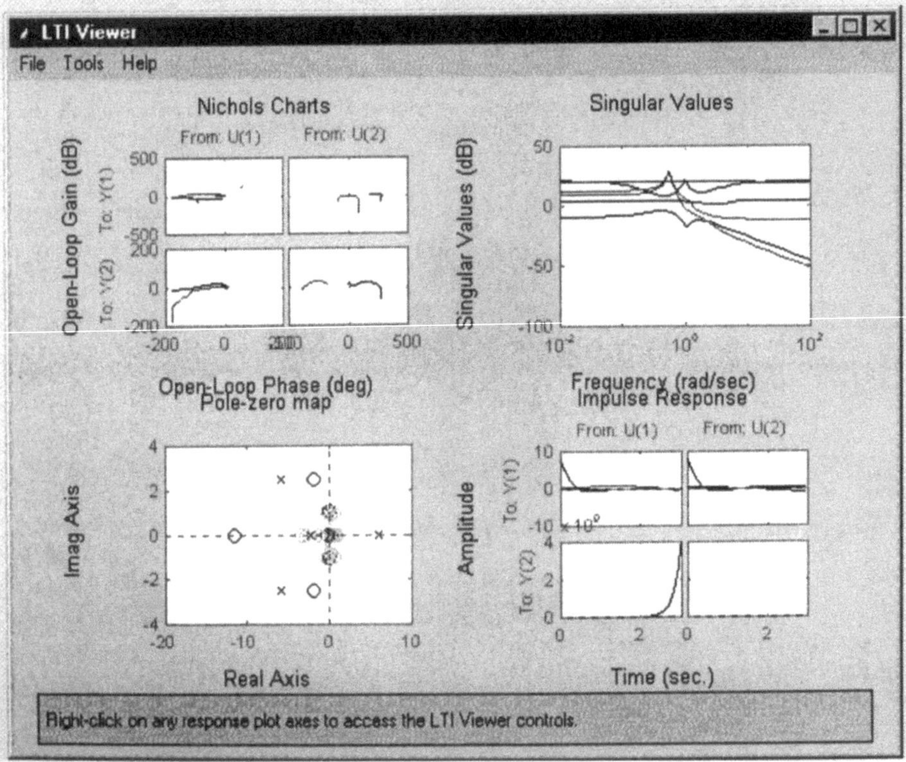

Let us now try out a few functions on the LTI array models and obtain responses for them. Let us choose modelA47 for which we have already obtained three responses in Figure 5.15 and 5.16. We shall get some other responses for this model by indexing or otherwise as follows. Observe carefully the way in which the indexes have been specified to obtain a particular response. If the index for an array model is specified by (a, b, c, d, e...), then, 'a' would denote the output (row) index, 'b' would denote the input (column) index, and the rest would denote the array dimensions (for example, if the response of 4th output to 2nd input of 2-by-3-by-1 array of a 3-by-3-by-3 array model with 5 inputs and 6 outputs is to be specified, then the index would be (4,2,2,3,1)).

The forthcoming examples will make things more clear.

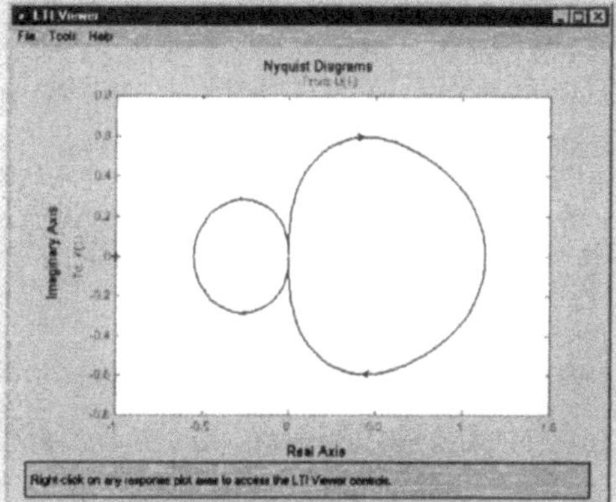

ltiview('nyquist',modelA47(3,2))
ltiview('nyquist',modelA47(3,2,:))
ltiview('nyquist',modelA47(3,2,:,:))

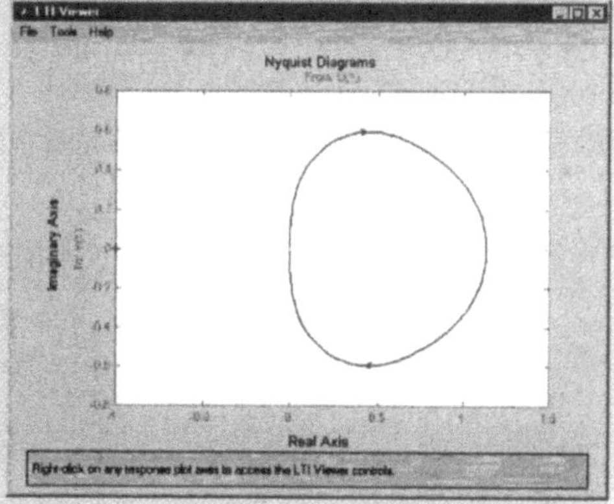

ltiview('nyquist',modelA47(3,2,1,1))

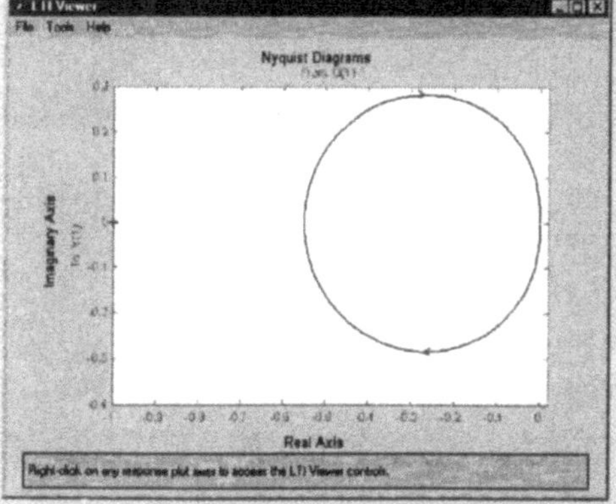

ltiview('nyquist',modelA47(3,2,1,2))

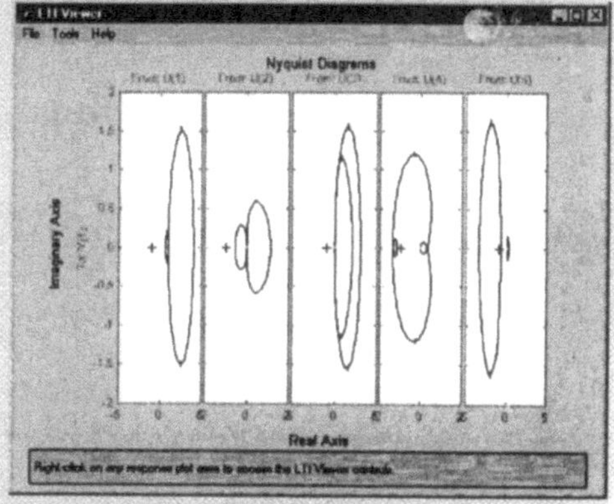

ltiview('nyquist',modelA47(3,:))
ltiview('nyquist',modelA47(3,:,:))
ltiview('nyquist',modelA47(3,:,:,:))

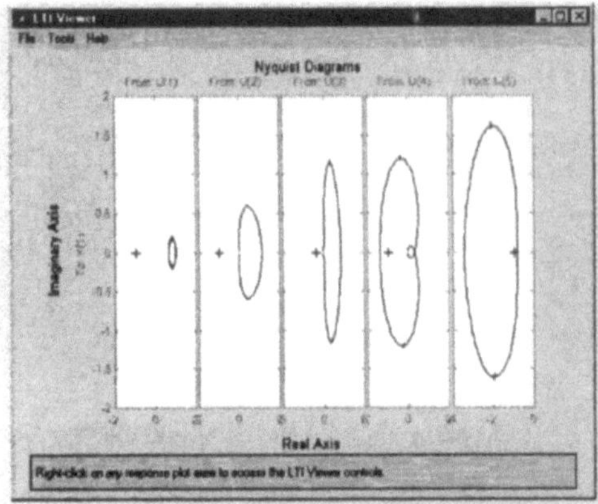

ltiview('nyquist',modelA47(3,:,1,1))

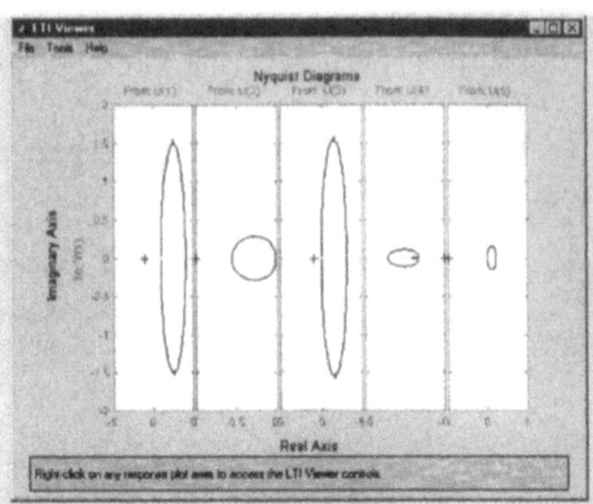

ltiview('nyquist',modelA47(3,:,1,2))

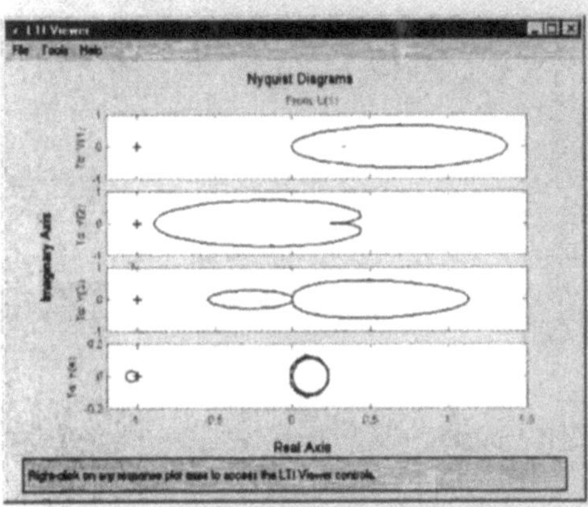

ltiview('nyquist',modelA47(:,2))
ltiview('nyquist',modelA47(:,2,:,:))

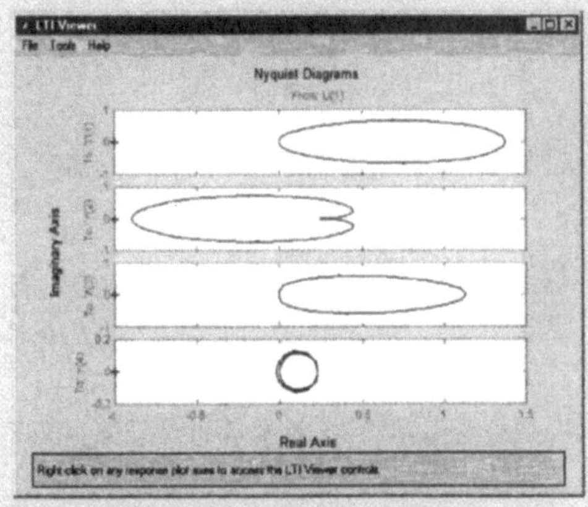

ltiview('nyquist',modelA47(:,2,1,1))

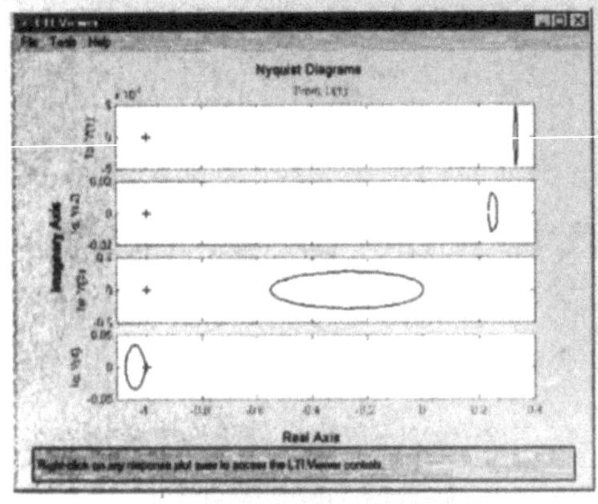

ltiview('nyquist',modelA47(:,2,1,2))

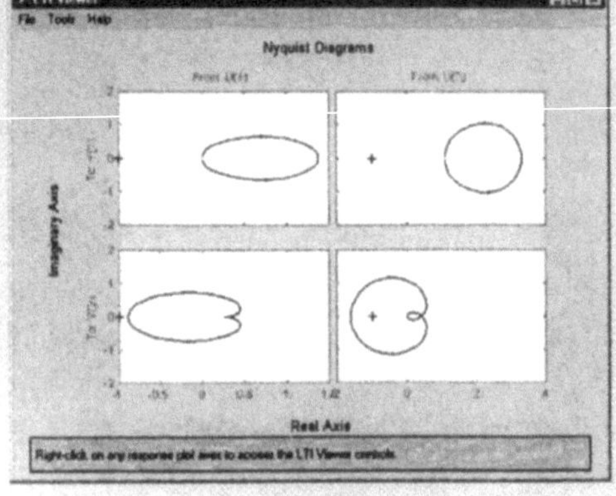

ltiview('nyquist',modelA47([1 2],[2 3],1,1))

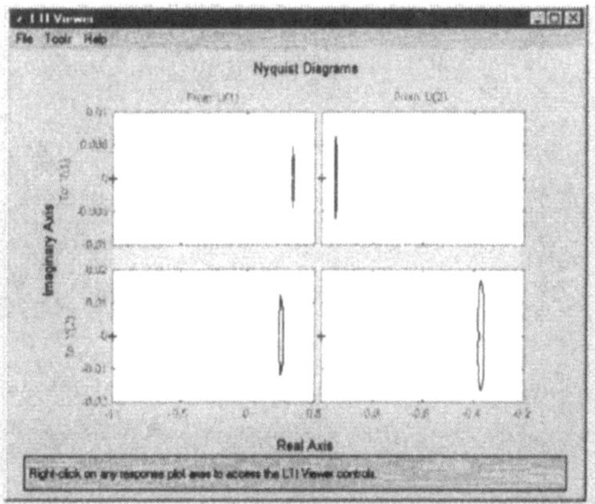

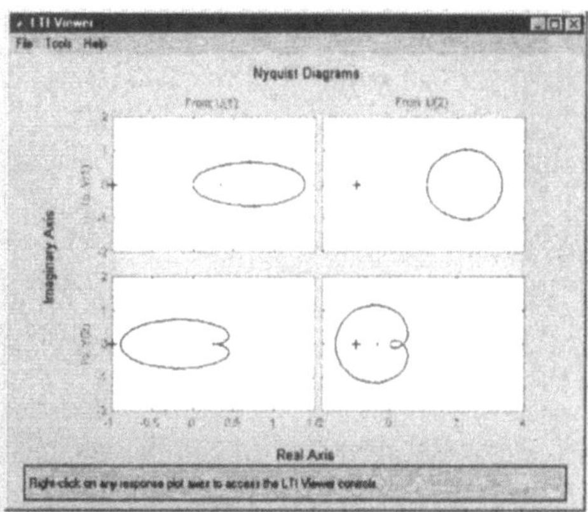

ltiview('nyquist',modelA47([1 2],[2 3],1,2))

ltiview('nyquist',modelA47([1 2],[2 3]))
ltiview('nyquist',modelA47([1 2],[2 3],:))
ltiview('nyquist',modelA47([1 2],[2 3],:,:))

You can also index into the Model Selector for LTI Arrays Window to obtain desired plots. It is also possible to obtain plots for certain conditions of plot characteristics (like Rise Time less than 15 seconds *etc.*, for time response plots and similar conditions for Gain and Phase Margins for frequency response plots) for an LTI array model using $ sign as the variable name as indicated in examples below for step response of modelA47:

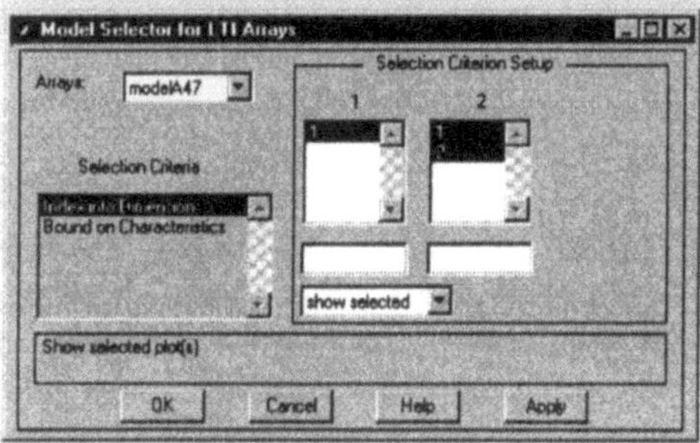

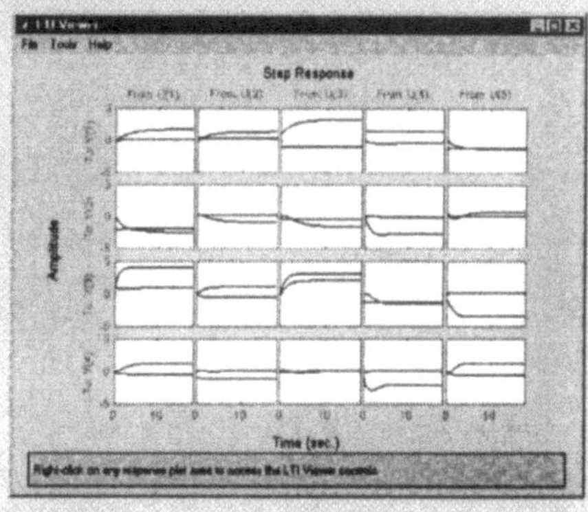

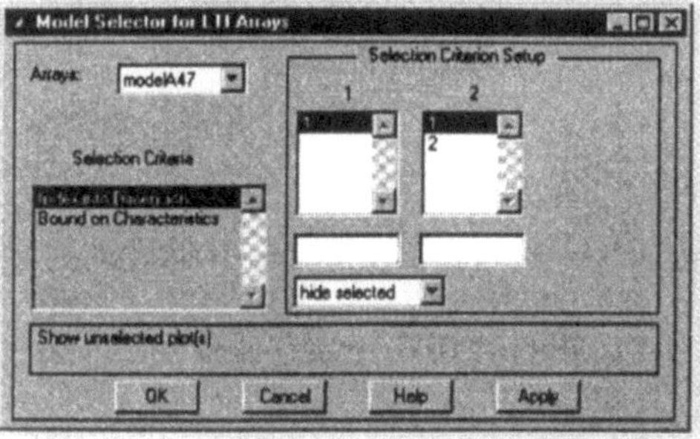

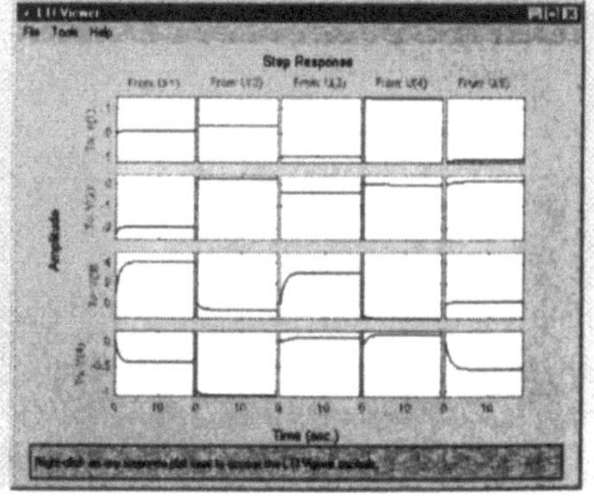

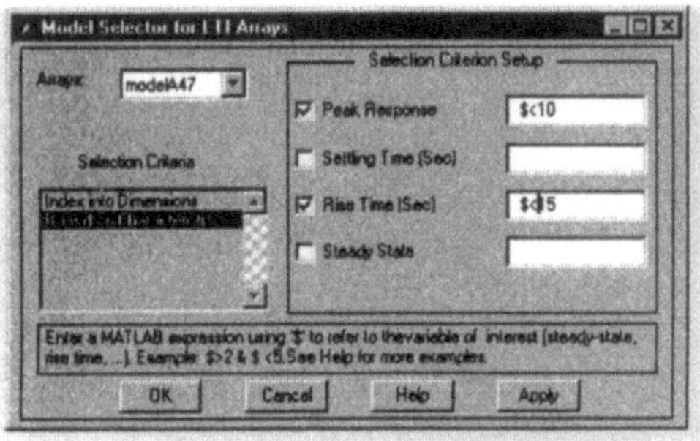

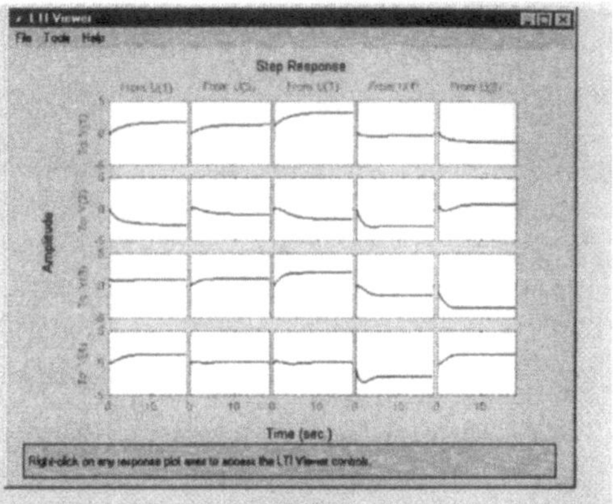

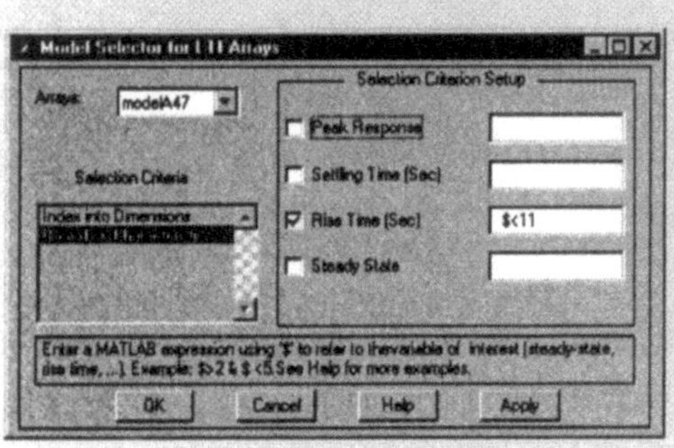

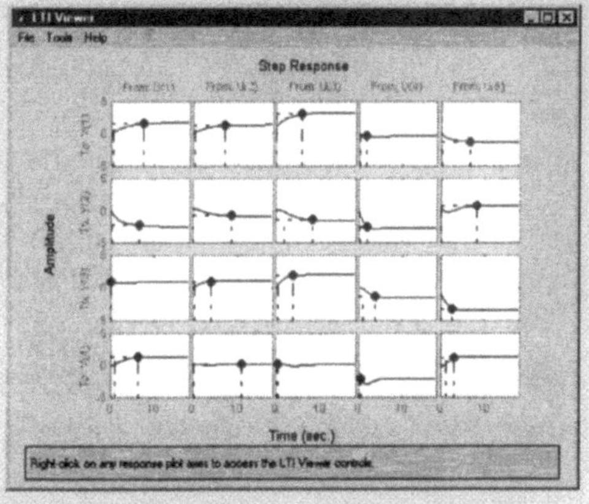

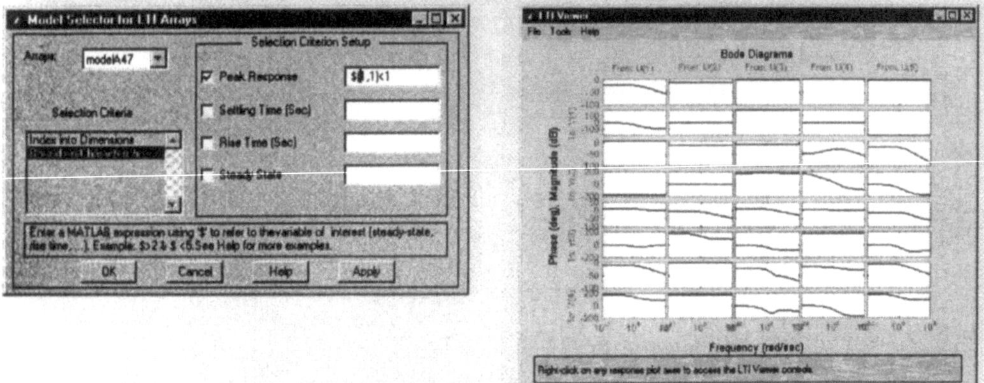

Similarly, frequency response plots for modelA47 can also be obtained.

Notice the 'Model Selector' for LTI Arrays for **PZmap** when 'Bound on Characteristics' is chosen in the 'Selection Criteria Window':

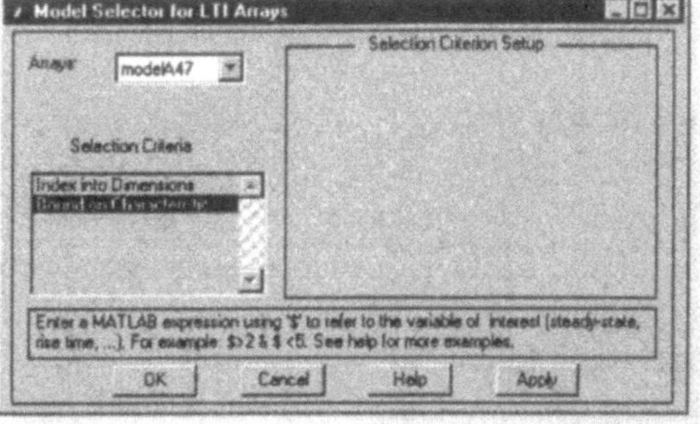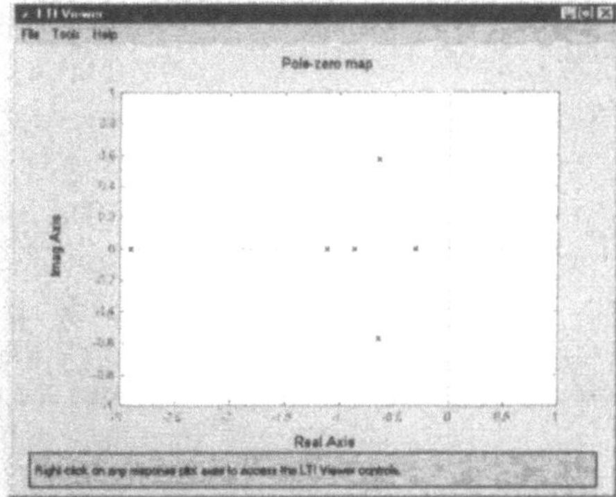

It has really been long since you took a practice test…

Practice Test 5.1.

1. Obtain all nine plots for modelA1 and modelA18.

2. Obtain all possible plot for modelA41.

3. Obtain step, Bode and Nichols plots for modelA47 for response of 4th input to 3rd output for 2nd array dimension.

4. Obtain impulse response plot for modelA47 for peak response greater than 2 seconds.

5.2 The Root Locus Design GUI

Control System Toolbox provides yet another Graphic User Interface for not only obtaining the root locus plot for SISO feedback models of zpk, tf and ss types but also to enable you to design compensators for the models. In the pages to follow, you will learn about this interactive GUI.

5.2.1 Initialising the Root Locus Design GUI

Just like the LTI Viewer, the Root Locus Design GUI can be initialised or invoked by any of the two methods described below:

- Obtaining an empty Root Locus Design GUI window:
By simply entering **rltool** at the command prompt of the MATLAB® command window. In this case, an empty LTI Viewer window appears on the screen. SISO LTI models with or without compensators whose root locus plots are desired to be studied are then loaded into the Root Locus Design GUI window.

- Obtaining a Root Locus Design GUI window with plots on it:
Invoking, Root Locus Design GUI, loading the models and compensators at desired location can be done simultaneously by using the same function **rltool** with the following syntax:

rltool(*ModelName, compensator, location, feedback*)

where,

ModelName .... specifies the names of the SISO LTI model in the workspace whose root locus plot is desired to be displayed in the Root Locus Design GUI window when it appears on the screen

compensator .... specifies the name of the SISO LTI compensator whose root locus plot is desired to be displayed in conjunction with the corresponding model (optional).

location .... specifies the location of the compensator (*i.e.*, in forward path or feedback path) as indicated in Table 5.4 (optional).

feedback .... specifies the sign of the feedback (*i.e.*, positive or negative feedback) of the compensator as indicated in Table 5.4 (optional).

Table 5.4. Root Locus Design GUI parameters

Parameter	Value	Description	Remarks
location	1	Places the compensator K in the forward path (default)	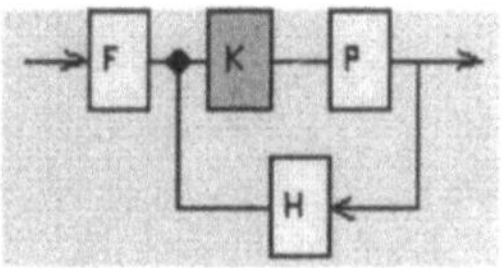 F → K → P, with H in feedback
	2	Places the compensator K in the feedback loop	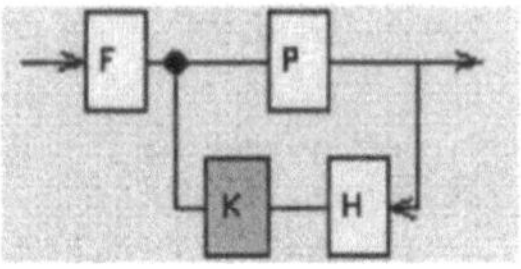 F → P, with K → H in feedback
feedback	-1	indicates negative feedback (default)	-
	1 or +1	indicates positive feedback	-

However, the initially displayed root locus plot can be changed by changing the model and/or the compensator that are currently loaded in the Root Locus Design GUI window. It is also possible to have more than one Root Locus Design GUI window open, for say, comparing the plot of a model with a different compensator.

5.2.2 The Root Locus Design GUI Environment

Just like the LTI Viewer window, the Root Locus Design GUI windows (Figure 5.20) have the following common features:

- a title bar with **Root Locus Design** or **Root Locus Design GUI:** *ModelName* written on it at the left hand corner and the familiar three command buttons for minimising, maximising and closing the Root Locus Design GUI window at the right hand corner, depending upon whether it was invoked by method one (which gives an empty Root Locus Design GUI window) or method two (which displays a plot in the Root Locus Design GUI window).

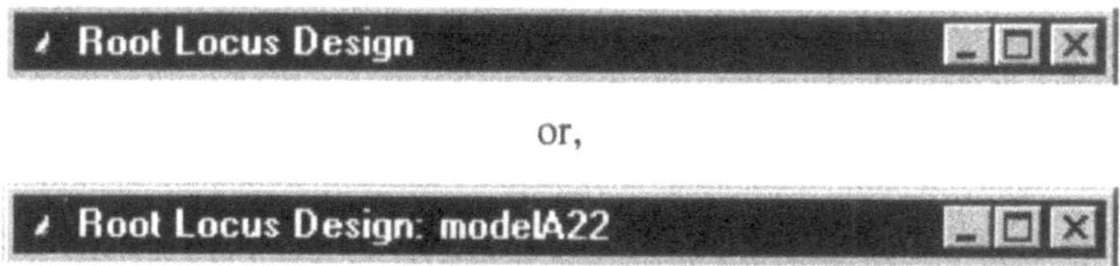

or,

* the top menu bar, displaying four words File, Tools, Window and Help. The details of the menu of the Root Locus Design GUI are summarised in Table 5.5.

* the compensator and feedback bar with two parts as indicated below. The compensator part indicates the zero/pole/gain model of the current compensator. Its default value is $\dfrac{\text{Gain}}{1}$ in case no compensator has been specified. The feedback part has a command button with +/- marked on it to toggle the feedback sign. Notice that the sign in the figure toggles when you click this button.

* the top toolbar as shown below with four tool buttons to 'drag' poles/zeros, 'add' poles, 'add' zeros and 'erase' poles/zeros respectively; a text box for entering 'compensator gain'; and a checkbox to 'add/delete grid' (starting from left hand side). The default value of gain is 1, which can be changed manually or by manipulating the closed loop poles on the root locus plot.

* the plot region that is either empty or displays the plot depending on the method that was used to initialise the LTI Viewer.

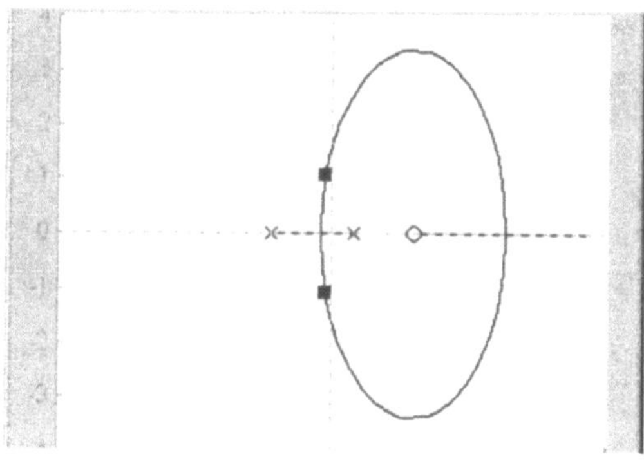

- the axes and zooming tool bar which has two parts as shown and described below:

    i. Axes settings -- with four command buttons to store axes limits, retrieve axes limits, make axes square and use equal axes aspect ratios (starting from the left hand side).
    ii. Zoom -- again with four tool buttons to zoom in X and Y direction, zoom in-X direction, zoom in-Y direction and for full view (starting from the left hand side respectively).

- the bottom plot bar with checkboxes to obtain the Step, Impulse, Bode, Nyquist and Nichols plots in the familiar LTI Viewer window (Figure 5.18) captioned **LTI Viewer for Root Locus Design:** *ModelName* at the top. Notice that the new LTI Viewer window right click menu for parameter **Characteristics** displays a new sub-menu **Stability Margins** for all frequency response plots, which was not available in the normal LTI Viewer.

- the bottom statusbar displaying a context sensitive information or help tip, which may be an instruction, a hint or an error message as you proceed through your design and analysis. The usual messages appearing as the help tip could be a message indicating that an unsupported function has been tried to be performed, a function has been completed or an additional information on the use of the Root Locus Design GUI or about the location of the newly placed compensator poles and zeros, as well as damping ratios, natural frequency including location of poles and zeros as you drag them.

Complete Root Locus Design GUI and response plots is shown in Figure 5.17 and Figure 5.18 respectively for modelA22 as main model P and modelA9 as the compensator K in the feedback path.

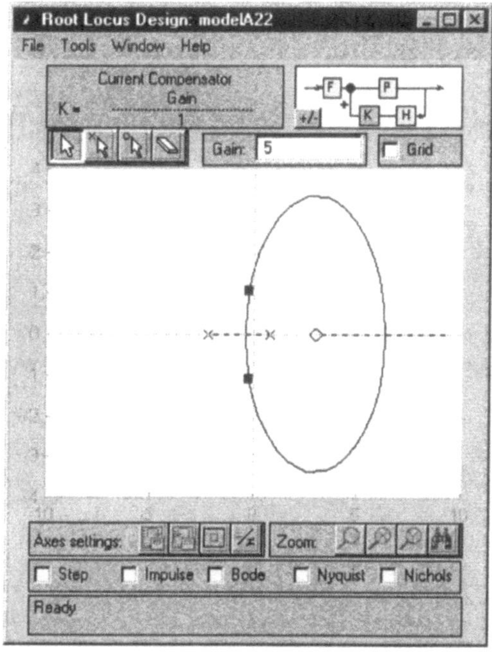

Figure 5.17. Complete Root Locus Design GUI (obtained by entering **rltool**(modelA22, modelA9, 2, +1) at the MATLAB® command prompt)

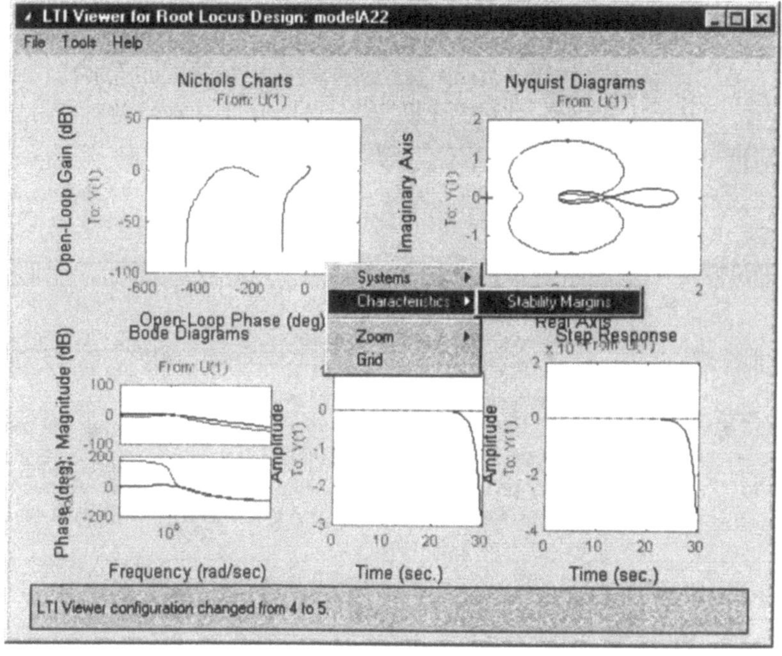

Figure 5.18. Response plot of Root Locus Design GUI shown in Figure 5.17

Table 5.5. The Root Locus Design GUI main menu

a. The File menu

Sub-menu	Description
Import Model (Ctrl+I)	opens the **Import LTI Design Model** window with the same caption in the title bar (Figure 5.19). It has four distinct parts: • the **Feedback Structure** part to decide the location of the compensator K (*i.e.,* in the forward path or in the feedback loop) by clicking the command button captioned 'Other…'. • the **Import From** part to import the model from the Workspace or saved/open MAT file or a SIMULINK® model by clicking their respective radio buttons or entering the *ModelName*, MAT file name or SIMULINK® model name respectively in the text box given at the bottom. • the **Workspace Contents/MAT File Contents/LTI Models** (depending upon the mode of selection of the model in the step given above) part to select the models for plant model P, sensor dynamics H and pre-filter F, from the list displayed in the window respectively . • the **Design Model** part to give a name to the system by entering it in the textbox titled **Name** and to select models for P, H and F from the list displayed in the above step by clicking the buttons with arrow marks adjacent to P, H and F after which, the *ModelName* is indicated in the text box adjacent to P, H and F. However, an appropriate zpk, tf or ss expression can also be entered directly into these text boxes to specify the models .
Import Compensator	opens the **Import LTI Design Model** window with the same caption in the title bar (Figure 5.20). It has three distinct parts: • the **Import From** part same as the one in Import LTI Design Model window explained above. • the **Workspace Contents/MAT File Contents/LTI Models** part same as the one in Import LTI Design Model window explained above. • the **Design Model** part similar to the one in Import LTI Design Model window explained above, except that here you import model for K *i.e.,* the compensator instead of P, H or F .
Export (Ctrl+E)	opens the **Export** window (Figure 5.21) captioned 'Export LTI Models/ Compensators' in the title bar displaying export list of the models, which can be exported from the Root Locus Design GUI's workspace to the MATLAB®'s workspace or a storage device.
Draw SIMULINK® Diagram	opens the SIMULINK® window captioned *ModelName* in the title bar displays the model's block diagram drawn on it. You can use the SIMULINK® features for further analysis of the system. (Figure 5.22).
Display History	opens an information window with no caption at the top in the title bar giving you information on the history of the system (fig 5.23).
Print Locus (Ctrl+P)	opens the old and familiar **Print** window to allow printing the root locus plot displayed in the Root Locus Design GUI.
Send Locus to Figure	sends the plot on the Root Locus Design GUI to the **Figure** window of MATLAB® such that the Figure window features are available to you to improve the appearance of root locus plot.
Exit Design (Ctrl+X)	closes the open **Root Locus Design GUI** window (clicking the cross on the top right side of the Root Locus Design GUI window does the same)

Table 5.5. (continued)

b. The Tools menu

Sub-menu	Description
Edit Compensator	opens the **Edit Compensator** window (Figure 5.24) captioned the same in the title bar to enable you to add, delete, change or edit compensator zeros and poles. Clicking on the 'Add Poles' or 'Add Zeros' command button produces empty text boxes with real and imaginary parts in which you can enter the values of poles and zeros respectively. Clicking on the checkbox under 'Delete' label removes the zeros or the poles.
List Model Poles/Zeros	opens a window captioned **Root Locus Plant: *ModelName*** in the title bar, which has three parts each for P, H and F . It displays the list of Poles/zeros of the system for P, H and F along with the names of the models (Figure 5.25). Each of the three parts has a command button named 'Show Model' which when clicked shows the model representation of P, H and F respectively of the system (Figure 5.26).
List Closed -loop Poles	produces a window captioned **Root Locus Design** in the title bar, displaying a list of the closed-loop poles of the system (Figure 5.27).
Convert Model/ Compensator	opens the **Convert Model/Compensator** window captioned the same in the title bar, which enables you to change a model/compensator from continuous to discrete or vice versa. It also allows you to choose between the methods (ZOH, FOH, Tutsin, Tutsin w/Prewarping and Matched pole-zero) and set the value of sampling time and critical frequency (Figure 5.28).
Clear Model	deletes the model currently loaded in the Root Locus Design GUI workspace.
Clear Compensator	deletes the compensator currently loaded in the Root Locus Design GUI workspace.
Add Grid/Boundary (Ctrl+G)	opens the **Grid and Constraint Options** captioned the same in the title bar, through which you can add grid lines and specify constraints for parameters as indicated in Figure 5.29 for the Root Locus plot. Specifying constraints results into a grey constraint line on the plot.
Set Axes Preferences (Ctrl+G)	opens the **Root Locus Axes Preferences** window captioned the same in the title bar, through which you can set the axes preferences (x and y axes limits, equal or square axes) and decide the colors and markers for Root Locus, Compensaator and Closed-loop poles (Figure 5.30).

c. The Window menu

Sub-menu	Description
0 MATLAB® Command Window	presents a list of the windows that are currently open, starting with MATLAB® command window which has serial number zero, and shall always have existence. Clicking on any of these makes that window active reducing others to the window's taskbar at the bottom of the screen. This action can also be accomplished by directly clicking the name of the window on the window's taskbar.
1 Root Locus Design	
2 Root Locus Design: *ModelName1*	
3 Root Locus Design: *ModelName2*	
4 LTI viewer window	

Table 5.5. (continued)

d. The Help menu

Sub-menu	Description
Main Help (Ctrl+H)	opens the MATLAB® **Help** window with the same caption. It has **Overview** written in the small text box at the top left corner of the window and the body of the window provides text explaining the features of the **Root Locus Design GUI** (Figure 5.31a).
Edit Compensator..	opens the MATLAB® **Help** window with the same caption. It has **Editing the Compensator** written in the small text box at the top left corner of the window and the body of the window provides text explaining the features of the **Edit Compensator** window (Figure 5.31b).
Convert Model..	opens the MATLAB® **Help** window with the same caption. It has **Convert Model** written in the small text box at the top left corner of the window and the body of the window provides text explaining the features of the **Convert Continuous/Discrete Model** window (Figure 5.31c).
Add Grid/Boundary..	opens the MATLAB® **Help** window with the same caption. It has **Grid and Constraints** written in the small text box at the top left corner of the window and the body of the window provides text explaining the features of the **Grid and Constraints Options** window (Figure 5.31d).
Set Axes Preferences..	opens the MATLAB® **Help** window with the same caption. It has **Axes Preferences** written in the small text box at the top left corner of the window and the body of the window provides text explaining the features of the **Axes Preferences** window (Figure 5.31e).

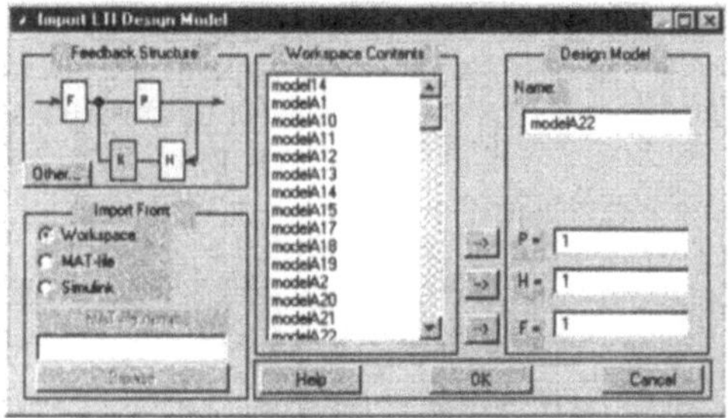

a. Import from workspace

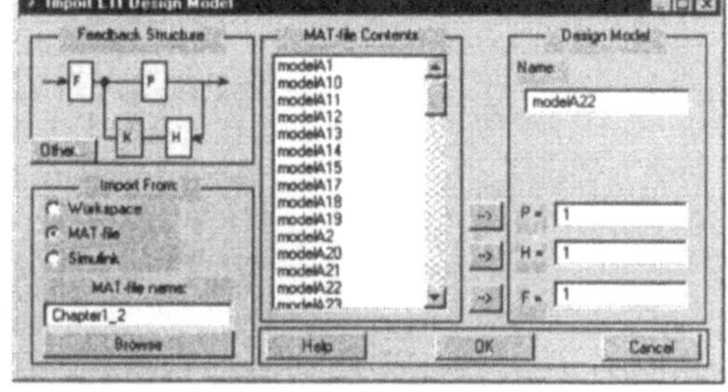

b. Import from MAT file

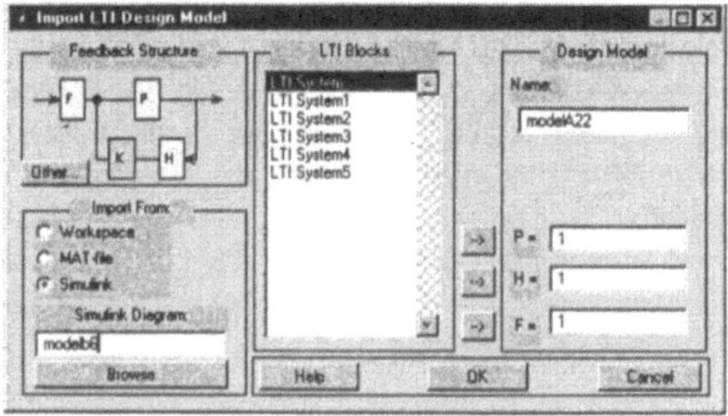

c. Import from SIMULINK®

Figure 5.19. The Import LTI Design Model Window

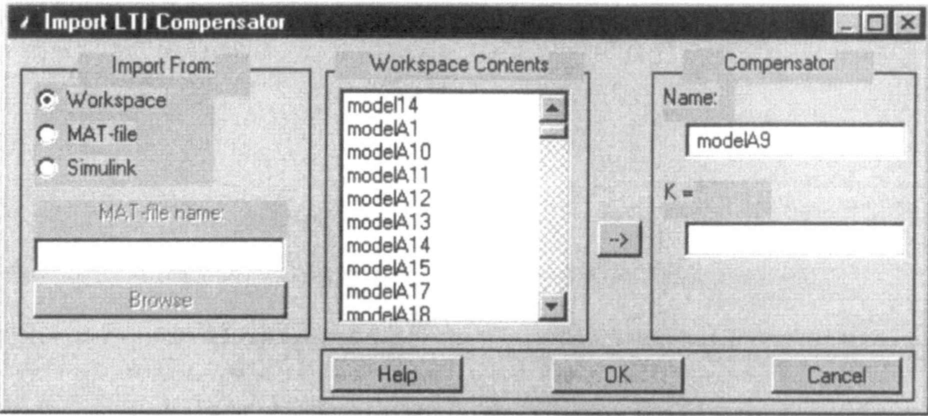

Figure 5.20. The Import LTI Compensator Window

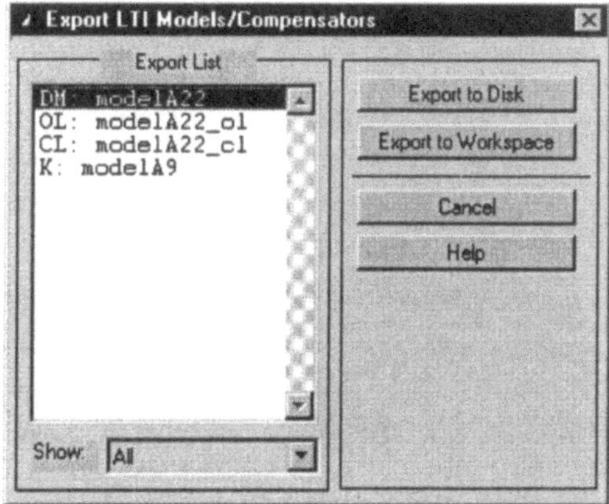

Figure 5.21. The Export LTI Model/Compensators Window

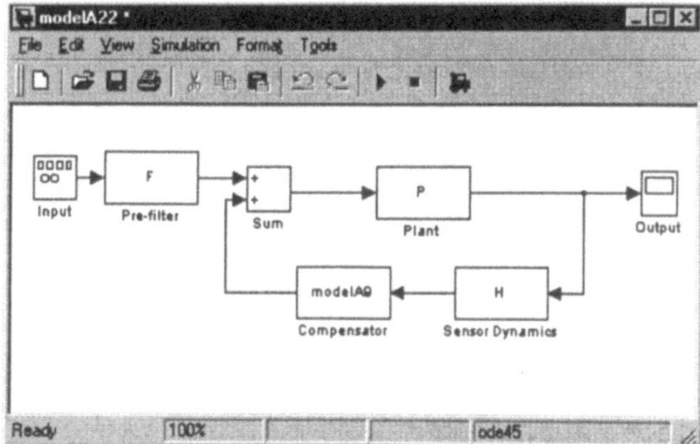

Figure 5.22. The SIMULINK® Window

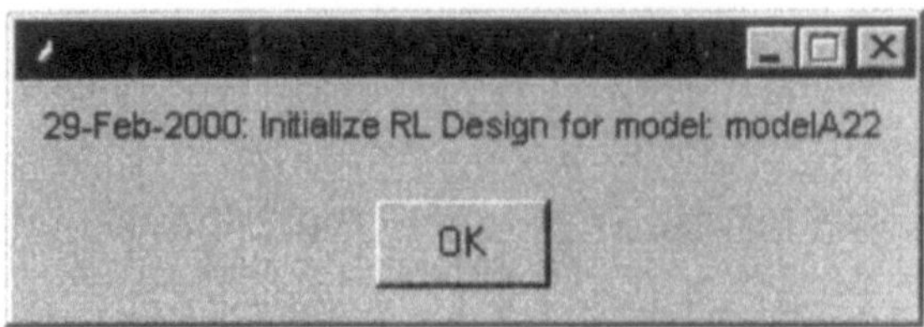

Figure 5.23. The Display History Window

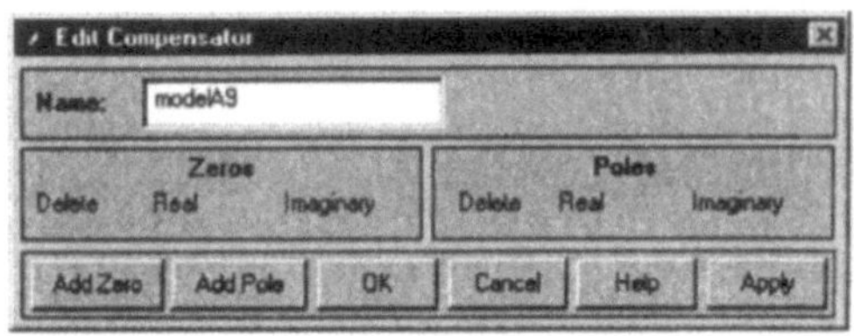

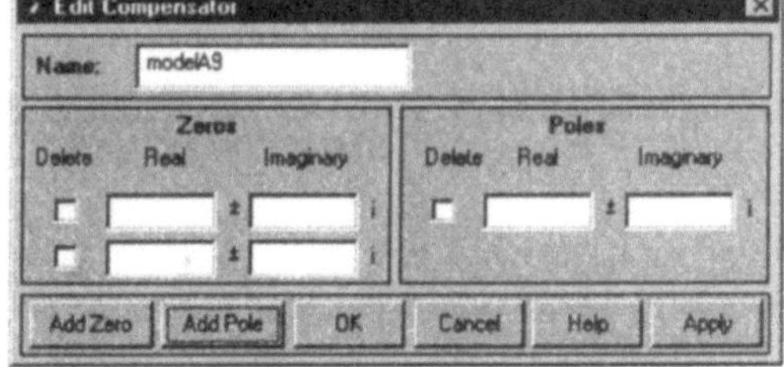

a. Empty Root Locus Design GUI

b. Edit Compensator window showing Empty Text Boxes for entering Additional Poles and Zeros (Delete Check-Box may be used to remove these Poles and Zeros)

Figure 5.24. The Edit Compensator Window

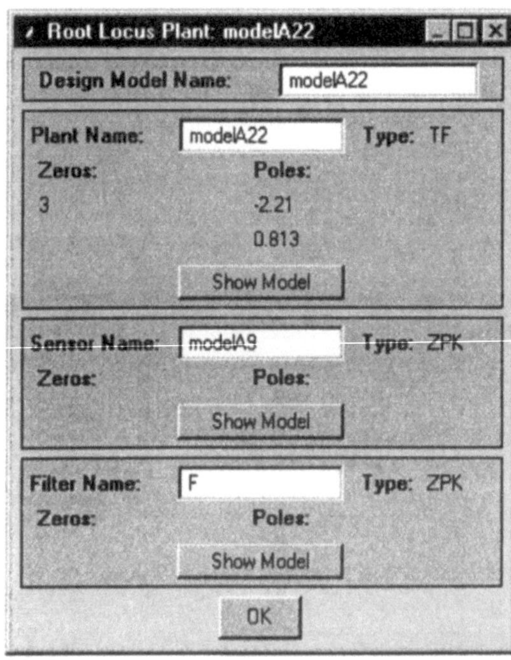

Figure 5.25. The Root Locus Plant Window showing the Model Poles/Zeros (obtained by entering **rltool**(modelA22, modelA9, 2, +1) at the MATLAB® command prompt)

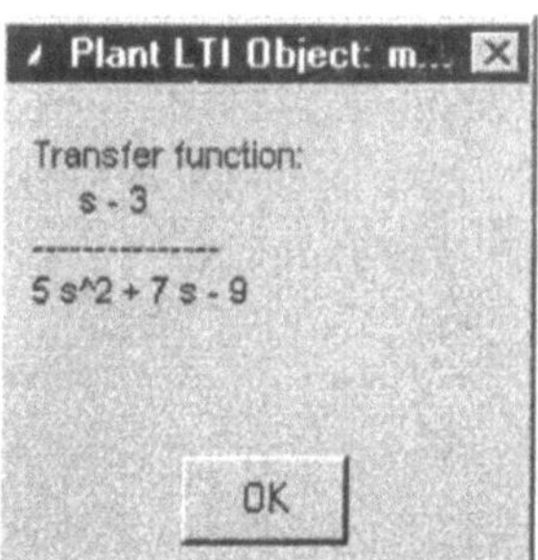

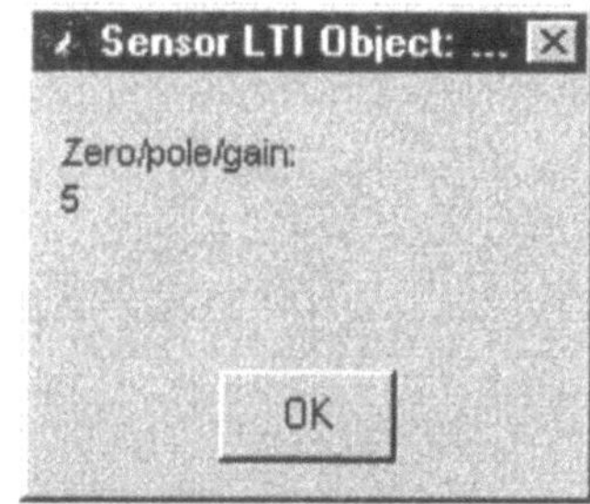

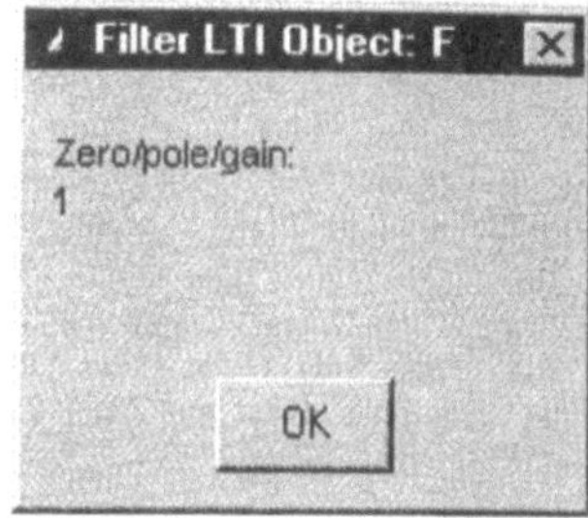

a. Plant Model   b. Sensor Dynamics Model   c. Pre-filter Model

Figure 5.26. Result of the clicking of Show Model command button of Figure 5.25 (obtained by entering **rltool**(modelA22,modelA9, 2, +1) at the MATLAB® command prompt)

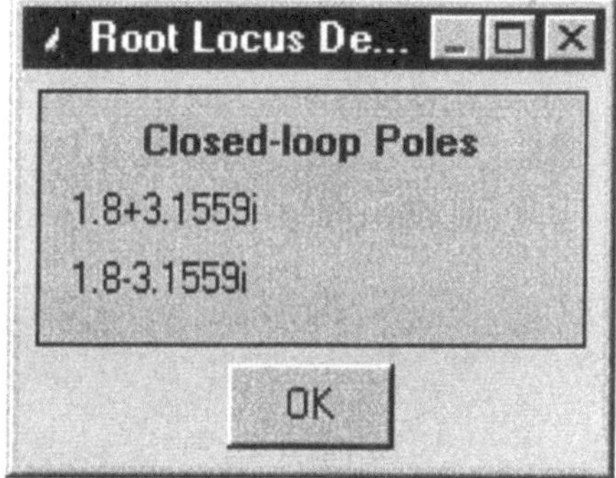

Figure 5.27. The Root Locus Design Window displaying a list of Closed-loop Poles (obtained by entering **rltool**(modelA22,modelA9, 2, +1) at the MATLAB® command prompt)

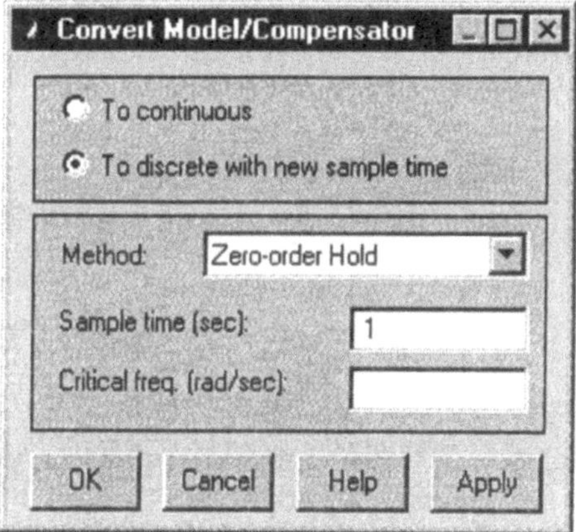

Figure 5.28. The Convert Model/Compensator Window

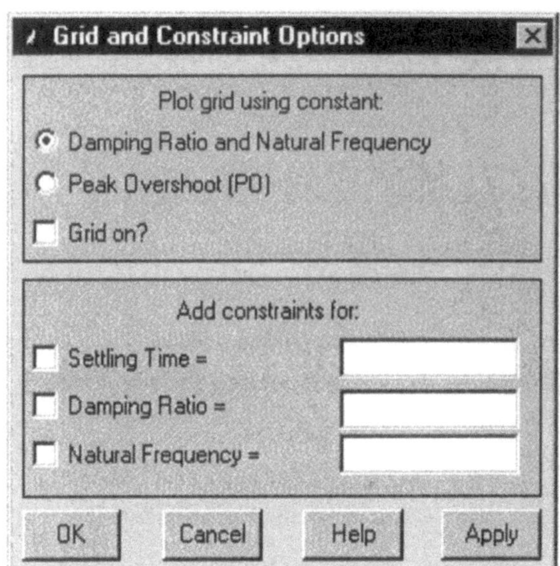

Figure 5.29. The Grid and Constraint Options Window

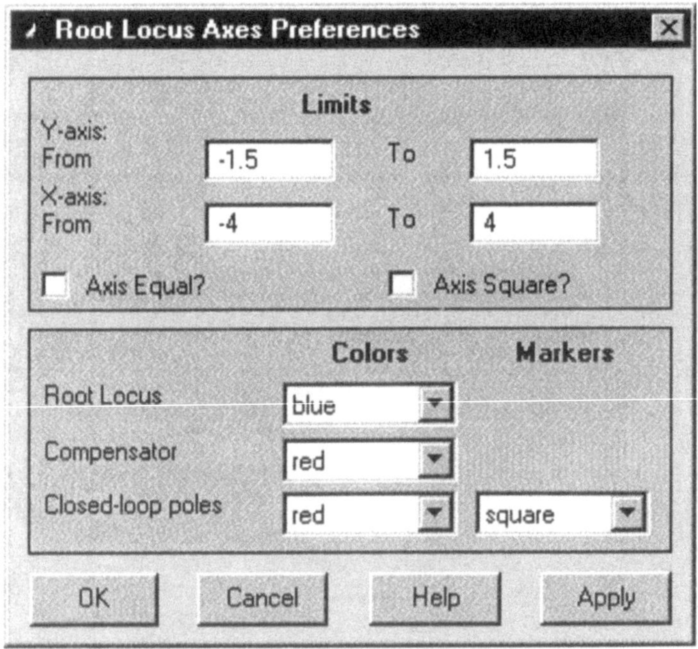

Figure 5.30. The Root Locus Axes Preferences Window

a. Main Help

b. Edit Compensator

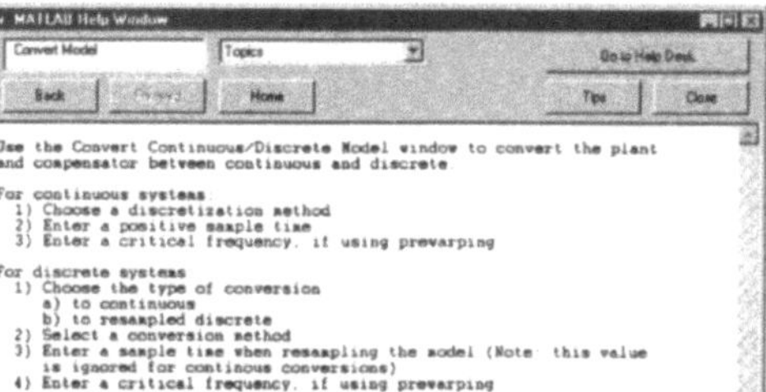

c. Convert Model

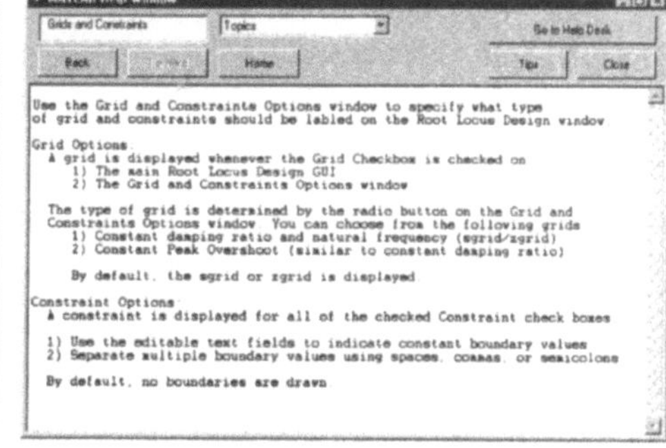

d. Add Grid/Boundary

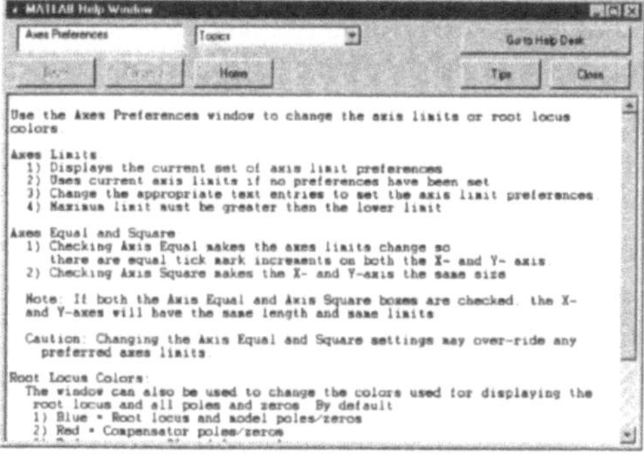

e. Set Axes Preferences

Figure 5.31. The Help Windows of Root Locus Design GUI

5.2.3 Using Root Locus Design GUI

As already mentioned, the Root Locus Design GUI, provides an interactive medium to design a suitable series (cascade) or feedback compensator for a SISO LTI model to obtain the desired performance. Invoking the Root Locus Design GUI with or without a model has already been explained in Article 5.2.1. If you have gone through Table 5.4 then you would be familiar with the environment of Root Locus Design GUI and also aware of many steps towards designing a compensator. You also know that there can be more than one method for achieving the same thing. However, the major steps involved in the design of a compensator are being enumerated below for your ready reference:

- Open the Root Locus Design GUI:
  i.   without a model;
  ii. with a model;
  iii. with a model and feedback compensator.
- If the Root Locus Design GUI was opened empty:
  i.   import a model using the Import Model Window or by entering a valid zpk, tf or ss expression for the model;
  ii. import a compensator using the Import Compensator Window or specifying a compensator by placing its poles and zeros manually.
- Display the response of the system.
- Change the gain set point by manually entering the values or by dragging the close loop poles represented by red squares on the plot to a new location using the mouse when the mouse cursor (currently an arrow) changes to a hand.
- Design the compensator:
  i.   to specify the design region boundaries on the root locus using the Grid and Constraint Options Window;
  ii. erase, edit or relocate the compensator poles and zeros using the toolbar, mouse or the Edit Compensator Window.
- List the poles and zeros.
- Open the LTI Viewer to observe the system response.
- Change the appearance of the Root Locus Design GUI Window and set axes preferences.
- Zoom in and out of the plot, especially, while trying to drag and place the closed loop poles of the system on the imaginary axis, or even place the poles and zeros of the compensator on the imaginary axis. This will enable you to place them exactly on desired location.
- Save the system to the workspace or a disk.
- Print the Root Locus.
- Draw the SIMULINK® diagram for the closed loop model.
- Convert between continuous and discrete models.
- Clear data if required.
- Get help if required.

Illustrations given below will help you in dealing with the intricacies involved in the procedure explained above.

Example:

rltool
returns

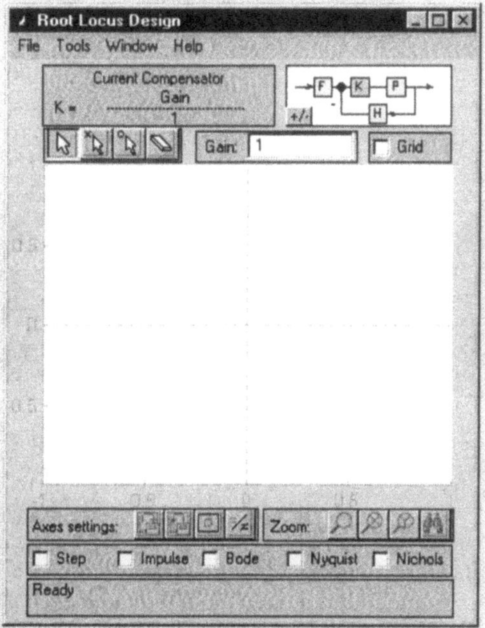

As our next example, let us examine the impulse response curves corresponding to different positions of poles in the s plane using the interactive features of Root Locus Design GUI.

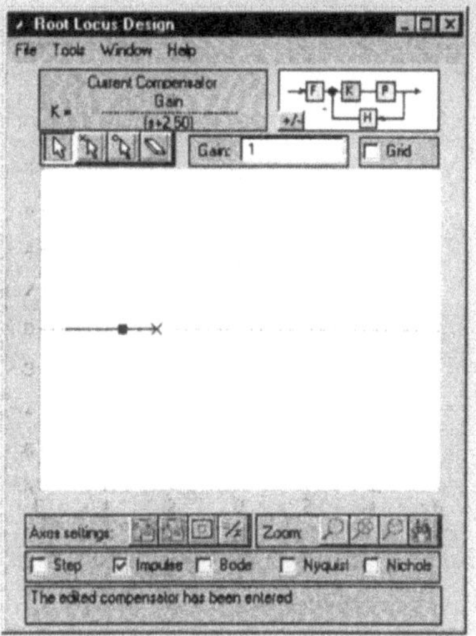

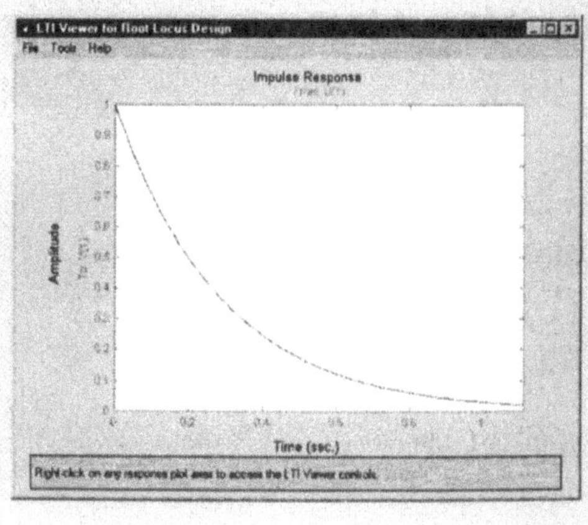

Single Pole on the Negative Real Axis *i.e.,* at s=-2.5

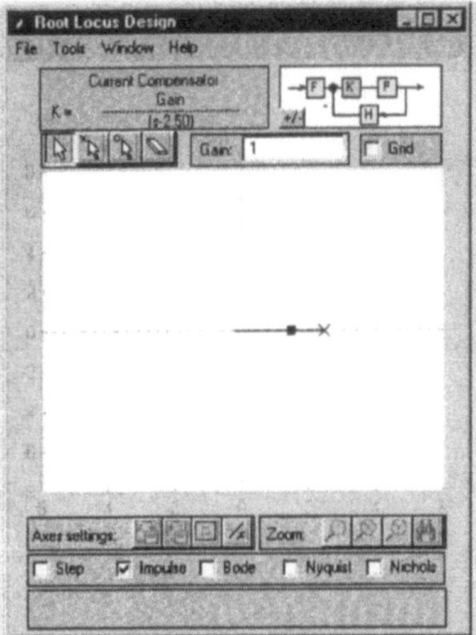

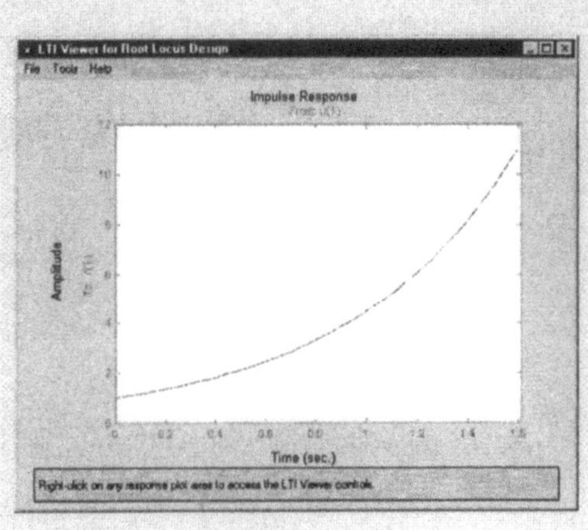

Single Pole on the Positive Real Axis *i.e.,* at s=2.5

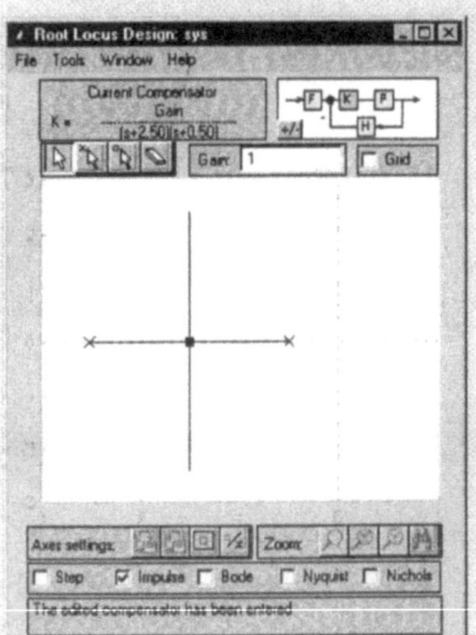

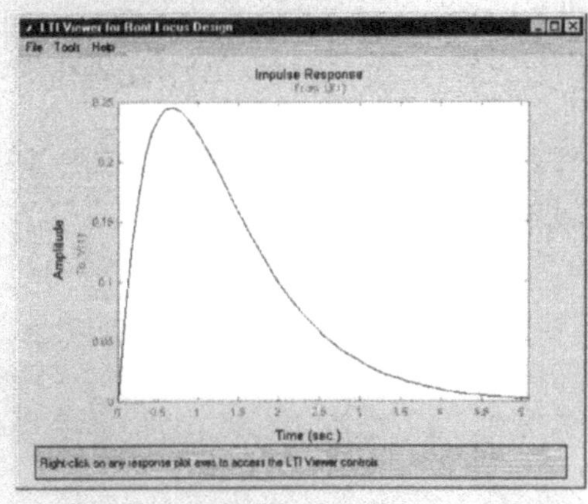

Pair of Poles on the Negative Real Axis *i.e.,* at s= -0.5 & -2.5

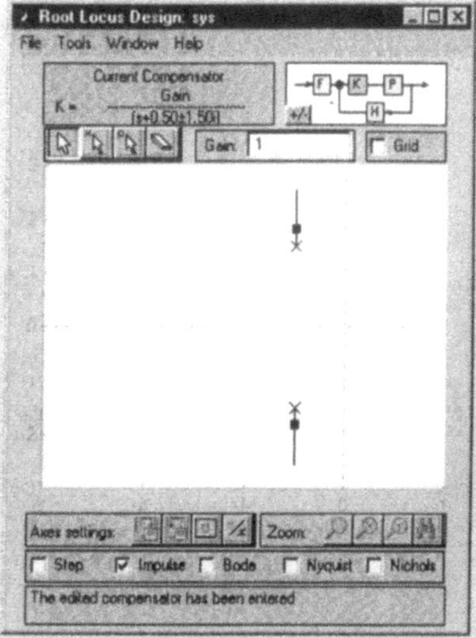
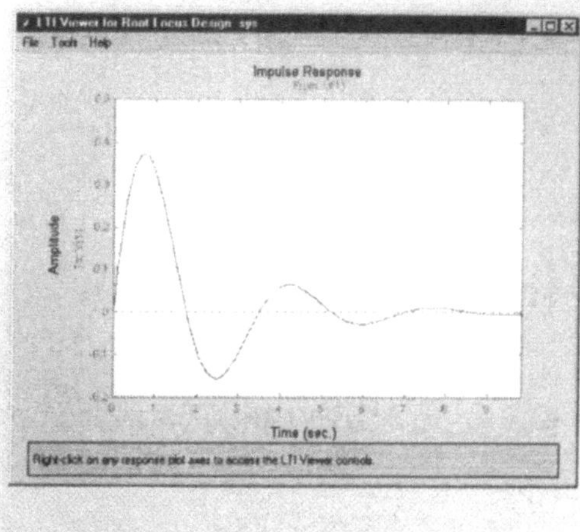

Pair of Complex Conjugate Poles on the Right Hand Side of the Imaginary Axis *i.e.*, at s= -0.5±1.5i

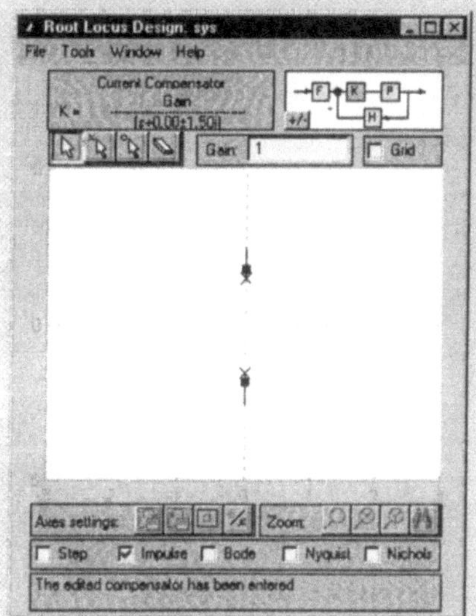
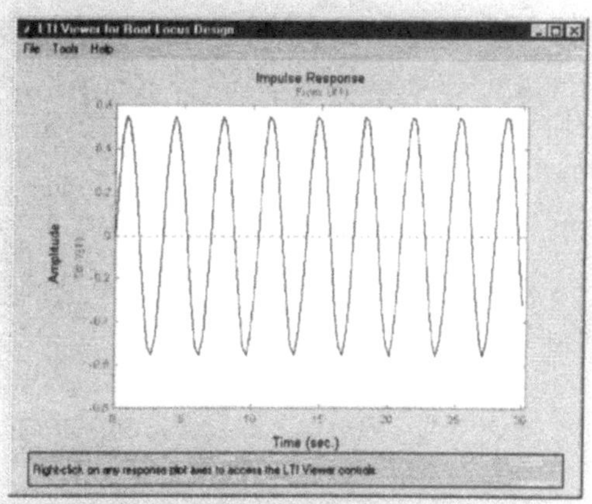

Pair of Poles on the Imaginary axis at s=±1.5i

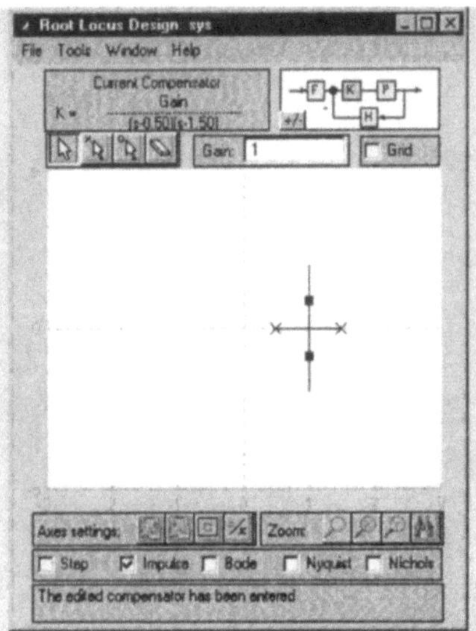
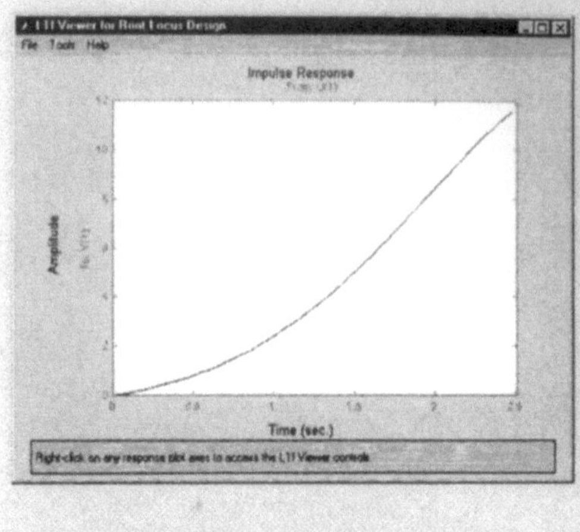

Pair of Positive Poles on the Real Axis *i.e.*, at s= 0.5 & 1.5

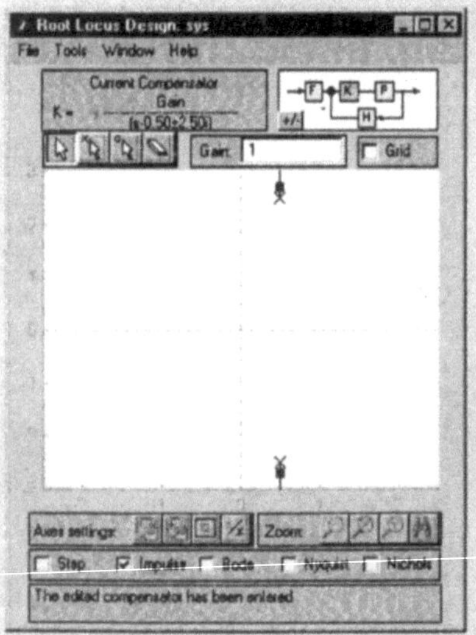
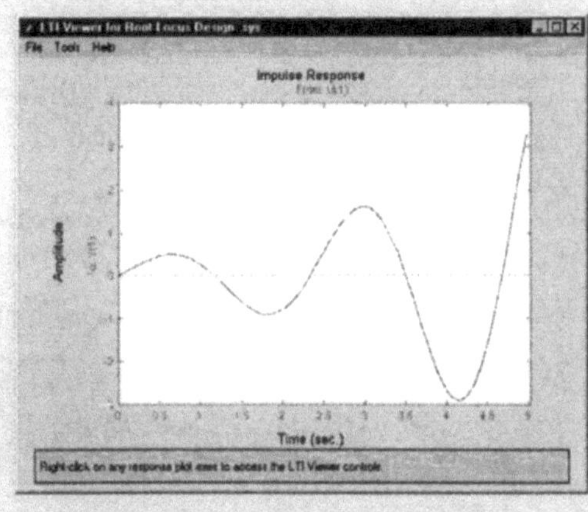

Pair of Complex Conjugate Poles on the Left Hand Side of the Imaginary Axis *i.e.*, at s=0.5 ± 1.5i

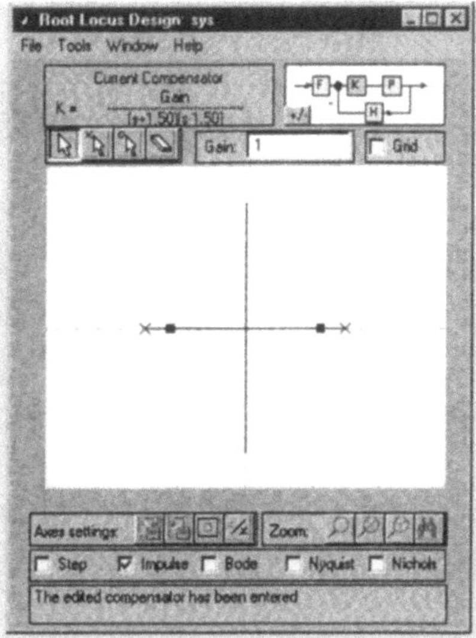
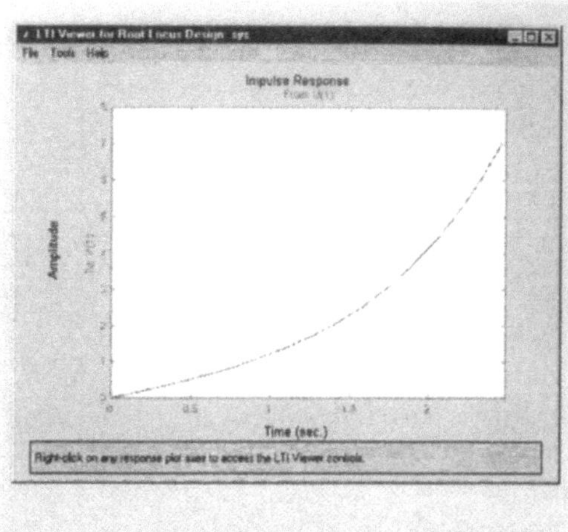

Single Pair of Poles at on the Real Axis *i.e.*, at s=±1.5

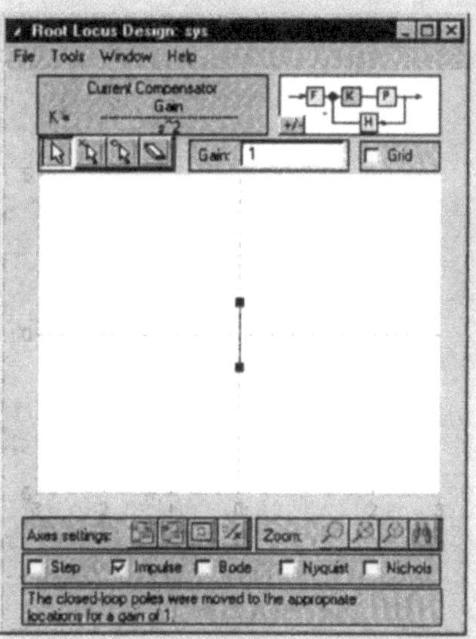
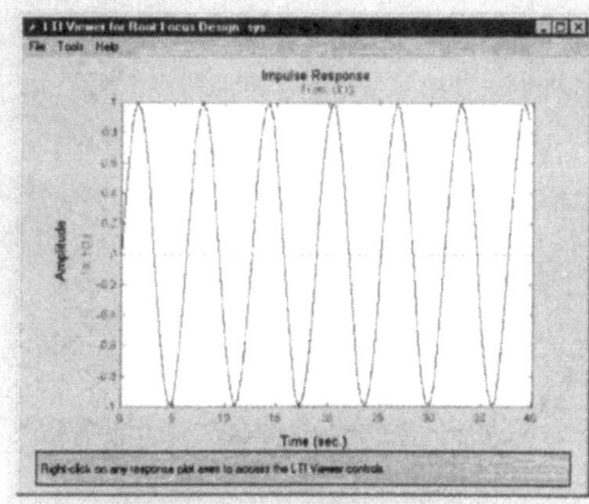

Double Poles at the Origin *i.e.*, at s=0

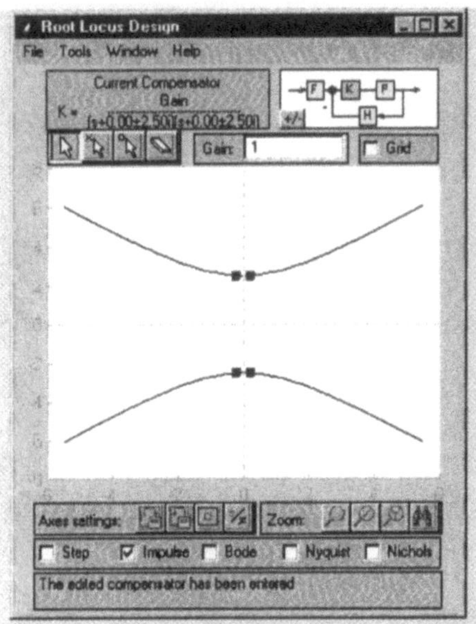
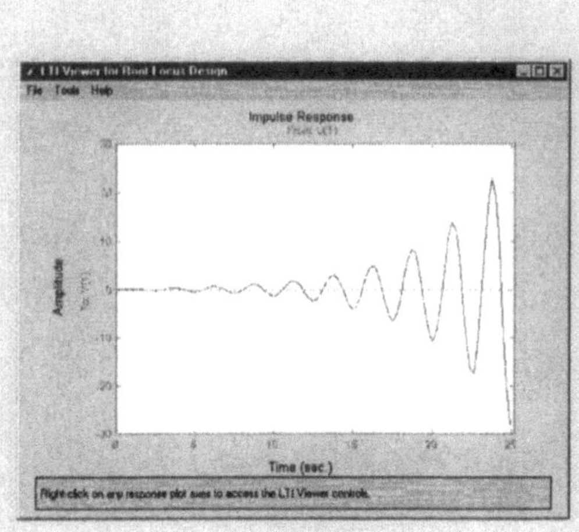

Double Poles on the Imaginary Axis *i.e.*, at s=±2.5

Let us next try out a few examples of design and compensation of LTI models for which the Root Locus Design GUI is actually designed.

Consider a system with unity feedback with the open loop transfer function as given below and design a series (cascade) compensator for it so as to increase the static velocity error constant K_v to about 5 sec^{-1} and a phase margin of about 60 degrees.

$$G(s) = \frac{K}{s(s+1)(s+2)}$$

Using the relationship,

$$K_v = s \xrightarrow[\text{}]{\lim} 0 \; sG_{(s)} = s \xrightarrow[\text{}]{\lim} 0 \frac{K}{s(s+1)(s+2)}$$

substituting $K_v = 5$,
we get the value of K, as K=10

Now enter the zpk([ ],[0 -1 -2], 10) in the P text box of Import Model Window of Root Locus Design GUI. The resultant Root Locus for the model is as shown in the forthcoming Figures.

Try adjuting the gain of the compensator by holding the closed loop poles represented by red squares by holding down the right mouse button (when the mouse cursor changes to a hand) and dragging the mouse so as to shift the poles to a new location where you release the mouse.

Observe that the value of gain changes in the gain window as you drag the mouse. You can also enter the value of the gain by typing into the gain text box directly.

Click the check box against Bode on the GUI window to get the Bode plot for the system. From the right click menu select the characteristics menu and margin sub-menu which marks the phase margin and gain margin for the system. Holding the mouse on these marks will display the values. Go on varying the value of the gain until you get the desired phase margin, which in this case is 60 degrees for the closed loop system at a gain of 0.0777.

Thus, connecting a series (or cascade) compensator with 0.0777 gain would result into desired system.

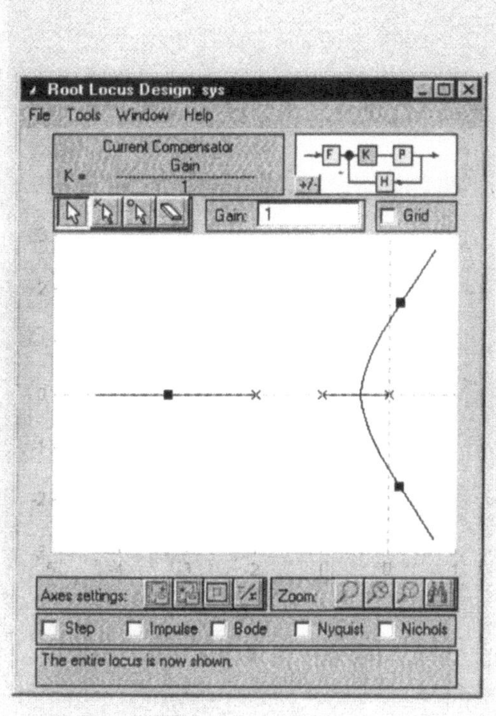

Root locus plot of the model

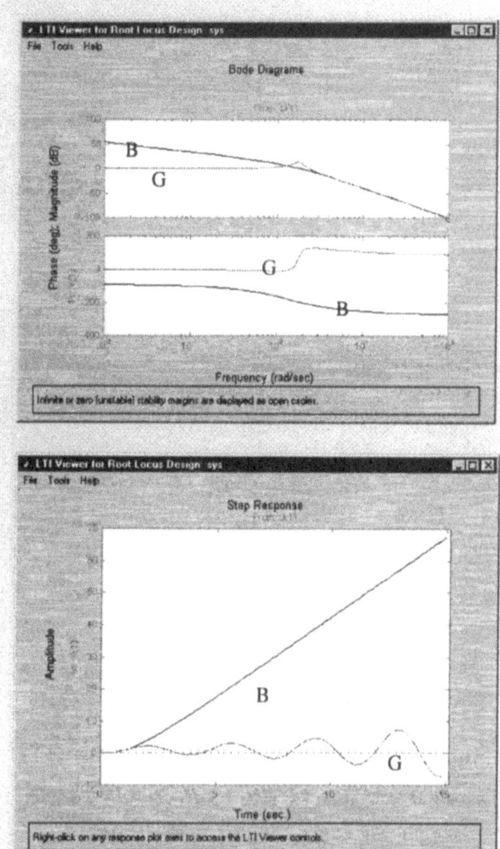

Bode plot and Step Response of the model
- open loop (blue, solid)
- closed loop (green, solid)

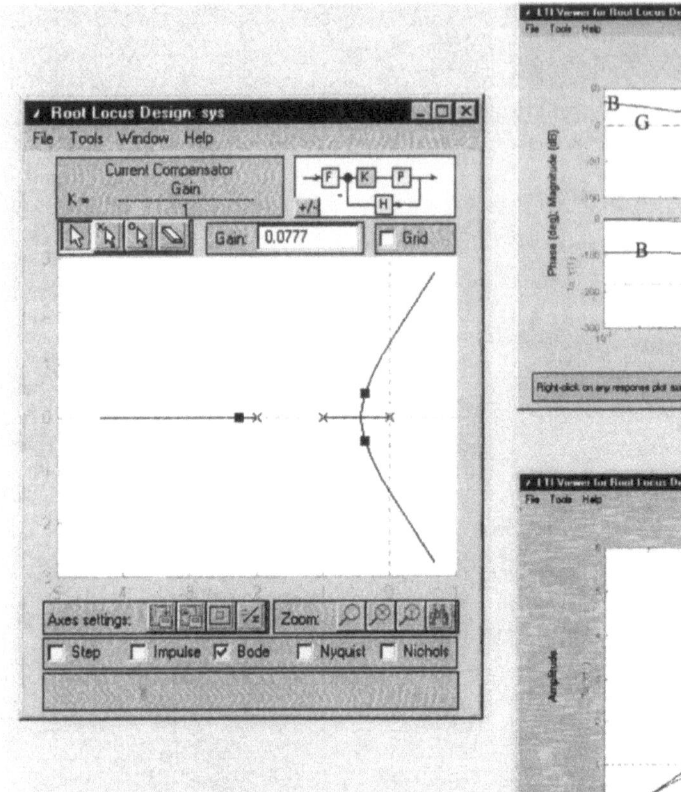

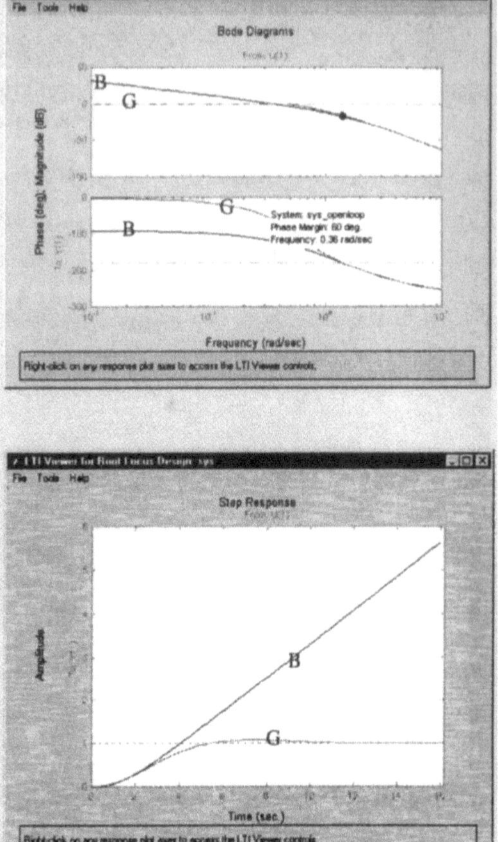

Root locus plot of the model after adjusting the gain

Bode plot and Step Response for the model after adjusting the gain
- open loop (blue, solid)
- closed loop (green, solid)

Get a SIMULINK® block diagram for the complete system. Observe the output response. Compare this output response to the response of the original system. You will see that after increasing the gain, the systems response has improved.

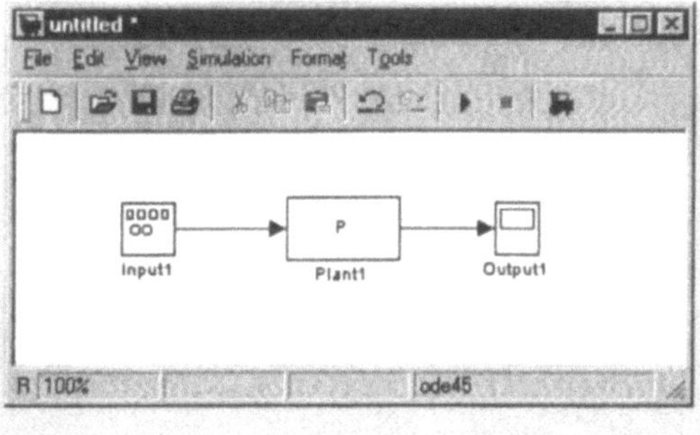

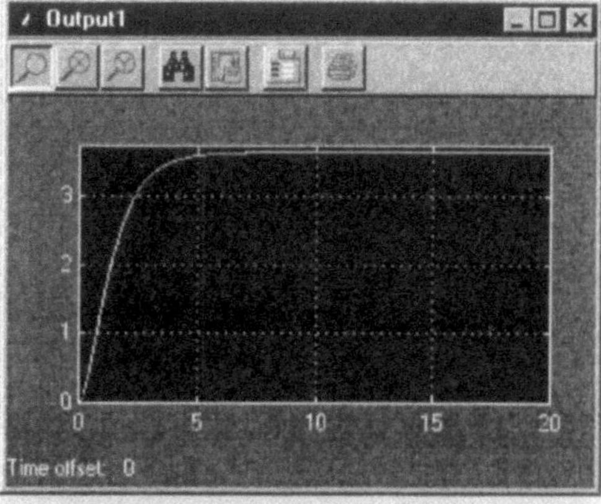

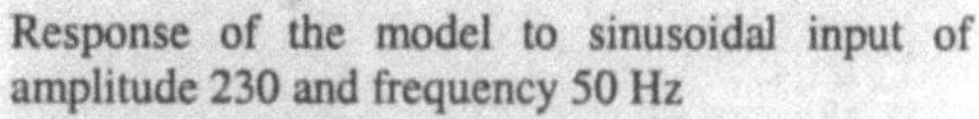

SIMULINK® diagram for the model

Response of the model to sinusoidal input of amplitude 230 and frequency 50 Hz

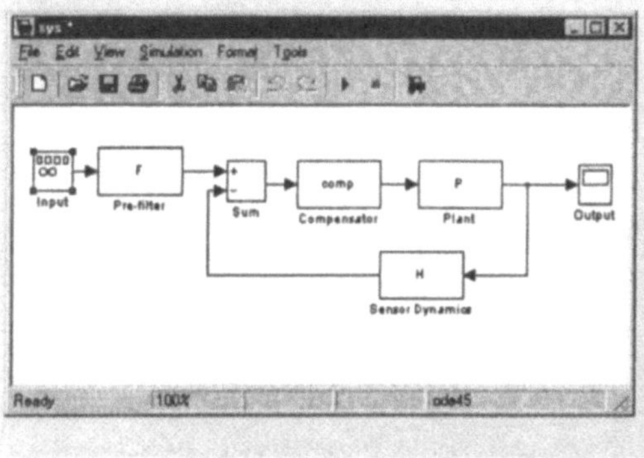

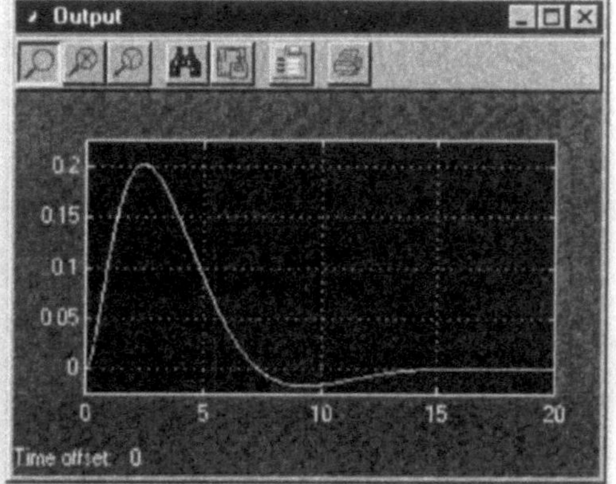

SIMULINK® diagram for the compensated model obtained by clicking Draw SIMULINK® Diagram sub-menu from the File menu.

Response of the compensated model to sinusoidal input of amplitude 230 and frequency 50 Hz. Notice that the response has improved.

The results are summarised in the table below for different values of gain parameters:

Gain = 1:

		Open Loop	Closed Loop
Bode Plot	**Peak Response**	54 dB at 0.01 rad/sec	14 dB at 1.72 rad/sec
	Gain Margin	unstable closed loop	unstable closed loop
	Phase Margin	unstable closed loop	unstable closed loop
Step Response	**Peak Response**	67.5 at 16 sec	-7.18 at 14.9 sec
	Settling Time	15	14.9
	Rise Time	-	0.596
	Steady State	dc gain 0	dc gain 1

Gain = 0.0777:

		Open Loop	Closed Loop
Bode Plot	**Peak Response**	31.8 dB at 0.01 rad/sec	0.151 dB at 0.281 rad/sec
	Gain Margin	17.8 dB at 1.41 rad/sec	16.5 dB at 1.41 rad/sec
	Phase Margin	120° at 0.359 rad/sec	60° at 0.36 rad/sec
Step Response	**Peak Response**	5.63 at 16 sec	1.08 at 7.41 sec
	Settling Time	16	10.6
	Rise Time	-	3.48
	Steady State	dc gain 0	dc gain 1

Let us take up another example, say, a discrete-time system.

Consider a unity feedback Control System with open loop transfer function as

$$G_{(s)} = \frac{K}{s(s + 2)}$$

We have to design a digital Control System with series (or cascade) compensator to meet the following specifications:

- the velocity error constant $K_v \geq 5$ sec^{-1}
- peak overshoot M_p to step input $\leq 16\%$
- settling time t_s(2% tolerance) ≤ 1.5sec
- Gain Margin ≥ 5dB
- Phase Margin $\geq 30°$

Consider sampling frequency 10 times the bandwidth and Zero-order Hold for discretization.

The system is as shown in the block diagram below:

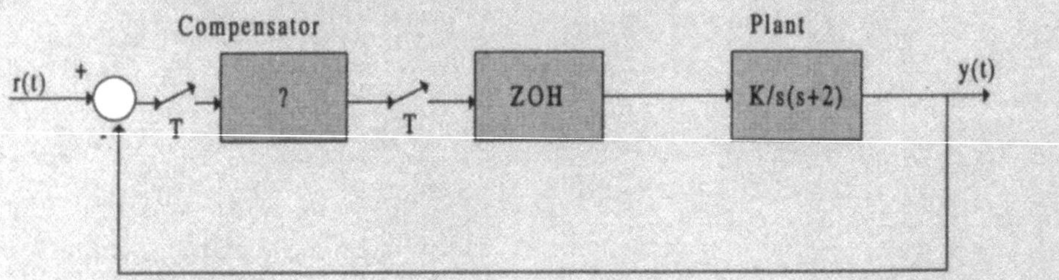

We have to design the compensator's parameters that are -- the system gain K, the compensator poles and zeros to fulfil our requirements, and the sampling time T.

Let us start with using the relationship,

$$K_v = \lim_{s \to 0} sG_{(s)} = \lim_{s \to 0} s\frac{K}{s(s + 2)}$$

substituting $K_v = 5$,
we get the value of K, as K=10

We next calculate some parameters as follows:
damping ratio,$\xi \cong 0.5$
natural frequency, $w_n \cong 5.3$ rad/sec
bandwidth, $w_b \cong 6.7$ rad/sec

Thus,
considering, sampling frequency = 10 times w_b

sampling time = $\dfrac{2\pi}{10w_b} \cong 0.1$ sec

use the relationships:

$$M_p = \exp^{\dfrac{-\pi\xi}{\sqrt{1-\xi^2}}}$$

$$t_s = \dfrac{4}{\xi w_n} \quad (\text{for 2\% tolerance})$$

$$W_b = w_n \sqrt{1 - 2\xi^2 + \sqrt{(2 - 4\xi^2 + 4\xi^4)}}$$

Now, open the Root Locus Design GUI and
- in the Import Model Window enter **zpk**([ ], [0 -2], 10) in the text box against P
- convert the model to discrete using the Convert Model/Compensator Window entering the sampling time as 0.1
- obtain the step response and Bode plot for the system for open loop (blue, solid line) and closed loop (green, solid line) and observe the characteristic values. These are shown in the forthcoming Figures and Tables. It is obvious that the system does not meet our design requirements.
- next, we design the compensator by intuitive guessing so as to meet our specifications by placing the compensator poles and zeros and adjusting the gain. We observe that the following compensator meets our requirements:

$$\frac{5.5(z - 0.74)}{(z - 0.25)}$$

- obtain the step response and the Bode plot of the system with compensator and observe the characteristics on the plots. These are shown in the figures and tables given below. Note that our design specifications are now met.
- obtain the SIMULINK® diagram for open loop as well as closed loop system and observe the response.

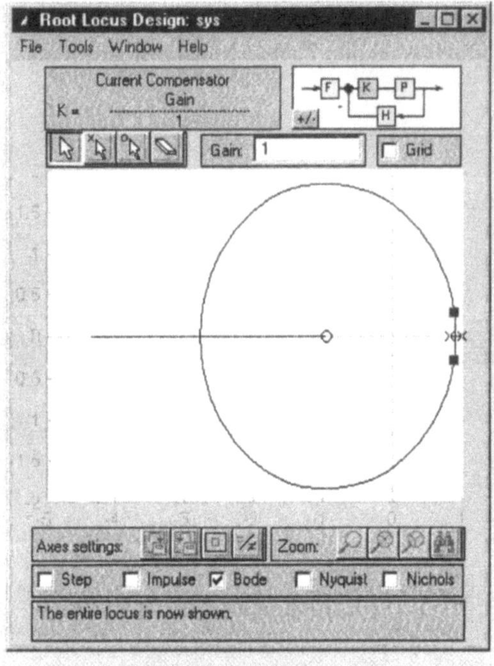

root locus plot for the model

zoomed root locus plot for the model

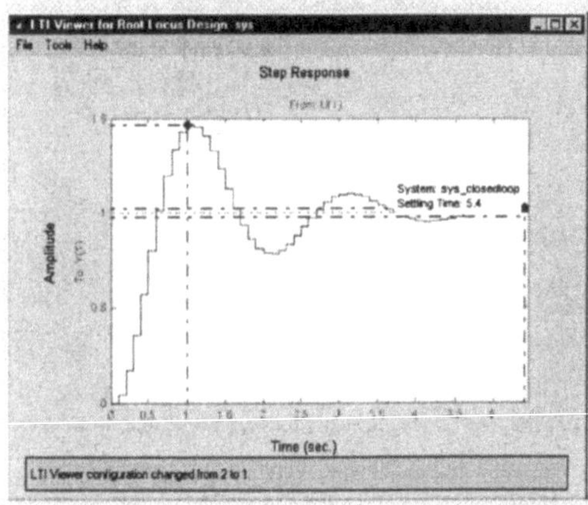

step response of the closed loop system for the model

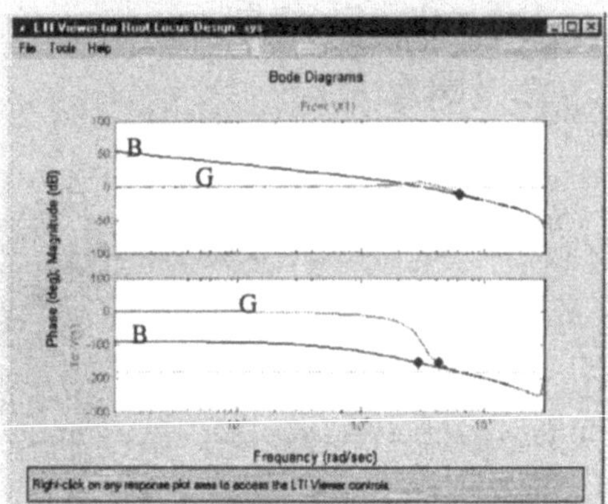

Bode plot of the closed loop (green, solid) and open loop (blue, solid) system for the model

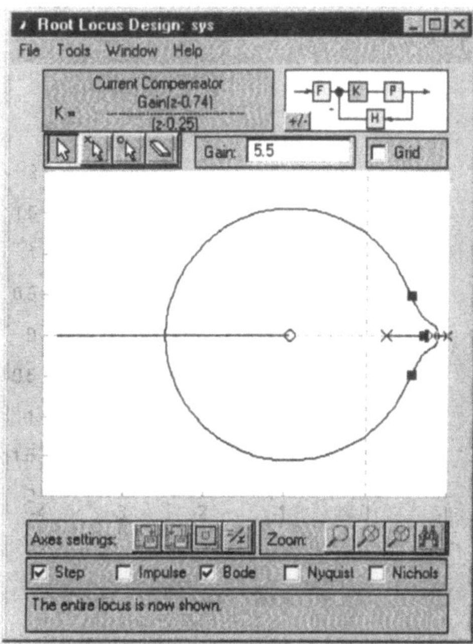

root locus plot for the compensated model

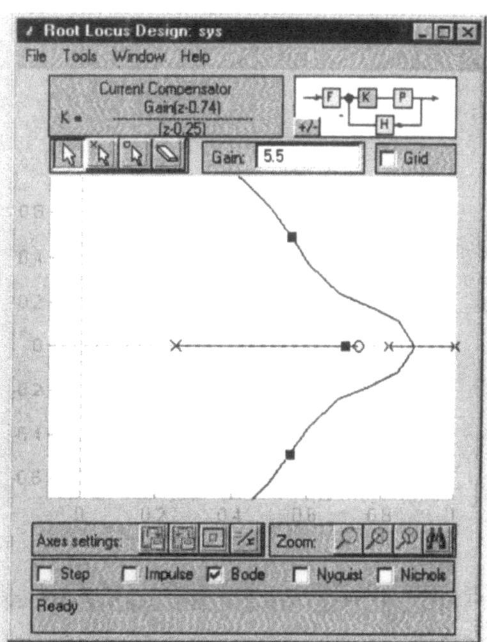

enlarged portion of the root locus plot for the compensated model obtained by zooming

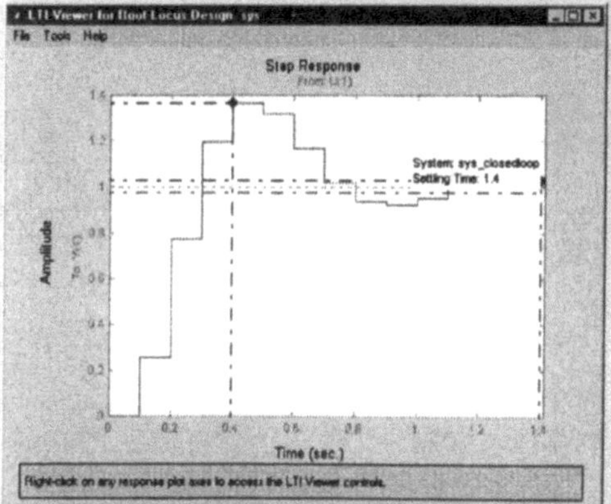

step response of the closed loop system for the compensated model

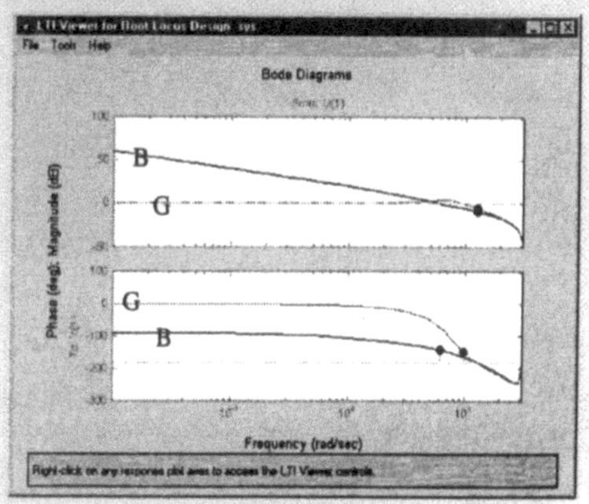

Bode plot of the closed loop (green, solid) and open loop (blue, solid) system for the compensated model

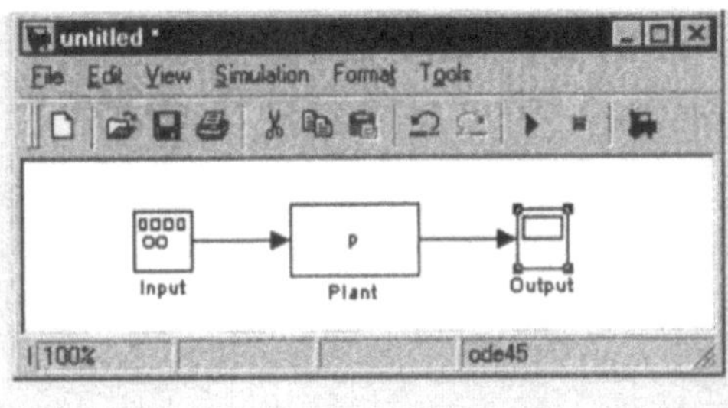

SIMULINK® diagram for the model (without feedback)

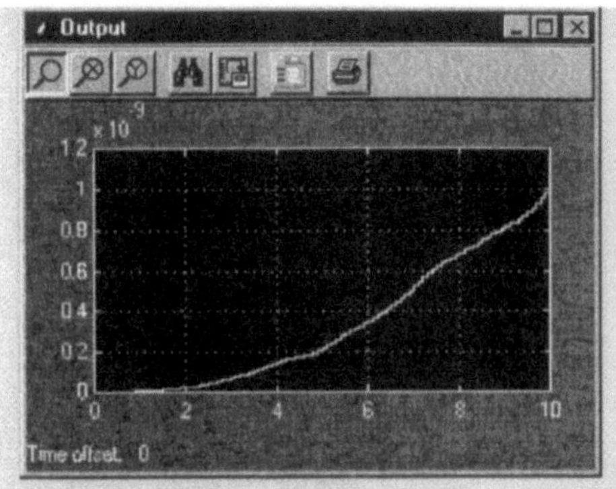

Response of the model (without feedback) to sinusoidal input of amplitude 230 and frequency 50 Hz

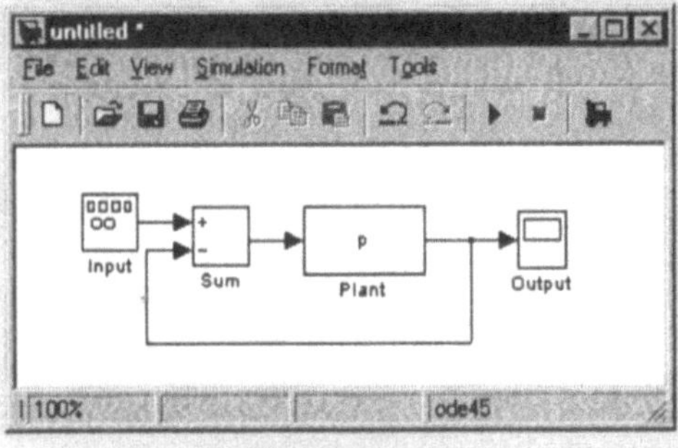

SIMULINK® diagram for the model (with feedback)

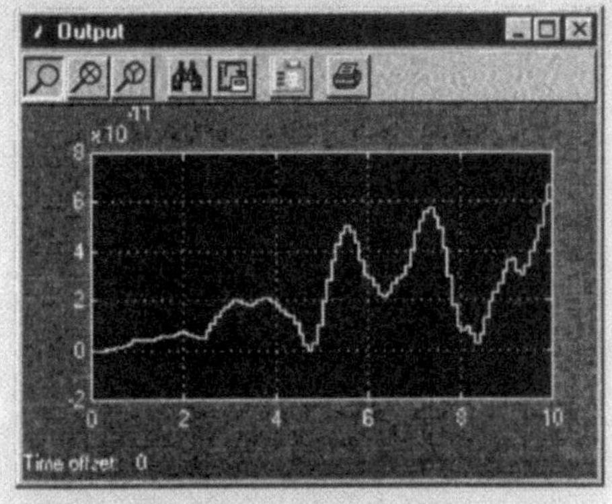

Response of the model (with feedback) to sinusoidal input of amplitude 230 and frequency 50 Hz

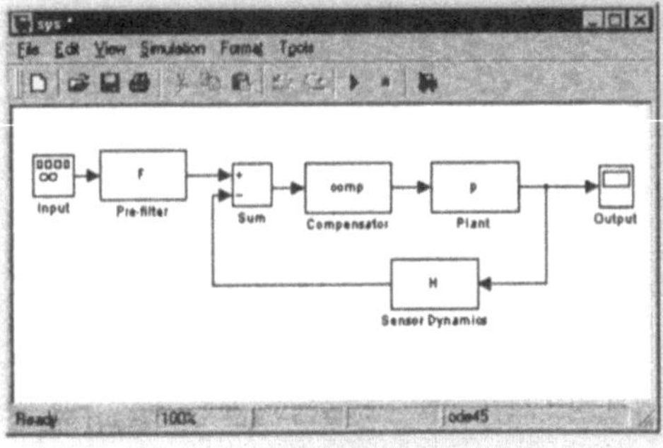

SIMULINK® diagram for the compensated model obtained by clicking Draw SIMULINK® Diagram sub-menu from the File menu.

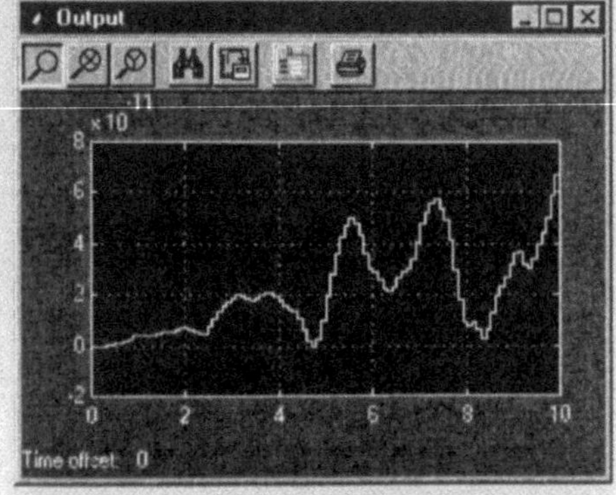

Response of the compensated model to sinusoidal input of amplitude 230 and frequency 50 Hz. Notice that the response has improved.

The results are summarised in the tables given below:

Uncompensated Model:

		Open Loop	Closed Loop
Bode Plot	Peak Response	54 dB at 0.01 rad/sec	6.74 dB at 2.96 rad/sec
	Gain Margin	9.93dB at 6.22 rad/sec	12.3 dB at 6.22 rad/sec
	Phase Margin	26.8° at 2.86 rad/sec	27.4° at 4.17 rad/sec
Step Response	Peak Response	37.5 at 8 sec	1.46 at 1 sec
	Settling Time	8	5.4
	Rise Time	-	0.4
	Steady State	dc gain 0	dc gain 1

Compensated Model:

		Open Loop	Closed Loop
Bode Plot	Peak Response	59.6 dB at 0.01 rad/sec	3.84 dB at 6.87 rad/sec
	Gain Margin	9.4 dB at 12.8 rad/sec	5.81 dB at 12.8 rad/sec
	Phase Margin	37.9° at 6.05 rad/sec	33.4° at 9.72 rad/sec
Step Response	Peak Response	45.3 at 5 sec	1.36 at 0.4 sec
	Settling Time	5	1.4
	Rise Time	-	0.2
	Steady State	dc gain 0	dc gain 1

Take up a practice test to see how much you have grasped…

Practice Test 5.2.

1. For the two examples considered above, place the compensator in the feedback path and compare the results with the ones obtained.

2. Consider a unity feedback Control System with open loop transfer function as

$$G_{(s)} = \frac{K}{s(s+5)}$$

Design a digital Control System with series (or cascade) compensator to meet the following specifications:

- the velocity error constant $K_v \geq 10$ sec^{-1}
- peak overshoot M_p to step input $\leq 25\%$
- settling time t_s(2% tolerance) ≤ 2.5sec
- Gain Margin ≥ 15dB
- Phase Margin $\geq 75°$

Consider sampling frequency 10 times the bandwidth and Zero-order Hold for discretization.

Exercise for Chapter 5:

1. Obtain all the response plots summarised in Table 5.3 for the models generated in exercise of Chapter 1 and Chapter 2 using LTI Viewer.

2. For the models and their responses obtained in Exercise 1 given above, display 1, 2, 3, 4, 5 and upto 6 response plots simultaneously at a time in the same LTI Viewer window.

3. For the models considered in Exercise 1 above, display any one type of response plot mentioned in Table 5.3 at a time for 1, 2, 3, 4, 5 and upto 6 models simultaneously. Choose these models on your own. Repeat this for other types of response plots mentioned in Table 5.3.

4. Obtain the root locus plot using Root Locus Design GUI for all the SISO models you have come across in this book so far.

5. For the models considered in Exercise 4 above, obtain the root locus plot using Root Locus Design GUI considering

   - unit positive feedback
   - unit negative feedback

Chapter 6

Control System Design through SIMULINK®

So far, you had been using the MATLAB® environment to solve your System Engineering problems. In this chapter, you will learn to solve the same problems and some more using the interactive environment of SIMULINK®.

We are all familiar with SIMULINK®, which is a software built atop MATLAB® environment for modeling, simulating and analysing dynamic systems. As we have seen in the previous chapters, systems with different sample times could not be operated upon together in MATLAB® environment. This shortcoming of Control System Toolbox of MATLAB® has been overcome in SIMULINK®, which supports multirate systems *i.e.,* different parts of a system can be sampled or updated at different rates. The SIMULINK® GUI provides easy click and drag-drop mouse operations for building models as block diagrams in an interactive environment. It has several libraries of ready to use blocks (see Appendix B) along with the facility for you to build your own blocks and save them in your personal library for subsequent use. Apart from these basic SIMULINK® blocks, there is a specialised library named Control System Toolbox with three blocks as shown in Table 6.1, which is especially tailor made to suit the requirements of System Engineering problems.

Table 6.1. Control System Toolbox library in SIMULINK®

Blocks	Symbol	Description
Input Point	>-+↓-> Input Point	This is the 'cstblocks/Input Point' block
LTI System	tf(1,[1 1]) LTI System	The LTI System block accepts both continuous and discrete LTI objects as defined in the Control System Toolbox. Transfer function, state-space, and zero-pole-gain formats are all supported in this block. Note: For multi-input systems, the input delays must be either all positive or all zero. Note: Initial states are only meaningful for state-space systems.
Output Point	>-◇-> Output Point	This is the 'cstblocks/Output Point' block

As this book does not intend to teach you the use of SIMULINK® and assuming that you already know how to use the blocks in the Model Window to represent a system, we will straight away proceed to use SIMULINK® to solve our System Engineering problems.

6.1 System Representation/Modeling

Representation of a System in SIMULINK® window is quite easy. SIMULINK® supports only three models -- the zpk model, the tf model and the ss model. These could be continuous-time or discrete-time models with the same or different sampling time. These models can be created by either of the following methods

- using SIMULINK® library
- using Control System Toolbox library

Note that frd model is not supported by SIMULINK®. In addition, the LTI arrays of models of any of the three types mentioned above, (*viz.,* zpk, tf and ss) are also not supported by SIMULINK®.

6.1.1 Modeling Using SIMULINK® Library

SIMULINK® library has sub-libraries titled Continuous and Discrete with eight blocks each as given in Appendix B. By setting the block parameters and by simple drag and drop action of the mouse you can use these blocks to represent the system models as zpk, tf or ss model of a continuous-time system or a discrete-time system.

The only drawback with the zpk or tf model using the blocks available in the Continuous or Discrete library is that for MIMO models the denominator (*i.e.,* the poles) should be same for all subsystem.

6.1.2 Modeling Using Control System Toolbox Library

The LTI System block available in the Control System Toolbox library can be used to represent any of the zpk, tf or ss models (barring LTI arrays of these). This block accepts

- valid expression for zpk, tf or ss model
- name of a zpk, tf or ss model loaded in the workspace
- valid expression involving name of zpk, tf or ss model loaded in the workspace

The block displays the expression in the first case and the model name or expression involving model names in second and third case respectively (Figure 6.1). There is however, no limitation for all the systems to have same denominator (*i.e.,* poles) in case of MIMO systems represented as zpk or tf model as against the case when blocks from the Continuous and Discrete libraries of SIMULINK® are used.

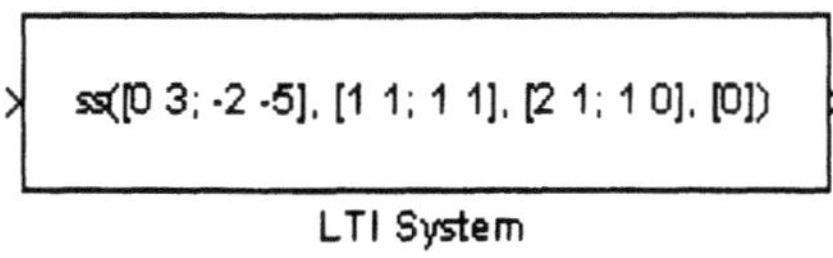

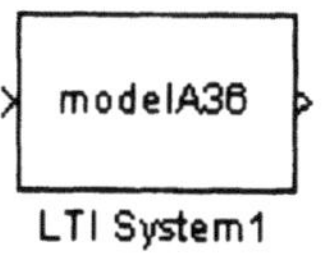

a. LTI System block using valid expression for ss model (*i.e.*, modelA36)

b. LTI System block using model name (*i.e.*, modelA36)

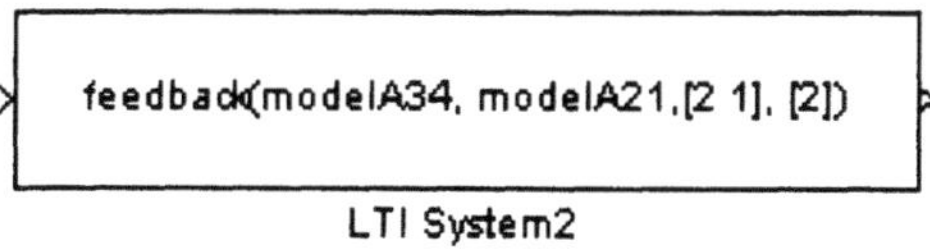

c. LTI System block using an expression containing model name (*i.e.*, **feedback**(modelA34,modelA21, [2 1],[2]))

Figure 6.1. The LTI System Block from the Control System Toolbox Library of SIMULINK®

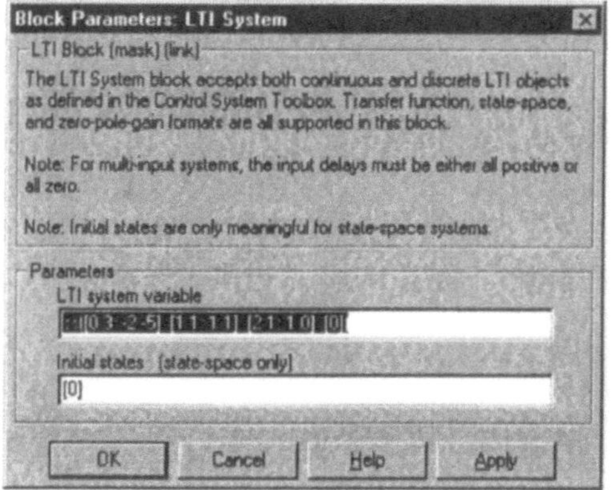

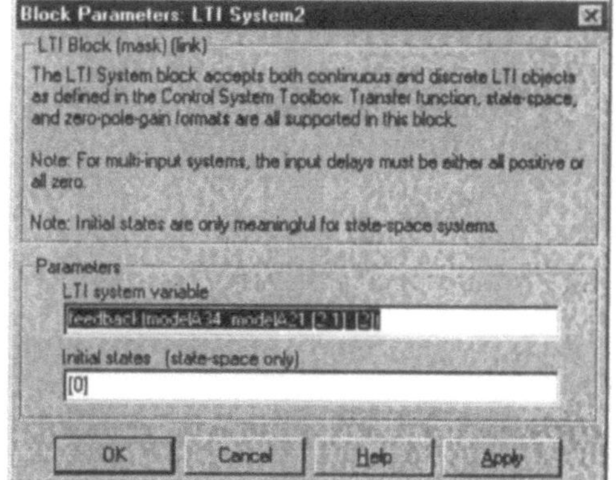

a. LTI System block indicating a valid expression for ss model

b. LTI System block indicating a valid expression involving names of model

Figure 6.2. The Block Parameters Window of LTI Syste Block of Control System Toolbox of SIMULINK®

6.2 Model Manipulation

In Chapter 2 you learnt how to use the built-in functions available with the Control System Toolbox of MATLAB® to manipulate the models representing small parts of a large system to obtain the complete representation of a bigger system. The principle of manipulation of the models remains same for SIMULINK® environment also. In the paragraphs to follow, you will learn how to do this.

6.2.1 Arithmetic Operations on the Models

SIMULINK® supports all arithmetic operations on the models that are mentioned in Table 2.2 of Chapter 2. You can perform arithmetic operations on models in several ways:

- you can perform arithmetic operation on the models using techniques you learnt in Chapter 2 and use the resulting model in SIMULINK® window.
- you can connect the individual models in SIMULINK® window to obtain equivalent representation of the system.
- you can straight away write the expression for arithmetic operation in the LTI system parameter box in the same way as you would do at the MATLAB® command prompt.

Going through examples considered below will further reinforce these concepts into your minds.

Example:

Let us first perform addition of models by all of the three ways explained above:

- by using resulting model of addition of modelA1 and modelA17 *i.e.,* modelB1

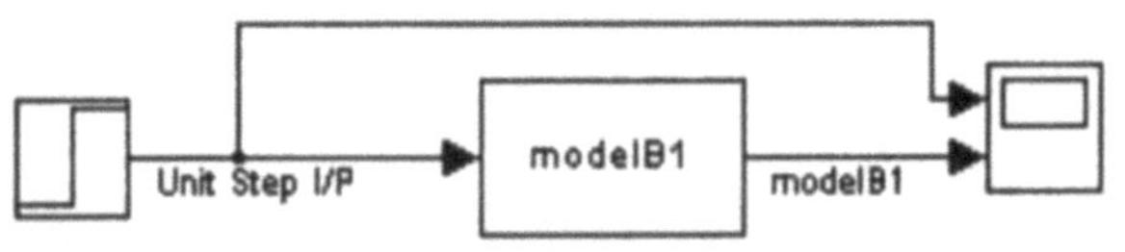

- by using MATLAB® expression for addition of modelA1 and modelA17

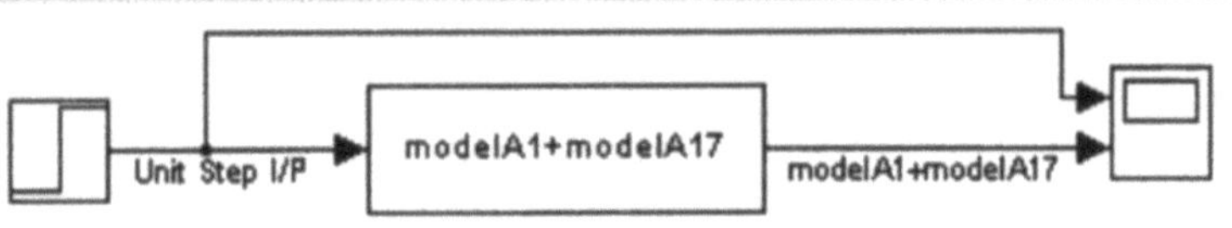

- by using equivalent representation for addition of modelA1 and modelA17

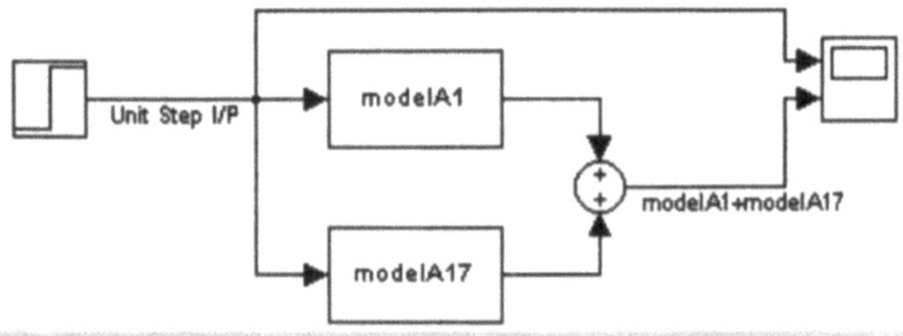

In all the three cases the result obtained for unit step input to the system is the same as displayed by the scope.

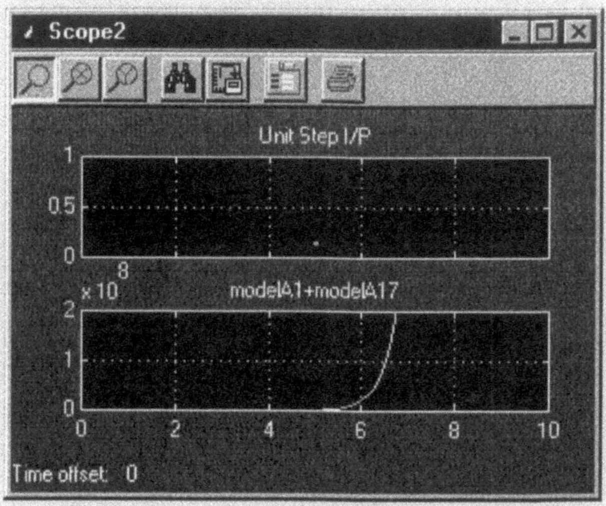

Hope you have understood how to perform arithmetic operations on the models in SIMULINK® environment. Better check if you really have...

Practice Test 6.1.

1. Try obtaining all the results of arithmetic operations you performed on models in Chapter 2 in SIMULINK® environment by all the methods you learnt recently.

6.2.2 Interconnection of Models

The procedure to be followed for interconnection of models in SIMULINK® window is the same as the one followed above for performing arithmetic operation on the models. Interconnection of models may be done in any of the following ways:

- using model from MATLAB® workspace obtained as a result of interconnection techniques you learnt in Chapter 2
- drawing interconnecting lines between the individual models with mouse
- straight away writing the expression for interconnection of models in the LTI system parameter box in the same way as you would do at the MATLAB® command prompt.

Following illustrations will make things more clear.

Example:

Let us obtain the Linear Fractional Transformation (lft) connection for modelA14 and modelA31

- by using resulting model of lft connection of modelA14 and modelA31 *i.e.*, modelB38

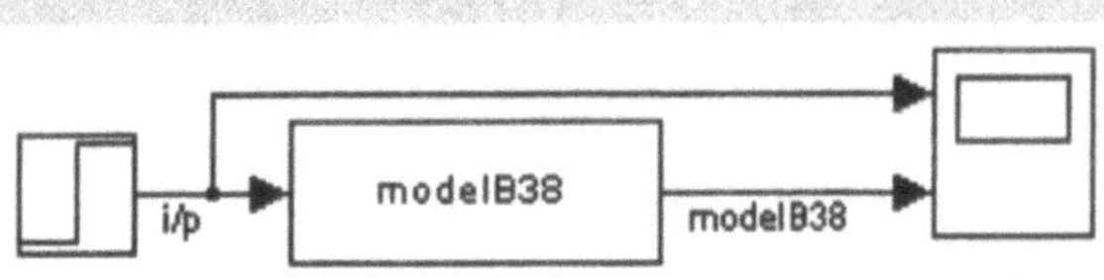

- by using equivalent representation for lft connection of modelA14 and modelA31

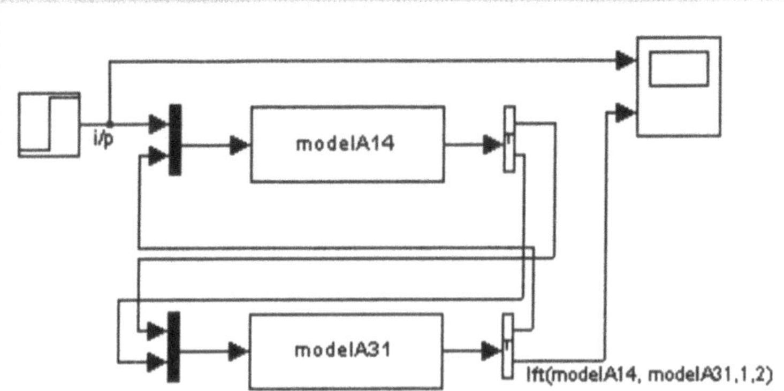

- by using MATLAB® expression for lft connection of modelA14 and modelA31

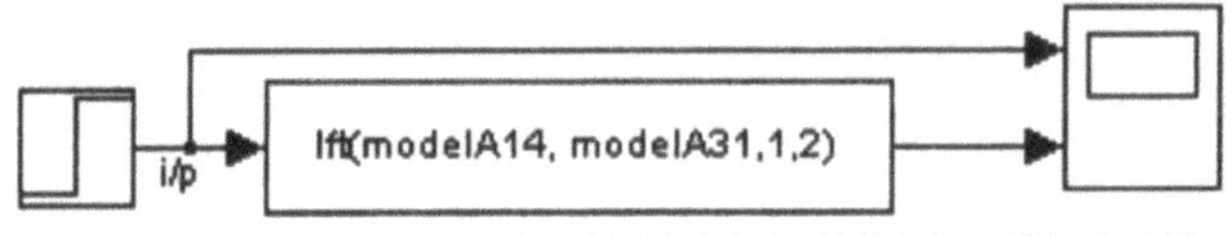

However, in all of the three cases the result obtained for unit step input to the system is the same as displayed by the scope:

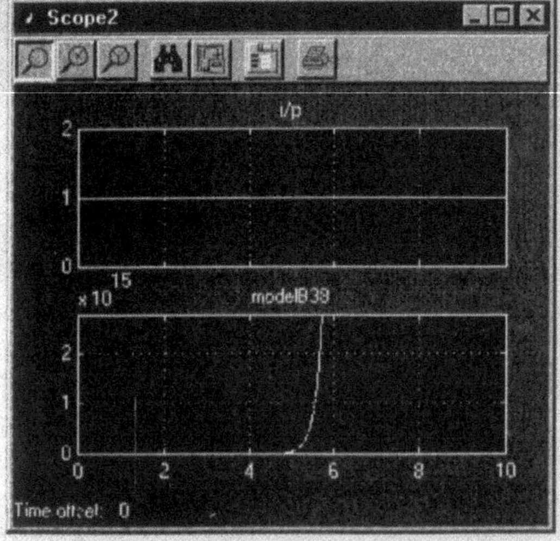

Next, let us obtain the feedback connection for modelA34 and modelA21,

- by using resulting model of feedback connection for modelA34 and modelA21 *i.e.,* modelB37

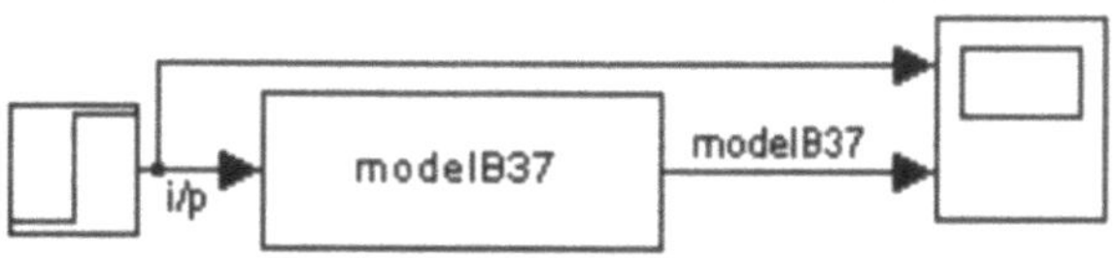

- by using equivalent representation for feedback connection for modelA34 and modelA21

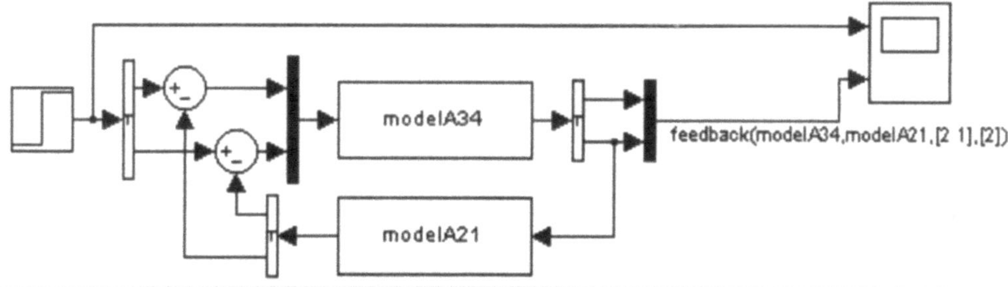

- by using MATLAB® expression for feedback connection for modelA34 and modelA21

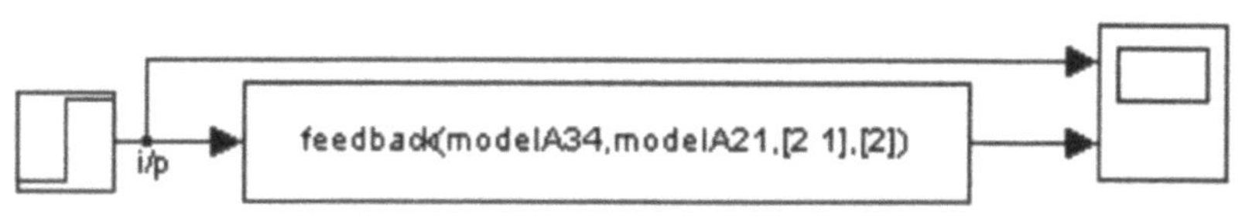

However, in all of the three cases the result obtained for unit step input to the system is the same as displayed by the scope:

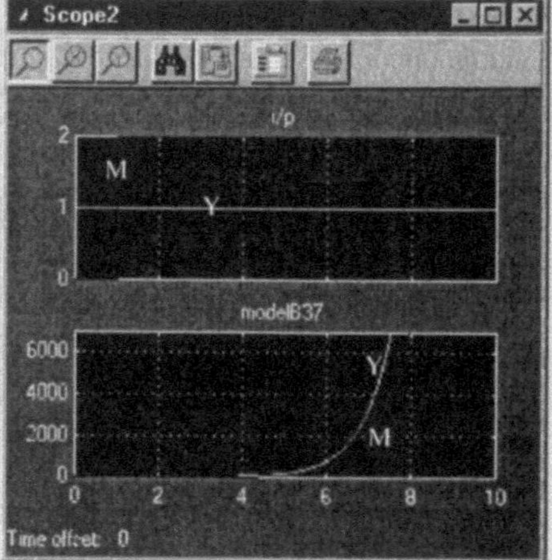

Check if you have really learnt the tricks of the trade…

Practice Test 6.2.

1. Try obtaining all the results of interconnection of models you did in Chapter 2 in SIMULINK® environment by all the methods you learnt recently.

6.3 Model Analysis

The next step after creating models, operating upon them and interconnecting them is to analyse the response of the system to some input. As you have already seen in the examples, you just went through, this is done quite easily by connecting the desired input from the Sources library to the system. The response of the system as a whole or at any point on the connecting lines can be observed on any of the blocks available in the Sinks library. However, the response so obtained is obviously the time response of the system. To obtain frequency response of the system, you have to take refuge to the SIMULINK® LTI Viewer, which is a GUI similar to the LTI Viewer you read about in Chapter 5. The SIMULINK® LTI Viewer can also be used to obtain the time response for the system.

6.4 The SIMULINK® LTI Viewer

The Control System Toolbox of MATLAB® provides you with yet another interactive GUI, to be used with SIMULINK®, similar in features to the LTI Viewer, you learnt to use in the previous chapter. This GUI which, you call as the SIMULINK® LTI Viewer can be used for performing the linear analysis like viewing, manipulating and analysing the response of the SIMULINK® models on any portion between two points. In the paragraphs to follow, you will learn how to invoke and use the SIMULINK® LTI Viewer.

6.4.1 Invoking and Using the SIMULINK® LTI Viewer

The steps involved in invoking and using the SIMULINK® LTI Viewer for analysis of a SIMULINK® model are

i. Load a model into the SIMULINK® window by either

   - opening an existing model in SIMULINK® window which was earlier saved as file with **.mdl** extension
   - creating a new model in a new window
   - changing an already existing model in an already existing window obtained from either of the two methods above.

ii. Click on the tools menu in the menu bar. A drop down sub-menu block appears as shown in Figure 6.3. Select the **Linear Analysis** sub-menu from this drop down menu.

iii. Two new windows open as follows:

- a SIMULINK® library window called **Model_Inputs_and_Outputs** having two blocks --
  Input Point and **Output Point** as shown in Figure 6.5. Note that these two blocks are the
  same as the ones existing in the Control System Toolbox library in SIMULINK® Library
  browser.
- an **LTI Viewer window**, as shown in Figure 6.6, with which you are already familiar.
  Note that the title bar displays the name of the SIMULINK® model to which it is linked and
  an additional menu called SIMULINK® appears in the menu bar with three submenus as
  described in Table 6.2.

Table 6.2. Main menu of SIMULINK® LTI Viewer

Sub-menu Item	Description
Get Lineariszed Model	Linearizes the SIMULINK® model portion which is to be analysed and imports the resulting linearized model into the LTI Viewer. Note that every time you select this menu item, the whole process of linearization is repeated and a new linearized analysis model is added to the SIMULINK® LTI Viewer workspace.
Set Operating Point	Opens the **Operating Point** window captioned the same also indicating the name of the SIMULINK® model to which it is linked, through which you can set or reset the operating conditions (Figure 6.8). It allows you to linearize states about: • Initial states in SIMULINK® diagram (default) • Zero state values • User-defined state values You can choose any of the above by clicking the option button placed next to them. The names of the models being analysed are displayed. Against this, a textbox called value is provided which is active when user defined 'state values' option is chosen into which you can enter the state values.
Remove Input/Output Points	Clears all the Input Point and the Output point blocks from the SIMULINK® link model in the SIMULINK® window.

iv. Specify the portion of the SIMULINK® model you want to analyse by placing the Input Point
and the Output Point block from the Model_Inputs_and_Outputs window at the desired
locations. The Input Point and Output Point blocks when dragged and dropped at the desired
location automatically get connected. More than one pair of Input Point and Output Point
blocks can be placed on one model which, could also be at different hierarchical levels of the
SIMULINK® model. In case, these Input Point and Output Point blocks are desired to be
removed, remove them one by one by selecting them and pressing the delete button on the
keyboard or all at once by selecting the Remove Input/Output Points sub-menu from
SIMULINK® main menu of the SIMULINK® LTI Viewer.
v. From the SIMULINK® LTI Viewer window, click the SIMULINK® main menu and select the
Set Operating Point sub-menu. Operating Point window displaying the name of the model
to which it is linked is shown in Figure 6.8. Through this window, you can set the state about
which you wish your model to be linearized.
vi. From the SIMULINK® main menu of the LTI Viewer, select **Get Linearized Model**. Step
Response of the portion of the SIMULINK® model is displayed in the SIMULINK® LTI Viewer
window. By the familiar method of right click menu, which is exactly same as for the LTI

viewer window you learnt about in Chapter 5, you can obtain other plots for the model. The rest of the procedure for obtaining different results with the SIMULINK® LTI Viewer and saving the model is the same as that with the LTI viewer window about which you read in Chapter 5 and about which you already know.

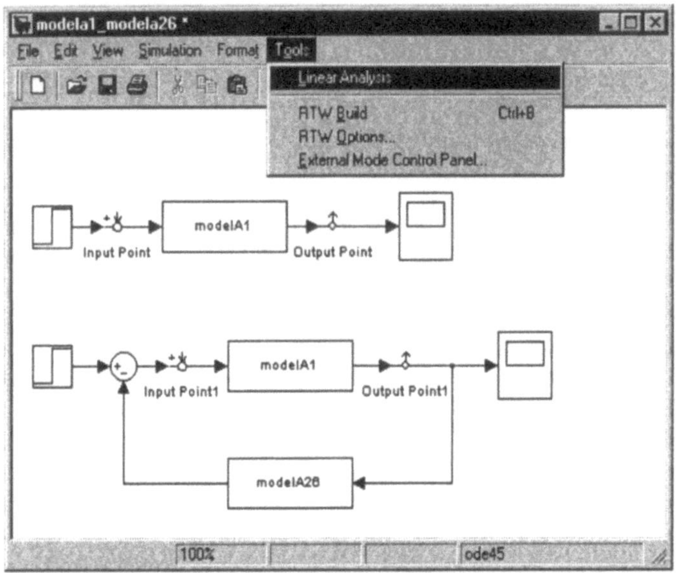

Figure 6.3. SIMULINK® Model Window showing the sub-menu items of the Tools main menu

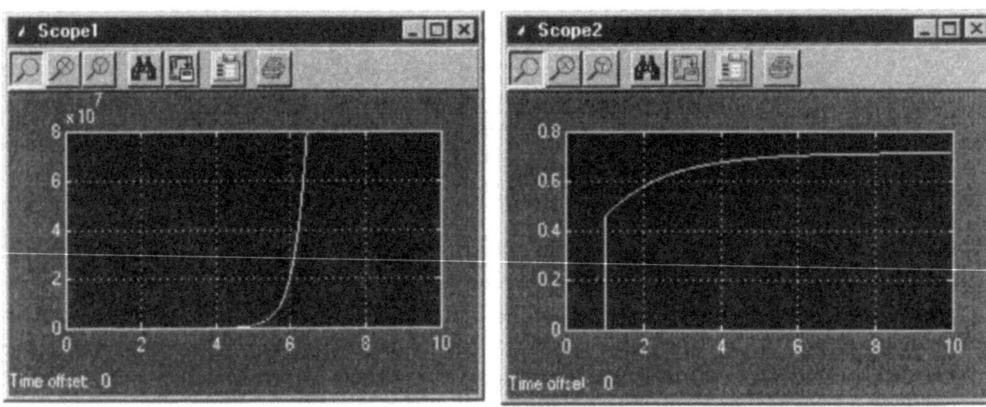

a. Time response of open-loop model shown in Figure 6.3

b. Time response of closed-loop model shown in Figure 6.3

Figure 6.4. Time response of SIMULINK® Model shown in Figure 6.3

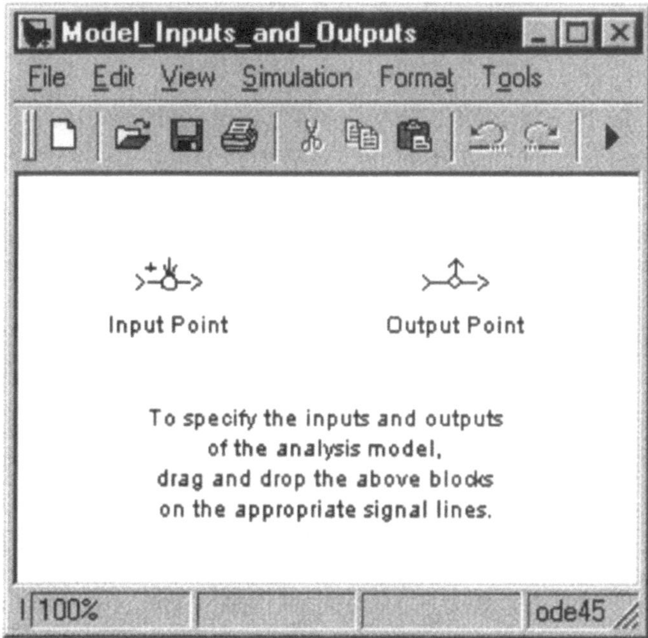

Figure 6.5. Model_Inputs_and_Outputs Window

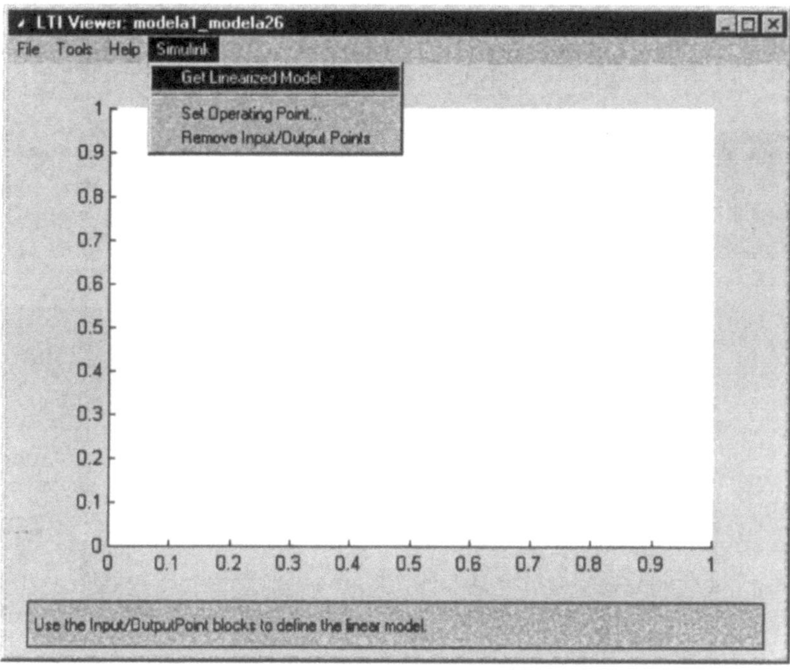

Figure 6.6. SIMULINK® LTI Viewer Window showing the sub-menu items of the SIMULINK® main menu

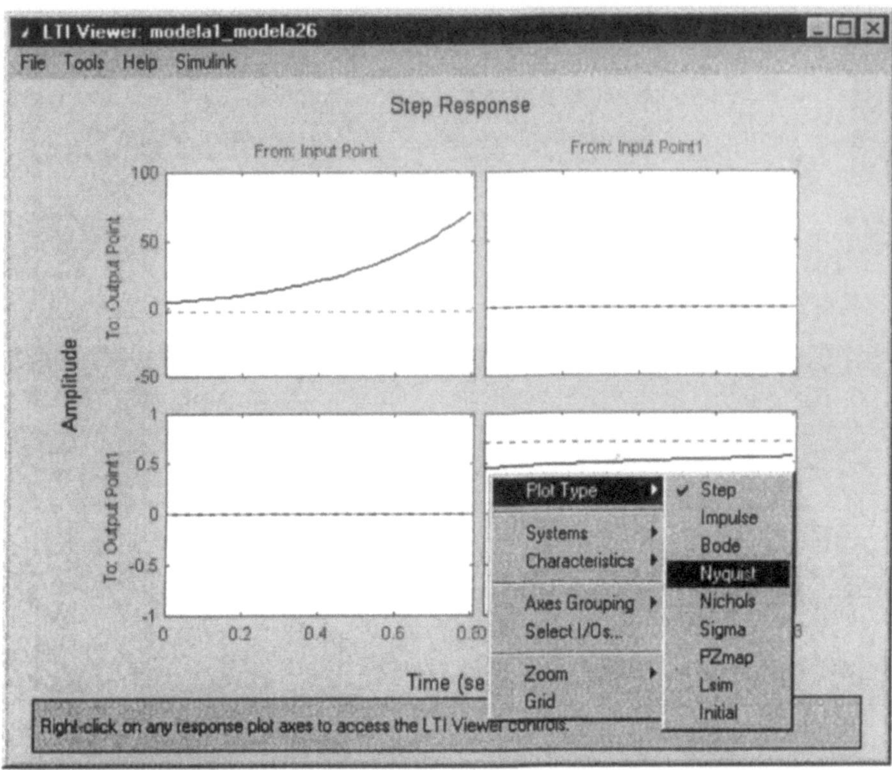

Figure 6.7. SIMULINK® LTI Viewer Window showing step response of the system shown in Figure 6.3

Figure 6.8. The Operating Point Window

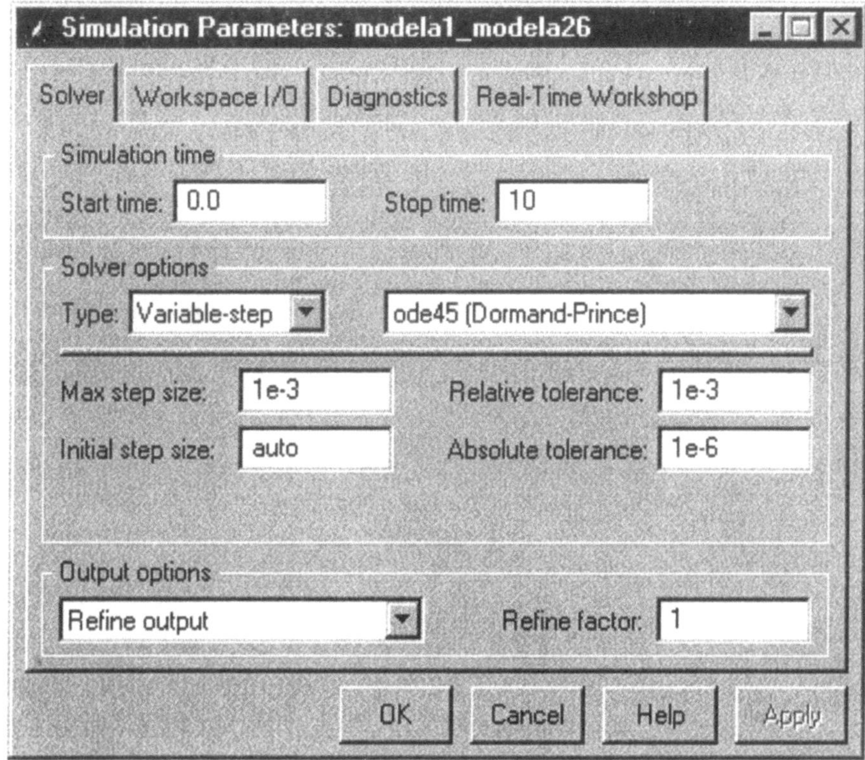

Figure 6.9. The Simulation Parameters Window

6.5 A Few Words of Caution

While working with SIMULINK® and SIMULINK® LTI Viewer, sometimes you may not get the expected results especially the waveforms. Here are a few tips that may prove to be of help to you in such trying moments:

- The sinusoidal waveform may appear to be triangular and so on. To fix this problem, open the Simulation Parameters window by clicking the Simulation main menu on the title bar of SIMULINK® window and then selecting **Parameters...**, sub-menu (Figure 6.9), try with different solver options setting suitable values for step sizes and tolerances till you obtain the desired result. In this book, the examples have been solved keeping the values of different parameters of the Parameter Window typically as shown in Figure 6.9 to produce sufficiently acceptable results.
- While selecting the Get Linearized Model sub-menu from the SIMULINK® main menu of the SIMULINK® LTI Viewer, keep in mind that a new model is added to the SIMULINK® workspace and the analysis is started afresh each time you select this.
- While analysing a closed loop model like the one shown in SIMULINK® window of Figure 6.3, you may place the Input_Point and Output_Point blocks as shown and expect the result to be that of a model in open loop. However, comparing this with the result you obtained for open loop by removing the closed loop links you observe that the two results

are not the same (observe the results in Figure 6.4 and Figure 6.7). Hence, be cautious while dealing with the closed loop system, in case you desire to analyse the model independently, remove the closing link paths.

Let us try solving few examples using SIMULINK®, some of which have already been solved in earlier chapters.

Example:

We are already familiar that as the value of damping factor varies, the response of a second-order system changes. We will model a system called zvariation with different values of damping factor and observe the response of the system to step input.

In the figures to follow, a subsystem is modeled with parameters as shown in the following Figure 6.10 and its response is observed on the scope. The response is again verified through SIMULINK® LTI Viewer.

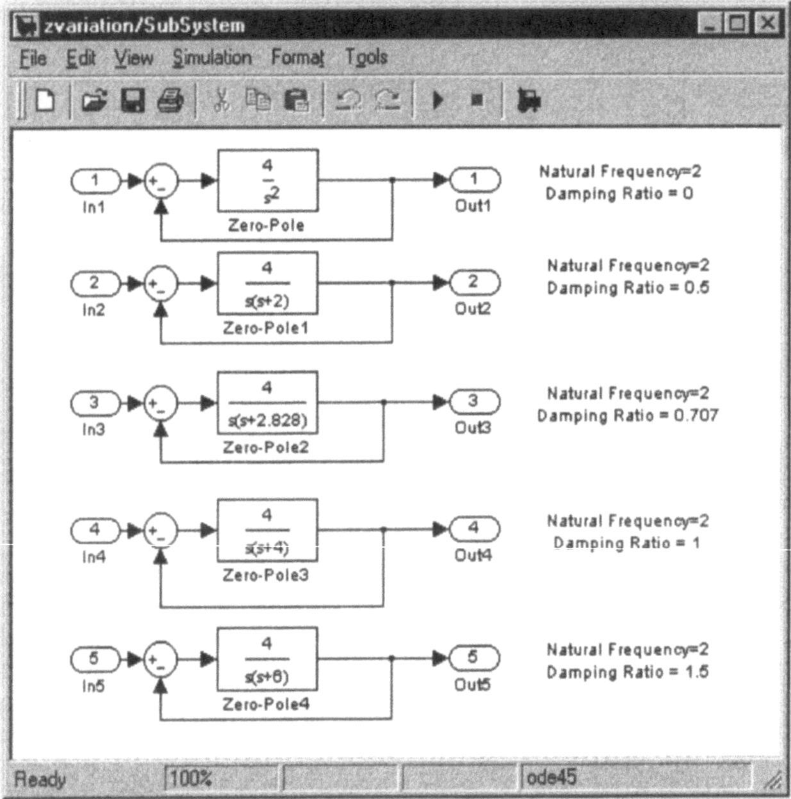

Figure 6.10. Subsystem for variation of damping factor for a second-order system

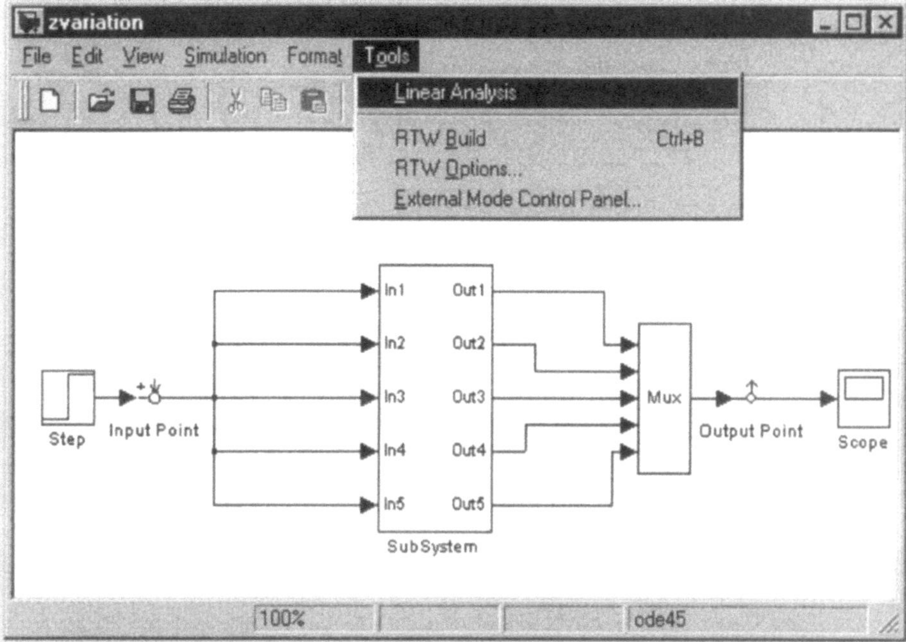

Figure 6.11. System for variation of damping factor for a second-order system

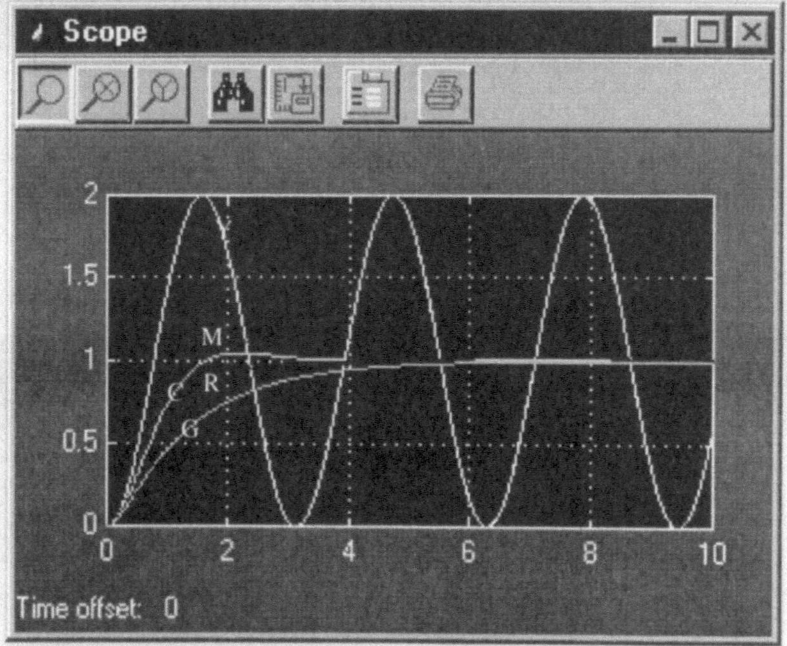

Figure 6.12. Response of the system to step input for variation of damping factor of a second-order system

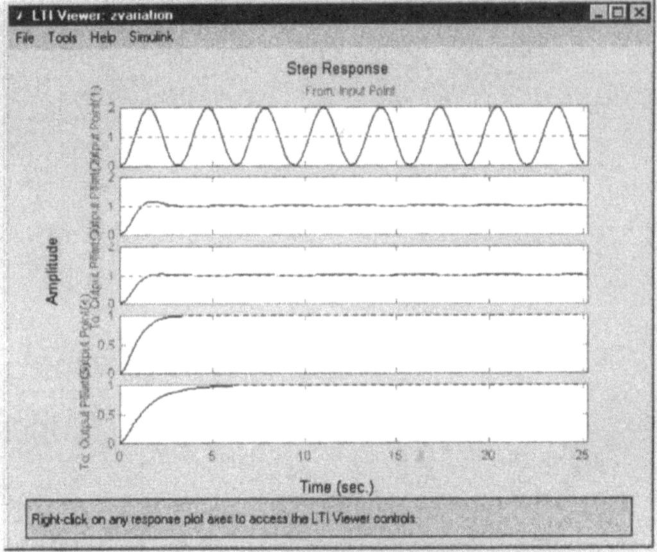

Figure 6.13. Response of the system on SIMULINK® LTI Viewer for variation of damping factor for a second-order system to step input, with axes ungrouped

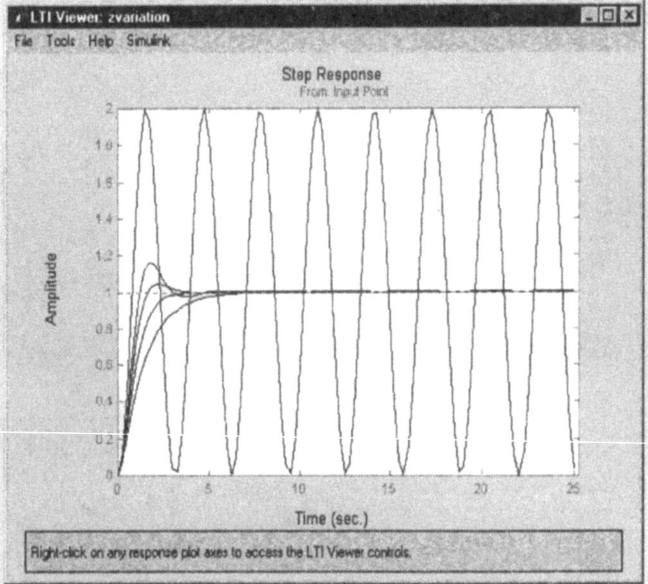

Figure 6.14. Response of the system on SIMULINK® LTI Viewer for variation of damping factor for a second-order system to step input, with axes grouped

Let us next model the system which, we have already analysed in Chapter 4 as an example of Article 4.2.2 using SIMULINK® and observe its response to sinusoidal input on scope.

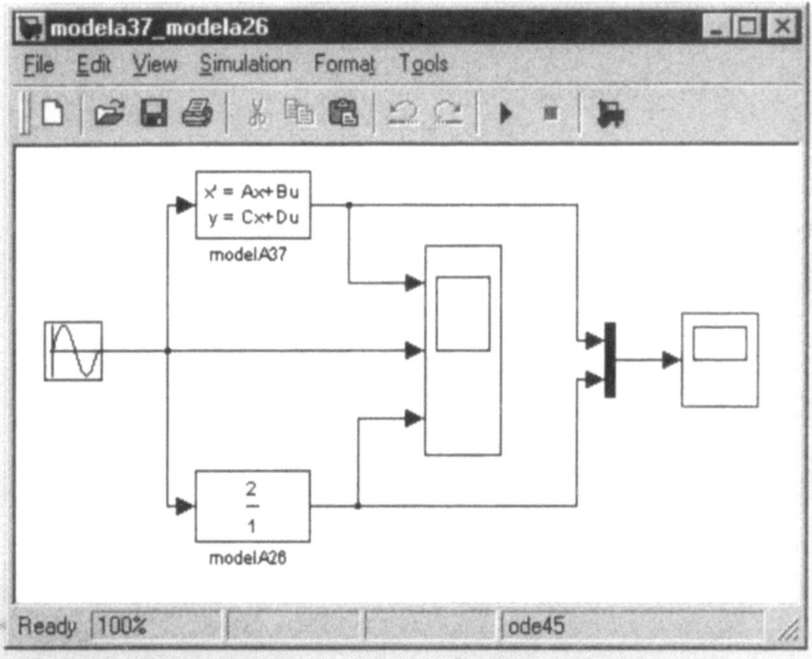

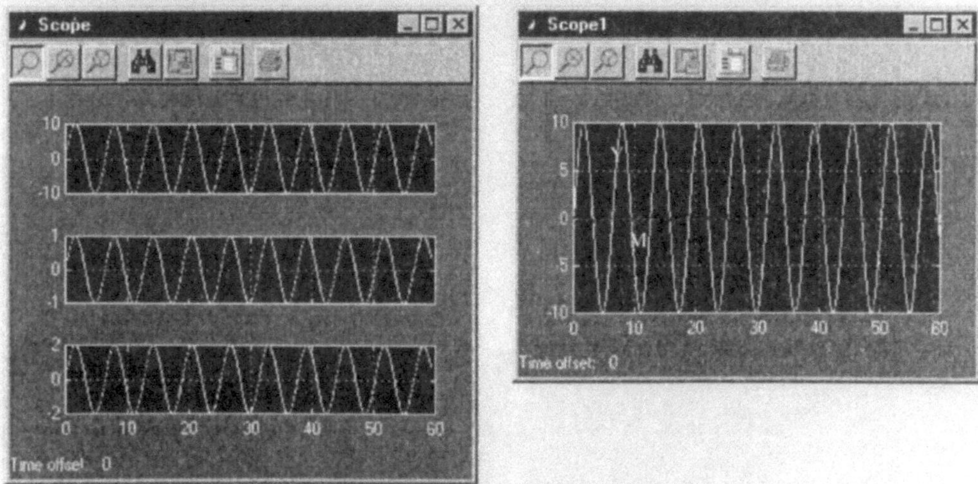

Figure 6.15. Response of the Model to Sinusoidal Input

Ready for last practice test for this chapter…

Practice Test 6.3.

1. Obtain response for all models of Chapter 4 using SIMULINK® by all the methods you can think of.

Exercise for Chapter 6:

1. Solve all the problems given in Exercise for Chapter 1, 2, 3 and 4 using SIMULINK®.

Chapter 7

Design of Compensators for Control Systems

In this chapter, you will learn to design compensators for LTI models using the classical Root Locus technique or the modern state space variable methods.

In previous chapters, we created models, manipulated them to get the complete representation of a system and obtained their responses for various inputs in time as well as frequency domain. We were least interested in whether the model gave the desired performance or not. In this chapter, we will learn to design System models, which behave in a certain controlled and desirable manner when subjected to a particular condition. To achieve this, the Control System Toolbox of MATLAB® offers two methods, well known to the Control System designers:

- classical or conventional designing techniques based upon Root Locus, Bode plot and Nichols chart
- modern methods involving pole placement and optimisation techniques

We will first have a bird's eye view of the classical method of designing and then switch over to learn the modern techniques in detail.

7.1 Classical Methods for Design

As already mentioned, any of the three plots *viz.,* Root Locus, Bode plot and Nichols chart can be used as a conventional design tool. We are already familiar with the methods of obtaining Bode plot and Nichols chart and know their use as design tools (see Chapter 4). Let us now learn the method to obtain the Root Locus plot of a model and manipulate it to make the model perform in a desirable manner.

7.1.1 Obtaining Root Locus Plot for a Model

As we all know, Root Locus plot describes the trajectories of closed-loop poles when the feedback gain k varies from zero to infinity. Syntax for obtaining a Root Locus plot for a SISO LTI model is summarised in Table 7.1.

Table 7.1. Syntax for obtaining Root Locus plot

Syntax	Results in...	Remarks
rlocus(*ModelName*)	a smooth plot of closed trajectories is displayed in the Figure window with the automatically generated positive gain values.	where, *ModelName*...... is the user-specified name of the model whose Root Locus plot is desired
rlocus(*ModelName,k*)	Root Locus plot for *ModeName* using user defined values for vector k of gains.	k... is the user-supplied vector of gains for which Root Locus plot of the model is desired
[r, k]= **rlocus**(*ModelName*)	a matrix r of complex root locations and associated gain k for the model.	r... is the matrix containing location of complex roots for gain k, such that the length of matrix r
r= **rlocus**(*ModelName*, k)	a matrix r of complex root locations for the user-specified gain k.	is k columns and i^{th} column lists the closed-loop roots for the gain k_i

Let us consider some examples to make things clear.

Example:

rlocus(modelA3)
returns

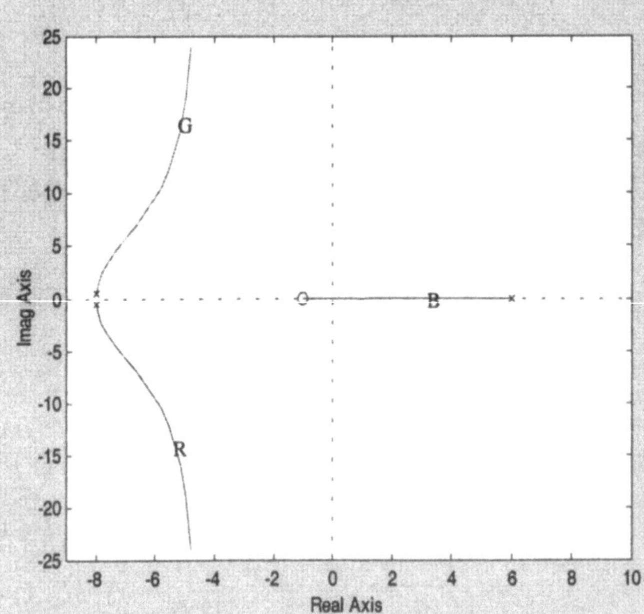

rlocus(modelB15)
returns

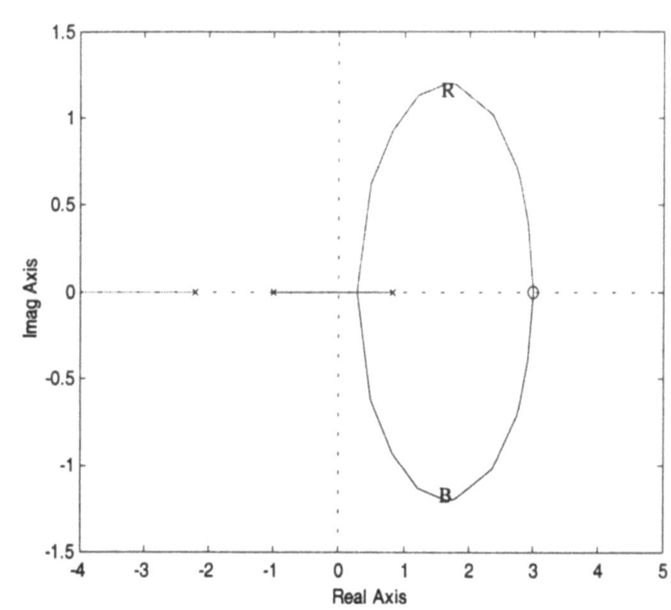

rlocus(modeB15,[0 1 2 5 10 50 100])
returns

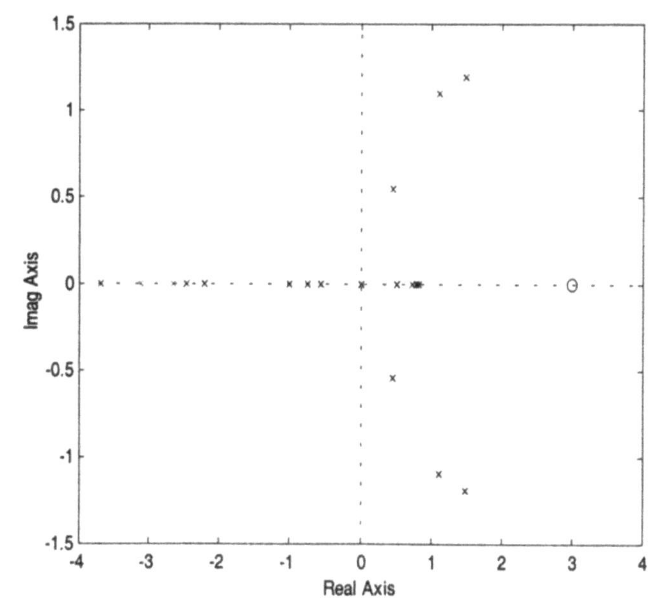

```
[r,k]=rlocus(modelB15)
returns

r =
 1.0e+002 *
 Columns 1 through 4
 -0.0100              -0.0098              -0.0095              -0.0088
 -0.0221              -0.0224              -0.0227              -0.0233
  0.0081               0.0081               0.0081               0.0080
 Columns 5 through 8
 -0.0075              -0.0050               0.0004               0.0022
 -0.0246              -0.0271              -0.0314              -0.0320
  0.0078               0.0072               0.0049               0.0036
 Columns 9 through 12
  0.0029               0.0029 - 0.0007i     0.0050 - 0.0062i     0.0082 - 0.0093i
 -0.0321              -0.0322              -0.0387              -0.0510
  0.0029               0.0029 + 0.0007i     0.0050 + 0.0062i     0.0082 + 0.0093i
 Columns 13 through 16
  0.0122 - 0.0113i     0.0167 - 0.0120i     0.0180 - 0.0120i     0.0238 - 0.0102i
 -0.0726              -0.1125              -0.1293              -0.2906
  0.0122 + 0.0113i     0.0167 + 0.0120i     0.0180 + 0.0120i     0.0238 + 0.0102i
 Columns 17 through 20
  0.0275 - 0.0071i     0.0279 - 0.0065i     0.0293 - 0.0039i      0.0300
 -0.7714              -0.9516              -2.8391               Inf
  0.0275 + 0.0071i     0.0279 + 0.0065i     0.0293 + 0.0039i      0.0300
k =
 1.0e+003 *
 Columns 1 through 7
      0      0.0001     0.0002     0.0004     0.0010     0.0023     0.0052
 Columns 8 through 14
  0.0057     0.0058     0.0058     0.0118     0.0267     0.0606     0.1375
 Columns 15 through 20
  0.1731     0.5474     1.7310     2.1792     6.8913      Inf

r=rlocus(modelB15,[0 1 2 5 10 50 100])
returns

r =
 Columns 1 through 4
 -1.0000              -0.7536              -0.5545              -0.0000
 -2.2133              -2.4624              -2.6582              -3.1138
  0.8133               0.7760               0.7327               0.5138
 Columns 5 through 7
  0.4454 - 0.5379i     1.1182 - 1.0912i     1.4937 - 1.1883i
 -3.6908              -6.6363              -9.3873
  0.4454 + 0.5379i     1.1182 + 1.0912i     1.4937 + 1.1883i
```

7.1.2 Selecting Feedback Gain from Root Locus Plot

Control System Toolbox of MATLAB® has a built-in function **rlocfind** for selecting feedback gain from Root Locus plot. The syntax for this function is:

[gain,poles]=**rlocfind**(*ModelName, p*)

where,

gain.................................	returns dc gain associated with the closed-loop poles
poles................................	returns poles associated with the closed-loop gain
p	is the user-specified vector of points at which the gain and poles are desired
ModelName	is the name of the model under consideration

The following illustrations will further clarify this concept.

Example:

```
[k,pole]=rlocfind(modelA3)
returns
```

Select a point in the graphics window

A figure window appears displaying the Root Locus plot for modelA3 alongwith a cross-hair controlled by the movement of the mouse. This is shown in figure given below.

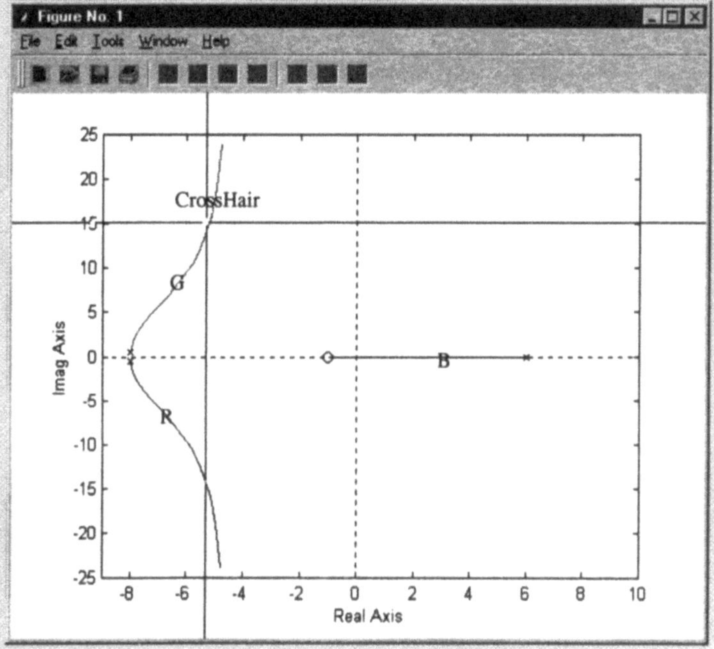

With the help of the cross-hair select a point on the Root Locus plot. MATLAB® now returns the following in the command window:

```
selected point =
 -5.2788 +14.1813i
k =
  50.9559
pole =
 -5.2853 +14.1819i
```

```
-5.2853 -14.1819i
 0.5707
```

In the Figure window, MATLAB® puts a cross (x) mark at the selected point (see the figure given below).

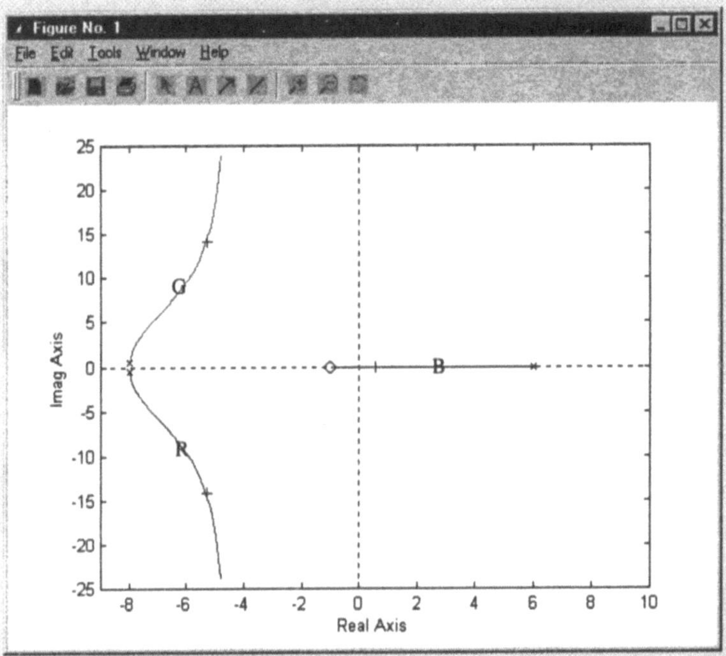

Instead of selecting the points using cross-hair you can specify the points in the from of a vector matrix *p* and obtain the gain and poles associated with these points also.

```
[gain,poles]=rlocfind(modelA3,[-1+j -1-j -2])
returns

gain =
  71.0502   71.0502   58.0000
poles =
  -5.0466 +17.2908i        -5.0466 +17.2908i        -5.1822 +15.3353i
  -5.0466 -17.2908i        -5.0466 -17.2908i        -5.1822 -15.3353i
   0.0932                   0.0932                   0.3645
```

7.1.3 Designing Compensator Using Root Locus Technique

You have already designed compensators for control systems using Root Locus technique in Chapter 5. There you used the Root Locus Design GUI as your tool. You can achieve the same results through MATLAB®'s command window too. Popular compensation techniques that you may be familiar with and which may be using Root Locus technique are summarised in Table 7.2.

Table 7.2. Types of conventional compensators

Type of Compensator	Compensator Connection
Series or Cascade Compensator	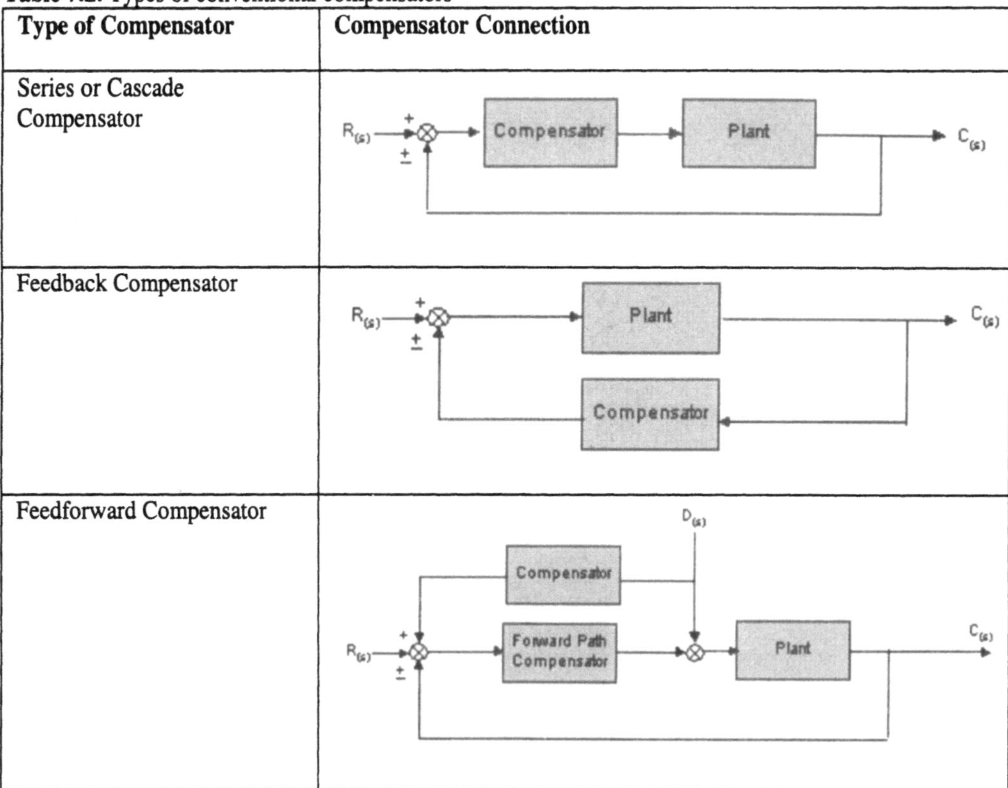
Feedback Compensator	
Feedforward Compensator	

In the examples of Chapter 5, what you designed was series (cascade) compensator for both continuous and discrete systems. You may design the same using the functions you just learnt to use using MATLAB® command window. We hope this is not a very difficult exercise for you!

Practice Test 7.1.

1. Solve all the problems you came across in Chapter 5 using Root Locus Design GUI using functions you just learnt in this chapter.

7.2 Solution of Lyapunov's Equation and Stability

Linear time invariant models have many methods like Bode plot, Nyquist plot, Nichols chart *etc.*, by which the stability of the model may be judged. However, when it comes to non-linear and/or invariant-time models, these methods which are popularly known as conventional methods of stability analysis do not apply. Stability of such systems are adjudged using solutions of Lyapunov's equations which are described in the paragraphs to follow.

A non-linear, time invariant model may be represented by the following state equation:

$\dot{x} = ax + bu$

$y = cx + du$

The Lyapunov's equation may be written for the model in any of the following two forms:

- $aP+Pa^T+Q=0$

for both continuous and discrete-time models.

Here, Q is an arbitrarily chosen positive-semidefinite real symmetric matrix of the same size as matrix a such that the solution to this may be obtained using built-in function **lyap** (for continuous models) or **dlyap** (for discrete models) that have the following syntax:

P=**lyap**(a, Q)
or,
P=**dlyap**(a, Q)

for continuous and discrete models respectively.

- $aP+bP+c=0$

which is a more generalised form of Lyapunov's equation, valid only for continuous-time models.

Its solution may be obtained using **lyap** function having a slightly different syntax than one described above:

P=**lyap**(a, b, c)

If the solution of Lyapunov's equation *i.e.,* value of P obtained is positive definite, the model may be said to be asymptotically stable.

Illustrations given below will further clarify this concept.

Example:

Let us assume the parameters for a continuous-time model as follows:

```
a=[0 1 0 0; 0 0 0 1;-1 -3 -6 -7; 0 1 2 3];
q=[0 0 0 0; 0 0 0 0; 0 0 0 0; 0 0 0 1];
```

then

```
P=lyap(a,q)
det(P)
returns
```

```
P =
    7.4167    0.0000   -4.5000    3.0000
    0.0000   -3.0000    1.4167   -0.0000
   -4.5000    1.4167    0.6250   -0.5000
    3.0000   -0.0000   -0.5000    0.1667
ans =
    5.3265
```

Since P is a positive definite matrix (since determinant of P is a positive value), the model is asymptotically stable.

Next, assume another model with the following parameters:

```
a=[0 1; -3 -4];
b=[ 0 -5;-10 2];
c=[1 0; 0 1];
```

then

```
P=lyap(a,b,c)
det(P)
returns
```

```
P =
   0.0317   0.1160
   0.1598  -0.0734
ans =
  -0.0209
```

Since determinant of P is negative, the model is unstable.

Let us next consider a discrete model as follows:

```
a=[0 1 0 0; 0 0 0 1;-1 -3 -6 -7; 0 1 2 3];
q=[0 0 0 0; 0 0 0 0; 0 0 0 0; 0 0 0 1];
t=0.1;
```

then

```
P=dlyap(a,q)
det(P)
returns
```

```
P =
   22.8000   20.7000  -31.2000   17.3000
   20.7000   22.8000  -33.8000   20.7000
  -31.2000  -33.8000   53.8000  -34.2000
   17.3000   20.7000  -34.2000   22.8000
ans =
 -110.4500
```

Since determinant of P is negative, model is not stable.

Ready to check yourself?

Practice Test 7. 2.

1. Find out the solution of Lyapunov's equation for ss models having parameters as given below:

a) $a = \begin{bmatrix} -1 & 0 & 0 \\ 0 & -2 & 0 \\ 0 & 0 & -3 \end{bmatrix}$; $q = \begin{bmatrix} 0 & 0 & 0 \\ 0 & 0 & 0 \\ 0 & 0 & 1 \end{bmatrix}$ for t=-1 and t=0.5

b) $a = \begin{bmatrix} 0 & 1 & -1 \\ -2 & 3 & 5 \\ 4 & 2 & 1 \end{bmatrix}$; $b = \begin{bmatrix} 0 & 1 & 1 \\ 1 & 0 & 0 \\ 0 & 1 & 1 \end{bmatrix}$; $c = \begin{bmatrix} 0 & 0 & 1 \\ 1 & 1 & 0 \\ 0 & 0 & 0 \end{bmatrix}$ for t=-1 and t=0.1

7.3 Modern Methods for Design

After trying out designing compensators by classical methods, I am sure, the amount of trial and error involved in the process must have irritated you. Moreover, you must have noticed that you have to make lot many compromises and perform several trade-offs to achieve desired system performance, say, like making the system stable. Your mind must have pondered over again and again, whether MATLAB® provides you with functions to design compensators using modern techniques which involves little amount of trial and error. Then, here is the big great news of relief... Yes, the Control System Toolbox of MATLAB® does support the modern methods of compensation techniques. In the forthcoming paragraphs, you will learn to design compensators for models using these techniques.

7.3.1 Design of State Feedback Controller

The simplest compensator, as you know, involves the pole placement technique. Consider a system represented by the following state space equations:

$$\dot{x} = ax + bu$$

$$y = cx + du$$

The Control System Toolbox of MATLAB® allows you to design the state feedback controller for this model, which could be a SISO model or a MIMO model. The state variable feedback for this system is a scalar function that takes the form:

$$kx = \begin{bmatrix} k_1 & k_2 & k_3 \cdots\cdots & k_n \end{bmatrix} \begin{bmatrix} x_1 \\ x_2 \\ x_3 \\ : \\ : \\ : \\ x_n \end{bmatrix}$$

The block diagram of the system with state variable feedback is shown in Figure 7.1. Observe that input u to the system is:

u = -kx + r

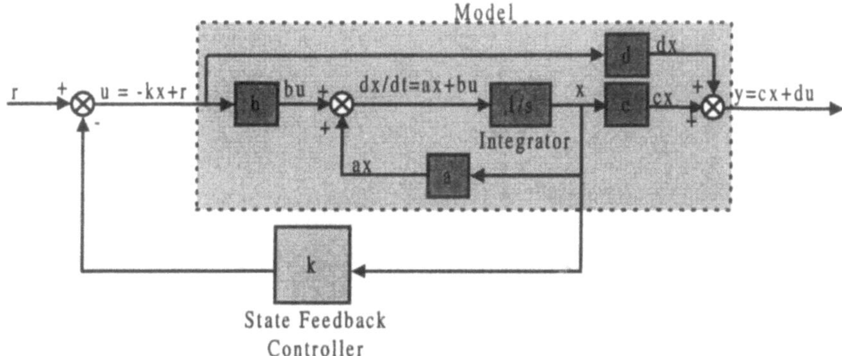

Figure 7.1. State Variable Feedback Controller

7.3.1.1 For SISO Models

Control System Toolbox has a built-in function, **acker,** based upon the ackermann's formula for designing State Feedback Controller for a SISO model. The syntax for this function is:

k=**acker**(a, b, p)

where,

k .... returns the feedback gain vector associated with the closed-loop poles p

a .... specifies 2D real valued system matrix of the SISO system under consideration, which may be continuous or discrete

b .... specifies 2D real valued input matrix of the SISO system under consideration, which may be continuous or discrete

p .... is the user-specified vector of desired closed-loop poles

The values of the feedback gain vector k so obtained, is such that the state feedback u= -kx places the closed-loop poles at the desired locations given by p, which is nothing but the eigenvalues of a-bk. When the feedback gain vector k is fedback along with the states x in the manner shown in Figure 7.1, the system exhibits desired performance.

Note that the states of the model are available for control of the model using feedback technique, only if the system is fully controllable. Hence, the name state feedback controller is given to this compensator.

Let us consider a few examples to have a better understanding.

Example:

```
a=[0 1 0; 0 0 1; -3 -5 -7];
b=[0;0;1];
c=[1 0 -1];
```

```
d=[0];
modelG1=ss(a,b,c,d);
```

Check whether the model is controllable

```
con=ctrb(modelG1);
rank(con)
returns

ans =
   3
```

Since the model is of third order and the rank of controllability matrix is also 3. Hence, the states of the model are fully controllable and hence available for state feedback control. Let us now design the State Feedback Controller for the model for desired location of closed-loop poles given by vector p.

```
p=[-2 -4 -6];
k=acker(a,b,p)
returns

k =
   45   39   5
```

For this system, let us compare the performance with and without state feedback for a zero step input and value of the initial states as x(1)=0, x(2)=0 and x(3)=0.5 using SIMULINK®. The result is as shown below. Observe how fast the response of the system with controller converges to zero.

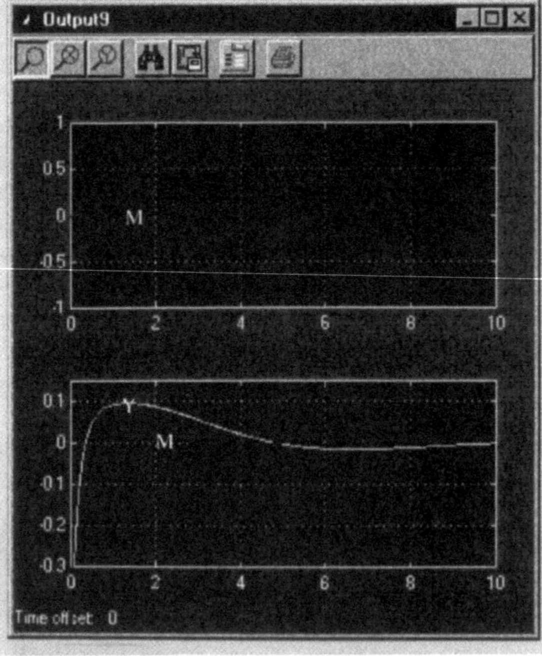

7.3.1.2 For MIMO Models

The function **acker** was meant for designing the State Feedback Controller only for SISO models. For MIMO models, Control System Toolbox of MATLAB® has yet another more versatile function **place** for designing State Feedback Controller. It acts equally well on SISO models too. The syntax for this function is:

[k, prec, message] = **place**(a, b, p)

where,

k .... returns feedback gain matrix associated with the closed-loop poles p

prec .... returns an estimate of closeness of eigenvalues of a-bk and p. It measures the accurate number of decimal digits present in actual closed-loop poles (optional)

message............................ returns a warning message if some non-zero closed-loop pole is 10% off the desired location (optional)

a .... specifies 2D real valued system matrix of the SISO model under consideration which may be continuous or discrete

b .... specifies 2D real valued input matrix of the SISO model under consideration which may be continuous or discrete

p .... is the user-specified vector of desired closed-loop poles

Similar to the **acker** function, the values of the feedback gain vector k so obtained, is such that the state feedback u= -kx places the closed-loop poles at the desired locations given by p, which is nothing but the eigenvalues of a-bk. When the feedback gain vector k is feedback along with the states x in the manner shown in Figure 7.1, the system exhibits desired performance.

For this function too, the model should have fully controllable states so that they are available for control.

Example:

```
a=[-2 1 0; 4 0 1; 0 0 0];
b=[0 1;1 0;-1 1];
c=[2 0 -1; 1 1 0];
d=[0];
modelG2=ss(a,b,c,d);
```

Check whether the model is controllable

```
con=ctrb(modelG2);
rank(con)
returns

ans =
   3
```

Since the model is of third order and the rank of controllability matrix is also 3 hence the states of the model are fully controllable and therefore available for state feedback control. Let us now design the State Feedback Controller for the model for desired location of closed-loop poles given by vector p.

```
p=[-0.5 -1+j*0.1 -1-j*0.1];
[k,prec,message]=place(a,b,p)
returns

k =
    2.0509    0.6863    0.4811
   -0.1297    0.9118    0.4246
prec =
    15
message =
    "
```

For this system, let us compare the performance with and without state feedback for a zero step input and value of the initial states as x(1)=0.5, x(2)=0 and x(3)=0 using SIMULINK®. The result is shown below. Observe how fast the response of the system with controller converges to zero.

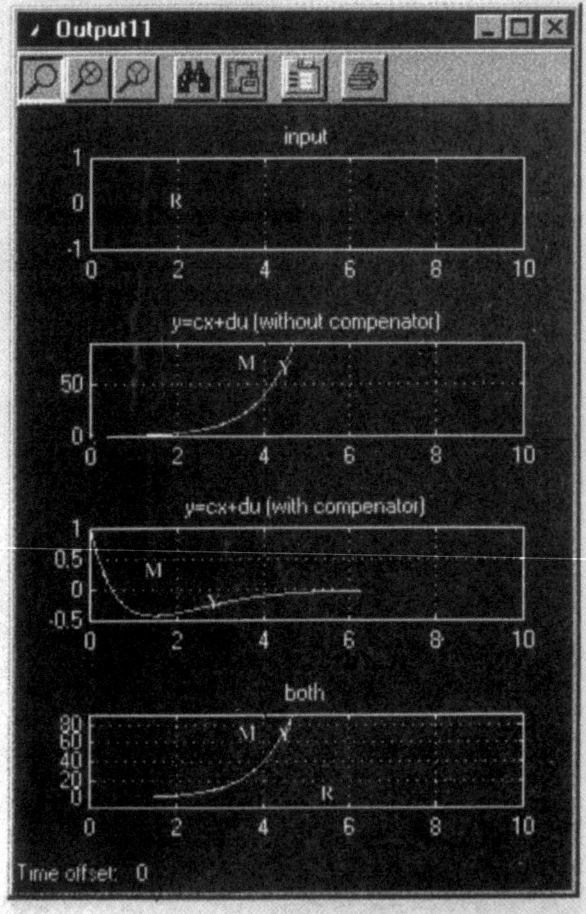

So, are you confident that you can design a State Feedback Controller for any given model?
Better check…

Practice Test 7.3.

1. For the ss models given below find out the feedback gain vector k of State Feedback
   Controller in order to have closed-loop poles at p as indicated. Compare the response of the
   two systems for various inputs assuming different initial values for the input states.

a) modelGp1=

$$\begin{bmatrix} \dot{x}1 \\ \dot{x}2 \\ \dot{x}3 \end{bmatrix} = \begin{bmatrix} -2 & 1 & 0 \\ 4 & 0 & 1 \\ 0 & 0 & -1 \end{bmatrix} \begin{bmatrix} x1 \\ x2 \\ x3 \end{bmatrix} + \begin{bmatrix} 0 & 1 \\ 1 & 0 \\ -1 & 1 \end{bmatrix} \begin{bmatrix} u1 \\ u2 \end{bmatrix}$$

$$\begin{bmatrix} y1 \\ y2 \end{bmatrix} = \begin{bmatrix} 2 & 0 & -1 \\ 1 & 1 & 0 \end{bmatrix} \begin{bmatrix} x1 \\ x2 \\ x3 \end{bmatrix} + \begin{bmatrix} 0 & 0 \\ 0 & 0 \end{bmatrix} \begin{bmatrix} u1 \\ u2 \end{bmatrix}$$

for p= -3, -4 , -5

b) modelGp2=

$$[\dot{x}] = \begin{bmatrix} 0 & 0 & 0 & 1 \\ 2 & 4 & 6 & -8 \\ 0 & 1 & 0 & 0 \\ 0 & 0 & 1 & 0 \end{bmatrix} [x] + \begin{bmatrix} 1 & 0 & 1 \\ 0 & 0 & 1 \\ 0 & 2 & -3 \\ 0 & 1 & 0 \end{bmatrix} [u]$$

for p= -1, -1+0.5j, -1-0.5j, -2

$$[y] = \begin{bmatrix} 1 & 0 & 2 \\ 2 & 0 & 1 \\ 3 & 3 & -3 \end{bmatrix} [x] + \begin{bmatrix} 1 & 0 & 0 \\ 0 & 0 & 1 \\ 0 & 1 & 1 \end{bmatrix} [u]$$

7.3.2 Design of State Estimator/Observer

In order to design a State Feedback Controller for a model it is necessary for the model to be
fully controllable. But in many systems, the states of the model are not fully controllable and
hence they cannot be taken out for feedback. It may be possible that the states of the models are
observable. Such models may also have some inputs which might be known while some inputs
may be entirely stochastic inputs whose values may be dependent upon probability of some
function. Similarly, only some of the outputs may be measurable which may be available for
feedback. For such models, State Feedback Controllers cannot be designed directly. The
observable states of such models have to be first estimated by an estimator and then for these
estimated states a state feedback controller may be designed. The arrangement for such models
is shown in Figure 7.2.

In the paragraphs to follow we will learn to design an estimator using **estim** function
available with Control System Toolbox of MATLAB®. Let us consider a model with all its inputs
w as stochastic and all outputs y as measurable, represented by the equation:

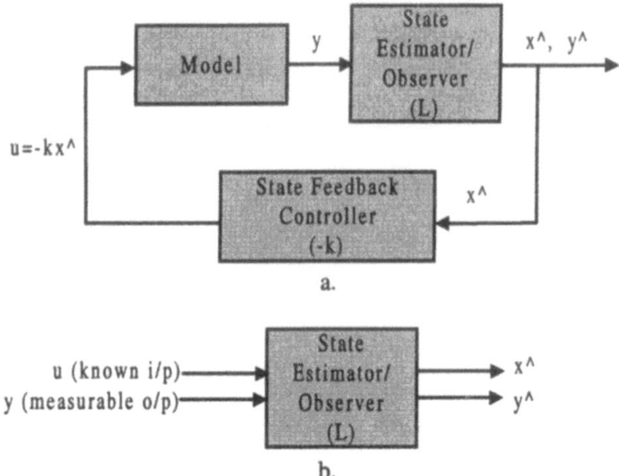

Figure 7.2. Estimator/Observer

$$\dot{x} = ax + bw$$

$$y = cx + dw$$

The Estimator/Observer can be designed for this model which may be represented by equation:

$$\dot{\hat{x}} = a\hat{x} + L(y - c\hat{x})$$

$$\begin{bmatrix} \hat{y} \\ \hat{x} \end{bmatrix} = \begin{bmatrix} c \\ I \end{bmatrix} \hat{x}$$

Such a model along with the Estimator/Observer connected in cascade is shown in Figure 7.3a.

However, for many models, all the inputs are not stochastic neither are all the outputs measurable. For such a model having u known inputs and w stochastic inputs, let there be y measurable outputs and z non-measurable outputs. Such a general model may be represented by the following equations:

$$\dot{x} = ax + b_1 w + b_2 u$$

$$\begin{bmatrix} z \\ y \end{bmatrix} = \begin{bmatrix} c_1 \\ c_2 \end{bmatrix} x + \begin{bmatrix} d_{11} \\ d_{21} \end{bmatrix} w + \begin{bmatrix} d_{12} \\ d_{22} \end{bmatrix} u$$

For such a model it is possible to design an Estimator/Observer having following equations:

$$\dot{\hat{x}} = a\hat{x} + b_2 u + L(y - c_2 \hat{x} - d_{22} u)$$

$$\begin{bmatrix} \hat{y} \\ \hat{x} \end{bmatrix} = \begin{bmatrix} c_2 \\ I \end{bmatrix} \hat{x} + \begin{bmatrix} d_{22} \\ 0 \end{bmatrix} u$$

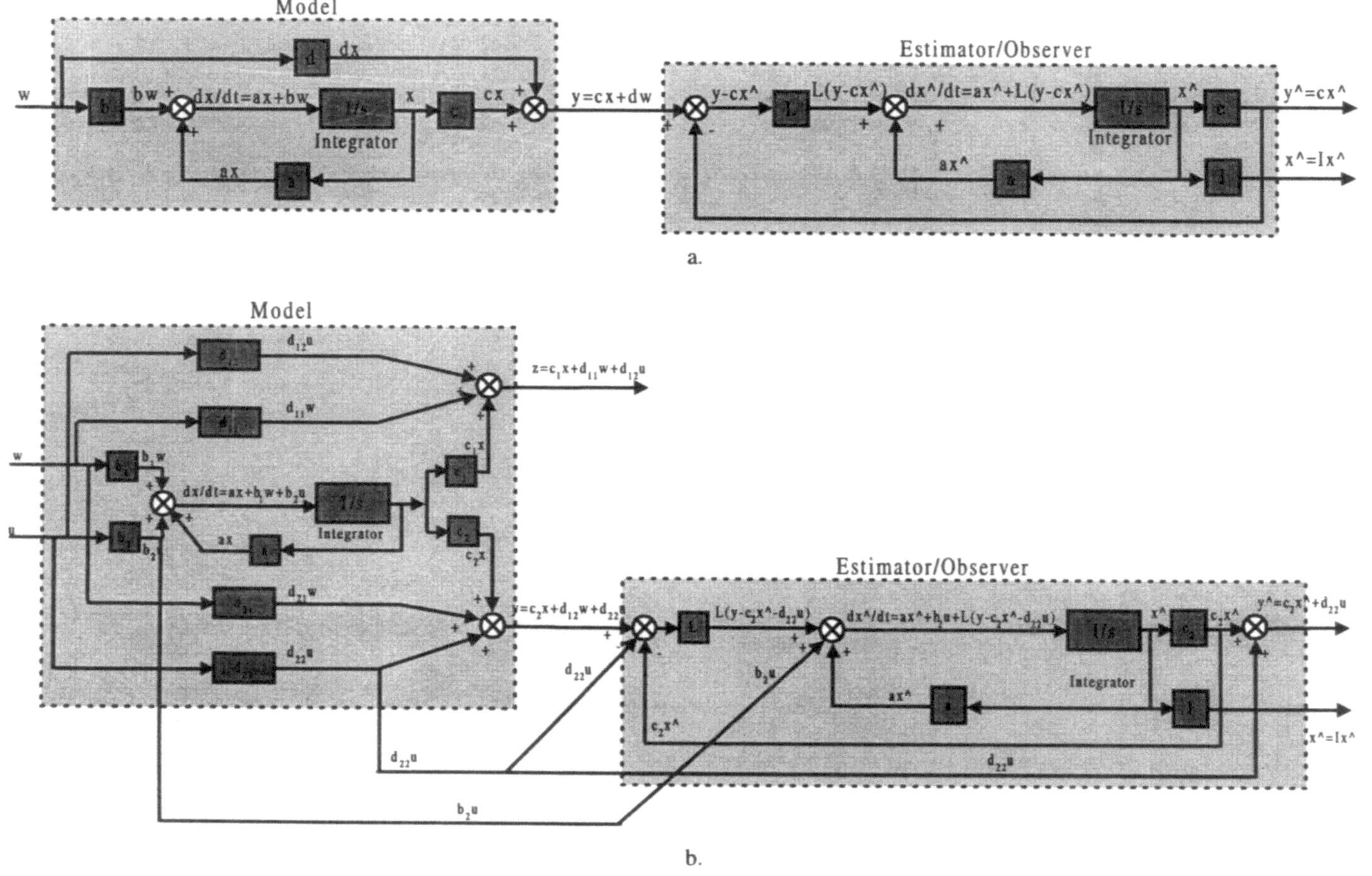

Figure 7.3. State Estimator/Observer

The model along with the Estimator/Observer connected in cascade is shown in Figure 7.3b.

Using Control System Toolbox of MATLAB® you can obtain the parameters of the Estimator/Observer using the **estim** function having the following syntax:

eo=**estim**(*ModelName, L, meas o/p sensors, known i/p*)

where,

eo..	returns the state/output Estimator/Observer
ModelName......................	specifies the model under consideration
L..	specifies estimator gain vector
meas o/p senors................	specifies the vector containing indices of the measurable output of the model *i.e.,* y (optional)
known i/p..........................	specifies the vector containing indices of the known inputs of the model *i.e.,* u (optional)

Note that if *meas o/p sensors* and *known i/p* vectors are not supplied, MATLAB® assumes that all inputs are stochastic and all outputs are non-measurable and estimator shown in Figure 7.2a is returned by MATLAB® in place of more general estimator shown in Figure 7.2b. Also note that L must have as many rows as states and as many columns as measurements y.

Like the state feedback controller, if you desire to design an estimator associated with some closed-loop poles given by vector p, you may calculate the estimator gain associated with closed-loop poles using **acker** and **place** functions for SISO and MIMO models respectively as follows:

L=**acker**(a', c', p).'
L=**place**(a', c', p).'

Let us consider some illustrations to further clarify this concept.

Example:

Let us first consider a SISO model with the following parameters:

```
a=[-1 0 0; 0 -5 0; 0 0 -10];
b=[0; 1; 1];
c=[2 1 3];
d=[0];
modelG3=ss(a,b,c,d,'InputName','w1','OutputName','y1')
returns
```

a =

	x1	x2	x3
x1	-1	0	0
x2	0	-5	0
x3	0	0	-10

b =

	w1
x1	0
x2	1
x3	1

modelG3=

$$\begin{bmatrix} \dot{x}1 \\ \dot{x}2 \\ \dot{x}3 \end{bmatrix} = \begin{bmatrix} -1 & 0 & 0 \\ 0 & -5 & 0 \\ 0 & 0 & -10 \end{bmatrix} \begin{bmatrix} x1 \\ x2 \\ x3 \end{bmatrix} + \begin{bmatrix} 0 \\ 1 \\ 1 \end{bmatrix} [w1]$$

$$[y1] = [2 \quad 1 \quad 3] \begin{bmatrix} x1 \\ x2 \\ x3 \end{bmatrix} + [0][w1]$$

```
c =
               x1        x2        x3
      y1        2         1         3

d =
               w1
      y1        0
```

Continuous-time model.

Let us check whether the model is controllable.

```
con=ctrb(modelG3);
rank(con)
```

returns
```
ans =
   2
```

Clearly, the model is not controllable. Hence, state feedback controller cannot be designed directly for this model. Let us now check if it is observable.

```
ob=obsv(modelG3);
rank(ob)
returns
```

```
ans =
   3
```

Model is observable and therefore, an Estimator/Observer can be designed for this model. Let the closed-loop poles be placed at -1, -2, -10. The gain L of the Estimator/Controller may be found out as follows:

```
p=[-1 -2 -10];
L=place(a',c',p).'
returns
```

```
place: ndigits= 15
L =
  -0.0000
  -3.0000
       0
```

The estimator may now be found out as follows assuming all inputs to be stochastic and all outputs to be measurable:

```
eoG3=estim(modelG3,L)
returns
```

$$eoG3=$$

$$\begin{bmatrix} \dot{\hat{x}}1 \\ \dot{\hat{x}}2 \\ \dot{\hat{x}}3 \end{bmatrix} = \begin{bmatrix} -1 & 0 & 0 \\ 0 & -5 & 0 \\ 0 & 0 & -10 \end{bmatrix} \begin{bmatrix} \hat{x}1 \\ \hat{x}2 \\ \hat{x}3 \end{bmatrix} + \begin{bmatrix} 0 \\ -3 \\ 0 \end{bmatrix} \left([y1] - [2 \quad 1 \quad 3] \begin{bmatrix} \hat{x}1 \\ \hat{x}2 \\ \hat{x}3 \end{bmatrix} \right)$$

$$\begin{bmatrix} \hat{y}1 \\ \hat{y}2 \\ \hat{y}3 \\ \hat{y}4 \end{bmatrix} = \begin{bmatrix} 2 & 1 & 3 \\ 1 & 0 & 0 \\ 0 & 1 & 0 \\ 0 & 0 & 1 \end{bmatrix} \begin{bmatrix} \hat{x}1 \\ \hat{x}2 \\ \hat{x}3 \end{bmatrix} + \begin{bmatrix} 0 \\ 0 \\ 0 \\ 0 \end{bmatrix} [y1]$$

```
a =

             x1            x2            x3
   x1        -1    4.9564e-016   1.4869e-015
   x2         6           -2             9
   x3         0            0           -10

b =

              y1
   x1 -4.9564e-016
   x2         -3
   x3          0

c =

             x1            x2            x3
   y1         2            1             3
   y2         1            0             0
   y3         0            1             0
   y4         0            0             1

d =

              y1
   y1          0
   y2          0
   y3          0
   y4          0

I/O groups:
    Group name        I/O    Channel(s)
    Measurement        I         1
    OutputEstimate     O         1
    StateEstimate      O        2,3,4

Continuous-time model.
```

The states $\hat{x}1, \hat{x}2, \hat{x}3$ estimated by the Estimator/Observer are compared with the states of the original model $x1, x2, x3$. The results are as indicated below. The states produced by the Estimator/Observer eoG3 follow the original states of the modelG3.

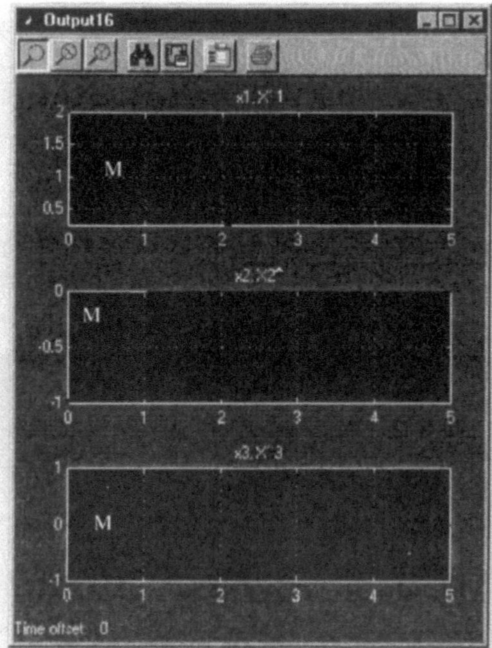

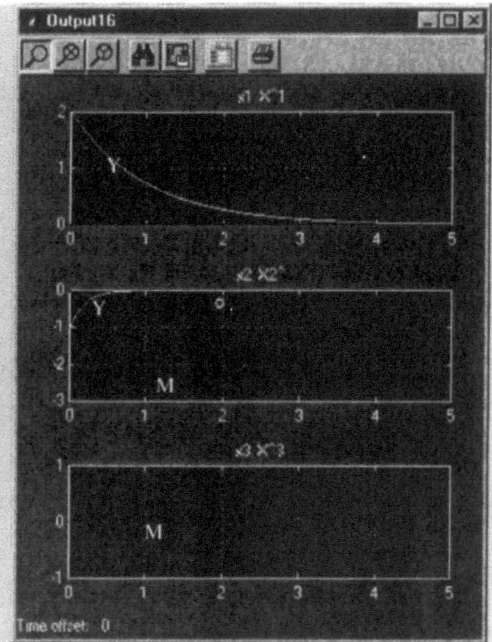

x1(0)=2, x2(0)= -1, x3(0)=0

$\hat{x}1(0) = 2, \hat{x}2(0) = -1, \hat{x}3(0) = 0$

States of the estimator follow the states of the model exactly for zero input.

x1(0)=2, x2(0)= -1, x3(0)=0

$\hat{x}1(0) = 0, \hat{x}2(0) = 0, \hat{x}3(0) = 0$

States of the estimator indicates some deviation initially since different initial states are specified for the model and the estimator, However, for zero input, after some time, states of the estimator follows the states of the model.

The state feedback compensator gain k may also be calculated using **place** function as before...

```
k=place(a,b,p)
returns
```

```
place: ndigits= 15
k =
   0.1567   -3.0000        0
```

Again, with the initial value of states of the model as x1(0)=2, x2(0)= -1 and x3(0)=0, and those of the estimator as $\hat{x}1(0) = 0, \hat{x}2(0) = 0, \hat{x}3(0) = 0$, we find out the response of the model to zero input. This is shown in the forthcoming Figure. Observe that with estimator, the response of the model improves.

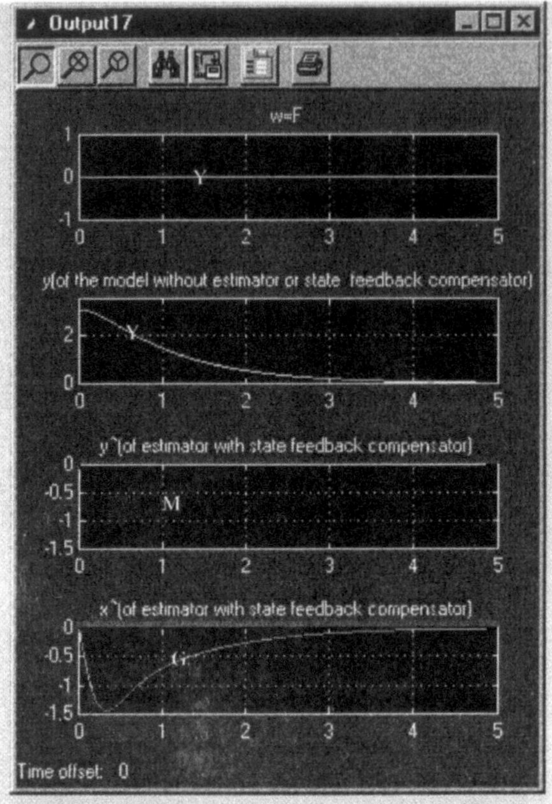

In a similar way, you can design state estimators for MIMO models too. The method to accomplish this is explained with an example:

```
a=[-2 1; 4 0 ];
b=[0 1;1 0];
c=[2 0 ; 1 1];
d=[0 0; 0 0];
modelG4=ss(a,b,c,d,'InputName', {'w' 'u'}, 'OutputName',{'z' 'y'})
returns
```

```
a =

            x1        ·x2
     x1     -2         1
     x2      4         0

b =

            w         u
     x1     0         1
     x2     1         0

c =

            x1        x2
     z      2         0
     y      1         1
```

$$
\begin{aligned}
\text{modelG4}= \\
\begin{bmatrix} x1 \\ x2 \end{bmatrix} &= \begin{bmatrix} -2 & 1 \\ 4 & 0 \end{bmatrix}\begin{bmatrix} x1 \\ x2 \end{bmatrix} + \begin{bmatrix} 0 & 1 \\ 1 & 0 \end{bmatrix}\begin{bmatrix} w \\ u \end{bmatrix} \\
\begin{bmatrix} z \\ y \end{bmatrix} &= \begin{bmatrix} 2 & 0 \\ 1 & 1 \end{bmatrix}\begin{bmatrix} x1 \\ x2 \end{bmatrix} + \begin{bmatrix} 0 & 0 \\ 0 & 0 \end{bmatrix}\begin{bmatrix} w \\ u \end{bmatrix}
\end{aligned}
$$

```
d =
                  w         u
        z         0         0
        y         0         0
```

Continuous-time model.

```
con=ctrb(modelG4);
rank(con)
returns

ans =
    2

obG4=obsv(modelG4);
rank(obG4)
returns

ans =
    2

p=[ -1+j*0.1 -1-j*0.1];
L=place(a',c',p).'
returns

place: ndigits= 15
L =
  -0.9500    0.9000
   1.5500    1.0000
```

Considering input1 as stochastic, input2 as known, output1 as non-measurable and output2 as measurable (*i.e.*, sensor) an estimator can be now designed for modelG4 as shown below:

```
eoG4=estim(modelG4, L(:, [2]), [2], [2])
returns

a =
                  x1        x2
        x1       -2.9       0.1
        x2        3         -1

b =
                  u         y
        x1        1         0.9
        x2        0         1

c =
                  x1        x2
        y1        1         1
        y2        1         0
        y3        0         1
d =
                  u         y
        y1        0         0
        y2        0         0
```

$$eoG4=$$

$$\begin{bmatrix} \dot{\hat{x}}1 \\ \dot{\hat{x}}2 \end{bmatrix} = \begin{bmatrix} -2 & 1 \\ 4 & 0 \end{bmatrix} \begin{bmatrix} \hat{x}1 \\ \hat{x}2 \end{bmatrix} + \begin{bmatrix} 0.9 \\ 1 \end{bmatrix} \left([y] - [1\ \ 1][\hat{x}] - [0][u] \right)$$

$$\begin{bmatrix} z \\ y \end{bmatrix} = \begin{bmatrix} 1 & 1 \\ 0 & 1 \\ 1 & 0 \end{bmatrix} \begin{bmatrix} \hat{x}1 \\ \hat{x}2 \end{bmatrix} + \begin{bmatrix} 0 \\ 0 \end{bmatrix} [u]$$

```
        y3           0          0
I/O groups:
   Group name      I/O     Channel(s)
   KnownInput       I         1
   Measurement          I        2
   OutputEstimate   O         1
   StateEstimate        O       2,3
```

Continuous-time model.

The states $\hat{x}1, \hat{x}2, \hat{x}3$ estimated by the Estimator/Observer are compared with the states of the original model x1, x2, x3. The results are as indicated below. The states produced by the Estimator/Observer eoG4 follow the original states of the modelG4.

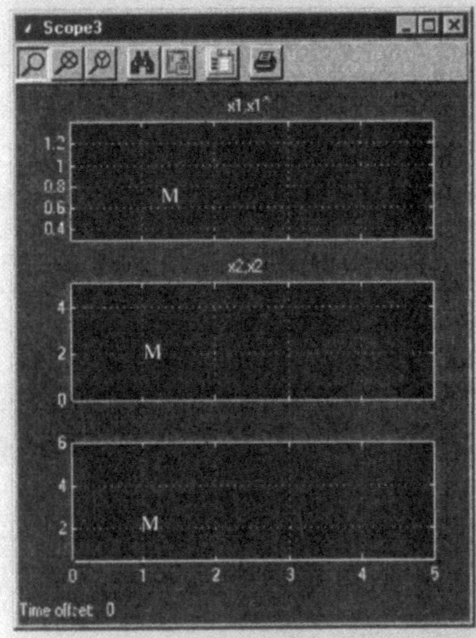

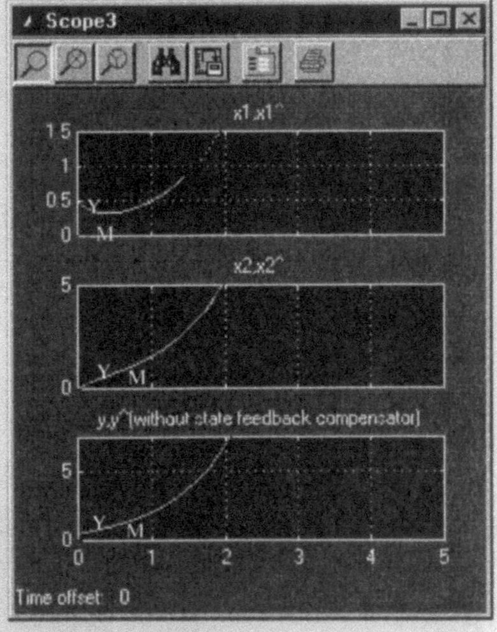

x1(0)=0.5, x2(0)= 0	x1(0)=0.5, x2(0)=0
$\hat{x}1(0) = 0$, $\hat{x}2(0) = 0$	$\hat{x}1(0) = 0$, $\hat{x}2(0) = 0$
States of the estimator follow the states of the model exactly for zero input.	States of the estimator indicates some deviation initially since different initial states are specified for the model and the estimator. However, for zero input, after some time, states of the estimator follows the states of the model.

The state feedback compensator gain k may also be calculated using **place** function as before...

```
[k,prec,message]=place(a,b,p)
returns


k =
    3.9000   1.0000
   -1.0000   1.1000
```

```
prec =
    15
message =
    ''
```

Again, with the initial value of states of the model as x1(0)=0.5, x2(0)= 0 and those of estimator as $\hat{x}1(0) = 0, \hat{x}2(0) = 0$, we find out the response of the model to zero input. This is shown in Figure given below. Observe that with estimator, the response of the model improves.

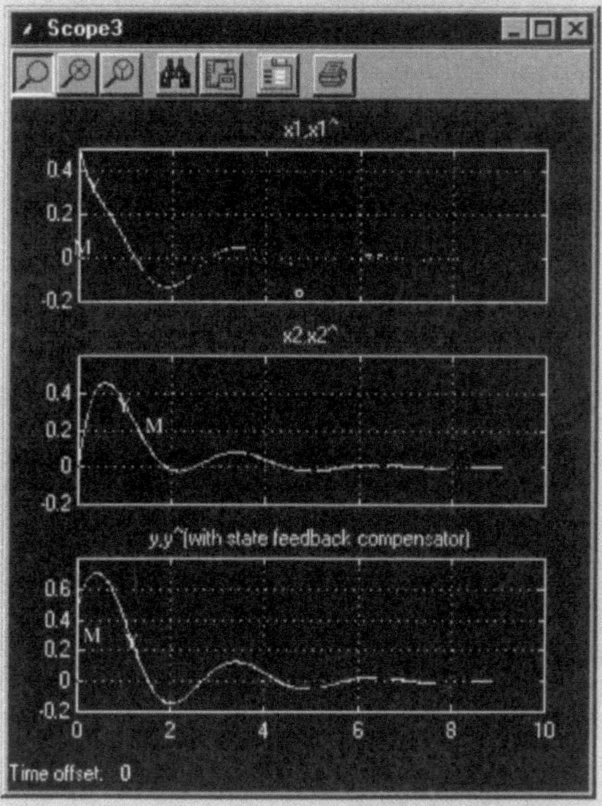

It is always good to practice whatever you have learnt immediately…

Practice Test 7.4.

1. Design an Estimator/Observer for the ss models given in Practice Test 7.3 *i.e.*, modelGp1 and modelGp2 for the closed-loop poles as indicated against them. Assume

a) all inputs to be stochastic and all outputs to be measured for both the models.

b) input1 as stochastic and output2 as measured for modelGp1.

c) input1 and input3 as stochastic and output2 and output3 as measured for modelGp2.

> Considering sampling time as -1 and 0.05, compare the estimated states so generated with the actual ones. Obtain the response of the system after connecting the gain feedback controller designed in Practice Test 7.3 and compare it with the response of the original models.

7.3.3 Design of State Regulator

The features of state feedback controller and Estimator/Observer may be combined to produce a complete compensator called state regulator as indicated in Figure 7.4 using the **reg** built in function of Control System Toolbox of MATLAB® which has the following syntax:

rg=**reg**(*ModelName, k, L, meas o/p sensors, known i/p, control i/p*)

where,
rg .... returns the state regulator
ModelName .... specifies the model under consideration which may be continuous or discrete
k .... specifies the feedback gain vector
L .... specifies estimator gain vector
meas o/p sensors .... specifies the vector containing indices of the measurable output of the model *i.e.*, y (optional)
known i/p .... specifies the vector containing indices of the known inputs of the model *i.e.*, u1 (optional) (it is the input to the state estimator)
control i/p .... specifies the vector containing indices of the control inputs of the model *i.e.*, u (optional) (it is the input to the controller having gain k)

Note that k must have as many rows as controls u and as many columns as states, while L must have as many rows as states and as many columns as measured output.

For a model given by:

$$\dot{x} = ax + bu$$

$$y = cx + du$$

The regulator can be designed arranging the state feedback compensator and the state Estimator/Observer as shown in Figure 7.4 to produce a regulator which can be represented by the following equations (Figure 7.5):

$$u = -k\hat{x}$$

$$\dot{\hat{x}} = a\hat{x} + bu + L(y - c\hat{x} - du)$$

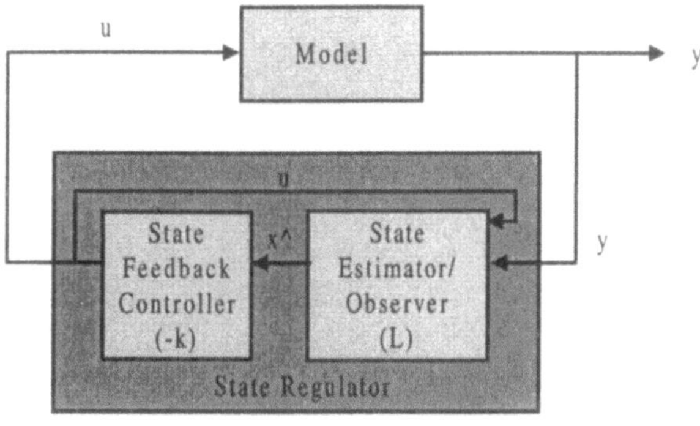

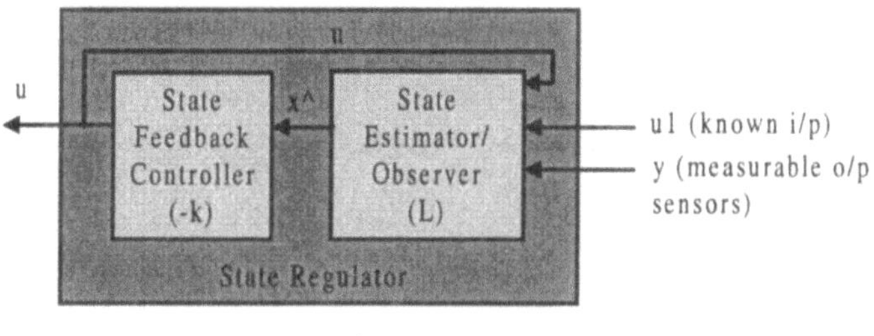

Figure 7.4. State Regulator

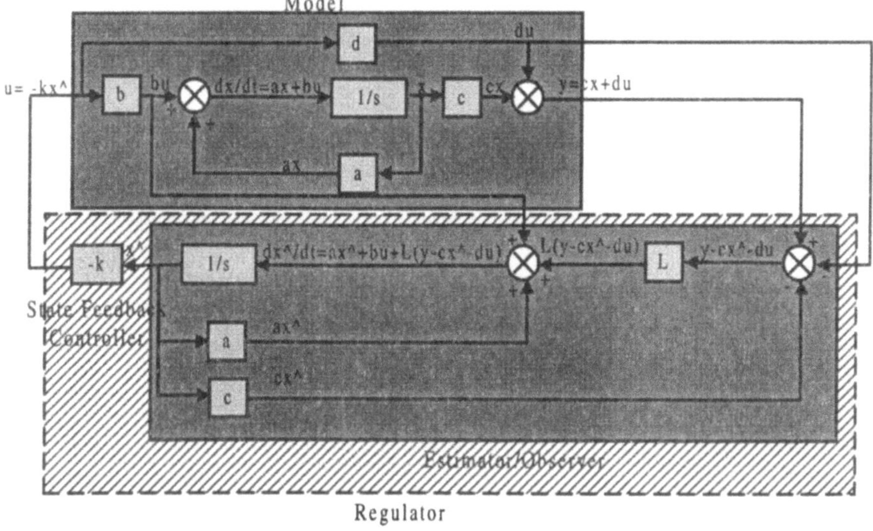

Figure 7.5. State Regulator

Note that the feedback is positive. Also, note that the input to the regulator is known inputs and measurable outputs. While the output of the regulator which forms feedback to the model follows the following precedence while being connected to the input of the model:

- control input of the model
- stochastic input to the model
- known input of the model

Let us consider some examples for better understanding.

Example:

Consider a model having following parameters...

```
a=[2 0 1; 2 -5 0; -2 1 -3];
b=[-1; 2; -3];
c=[-1 2 3];
d=[0];
modelG5=ss(a,b,c,d,'InputName','w1','OutputName','y1')
returns
```

```
a =
            x1      x2      x3
     x1      2       0       1
     x2      2      -5       0
     x3     -2       1      -3

b =
            w1
     x1     -1
     x2      2
     x3     -3

c =
            x1      x2      x3
     y1     -1       2       3

d =
            w1
     y1      0

Continuous-time model.
```

```
p=[-1 -2 -1.5];
k=place(a,b,p)
returns
```

```
place: ndigits= 15
k =
   -0.3077   -1.5385   -0.4231
```

```
L=place(a',c',p).'
returns
```

```
place: ndigits= 15
L =
  -1.0556
  -5.2778
   2.6667
```

With these values let us now design a regulator for the model assuming all inputs as stochastic and all outputs as measured.

$$
\begin{aligned}
&\text{rgG5=} \\[4pt]
&w1 = -[-0.3077 \;-1.5385\; -0.4231]\begin{bmatrix} \hat{x}1 \\ \hat{x}2 \\ \hat{x}3 \end{bmatrix} \\[8pt]
&\begin{bmatrix} \dot{\hat{x}}1 \\ \dot{\hat{x}}2 \\ \dot{\hat{x}}3 \end{bmatrix} = \begin{bmatrix} 2 & 0 & 1 \\ 2 & -5 & 0 \\ -2 & 1 & -3 \end{bmatrix}\begin{bmatrix} \hat{x}1 \\ \hat{x}2 \\ \hat{x}3 \end{bmatrix} + \begin{bmatrix} -1 \\ 2 \\ -3 \end{bmatrix}[w1] + \begin{bmatrix} -1.0556 \\ -5.2778 \\ 2.6667 \end{bmatrix}\left([y1] - [-1 \;\; 2 \;\; 3]\begin{bmatrix} \hat{x}1 \\ \hat{x}2 \\ \hat{x}3 \end{bmatrix} + [0][w1] \right)
\end{aligned}
$$

```
rgG5=reg(modelG5,k,L)
returns


a =
                x1          x2          x3
       x1     0.63675     0.57265      3.7436
       x2    -2.6624      8.6325       16.679
       x3    -0.25641    -8.9487     -12.269

b =
                y1
       x1    -1.0556
       x2    -5.2778
       x3     2.6667

c =
                x1          x2          x3
       w1     0.30769     1.5385      0.42308

d =
                y1
       w1       0

I/O groups:
   Group name        I/O    Channel(s)
   Measurement        I         1
   Controls           O         1

Continuous-time model.
```

Choosing initial states of modelG5 as x1(0)=1, x2(0)=x3(0)=0, and those of regulator as all zero, compare the response of the system with and without feedback regulator. From the forthcoming figure, it can be concluded that incorporation of the regulator has made the model stable.

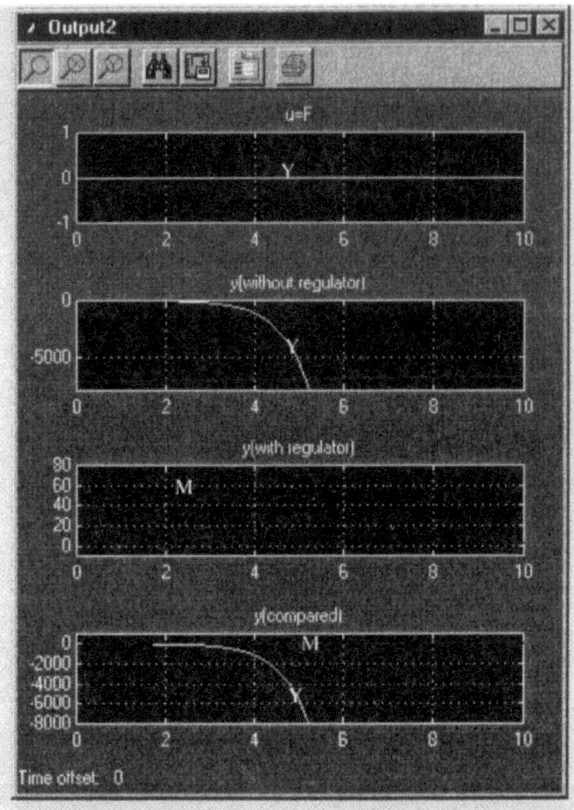

Practice a few more problems to get a better understanding…

Practice Test 7.5.

1. Design a regulator for the ss models given in Practice Test 7.3 *i.e.,* modelGp1 and modelGp2 for the closed-loop poles as indicated against them. Assume

a) all inputs to be stochastic and all outputs to be measured for both the models.

b) input1 as stochastic, input2 as known and output2 as measured for modelGp1.

c) input1 and input3 as stochastic, input2 as controlled, input4 as known; output2 and output3 as measured for modelGp2.

Considering sampling time as -1 and 0.05, compare the response of the system with those of the model alone. Compare the result with those obtained for the same models in Practice Test 7.4. Comment on the results.

7.4 Design of Optimal Compensators

While designing compensators using classical or modern technique explained in the previous articles, you must have realised that for a particular model, several different compensators can be designed. These compensators may or may not result into the best possible performance. Compensators based on classical methods were meant to satisfy specific requirements like steady state error, transient response, stability margins *etc., However,* all requirements could rarely be met due to limitations of design techniques and several trade-offs involved. While modern technique allowed arbitrary pole placement, the stability requirements were however, difficult to meet. Thus, after designing a compensator using these methods you would never feel sure that what you designed has resulted into the best compensator for the model. To eliminate this uncertainty, optimal controllers were developed with which you can always be sure that you are getting the best possible results. Before you learn to design the optimal controllers with the functions available with the Control System Toolbox of MATLAB®, let us first learn to solve Riccati equations, the solution of which is useful for designing such controllers.

7.4.1 Riccati Equations

Modern Control System Techniques based on state space representation of systems uses solution of Riccati equations for design of optimal compensators for the model. Two built-in functions **care** (continuous algebraic Riccati equation) and **dare** (discrete algebraic Riccati equation) are available with Control System Toolbox of MATLAB® for solving these equations for continuous-time and discrete-time models respectively.

7.4.1.1 Solution of Continuous-time Algebraic Riccati Equations

Continuous-time algebraic Riccati equations are mostly written in two forms. Depending upon the form of Riccati equation, the syntax used to obtain the solution of the equation varies. Table 7.3 summarises various syntax of the function **care** to obtain the solution of Riccati equation depending upon its form.

Let us try solving Riccati equations of different forms to see if the function really works.

Example:

Let us consider a state space model whose a, b, c, d parameters are given by:

```
a=[0 1; 0 0];
b=[0;1];
c=[1 0];
d=[0];
modelG7=ss(a,b,c,d);
```

Let the Riccati equation have the form as indicated below:

```
Riccati(P) = a'P+Pa-Pbb'P+Q = 0
```

Choosing the value of matrix Q as shown below, let us find out the solution of the equation....

```
Q=[2 0; 0 0];
[P, L, g, rr]=care(a,b,Q)
```

Table 7.3. Syntax for solving Riccati equations for continuous-time models

Form of Riccati Equation	Syntax	Results into	Remarks
Riccati(P)=0 or, a'P+Pa- Pbb'P+Q=0	[P,L,g,rr]= **care** (a,b,Q)	P... a symmetric matrix called stabilising solution of Riccati equation L... eigenvalues of $(a\text{-}bb'P)$ g... gain matrix $(b'P)$ rr... relative residual $\dfrac{\mid Riccati(P)\mid_F}{\mid P\mid_F}$	all eigenvalues of $(a\text{-}bb'P)$ should be in the left half plane
Riccati(P)=0 or, a'Pe+e'Pa- $(e'Pb+s)r^{-1}$ $(b'Pe+s')+$ Q=0	[P,L,g,(rr or report)]= **care** $(a,b,Q,r,s,e,'report')$	P... a symmetric matrix called stabilising solution of Riccati equation L... eigenvalues of $(a\text{-}b*g,e)$ g... gain matrix $r^{-1}(b'Pe+s')$ report... (optional) turns off the error messages when the solution P fails to exist and returns a failure report instead which has values: • -1... indicating associated hamiltonian pencil has eigenvalues on or very near the imaginary axis *i.e.*, failure. • -2... indicating there is no finite solution *i.e.*, $P=P1P2^{-1}$ has P1 singular, failure • rr... relative residual $\dfrac{\mid Riccati(P)\mid_F}{\mid P\mid_F}$ when solution exists	
	[P1,P2,L,report]= **care** $(a,b,Q,r,s,e,'implicit')$	• same as above • turns off error message • P in implicit form *i.e.*, $P=P2P1^{-1}$ when successful • zero (0) in case of failure	

```
returns

P =
   2.3784    1.4142
   1.4142    1.6818
L =
  -0.8409 + 0.8409i
  -0.8409 - 0.8409i
g =
   1.4142    1.6818
rr =
  1.3953e-015
```

The solution has resulted into a positive definite matrix P, which is indeed the stabilising solution of the Riccati equation.

We can also design an optimal state feedback compensator for the model. Using the solution of Riccati equation. For the same model, choosing $r=2$, and the value of initial states x1(0)=1 and x2(0)=-2, lets try out designing...

```
r=2;
[P, L, g, rr]=care(a,b,Q,r)
returns

P =
   2.8284    2.0000
   2.0000    2.8284
L =
  -0.7071 + 0.7071i
  -0.7071 - 0.7071i
g =
   1.0000    1.4142
rr =
  1.7971e-015
```

For the model the optimal state feedback compensator input is u=-kx where the gain k can be calculated using expression $k = r^{-1} b'P$,

```
k=inv(r )*b'*P
returns

k =
   1.0000    1.4142
```

The output of the model with and without optimal state feedback controller is shown in the forthcoming Figure for zero input. Observe how the output converges to zero if the optimal state feedback controller is used.

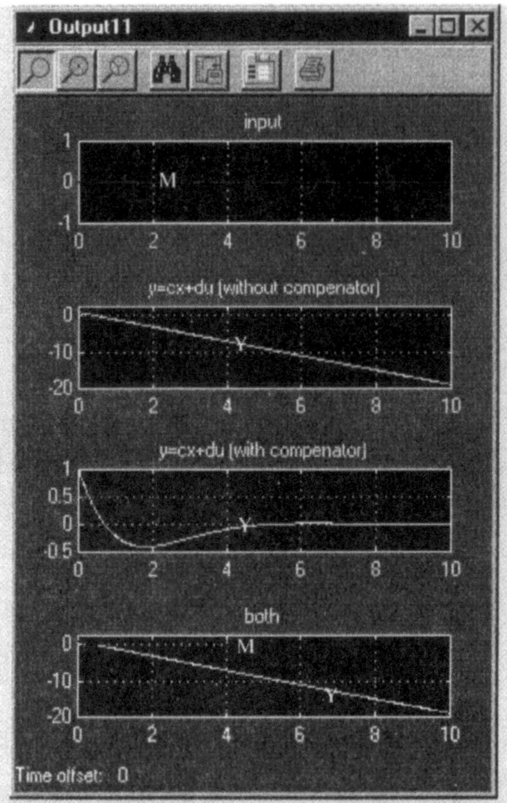

Let us try using other syntax of **care** function for the same model.

```
[P, L, g, report]=care(a,b,Q,'report')
returns

P =
    2.3784    1.4142
    1.4142    1.6818
L =
  -0.8409 + 0.8409i
  -0.8409 - 0.8409i
g =
    1.4142    1.6818
report =
  1.3953e-015


[P1,P2,L,report]=care(a,b,Q,'implicit')
returns

P1 =
  -0.2259 - 0.2950i   0.3003 + 0.3241i
   0.4380 + 0.0581i  -0.1338 - 0.5310i
P2 =
   0.0822 - 0.6195i   0.5251 + 0.0200i
   0.4172 - 0.3195i   0.1997 - 0.4346i
```

```
L =
  -0.8409 + 0.8409i
  -0.8409 - 0.8409i
report =
    0
```

Now let us try out using **care** function to obtain solution for the more general form of Riccati equation given by:

$$\text{Riccati}(P) = a'Pe + e'Pa - (e'Pb+s)r^{-1}(b'Pe+s') + Q = 0$$

Using the same values of a, b, c, d let us assume the value of additional parameter s (same size as b) and e (same size as a) as follows:

```
s=[0; 0.5];
e=[ 2 1; 3 4];
[P, L, g, rr]=care(a,b,Q,r,s,e)
returns

P =
   1.0642    0.2225
   0.2225    0.5183
L =
  -0.3064
  -0.6528
g =
   1.0000    1.3979
rr =
 1.0009e-014

[P, L, g, report]=care(a,b,Q,r,s,e,'report')
returns

P =
   1.0642    0.2225
   0.2225    0.5183
L =
  -0.3064
  -0.6528
g =
   1.0000    1.3979
report =
 1.0009e-014

s=[0; 0.5];
e=[ 0 1; 1 0];
[P, L, g, report]=care(a,b,Q,r,s,e,'report')
returns

P =
   []

L =
   []
```

```
g =
    []
report =
   -2
```

report = -2 indicates no finite solution with these values of *s* and *e* matrix *i.e.*, failure.

```
[P1,P2, L,report]=care(a,b,Q,r,s,e,'implicit')
returns
```

```
P1 =
       0 - 0.0000i          0.0000 - 0.0000i
  0.0000 + 0.0000i         -0.4472 + 0.0000i
P2 =
 -0.0000 + 1.0000i          0.0000 + 0.0000i
  0.0000 + 0.0000i         -0.8944 - 0.0000i
L =
 -1.0000 + 0.0000i
 -1.0000 - 0.0000i
report =
    0
```

report =0 indicates success.

7.4.1.2 Solution of Discrete-time Algebraic Riccati Equations

Similar to Riccati equations for continuous-time models, Riccati equations for discrete-time models also have two main forms of representation. The built-in function **dare** is used to solve these equations. The solution of Riccati equations based upon its form is summarised in Table 7.4.

Let us consider some illustrations to further elucidate this concept.

Example:

Let us consider the same state space model that we considered for the continuous-time model. Let the model have an additional parameter t (sampling time) in addition to *a*, *b*, *c*, *d* making it a discrete model. Let the parameters have values as given below:

```
a=[0 1; 0 0];
b=[0;1];
c=[1 0];
d=[0];
t=0.1;
modelG8=ss(a,b,c,d,t);
```

Let the Riccati equation have the form as indicated below:

Riccati(P) = a'Pa-P- a'Pb(b'Pb+r)$^{-1}$b'Pa +Q=0

Choosing the value of matrix *Q* as given below, let us find out the solution of the equation...

Table 7.4. Syntax for solving Discrete-time type of Riccati equations

Form of Riccati Equation	Syntax	Results into	Remarks
Riccati(P)=0 or, a'Pa-P-a'Pb(b'Pb+r)$^{-1}$b'Pa +Q=0	[P,L,g,rr]= **dare** (a,b,Q,r)	P... a symmetric matrix called stabilizing solution of Riccati equation L... eigenvalues of a_{cl} g... gain matrix $((b'Pb+r)^{-1}b'Pa)$ rr... relative residual $\dfrac{\|\,\mathrm{Riccati(P)}\,\|_F}{\|P\|_F}$	all eigenvalues of $a_{cl}=a-b(b'Pb+r)^{-}b'Pa$ should be in the unit circle
Riccati(P)=0 or, a'Pa+e'Pe-(a'Pb+s)(b'Pb+r)$^{-1}$(b'Pa+s') +Q=0	[P,L,g,(rr or report)]= **dare** $(a,b,Q,r,s,e,'report')$	P... a symmetric matrix called stabilizing solution of Riccati equation L... eigenvalues of a_{cl} g... gain matrix $((b'Pb+r)^{-1}(b'Pa+s'))$ report... (optional) turns off the error messages when the solution P fails to exist and returns a failure report instead which has values • -1... indicating associated hamiltonian pencil has eigenvalues on or very near the imaginary axis *i.e.*, failure. • -2... indicating there is no finite solution *i.e.*, P=P1P2^{-1} has P1 singular, failure • rr... relative residual $\dfrac{\|\,\mathrm{Riccati(P)}\,\|_F}{\|P\|_F}$ when solution exists	
	[P1,P2,L,report]= **dare** $(a,b,Q,r,s,e,'implicit')$	• same as above • turns off error message • returns P in implicit form *i.e.*, P=P2P1^{-1} when successful • returns zero (0) in case of failure	

```
Q=[2 0; 0 0];
[P, L, g, rr]=dare(a,b,Q)
returns


P =
    2    0
    0    2
L =
  1.0e-016 *
  -0.8635
        0
g =
    0    0
rr =
    0
```

The solution has resulted into a positive definite matrix P, which is indeed the stabilising solution of the Riccati equation. We can also design an optimal state feedback compensator for the model using the solution of Riccati equation. For the same model, choosing $r=2$, and the value of initial states $x1(0)=1$ and $x2(0)= -2$, lets try out designing…

```
r=2;
[P, L, g, rr]=dare(a,b,Q,r)
returns


P =
    2.0000    0.0000
    0.0000    2.0000
L =
  1.0e-016 *
  -0.8882
        0
g =
  1.0e-047 *
        0    0.2737
rr =
  7.8505e-017
```

For the model the optimal state feedback compensator input is u=-kx where the gain k can be calculated using expression $k= r^{-1}\, b'P$,

```
k=inv(r )*b'*P
returns


k =
    0.0000    1.0000
```

You may compare the outputs of the models with and without optimal state feedback controller for zero input in the same way as you did for the continuous modelG7 above to see if you have indeed designed an optimal controller.

Let us try using other syntax of **dare** function for the same model.

```
[P, L, g, report]=dare(a,b,Q,'report')
returns

P =
   2    0
   0    2
L =
  1.0e-016 *
  -0.8635
       0
g =
   0    0
report =
   0


[P1,P2,L,report]=dare(a,b,Q,'implicit')
returns

P1 =
  -0.8321         0
       0  -0.8321
P2 =
  -1.6641         0
       0  -1.6641
L =
  1.0e-016 *
  -0.8635
       0
report =
   0
```

Now let us try using **dare** function to obtain solution for the more general form of Riccati
equation given by:

$$\text{Riccati}(P)=a'Pa+e'Pe-(a'Pb+s)(b'Pb+r)^{-1}(b'Pa+s')+Q=0$$

using the same values of a, b, c, d let us assume the value of additional parameter s (same size as
b) and e (same size as a) as follows:

```
s=[0; 0.5];
e=[ 2 1; 3 4];
[P, L, g, rr]=dare(a,b,Q,r,s,e)
returns

P =
   1.9806   -0.7871
  -0.7871    0.3914
L =
  -0.5520
   0.0000
g =
   0    -0.1200
rr =
  1.3315e-015
```

```
[P, L, g, report]=dare(a,b,Q,r,s,e,'report')
returns

P =
   1.9806   -0.7871
  -0.7871    0.3914
L =
  -0.5520
   0.0000
g =
      0  -0.1200
report =
  1.3315e-015

s=[0; 0.5];
e=[ 0 1; 1 0];
[P, L, g, report]=dare(a,b,Q,r,s,e,'report')
returns

P =
   []
L =
   []
g =
   []
report =
   -1
```

report = -1 indicates eigenvalues on or near unit circle *i.e.*, failure.

```
[P1,P2, L,report]=dare(a,b,Q,r,s,e,'implicit')
returns

P1 =
   []
P2 =
   []
L =
   []
report =
   -1
```

report =-1 indicates failure.

Ready for a test?

Practice Test 7. 5.

1. Solve all forms of Riccati equations for modelGp1 and modelGp2 using syntax summarised in Table 7.2 and Table 7.3. Choose the value of other relevant parameters on your own. Assume sampling time as ts=-1 and 0.1 respectively. Also design an optimal compensator for the two models using the solution so obtained. Comment on the response.

7.4.2 Design of Kalman State Estimator

Models that were considered previously in this chapter had certain uncertainties involved in the form of stochastic inputs and nonmeasurable outputs. But even these can be estimated. Sometimes the model has inputs and outputs entirely random in nature depending upon multitudes of factors which can never be computed. Such inputs as you already know are called noise. You must also be aware of the fact that there is almost no system that is not amenable to noise. Therefore, it becomes necessary to design compensators which take noise into consideration too. Kalman filtering technique gives design of compensators for such models.

7.4.2.1 Kalman Estimator for Continuous-/Discrete-time Models

Kalman state estimator can be designed for continuous-time and discrete-time state space models with the help of built-in function **kalman** available with Control System Toolbox of MATLAB®, which takes into consideration the process and measurement noise covariance data. Details of such design process are summarised in Table 7.5. Syntax for achieving this is:

[kest, L, P, M, Z]=**kalman**(*ModelName, Qn, rn, Nn, meas o/p sensors, known i/p*)

where,

kest	returns gain of discrete Kalman State Estimator
L	returns Kalman gain for continuous model and filter gain for discrete model
M	returns innovations gain M
P	returns solution of associated Riccati equation for continuous models (steady state error covariance matrices)
Z	returns steady state error covariance matrices for discrete models (does not exist for continuous models)
ModelName	specifies the state space model under consideration for which the estimator has to be designed
Qn,rn,Nn	specifies the noise covariance data matrices (default value of N=0 if not otherwise specified or when N is omitted)
meas o/p sensors	specifies the vector containing indices of the measurable output of the model *i.e.,* y (optional)
known i/p	specifies the vector containing indices of the known inputs of the model *i.e.,* u (optional)

Note that *rn* must be square with as many rows as measured outputs while *Nn* must have as many rows as *Qn* and as many columns as *rn*.

Table 7.5. Kalman State Estimator

State Space Equation of Model	Noise Covariance Data	Optimal Kalman State Estimator	Remarks
For Continuous Model:			
$\dot{x} = ax + bu + gw$ $y_v = cx + du + hw + v$	$E(w)=E(v)=0,$ $E(ww')=Q,$ $E(vv')=r,$ $E(wv')=N$	$x^\wedge = ax^\wedge + bu + L(y_v - cx^\wedge - du)$ $\begin{bmatrix} y^\wedge \\ x^\wedge \end{bmatrix} = \begin{bmatrix} c \\ I \end{bmatrix} x^\wedge + \begin{bmatrix} d \\ 0 \end{bmatrix} u$	Steady State error covariance P is minimised
For Discrete Model:			
$\dot{x}[n+1] = ax[n] + bu[n]$ $\qquad + gw[n]$ $y_v[n] = cx[n] + du[n]$ $\qquad + hw[n] + v[n]$	$E(w[n]w[n]')=Q,$ $E(v[n]v[n]')=r,$ $E(w[n]v[n]')=N$	$x^\wedge[n+1\mid n] = ax^\wedge[n+1\mid n] + bu[n]$ $\qquad + L(y_v[n] - cx^\wedge[n$ $\qquad +1\mid n] - du[n])$ $\begin{bmatrix} y^\wedge[n] \\ x^\wedge[n] \end{bmatrix} = \begin{bmatrix} c(I - Mc) \\ I - Mc \end{bmatrix} x^\wedge[n+1\mid n]$ $\qquad + \begin{bmatrix} (I - cM)d & cM \\ -Md & M \end{bmatrix} \begin{bmatrix} u[n] \\ y_v[n] \end{bmatrix}$	M is used to update prediction $x^\wedge[n\mid n-1]$ using new measurement $x^\wedge[n\mid n]$

Note:

- $P = s \xrightarrow{\lim} \infty \, (E(x - x^\wedge)(x - x^\wedge)')$
- $x^\wedge[n\mid n] = x^\wedge[n\mid n-1] + M(y_v[n] - cx^\wedge[n\mid n-1] - Du[n])$

Kalman Estimator

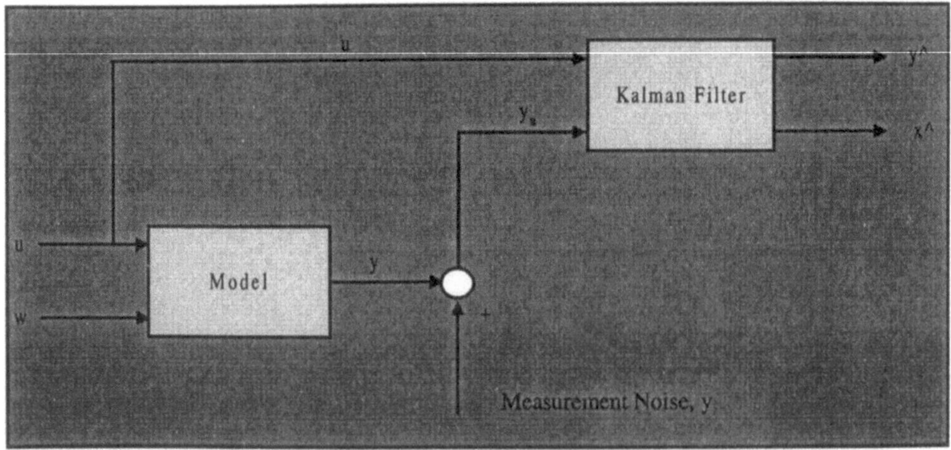

Figure 7.6. Kalman Estimator

Forthcoming illustrations will further elucidate this concept.

Example:

Consider the modelG3 for which we designed a simple estimator earlier. For this model and
Qn=[1] and rn=2, we will design a Kalman estimator and observe its response.

```
a=[-1 0 0; 0 -5 0; 0 0 -10];
b=[0; 1; 1];
c=[2 1 3];
d=[0];
modelG3=ss(a,b,c,d,'InputName','w1','OutputName','y1');
Qn=[1];
rn=2;
[kestG3,L,P,M,Z]=kalman(modelG3,Qn,rn)
returns
```

a =

	x1_e	x2_e	x3_e
x1_e	-1	-1.3086e-018	-3.9258e-018
x2_e	-0.28968	-5.1448	-0.43453
x3_e	-0.21128	-0.10564	-10.317

b =

	y1
x1_e	1.3086e-018
x2_e	0.14484
x3_e	0.10564

c =

	x1_e	x2_e	x3_e
y1_e	2	1	3
x1_e	1	0	0
x2_e	0	1	0
x3_e	0	0	1

d =

	y1
y1_e	0
x1_e	0
x2_e	0
x3_e	0

e =

	x1_e	x2_e	x3_e
x1_e	1	0	0
x2_e	0	1	0
x3_e	0	0	1

I/O groups:

Group name	I/O	Channel(s)
Measurement	I	1
OutputEstimate	O	1
StateEstimate	O	2,3,4

Continuous-time model.

Note that:

- Rn must be square with as many rows as measured outputs.
- Nn must have as many rows as Qn and as many columns as Rn.
- Length of KNOWN and size of Qn should add up to the number of inputs in SYS.
- The matrix [G 0;H I]*[Qn Nn;Nn' Rn]*[G 0;H I]' should be positive semi-definite.

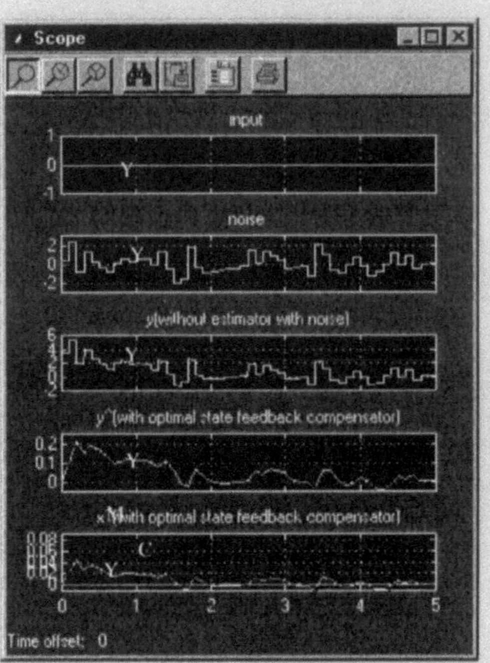

a.

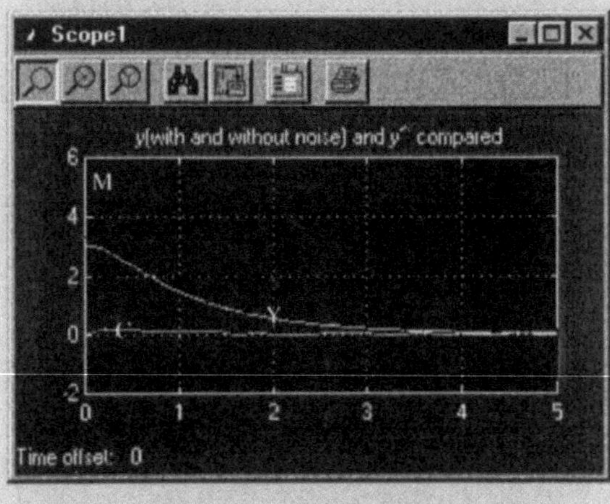

b.

Figure 7.7. Response of modelG3 with and without Kalman estimator and optimal State Feedback Controller

```
L =
    0.0000
    0.1448
    0.1056
P =
         0    0.0000    0.0000
    0.0000    0.0958    0.0646
```

```
  0.0000   0.0646   0.0489
M =
   []
Z =
   []
```

Next we design a state feedback controller k using Riccati solution P in the same way as we did earlier.

```
k=inv(rn)*b'*P
returns

k =
  0.0000   0.0802   0.0568
```

With the Kalman estimator kest, state covariance matrices P, a state feedback controller gain k, initial value of the states x(1)=2, x(2)=-1, x(3)=0 and zero input signal, the response of the model mod with and without compensator is compared as shown below with bandlimited white noise signal as the noise input. Observe that with compensator the output is least affected by the noise and maintains well around zero.

Take up a practice test for a change…

Practice Test 7. 6.

1. Assuming suitable values of other relevant parameters, design an optimal compensator using Kalman estimator and optimal state feedback controller for modelGp1 and modelGp2 for sampling time ts= -1 and 0.1 respectively considering,

a) all inputs are stochastic and all outputs are measured for both the models.

b) input1 is stochastic, input2 is known and output2 is measured for modelGp1.

c) input1 and input3 are stochastic, input2 and input4 are known; output2 and output3 are measured for modelGp2.

Comment on the response.

7.4.2.2 Discrete Kalman State Estimator for Continuous-time Models

Discrete Kalman state estimator can be designed for a continuous model. Let us consider a model with process noise w and measurement noise v represented by the equation given as:

$$\dot{x} = ax + bu + gw$$

$$y_v = cx + du + v$$

where, w and v are such that the following conditions are satisfied:

$$E(w)=E(v)=0, \quad E(ww')=Qn, \quad E(vv')=rn, \quad E(wv')=0$$

then Control System Toolbox of MATLAB® computes the discrete Kalman estimator for the continuous-time model by:

- discretizing the model using zero-order hold with sample time Ts
- replacing continuous noise covariance matrices Qn and rn by their discrete counterparts Qd and rd where:

$$Qd = \int_0^T e^{A\tau} gQg' e^{A'\tau} d\tau$$

$$rd = \frac{R}{Ts}$$

- computing the integral by using matrix exponential formulas
- designing a discrete-time estimator for the discretized model and noise

The function achieving all this is **kalmd** for which the syntax is:

[kest, L, P, M, Z]=**kalmd**(*ModelName, Qn, rn, Ts*)

where,

kest	returns gain of discrete Kalman state estimator
L,M	returns estimator gain
P,Z	returns discrete error covariance matrices P and Z
ModelName	specifies the continuous state space model under consideration for which the estimator has to be designed
Qn, rn	specifies the continuous noise covariance data matrices
Ts	specifies Sampling Time of Kalman State Estimator

Note that covariance matrix *rn* is a positive definite square matrix with as many rows as measured outputs.

Let us take up a few examples to get a better understanding.

Example:

```
a=[-1 0 0; 0 -5 0; 0 0 -10];
b=[0; 1; 1];
c=[2 1 3];
d=[0];
modelG3=ss(a,b,c,d);
ts=0.5;
Qn=[1];
rn=2;
[kestG10,L,P,M,Z]=kalmd(modelG3,Qn,rn,ts)
returns
```

```
a =
                x1_e        x2_e          x3_e
    x1_e       0.60653        0             0
    x2_e     -0.0099452    0.077112     -0.014918
    x3_e    -0.00058985  -0.00029493    0.0058532

b =
                y1
    x1_e         0
    x2_e     0.0049726
    x3_e     0.00029493

c =
                x1_e        x2_e         x3_e
    y1_e        1.6162      0.80811      2.4243
    x1_e          1           0            0
    x2_e       -0.12116     0.93942     -0.18174
    x3_e      -0.087542    -0.043771     0.86869

d =
                y1
    y1_e      0.19189
    x1_e         0
    x2_e      0.060579
    x3_e      0.043771

e =
                x1_e        x2_e        x3_e
    x1_e          1           0           0
    x2_e          0           1           0
    x3_e          0           0           1

I/O groups:
    Group name      I/O     Channel(s)
    Measurement      I          1
    OutputEstimate   O          1
    StateEstimate    O        2,3,4

Sampling time: 0.5
Discrete-time model.

L =
     0
   0.0050
   0.0003
P =
     0        0        0
     0     0.0999   0.0667
     0     0.0667   0.0500
M =
     0
   0.0606
   0.0438
Z =
     0        0        0
     0     0.0817   0.0535
     0     0.0535   0.0405
```

```
k=inv(rn)*b'*P

k =
     0    0.0833    0.0583
```

With the discrete Kalman estimator kest10, state covariance matrices P, a state feedback controller gain k, initial value of the states x(1)=2, x(2)=-1, x(3)=0 and zero input signal, the response of the continuous model modelG3 with and without compensator is compared as shown below with bandlimited white noise signal as the noise input. Observe that with compensator the output is least affected by the noise and maintains well around zero.

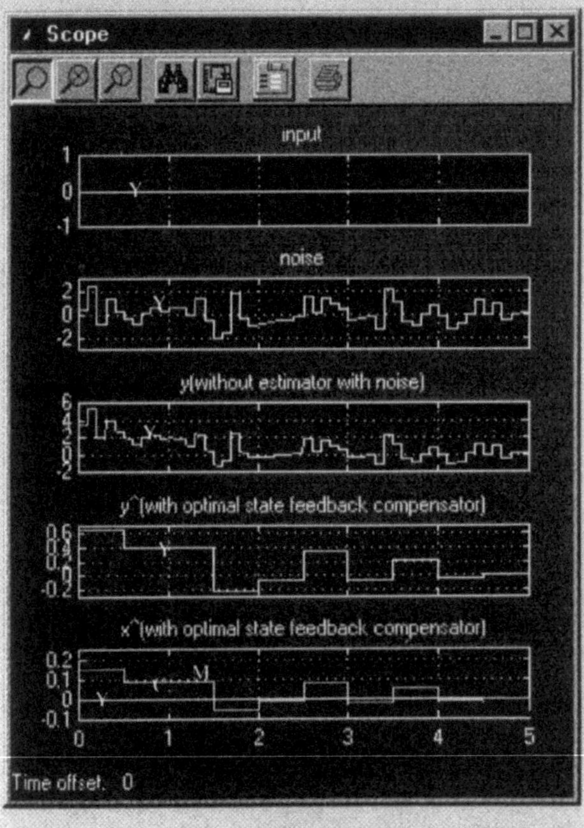

a.

Figure 7.8. Response of modelG3 with and without discrete Kalman estimator and optimal state feedback controller (Panel b. facing page)

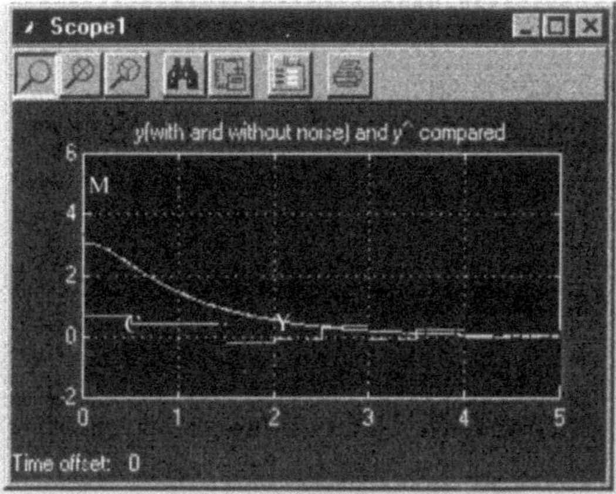

Figure 7.8b.

Take up a practice test for a change…

Practice Test 7. 7.

1. Assuming suitable values of other relevant parameters, design an optimal compensator using discrete Kalman estimator and optimal state feedback controller for modelGp1 and modelGp2 considering

a) all inputs are stochastic and all outputs are measured for both the models.

b) input1 is stochastic, input2 is known and output2 is measured for modelGp1.

c) input1 and input3 are stochastic, input2 and input4 are known; output2 and output3 are measured for modelGp2.

    Comment on the response.

7.4.3 Design of Linear Quadratic State Feedback Regulator

We know that optimisation refers to the science of maximising or minimising objectives based upon measure of performance popularly known as objective or cost function. Control System Toolbox of MATLAB® has two built-in functions **lqr** and **dlqr** for designing such optimal controllers for continuous-time and discrete-time systems respectively.

7.4.3.1 Linear Quadratic State Feedback Regulator for Continuous Models
For a continuous-time model represented by the state space equations:

$$\dot{x} = ax + bu$$

$$y = cx + du$$

the design of optimal compensator requires finding a control function u(t) that minimises the cost function J given by:

$$J = \int_{0}^{\infty} (x'Qx + u'ru + 2x'Nu)dt$$

This can be achieved by using built-in function **lqr** of Control System Toolbox of MATLAB® for which the syntax is:

$$[k, s, e] = \mathbf{lqr}(a, b, Q, r, N)$$

where,

k	returns the optimal feedback gain matrix such that control function u(t)=-kx
s	returns solution of associated Riccati equation: a's+sa-(sb+N)r⁻¹(b's+N')+Q=0 (default value of N=0 if not otherwise specified or N is omitted)
e	returns the eigenvalues eig(a-bk) where k is derived as k=r⁻¹(b's+N')
a	specifies 2D real valued system matrix of the model under consideration
b	specifies 2D real valued input matrix of the model under consideration
Q, r, N	specifies quantities associated with Riccati equation a's+sa-(sb+N)r⁻¹(b's+N')+Q=0

Note that a and Q matrices must be the same size. Similarly, b and N matrices must be the same size.

Let us take up a few illustrations to clarify the concept.

Example:

```
a=[-2 1 0; 4 0 1; 0 0 0];
b=[0; 0; 1];
c=[2 0 -1; 1 1 0];
d=[0];
modelG10=ss(a,b,c,d);
Q=[1 0 0;0 0 0;0 0 0];
r=2;
N=[1;0 ;0];
[k,s,e] = lqr(a,b,Q,r,N)
returns

k =
    4.3610    3.3498    2.5883

s =
   13.1830   11.2210    7.7219
   11.2210    9.6188    6.6995
    7.7219    6.6995    5.1767
```

```
e =
  -3.2704
  -1.1259
  -0.1920
```

For this system, let us compare the performance with and without state feedback for a zero step input and value of the initial states as $x(1)=0.5$, $x(2)=0$ and $x(3)=0$ using SIMULINK®. The result is as shown below. Observe how fast the response of the system with controller converges to zero.

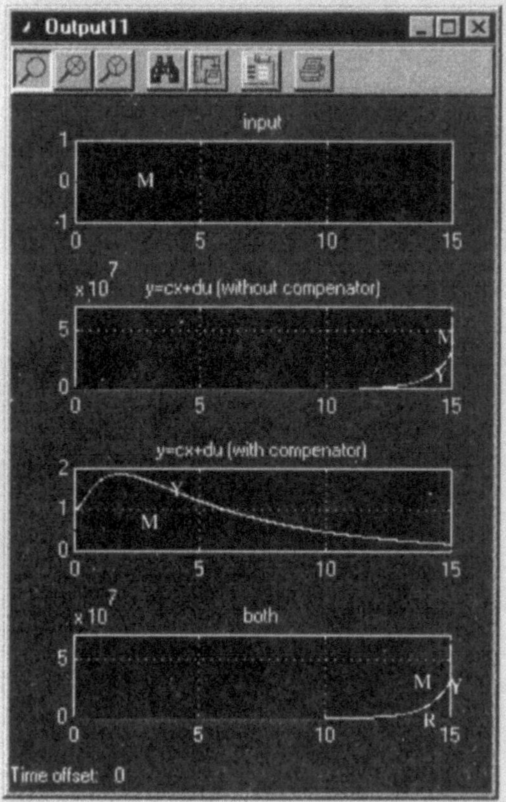

Take up a practice test for a change…

7.4.3.2 Linear Quadratic State Feedback Regulator for Discrete models

For a discrete-time model represented by the state space equations:

$$\dot{x}_{(n+1)} = ax_{(n)+} + u_{(n)}$$

$$y_{(n)} = cx_{(n)} + du_{(n)}$$

the design of optimal compensator requires finding a control function u(t) that minimises the cost function J given by:

$$J = \sum_{n=1}^{\infty} \left(x[n]'Qx[n] + u[n]'Ru[n] + 2x[n]'Nu[n] \right)$$

This can be achieved by using built-in function **dlqr** of Control System Toolbox of MATLAB® for which the syntax is:

[k, s, e] = **dlqr**(*a, b, Q, r, N*)

where,

k .... returns the optimal feedback gain matrix such that control function u(n)=-kx[n]

s.. returns solution of associated Riccati equation
a'sa-s-(a'sb+N)(b'xb+r)$^{-1}$(b'sa+N')+Q=0
(default value of N=0 if not otherwise specified or N is omitted)

e.. returns the eigenvalues eig(a-bk) where k is derived as
k=(b'xb+r)$^{-1}$(b'sa+N')

a .... specifies 2D real valued system matrix of the model under consideration

b .... specifies 2D real valued input matrix of the model under consideration

Q, r, N.. specifies quantities associated with Riccati equation
a'sa-s-(a'sb+N)(b'xb+r)$^{-1}$(b'sa+N')+Q=0

Let us take up a few examples to get a better understanding.

Example:

```
a=[-2 1 0; 4 0 1; 0 0 0];
b=[0; 0; 1];
c=[2 0 -1; 1 1 0];
d=[0; 0];
ts=0.1
modelG11=ss(a,b,c,d,ts);
Q=[1 0 0;0 0 0;0 0 0];
r=2;
N=[1;0 ;0];
[k,s,e] = dlqr(a,b,Q,r,N)
returns
```

```
ts =
   0.1000
k =
 -28.9241    9.1479   -2.6656
s =
  1.0e+003 *
    2.6687   -0.7871    0.2662
   -0.7871    0.2338   -0.0776
    0.2662   -0.0776    0.0271
e =
   -0.3050
    0.8358
    0.1348
```

For this system, compare the performance with and without state feedback for a zero step input and value of the initial states as x(1)=0.5, x(2)=0 and x(3)=0 using SIMULINK® as before. Observe that the response of the system with compensator converges to zero when optimal state feedback controller with gain k is used.

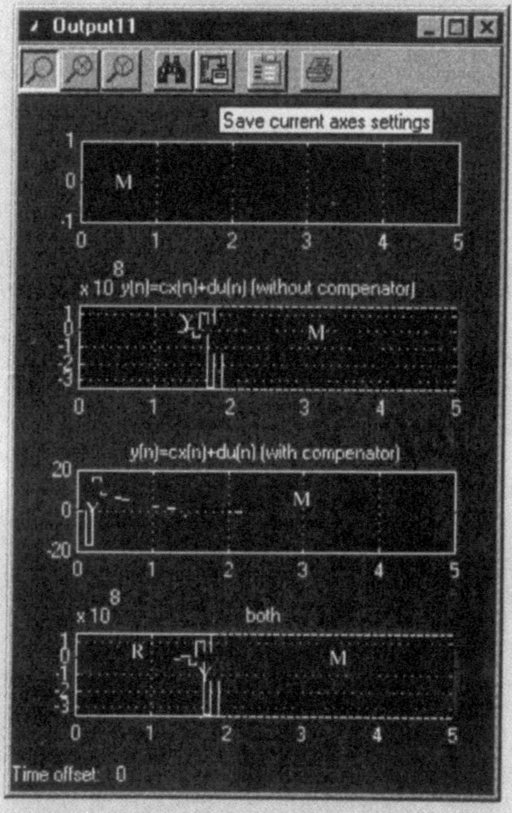

Take up a practice test for a change…

7.4.3.3 Linear Quadratic State Feedback Regulator with Output Weighting

When the quadratic cost function has output weighting, the linear quadratic state feedback regulator can still be designed using **lqry** function available with Control System Toolbox of MATLAB®.

For a continuous-time model represented by the state space equations:

$$\dot{x} = ax + bu$$

$$y = cx + du$$

the design of optimal Linear Quadratic State Feedback Regulator with Output Weighting requires finding a control function u=-kx. that minimises the cost function J with output weighting given by:

$$J = \int_{0}^{\infty} (y'Qy + u'ru + 2y'Nu)dt$$

Syntax of **lqry** function is:

[k, s, e] = **lqry**(*ModelName, Q, r, N*)

where,

k	returns the optimal feedback gain matrix such that control function u=-kx
s	returns solution of associated Riccati equation a's+sa-(sb+N)r^{-1}(b's+N')+Q=0 (default value of N=0 if not otherwise specified or N is omitted)
e	returns the eigenvalues eig(a-bk) where k is derived as k=r^{-1}(b's+N')
ModelName	specifies the name of the model under consideration
Q, r, N	specifies quantities associated with the cost function J

In a similar way Linear Quadratic State Feedback Regulator with Output Weighting can also be designed for discrete-time models. In fact, function **lqry** is equivalent to **lqr** or **dlqr** with weighting matrices given by:

$$\begin{bmatrix} \overline{Q} & \overline{N} \\ \overline{N}' & \overline{r} \end{bmatrix} = \begin{bmatrix} c' & 0 \\ d' & I \end{bmatrix} \begin{bmatrix} Q & N \\ N' & R \end{bmatrix} \begin{bmatrix} c & d \\ 0 & I \end{bmatrix}$$

Note that Q must be symmetric with as many rows as C, D and N matrices must be of the same size, while the R matrix must be square with as many columns as B.

Let us take up a few examples to elucidate this concept.

Example:

```
a=[9 1 2 1; 1 2 3 1;1 3 6 5; 7 1 2 3];
b=[1 4 ; 1 2; 3 1; 2 1];
c=[2 4 6 8;9 7 5 3];
d=[1 0; 0 0];
modelG12=ss(a,b,c,d);
Q=[1 0;0 0];
r=[ 4 0; 0 2];
N=[0];
[k,s,e] = lqry(modelG12,Q,r,N)
returns


k =
   -0.6944    3.1683    4.2401    4.1861
    6.8932   -0.9734    1.2189    2.6931
s =
    7.2958   -6.3545   -1.0379   -1.6497
   -6.3545    9.7365    0.4629    3.5354
   -1.0379    0.4629    4.4478    1.2161
   -1.6497    3.5354    1.2161    3.6982
e =
  -25.8141
   -4.5823
   -2.1810
   -0.5267
```

For this system, compare the performance with and without state feedback for a zero step input and value of the initial states as x(1)=1, x(2)=2, x(3)=3 and x(4)=4 using SIMULINK® as before. Observe that the response of the system with compensator converges to zero when optimal state feedback controller with gain k is used.

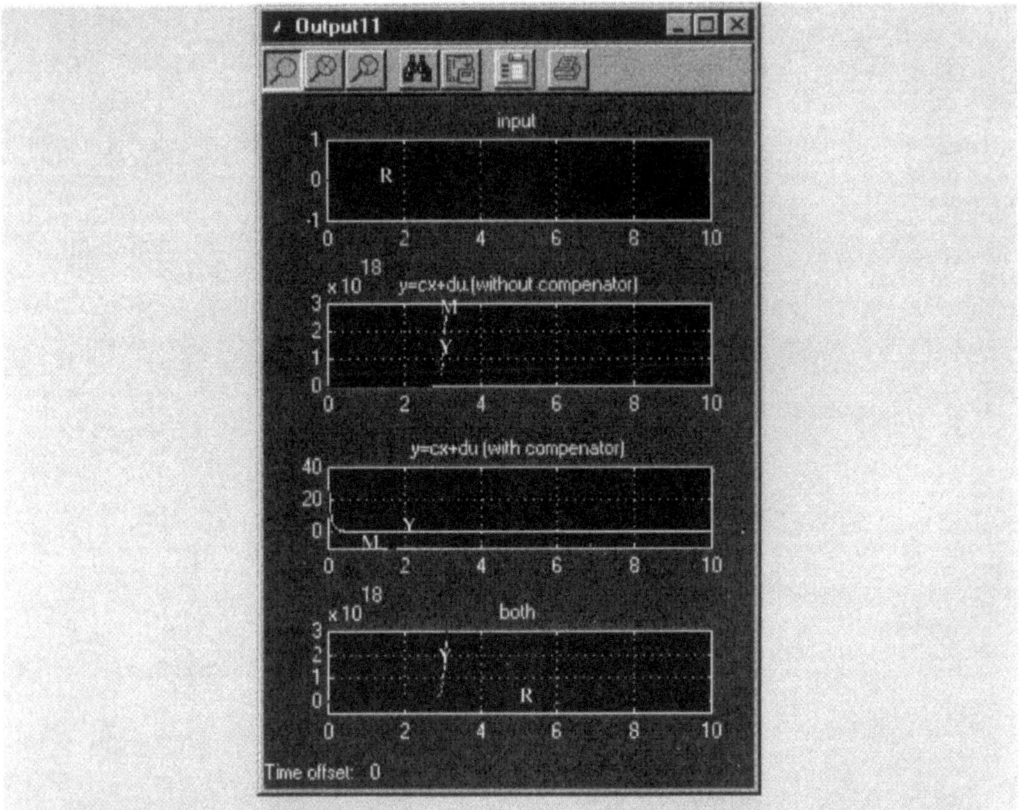

Take up a practice test for a change…

Practice Test 7.10.

1. Assuming suitable values of other relevant parameters, design a Linear Quadratic State Feedback regulator with Output Weighting for modelGp1 and modelGp2 considering sampling time as -1 and 0.01. Comment on the response.

7.4.3.4 Discrete Linear Quadratic State Feedback Regulator for Continuous Model

For a continuous-time model represented by the following state space equations:

$$\dot{x} = ax + bu$$

$$y = cx + du$$

design of discrete Linear Quadratic State Feedback Regulator with sample time Ts is also possible. It requires finding a control function $u[n] = -k_d x[n]$ that minimises the cost function J having a cross-coupling term as given below:

$$J = \int_{0}^{\infty} (x'Qx + u'ru + 2x'Nu)dt$$

This can be achieved by using built-in function **lqrd** of Control System Toolbox of MATLAB® for which the syntax is:

[kd, s, e] = **lqrd**(a, b, Q, r, N, ts)

where,

kd	returns the optimal feedback gain matrix such that control function u[n]=-k_dx[n] (kd is determined by discretizing the continuous model and weighting matrices using the sample time Ts an the zero-order hold approximation).
s	returns solution of associated Riccati equation
e	returns the eigenvalues eig(ad-bd*kd)
a	specifies 2D real valued system matrix of the model under consideration
b	specifies 2D real valued input matrix of the model under consideration
Q, r, N	specifies quantities associated with the cost function J
ts	specifies the sampling time for the controller with gain kd

Let us take up a few examples to get a better understanding.

Example:

```
a=[-2 1 0; 4 0 1; 0 0 0];
b=[0; 0; 1];
c=[2 0 -1; 1 1 0];
d=[0];
ts=0.1;
modelG13=ss(a,b,c,d);
Q=[1 0 0;0 0 0;0 0 0];
r=2;
N=[1;0 ;0];
[kd,s,e] = lqrd(a,b,Q,r,N,ts)
returns

kd =
    4.0537    3.1466    2.4314
s =
   13.2082   11.2395    7.7357
   11.2395    9.6325    6.7097
    7.7357    6.7097    5.1842
e =
    0.7210
    0.8935
    0.9810
```

For this system, compare the performance with and without state feedback for a zero step input and value of the initial states as x(1)=0.5, x(2)=-1 and x(3)=2 using SIMULINK® as before. Observe that the response of the system with compensator converges to zero when optimal state feedback controller with gain kd is used.

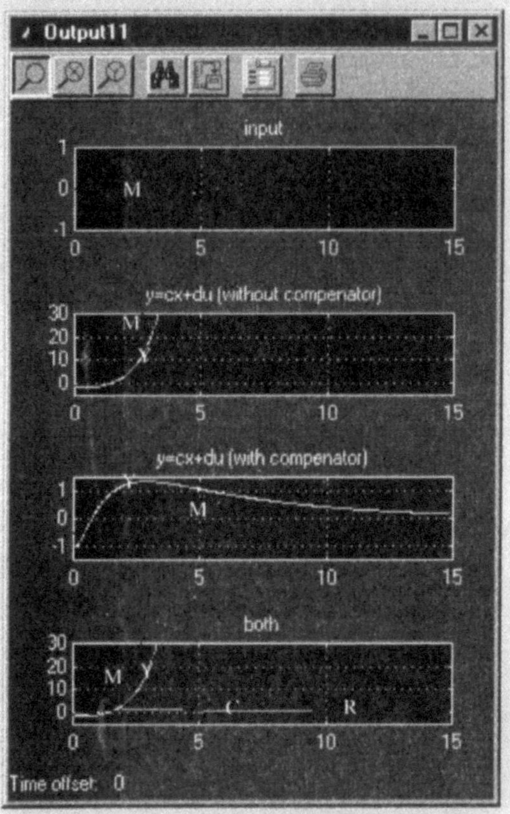

Take up a practice test for a change…

Practice Test 7.11.

1. Assuming suitable values of other relevant parameters, design a Discrete Linear Quadratic State Feedback regulator with Output Weighting for modelGp1 and modelGp2. Comment on the response.

7.4.4 Design of Linear Quadratic Gaussian Regulator

You know that the State Regulator designed earlier, had the drawback of not being optimal compensator for a model. Using similar principles an optimal regulator known as Linear Quadratic Gaussian Regulator may be designed by connecting the Kalman estimator (designed using **kalman** function) and the optimal state-feedback gain (designed using **lqr**, **dlqr** or **lqry** functions). This is shown in Figure 7.9. A continuous regulator for a continuous plant is produced using **lqr**, **lqry** and **kalman** functions. A discrete regulator for a discrete plant is produced using **dlqr**, **lqry** and **kalman** functions. A discrete regulator for continuous plant is produced using **lqrd** and **kalman** functions. This regulator is dynamic and produces the regulating command depending on the noisy output measurements of the model. It trades off between regulation performance and control effort to minimise the quadratic cost function.

For a continuous model, the regulator generates the command u=-kx^, where, x^ is the Kalman state estimate. The regulator is represented by the following state space equations

$$u = -kx^\wedge$$

$$\dot{\hat{x}} = a\hat{x} + bu + L(y_V - c\hat{x} - du)$$

For a discrete model, the regulator generated is:

u[n]=-kx^[n|n-1] (by default)
or,
u[n]=-kx^[n|n] (that is the current estimator when otherwise specified)

where,
x^[n] based on yv[n-1] (or x^[n|n] baed on yv[n]) is the Kalman state estimate.

This linear Quadratic Gaussian Regulator is designed using the built-in function, **lqgreg** available with the Control System Toolbox, which has following syntax:

lqgrg = **lqgreg**(kest, k, controls, 'current')

where,

lqgrg	returns the state space model of the linear quadratic Gaussian regulator which for discrete time is u[n]=-kx^[n|n-1] (default)
kest	specifies Kalman state estimator gain
k	specifies state feedback gain matrix
controls	is vector specifying indices of known inputs ud of the model (the resulting regulator has ud and yv as its input)
current	when specified, returns current regulator u[n]=-kx^[n|n] (instead of u[n]=-kx^[n|n-1] that you obtain by default)

Note the positive feedback for connection.
Let us try designing some of these regulators for some models to get better concepts about their design principles.

Example:

```
a=[-1 0 0; 0 -5 0; 0 0 -10];
b=[0; 1; 1];
c=[2 1 3];
d=[0];
modelG3=ss(a,b,c,d,'InputName','w1','OutputName','y1');
Qn=[2];
rn=1;
```

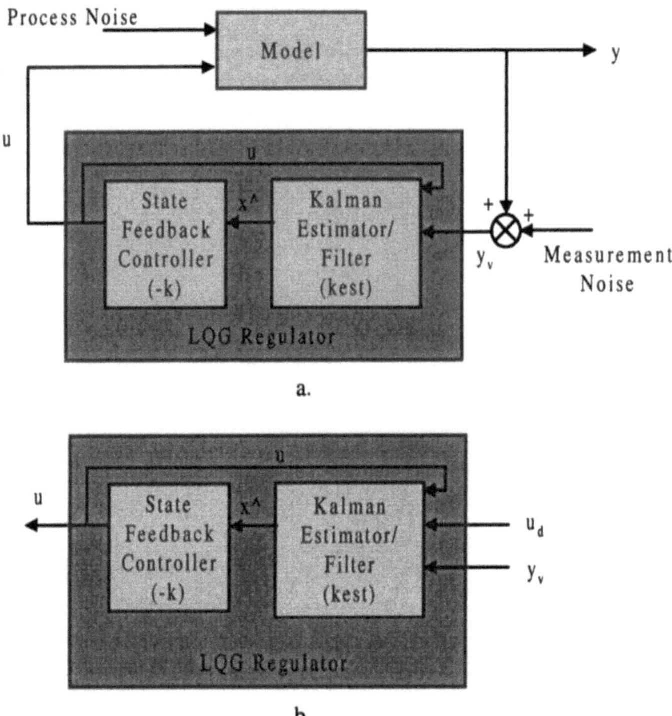

Figure 7.9. Linear Quadratic Gaussian Regulator

```
[kestG14,L,P,M,Z]=kalman(modelG3,Qn,rn)
k=inv(rn)*b'*P
lqgrgG14 = lqgreg(kestG14,k,[1])
returns

a =

              x1_e          x2_e           x3_e
    x1_e       -1         1.2312e-016    3.6937e-016
    x2_e     -1.0599       -5.53         -1.5899
    x3_e     -0.79169     -0.39584       -11.188

b =

              y1
    x1_e    -1.2312e-016
    x2_e     0.52996
    x3_e     0.39584

c =

            x1_e    x2_e    x3_e
    y1_e     2       1       3
    x1_e     1       0       0
    x2_e     0       1       0
    x3_e     0       0       1
```

```
d =
                y1
    y1_e         0
    x1_e         0
    x2_e         0
    x3_e         0

e =
               x1_e        x2_e        x3_e
    x1_e         1           0           0
    x2_e         0           1           0
    x3_e         0           0           1

I/O groups:
   Group name          I/O    Channel(s)
   Measurement          I         1
   OutputEstimate       O         1
   StateEstimate        O       2,3,4

Continuous-time model.

L =
  -0.0000
   0.5300
   0.3958
P =
      0      0.0000   -0.0000
   0.0000   0.1719    0.1193
  -0.0000   0.1193    0.0922
M =
   []
Z =
   []
k =
  -0.0000    0.2913    0.2115

a =
               x1_e         x2_e         x3_e
    x1_e         -1       1.5898e-016  3.9541e-016
    x2_e      -1.0599      -5.6843      -1.702
    x3_e      -0.79169     -0.51114     -11.271

b =

   Empty matrix: 3-by-0

c =
               x1_e         x2_e         x3_e
     y1     2.9524e-017   -0.29126     -0.21151

d =

   Empty matrix: 1-by-0

I/O groups:
   Group name    I/O    Channel(s)
   Controls        O        1
```

Continuous-time model.

For this system, compare the performance with and without state feedback for a zero step input and value of the initial states as x(1)=2, x(2)=-1 and x(3)=0 using SIMULINK® as before. Observe that the response of the system with compensator converges to zero when LQG regulator is used despite noise input to the system.

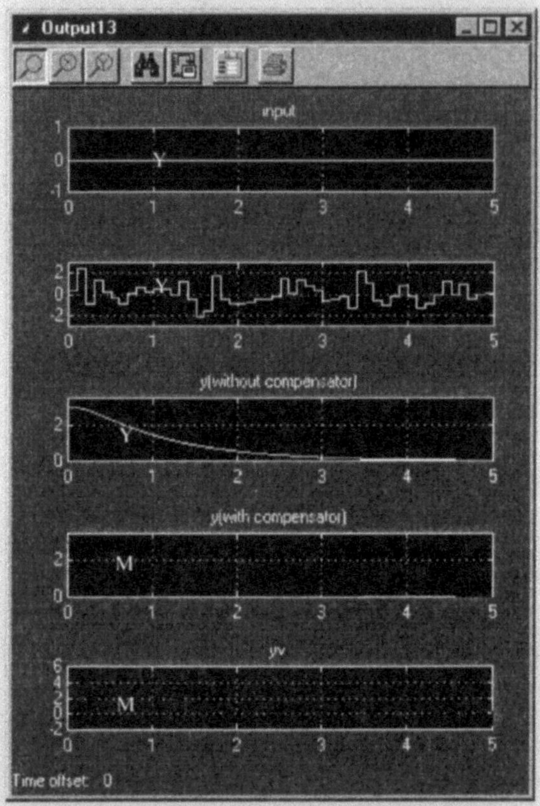

The same treatment can be extended to discrete models too.

```
a=[-2 1 0; 4 0 1; 0 0 0];
b=[0; 0; 1];
c=[2 0 -1];
d=[0];
ts=0.1;
modelG15=ss(a,b,c,d,ts);
Qn=[1];
rn=[2];
[kestG15,L,P,M,Z]=kalman(modelG15,Qn,rn)
k=inv(rn)*b'*P
lqgrgG15 = lqgreg(kestG15,k,[1])
returns
```

```
a =
                x1_e            x2_e          x3_e
      x1_e     0.39879            1         -1.1994
      x2_e     0.20241            0          2.8988
      x3_e           0            0               0

b =
                  y1
      x1_e     -1.1994
      x2_e      1.8988
      x3_e           0

c =
                x1_e            x2_e          x3_e
      y1_e     0.057832           0        -0.028916
      x1_e     0.043374           0          0.47831
      x2_e      0.48554           1         -0.24277
      x3_e     0.028916           0          0.98554

d =
                  y1
      y1_e      0.97108
      x1_e      0.47831
      x2_e     -0.24277
      x3_e     -0.014458

e =
                x1_e            x2_e          x3_e
      x1_e        1              0             0
      x2_e        0              1             0
      x3_e        0              0             1

I/O groups:
   Group name        I/O     Channel(s)
   Measurement        I          1
   OutputEstimate     O          1
   StateEstimate      O         2,3,4

Sampling time: 0.1
Discrete-time model.

L =
   -1.1994
    1.8988
        0
P =
   16.5416   -8.3958   -0.0000
   -8.3958   16.2916   -0.0000
   -0.0000   -0.0000    1.0000
M =
    0.4783
   -0.2428
   -0.0145
Z =
    0.7175   -0.3642    0.4783
   -0.3642   12.2151   -0.2428
    0.4783   -0.2428    0.9855
```

```
k =
  -0.0000  -0.0000   0.5000

a =
                  x1_e            x2_e         x3_e
     x1_e       0.39879            1         -0.5997
     x2_e       0.20241       1.3668e-017    1.9494
     x3_e          0              0             0
b =

   Empty matrix: 3-by-0

c =
                  x1_e            x2_e         x3_e
     y1        4.7471e-017    7.1984e-018     -0.5

d =

   Empty matrix: 1-by-0

I/O groups:
   Group name    I/O    Channel(s)
   Controls       O         1

Sampling time: 0.1
Discrete-time model.

lqgrgG16 = lqgreg(kestG15,k,'current')
returns

a =
                  x1_e            x2_e         x3_e
     x1_e       0.41626            1         -0.60407
     x2_e       0.17476       1.3768e-017    1.9563
     x3_e          0              0             0

b =

   Empty matrix: 3-by-0

c =
                  x1_e            x2_e         x3_e
     y1        -0.014563    7.2508e-018     -0.49636

d =

   Empty matrix: 1-by-0

I/O groups:
   Group name    I/O    Channel(s)
   Controls       O         1

Sampling time: 0.1
Discrete-time model.
```

Let us take up a practice test to reinforce your learning…

Exercise for Chapter 7:

1. For all the models generated in the exercise for Chapter 1 to 6, design all types of compensators about which you learnt in this chapter. Assume suitable values of relevant parameters to obtain the best response. Compare the results so obtained. (Remember to first convert the models into type suitable for such design).

Chapter 8

Some Simple Applications

So far you have learnt to use various features of the Control System Toolbox. In this chapter, you will apply all the learning acquired so far, to simulate some simple real physical systems that you might have come across in your day to day life.

In previous chapters, you learnt the use of built-in functions provided with Control System Toolbox of MATLAB® and SIMULINK® for computation, analysis and simulation of various Systems. But all the skills acquired so far can be appreciated only if they can be applied to study some real physical systems, which we come across in our day-to-day life. In this chapter, you will simulate, model, design and analyse the response of some real physical systems of simple nature.

8.1 Which Method to Choose

The moment you are given a problem to solve, your mind starts thinking about what should be done next and how it should be done. While going through previous chapters, you must have noticed that a problem can be solved by several different methods to give the same result. These methods could be summarised as follows:

- through MATLAB® environment using built-in functions provided with the Control System Toolbox
- through SIMULINK® using the block libraries of SIMULINK® or Control System Toolbox
- through the GUIs provided with the Control System Toolbox
- a combination of all of the above

You may wonder at times about which method you should choose to solve a particular problem. In such situation, follow the golden rules:

- choose a method you are good at
- choose a method that is appropriate and most suitable for a particular problem

The first choice mentioned above, however, needs no explanation as it depends entirely upon your judgement and self-evaluation. However, tips given below will to help you to decide the appropriateness of a suitable method for a particular problem:

- SIMULINK® does not support frd models, the models of LTI arrays and the descriptor state space models. Hence, for systems involving these, MATLAB® environment is the only refuge.
- Think of a problem where impulse response of a system is required. Try searching out the impulse signal in the block library of SIMULINK® or Control System Toolbox. It certainly does not exist. Hence, for such problems, you have two options:

    i. go to the MATLAB® command prompt and use the **impulse** function of Control System Toolbox. However, if you still insist on using SIMULINK®, then use SIMULINK® LTI Viewer for solving the problem.
    ii. obtain the impulse signal by differentiating the step signal in the SIMULINK® window. However, if you resort to this method, make sure that you do not set the step time parameter of step block to zero, else your scope will display a zero signal instead of an impulse signal. Also be prepared for warning messages like one given below, when you use a step signal with step time 1 (for example) for obtaining impulse signal:

    Warning: Unable to meet integration tolerances without violating a minimum machine step size of 3.552713678800526e-015 at time 1.000000000000011 for model 'signals'. It appears that there is a singularity in the solution. If not, try reducing the step size (either by reducing the fixed step size or by tightening the error tolerances).

- While doing interconnection of systems in SIMULINK® environment, you can assign different signals to different inputs of the MIMO model, try doing the same in MATLAB® environment! What conclusion do you arrive at as far as ease is concerned?
- Synthesising signals like pulse width modulated signal is quite natural and simple in SIMULINK® environment, giving you the ease of checking the resulting signal at each stage. Check out if it is so easy in MATLAB® environment too.
- SIMULINK® environment has yet another advantage over the MATLAB® environment. SIMULINK® supports multirate systems. Hence, while working with models involving these, SIMULINK® is the only refuge.

These were just a few examples out of many explained in order to bring out some of the limitations of SIMULINK® for solving Control System problems. As you practice more and more, you will develop an insight regarding the appropriateness of a suitable method for a particular problem. However, instead of choosing MATLAB® environment or SIMULINK® environment exclusively for a particular problem, the best policy is to work in mixed environment extracting the best out of both.

8.2 Electrical Systems

In order to learn simulation of physical systems through Control System Toolbox of MATLAB® and SIMULINK®, let us first learn to simulate electrical circuits. You must be familiar with the fact that analogous electrical circuits can be drawn for many other physical systems. Thus, if you master simulation of electrical circuits, working with other physical systems is going to be quite easy.

To start with, we will use SIMULINK® as our simulation environment. Then we will switch over to examples simulated in MATLAB® environment and finally, we will take up examples using both.

Let us start with obtaining response of the basic elements of electrical circuits to different signals.

8.2.1 Purely Resistive Circuit

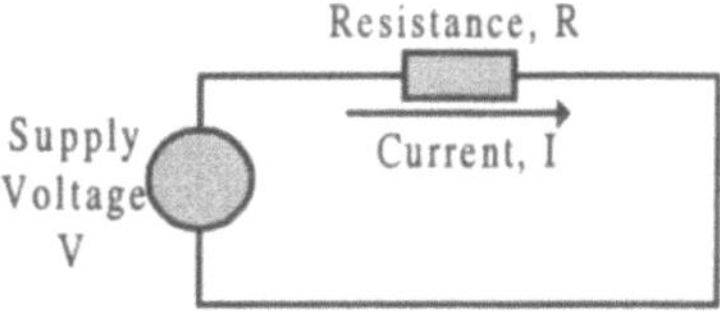

Figure 8.1. Purely Resistive Circuit

Consider a circuit having pure resistance (say a small piece of constantan wire from a heating element) as shown in the Figure 8.1. This circuit can be simulated using equations obtained by applying Ohm's Law to the circuit giving relationship between current and voltage as follows:

$$V = RI \Leftrightarrow I = \frac{V}{R}$$

Power consumed and energy dissipated in the resistance are given by the following relationships:

Power, $P = VI$

Energy, $W = \int Pdt$

The block diagram for this can be drawn as shown in Figure 8.2. The response of the circuit with resistance of 10 ohms to step input and sinusoidal input of amplitude 100 volts respectively is shown in Figure 8.3.

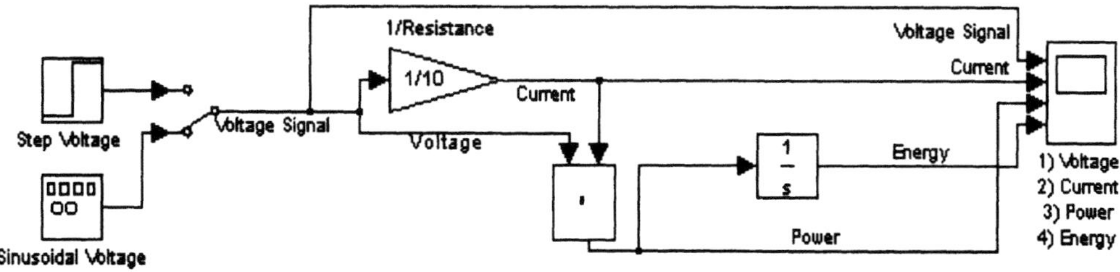

Figure 8.2. Simulation of a Purely Resistive Circuit

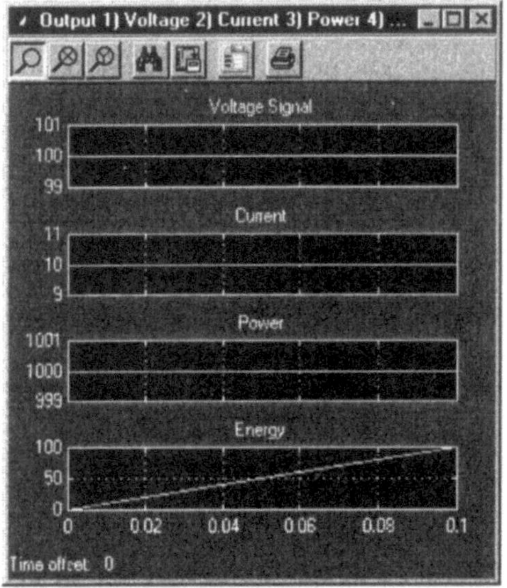
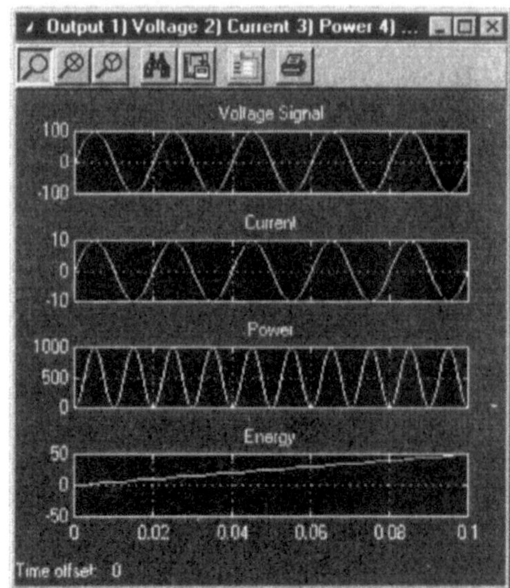

a. Response of Purely Resistive Circuit to step input

b. Response of Purely Resistive Circuit to sinusoidal input

Figure 8.3. Response of Purely Resistive Circuit to different inputs

Practice Test 8.1.

1. Three incandescent lamps are connected as shown in the figure given below:
    Simulate the circuit in SIMULINK® environment for the following situations:

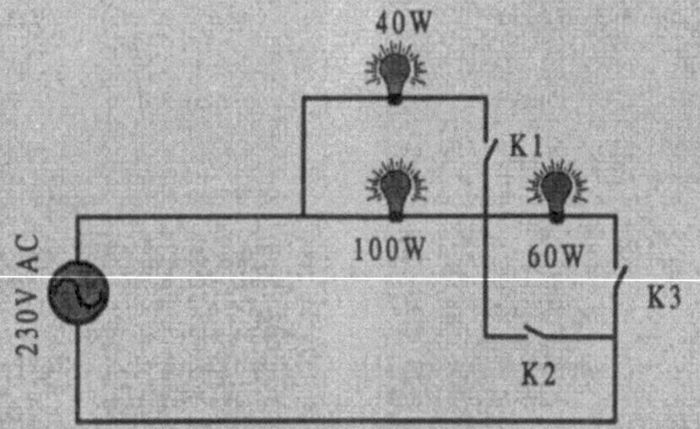

- K1=Closed; K2= Open; K3=Closed
- K1= Open; K2= Closed; K3= Open
- K1= Open; K2= Closed; K3= Closed
- K1= Closed; K2= Open; K3= Open

   In which of the above situations bulb of 100 Watt will have maximum brightness?

8.2.2 Purely Inductive Circuit

Replace the resistance in Figure 8.1 with an inductance of 100 mH, as shown in Figure 8.4.

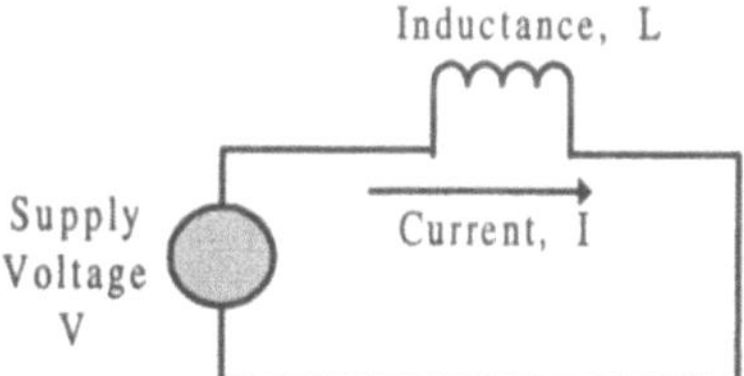

Figure 8.4. Purely Inductive Circuit

This circuit is governed by the following relationships:

$$V = L\frac{dI}{dt} \Leftrightarrow I = \frac{1}{L}\int V dt$$

$$P = VICos\Phi$$

$$W = \int P\, dt$$

Figure 8.5 shows the simulation diagram for the circuit using block diagram approach. The response of the circuit to ramp, sinusoidal, square and sawtooth inputs are shown in Figure 8.6. You may try obtaining response of the circuit to some different input also.

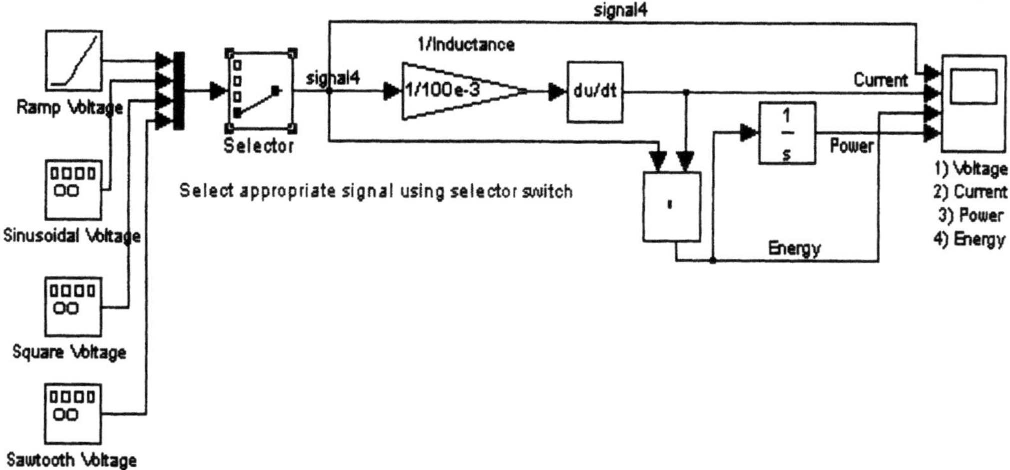

Figure 8.5. Simulation of a Purely Inductive Circuit

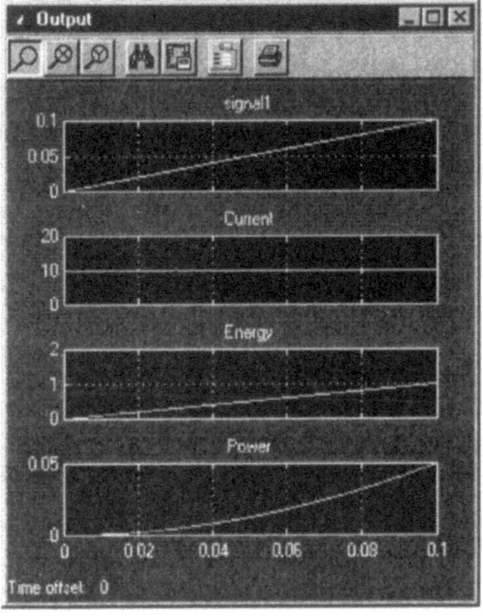

a. Response of Purely Inductive Circuit to ramp input

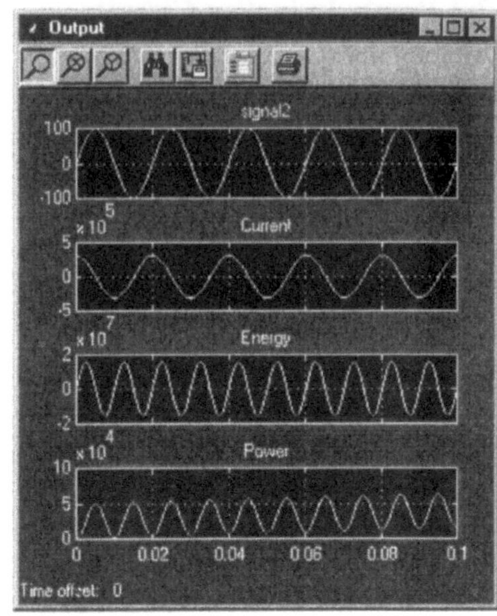

b. Response of Purely Inductive Circuit to sinusoidal input

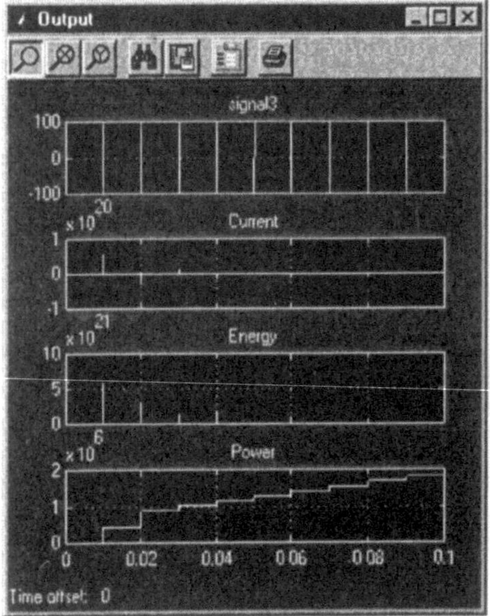

c. Response of Purely Inductive Circuit to square input

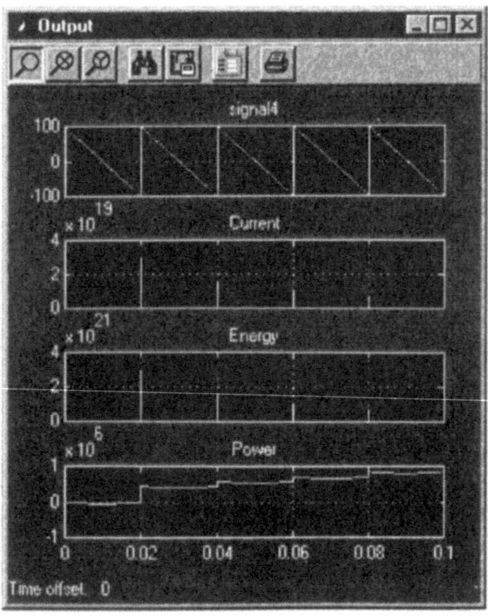

d. Response of Purely Inductive Circuit to Sawtooth Input

Figure 8.6. Response of Purely Inductive Circuit to different inputs

Practice Test 8.2.

1. In the inductive circuit given above, use full wave and half wave rectified sinusoidal input signals. Explain the reason for nature of the waveform so obtained.

2. Simulate the circuit shown in Figure given below and obtain the value of I1, I2, I3 for K open and closed respectively for single triangular pulse and single rectangular pulse voltage signals as indicated.

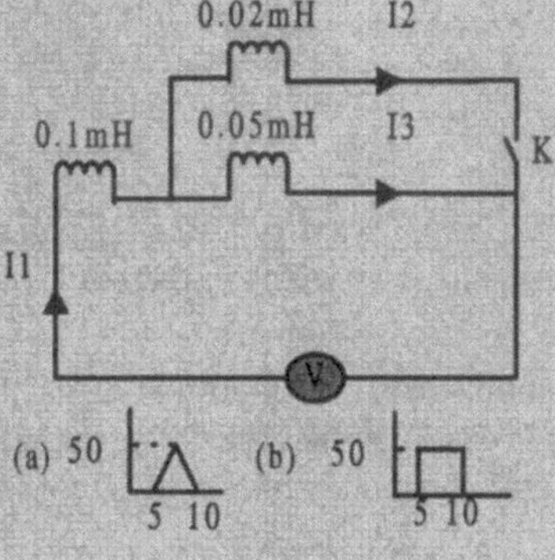

8.2.3 Purely Capacitive Circuit

Replace the resistance in Figure 8.1 with a Capacitance of 1μF as shown in Figure 8.7.

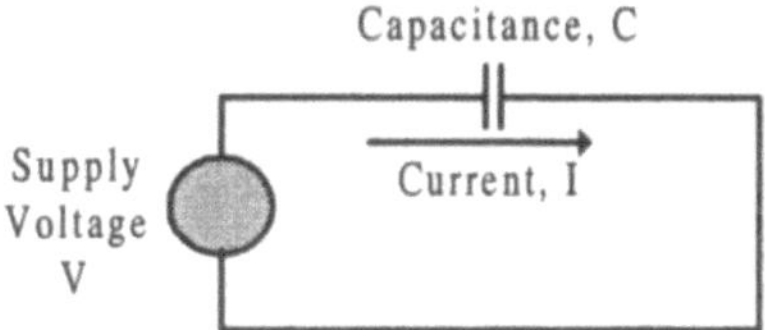

Figure 8.7. Purely Capacitive Circuit

This circuit is governed by the following relationships:

$$V = \frac{1}{C}\int I\,dt \Leftrightarrow I = C\frac{dV}{dt}$$
$$P = VI\mathrm{Cos}\Phi$$
$$W = \int P\,dt$$

Figure 8.8 shows the block diagram for simulation of the circuit. The responses of the circuit to step, sinusoidal, square and sawtooth input signals are shown in Figure 8.9. You may try obtaining response of the circuit to some different input also.

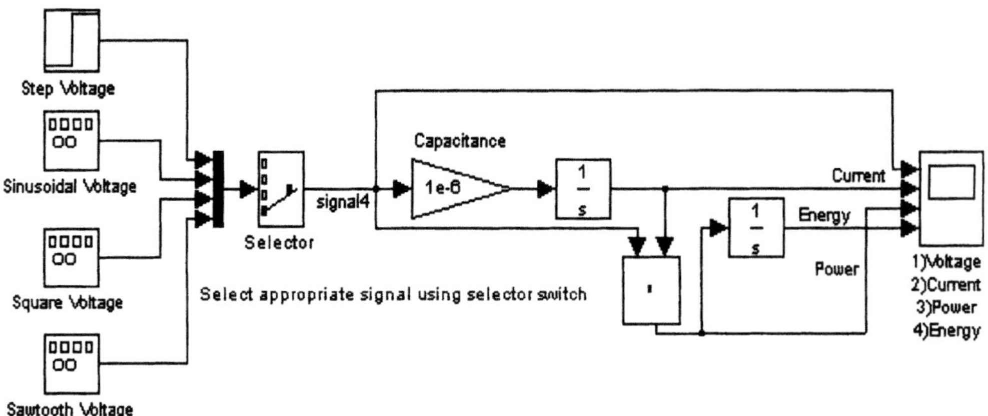

Figure 8.8. Simulation of a Purely Capacitive Circuit

Little perplexed! Is SIMULINK® the only answer for simulation of a problem? No, it is not. As already mentioned time and again, you can simulate the same problem in several ways and yet arrive at the same result. Let us simulate the next problem in all the possible ways we can think of and see for ourselves.

Practice Test 8.3.

1. In the capacitive circuit explained above, use full wave and half wave rectified sinusoidal input signals. Explain the reason for nature of the waveform.

2. Simulate the circuit shown in Figure given below and obtain the value of I1, I2, I3 for K open and closed respectively for single triangular pulse and single rectangular pulse voltage signals as indicated.

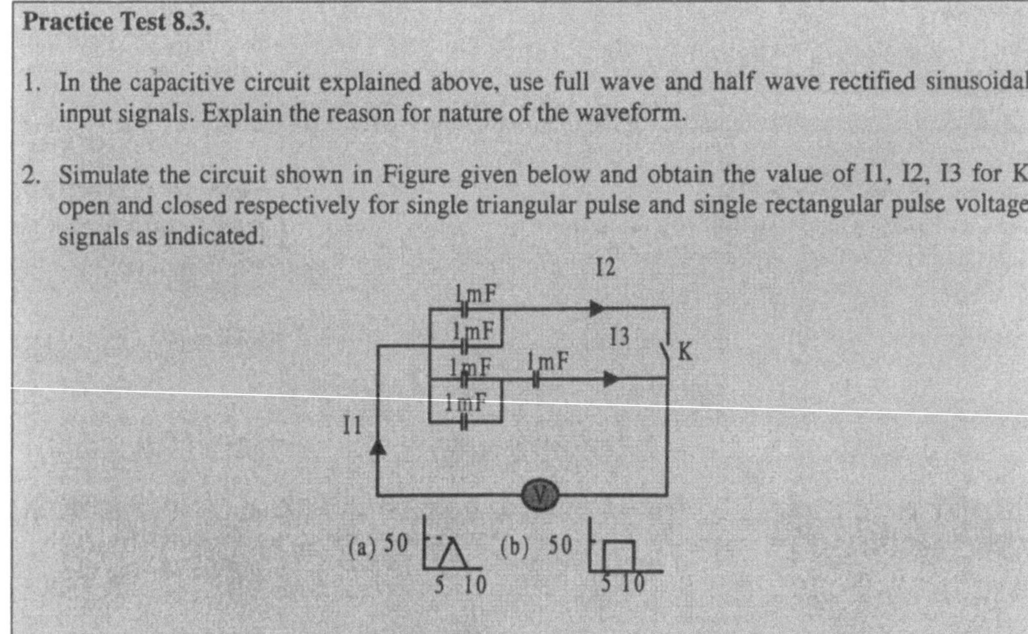

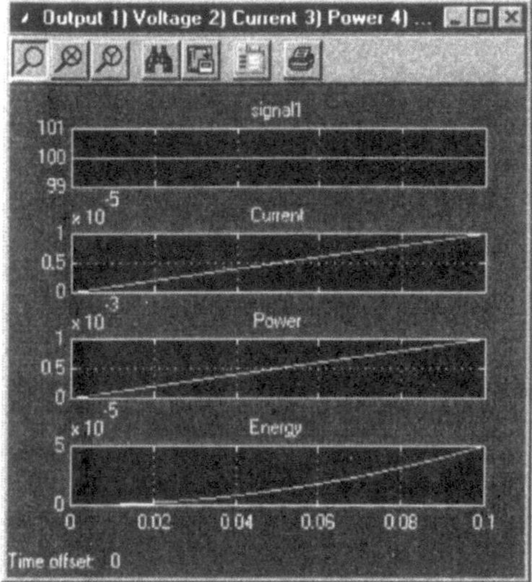

a. Response of Purely Capacitive Circuit to step input

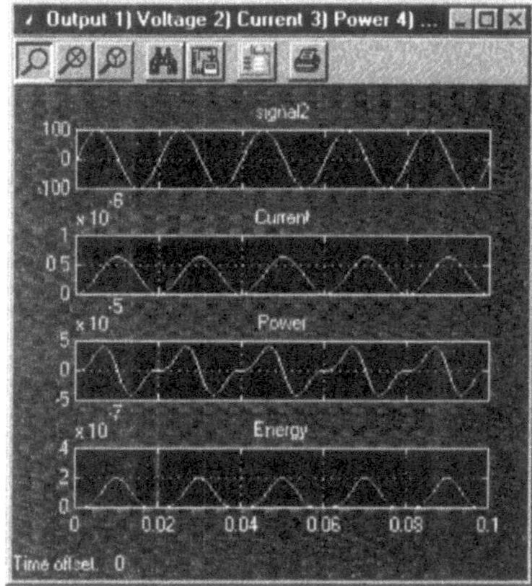

b. Response of Purely Capacitive Circuit to sinusoidal input

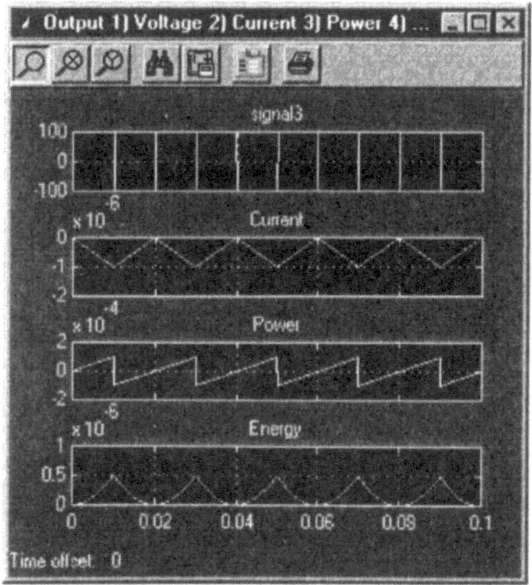

c. Response of Purely Capacitive Circuit to Square Input

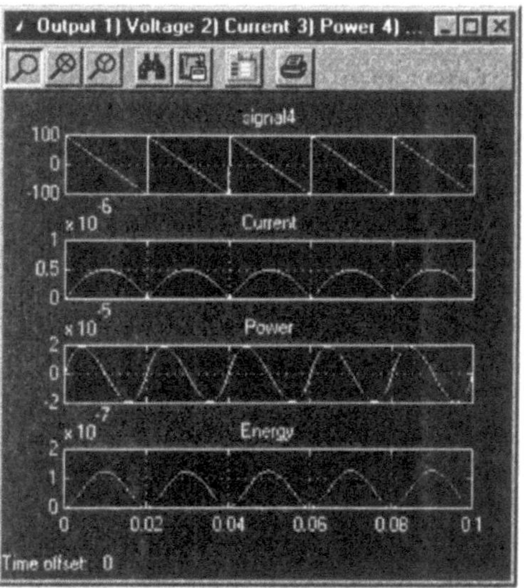

d. Response of Purely Inductive Circuit to Sawtooth Input

Figure 8.9. Response of Purely Capacitive Circuit to Different Inputs

8.2.4 Series RL Circuit

Let us consider a simple RL series circuit with resistance, R equal to 10 Ω and inductance, L equal to 100 mH. We will simulate the circuit to obtain its response to 100 volts step input.

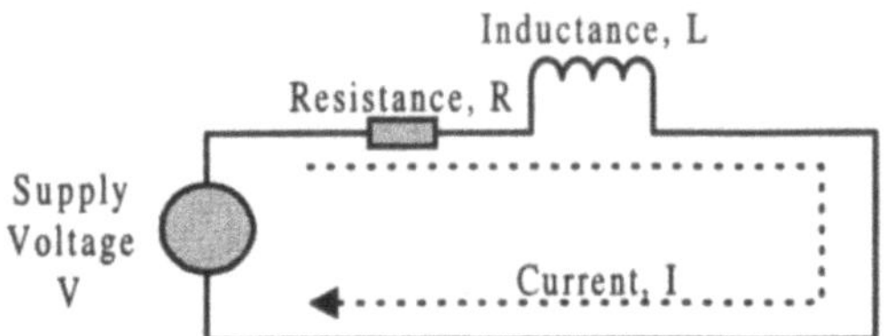

Figure 8.10. Series RL Circuit

The equations governing this circuit are as follows:

$$V = RI + L\frac{dI}{dt} \Leftrightarrow I = \frac{V}{R} + \frac{1}{L}\int V dt$$

$$P = VI \, Cos\Phi$$

$$W = \int P \, dt$$

Laplace transform of the voltage equation given above is:

$$V_{(s)} = (R+sL) \, I_{(s)}$$

from which we can obtain the transfer function as:

$$\frac{I(s)}{V(s)} = \frac{1}{sL + R}$$

We can also obtain state space variable representation of the circuit as follows:

Since this circuit contains only one inductor, hence it is a first-order system. Obviously, it will have only one state. Let us choose the state, x_1 as current through the inductor, I. It is also the output, y of the circuit, while input, u of the circuit is voltage, V.

Then,

$$x_1 = I$$

$$\dot{x}_1 = \frac{dI}{dt} = -\frac{R}{L}x_1 + \frac{1}{L}V$$

$$y = x_1$$

In vector matrix form,

$$\left[\dot{x}_1\right]=\left[-\frac{R}{L}\right]\left[x_1\right]+\left[\frac{1}{L}\right]\left[v\right]$$

$$\left[y\right]=\left[1\right]\left[x_1\right]+\left[0\right]\left[v\right]$$

You can use the equations directly, or use the transfer function and state space equations to simulate the circuit. This can be achieved from the MATLAB® command window or through the SIMULINK® window. In the paragraphs to follow, we shall try both.

8.2.4.1 Simulation from the MATLAB® Window

Substitute the values of L and R in the expression for transfer function and state space representations. Enter the following at the MATLAB® command prompt:

```
% considering transfer function representation
rltf=tf([1], [100e-3 10]);
x=100*step(rltf);
plot(x)
```

or,

```
% considering state space representation
rlss=ss([-10/100e-3],[1/100e-3],[1],[0]);
x=100*step(rlss);
plot(x)
```

both of the above return the same plot in the Figure window as shown in Figure 8.11.

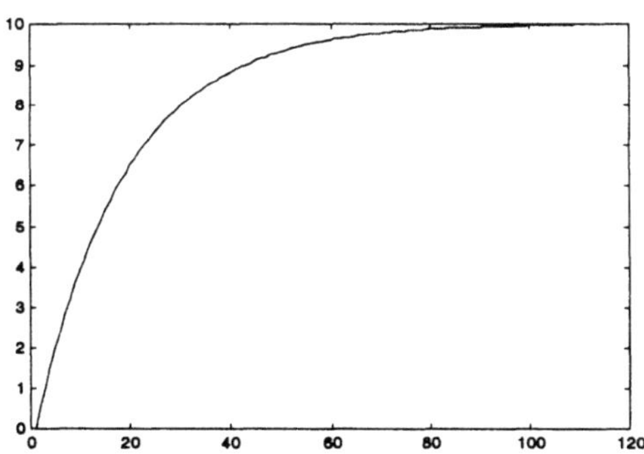

Figure 8.11. Response of Series RL Circuit to 100 V step input

8.2.4.2 Simulation from the SIMULINK® Window

Same results, as obtained above can also be obtained form the SIMULINK® window. You have in fact three options:

- You can simulate basic circuit equations relating voltage and current as a block diagram. The equation can be written down in three forms as follows:

$$V - L\frac{di}{dt} = Ri$$

$$V - Ri = L\frac{di}{dt}$$

$$V - Ri + L\frac{di}{dt} = 0$$

The three forms of equation are simulated as shown in Figure 8.12.

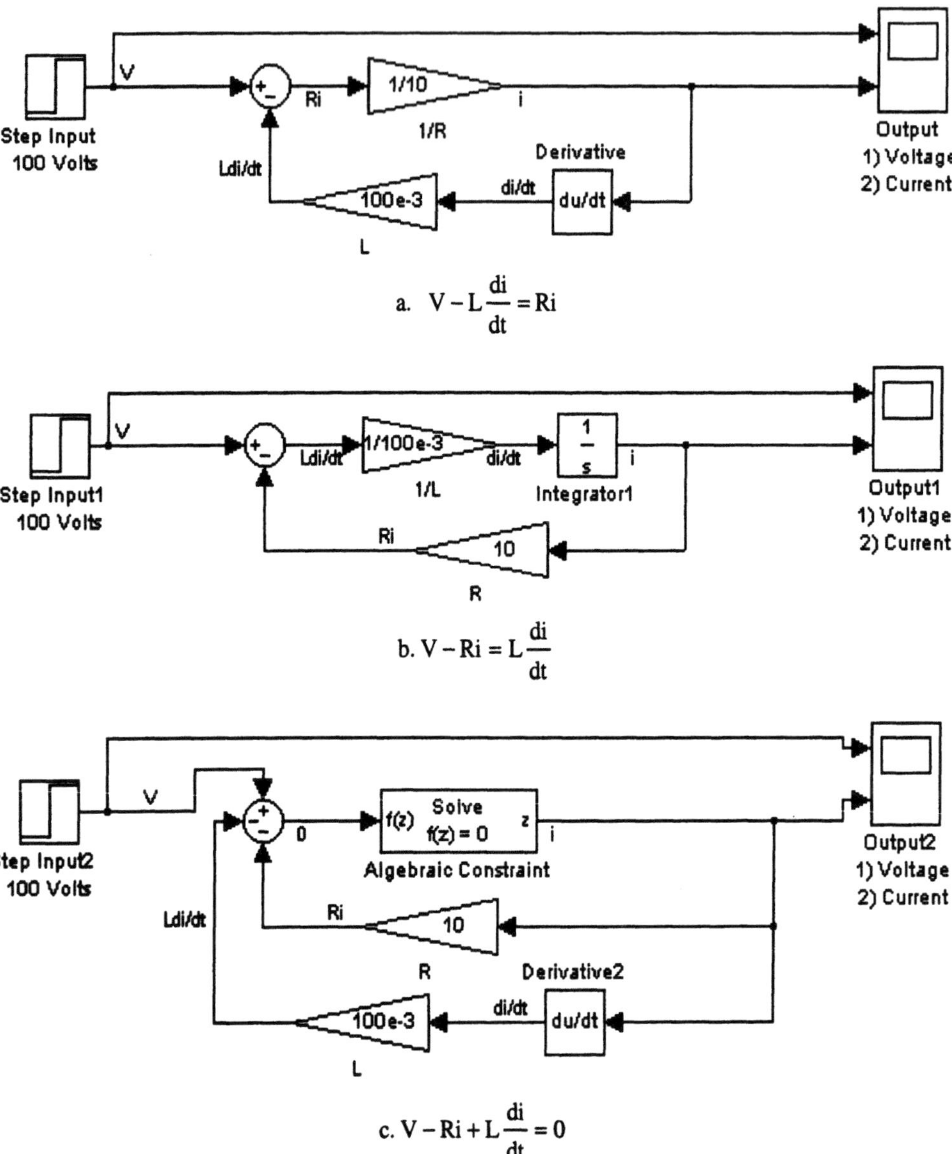

$$\text{a.} \quad V - L\frac{di}{dt} = Ri$$

$$\text{b.} \; V - Ri = L\frac{di}{dt}$$

$$\text{c.} \; V - Ri + L\frac{di}{dt} = 0$$

Figure 8.12. Series RL Circuit representation using basic equations using block diagram approach

- You can simulate the transfer function expression. The transfer function can be expressed in two ways:

i. Using the transfer function block available in SIMULINK® Continuous Library directly as a first-order system with unity feedback loop.

ii. Using the transfer function block available in SIMULINK® Continuous Library directly in the forward path without any feedback loop.

The two styles of transfer function representation are simulated as shown in Figure 8.13.

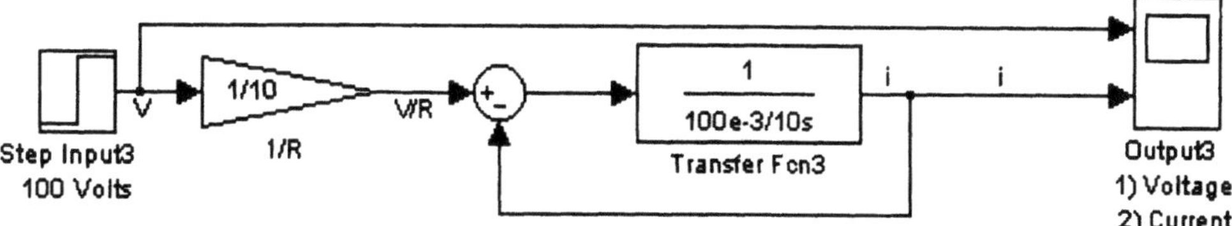

a. Series RL Circuit representation as first-order system with unity feedback

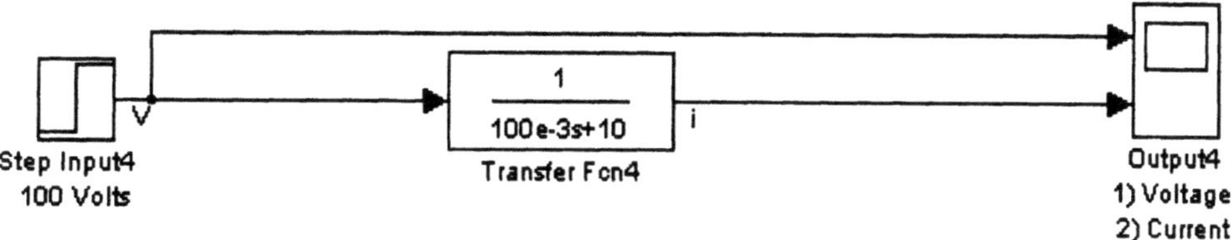

b. Series RL Circuit representation as second-order system

Figure 8.13. Series RL Circuit representation using transfer function approach

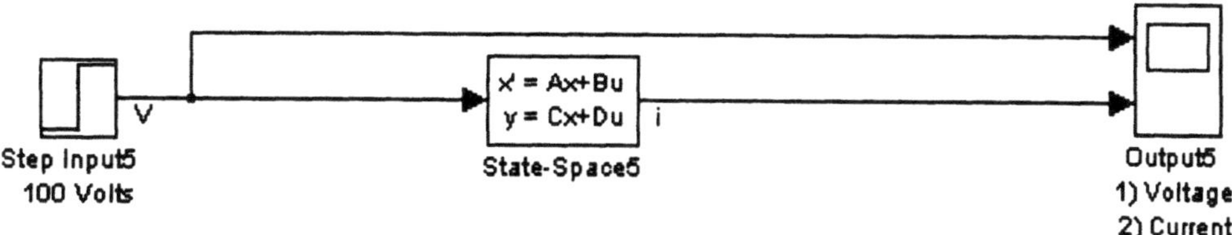

a. Series RL Circuit representation using state space variable approach

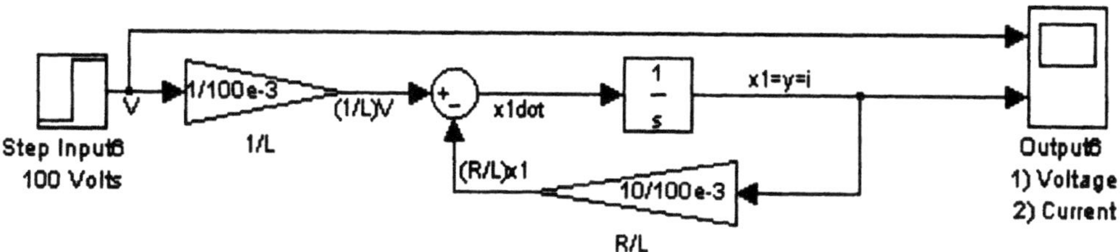

b. Series RL Circuit representation as block diagram of state space model

Figure 8.14. Simulation of RL Series Circuit using state space variable approach

- You can simulate the state space relation. The state variables can be expressed in two ways:

i). Using the state space block available in SIMULINK® Library directly.
ii). Representing the state space by block diagram.

The two styles of state space representation are simulated as shown in Figure 8.14.
However, the response obtained for 100V step input to any of the above simulation model is the same which is shown in Figure 8.15.

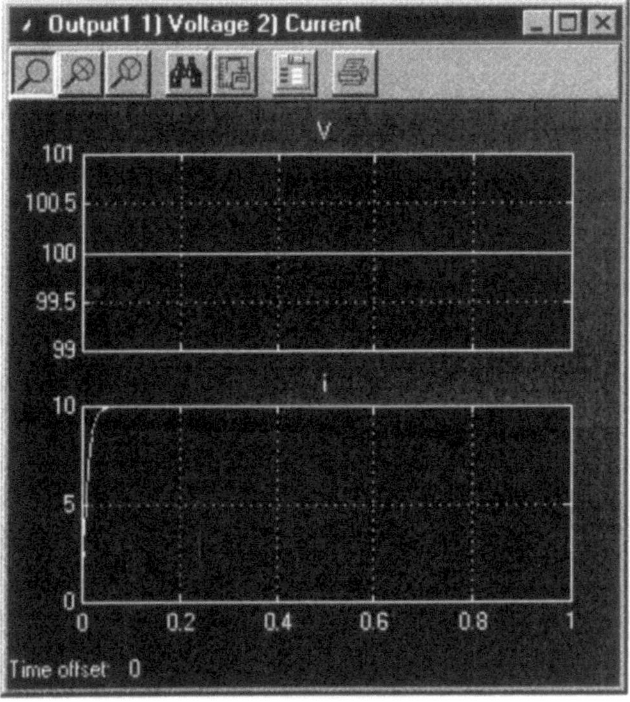

Figure 8.15. Simulation result of RL Series Circuit to step input

This exercise must have given you an insight into different approaches, which can be adopted to simulate any problem. You may obtain the power and energy waveforms in the same manner as in earlier articles.

Practice Test 8.4.

1. Repeat the above for
- simple parallel RL circuit with same parameters
- for series RC circuit with value of R=10 Ω, C= 0.25 Farad
- for parallel RC circuit with value of R=10 Ω, C= 0.25 Farad

Next, let us take an example of pnp transistor and simulate it in mixed environment *i.e.,* we shall use SIMULINK® as well as MATLAB® windows for simulation.

8.2.5 Output Characteristics of a PNP Transistor (Linear Portion in Common Emitter Configuration)

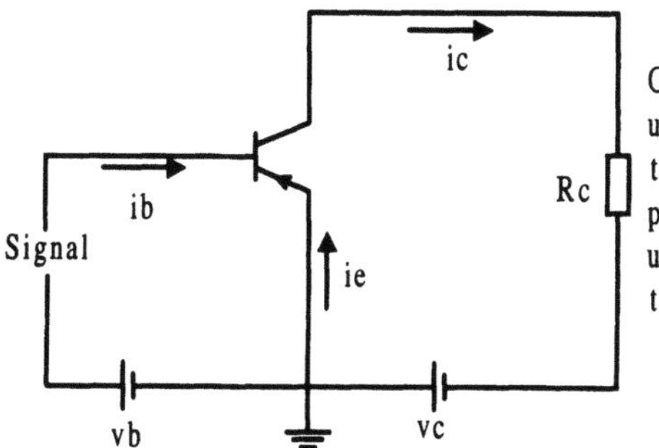

a. Connection schematic of PNP Transistor in common emitter configuration

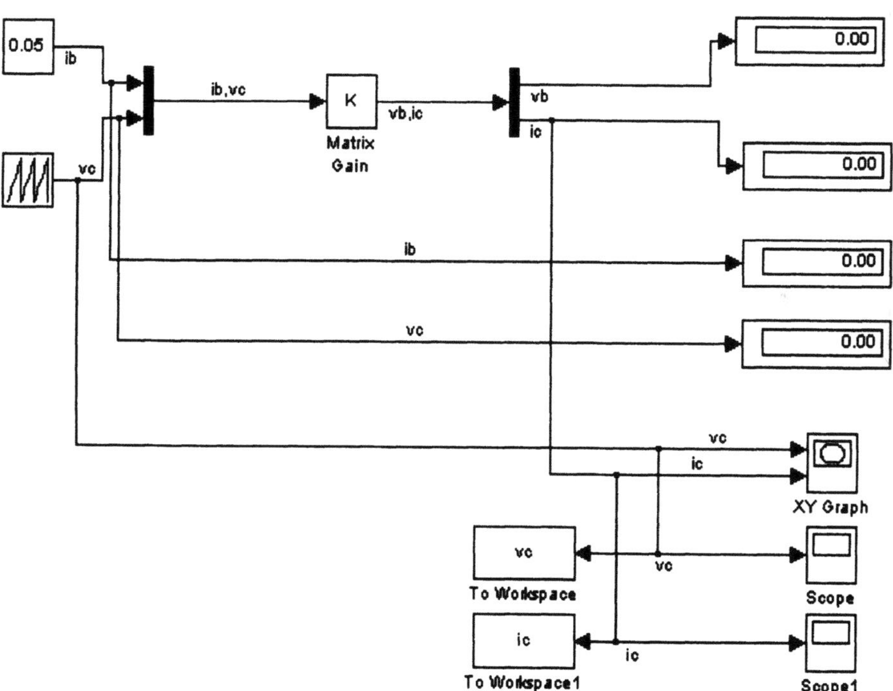

b. Simulation diagram of a PNP Transistor hybrid model for obtaining its output characteristics

Figure 8.16. PNP Transistor in common emitter configuration

The most popular way of representing a transistor is by its hybrid model. For the pnp transistor, the hybrid model is given as follows:

$$\begin{bmatrix} v_b \\ i_c \end{bmatrix} = \begin{bmatrix} h_{11} & h_{12} \\ h_{21} & h_{22} \end{bmatrix} \begin{bmatrix} i_b \\ v_c \end{bmatrix}$$

let,

$$\begin{bmatrix} h_{11} & h_{12} \\ h_{21} & h_{22} \end{bmatrix} = \begin{bmatrix} 1000 & 0.025 \\ 75 & 1.36 \end{bmatrix}$$

using these values the linear portion of the output characteristics of a PNP transistor in common emitter configuration (Figure 8.16a) is modelled using SIMULINK® (Figure 8.16b). The response of the model *i.e.,* the values vc (collector to emitter voltage) and ic (collector current) is stored in the workspace for different values of ib (base current). Let the variables storing these values be specified by vc, vc1, vc2..., and ic, ic1, ic2... By entering the following at the command prompt, response plot is obtained as shown in Figure 8.17:

```
plot(Vc,ic)
plot(Vc1,ic1)
|
|
|
plot(vc10,ic10)
```

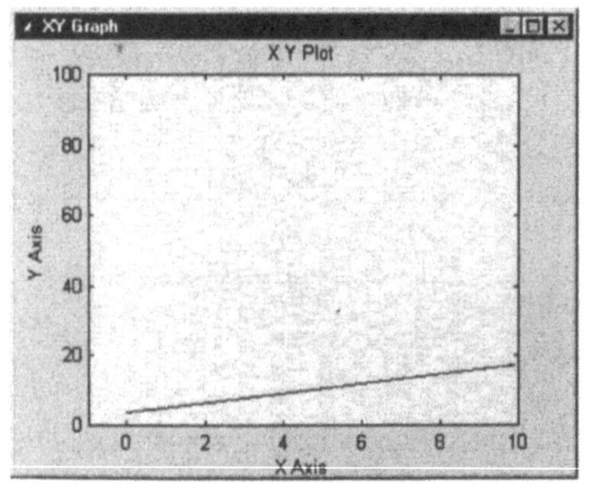

a. V_c versus I_c for Ib= 0.05

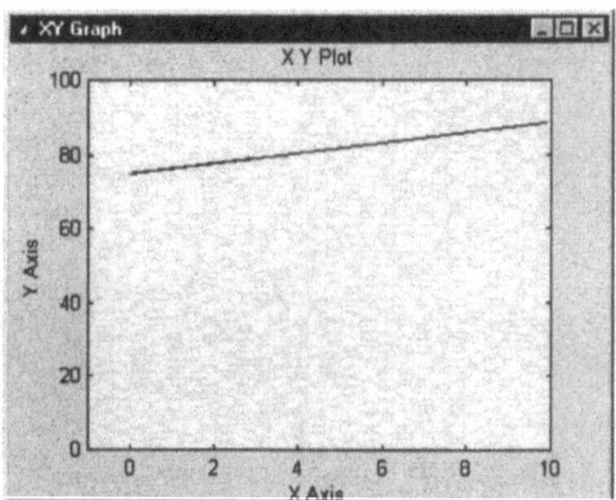

b. V_c versus I_c for Ib= 1

Figure 8.17. Response of pnp transistor hybrid model to different inputs

At the command prompt in MATLAB® window, enter

plot(vc,ic,vc1, ic1, vc2, ic2, vc3, ic3, vc4, ic4, vc5, ic5, vc8, ic8, vc10, ic10)

This returns a Figure Window displaying all the plots as shown in Figure 8.18.

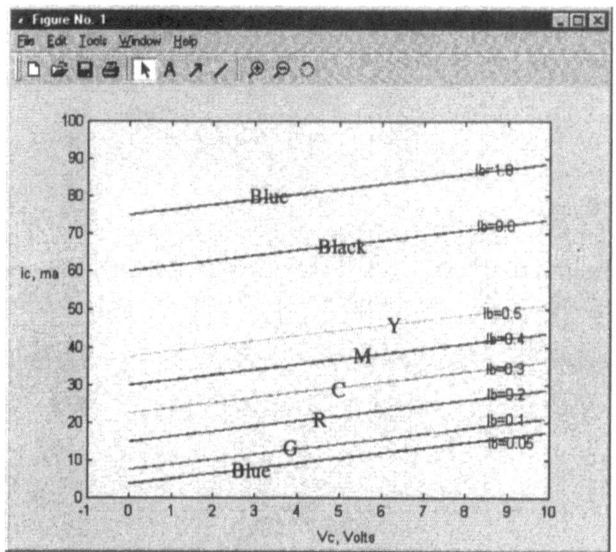

Figure 8.18. Combined response plot of the PNP Transistor hybrid model for different inputs

Practice Test 8.5.

1. Obtain the input characteristics (linear portion) for the same pnp transistor used above in common emitter configuration.

8.2.6 Series RLC Circuit

Let us, now, simulate a simple RLC series circuit shown in Figure 8.19.

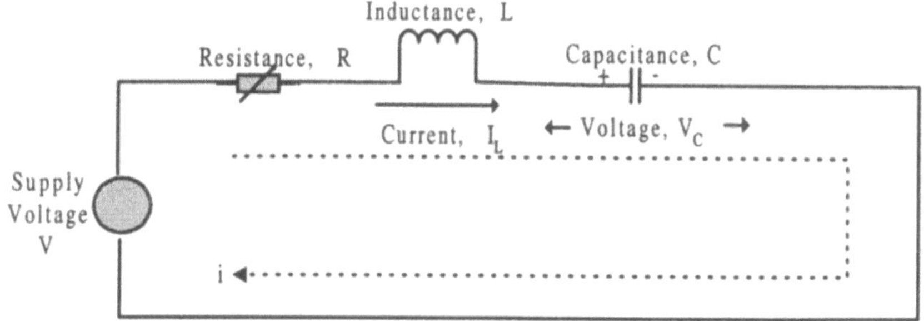

Figure 8.19. RLC Series Circuit

The equation governing the circuit can be written down as:

$$I_L = i$$

$$V = Ri + L\frac{di}{dt} + C\int idt$$

The transfer function of the system that can be easily derived from the equations above is:

$$\frac{I_{(s)}}{V_{(s)}} = \frac{sC}{s^2LC + sRC + 1}$$

Similarly, the state space equations can also derived from above by choosing V as input, i as output and states as the current through inductor, I_L (= i) and voltage across the capacitor V_C.

$$\frac{d}{dt}\begin{bmatrix} i_L \\ V_C \end{bmatrix} = \begin{bmatrix} -\dfrac{R}{L} & \dfrac{-1}{L} \\ \dfrac{1}{C} & 0 \end{bmatrix}\begin{bmatrix} i_L \\ V_C \end{bmatrix} + \begin{bmatrix} \dfrac{-1}{L} \\ 0 \end{bmatrix}[V]$$

$$\begin{bmatrix} i_L \end{bmatrix} = \begin{bmatrix} 1 & 0 \end{bmatrix}\begin{bmatrix} i_L \\ V_C \end{bmatrix}$$

The circuit is now simulated with following circuit parameters using the block diagram approach for representing the basic state space equations as shown in Figure 8.20.

V= Sinusoidal Voltage of 100 V amplitude and 50 Hz frequency
R= variable from 0 Ω to 10000 Ω
L= 1mH
C= 0.25 Farad

The response of the circuit with respect to time for different values of R to sinusoidal input voltage is shown in Figure 8.21.

Let us now study the frequency response of the circuit. The resonance frequency for the circuit can be readily calculated using the following relationship:

$$f_0 = \frac{1}{2\sqrt{LC}} = 10Hz \text{ (approx.)}$$

Bode plot for the circuit is plotted for different values of R. It is interesting to note that as the value of R increases from 0 to higher value, the magnitude curve flattens down indicating reduction in the sharpness of the curve which implies lower selectivity or quality factor (Figure 8.21).

The system is clearly of second order (the order of an electrical circuit being judged by the number of inductors and/or capacitors present in the circuit, which in this case is two) and hence the step response of the system indicates typical second-order response varying from sustained oscillations for R=0 to damped oscillations for higher values of R (Figure 8.22). Clearly, R affects the damping ratio of the system, which is also indicated by the transfer function expression of the system.

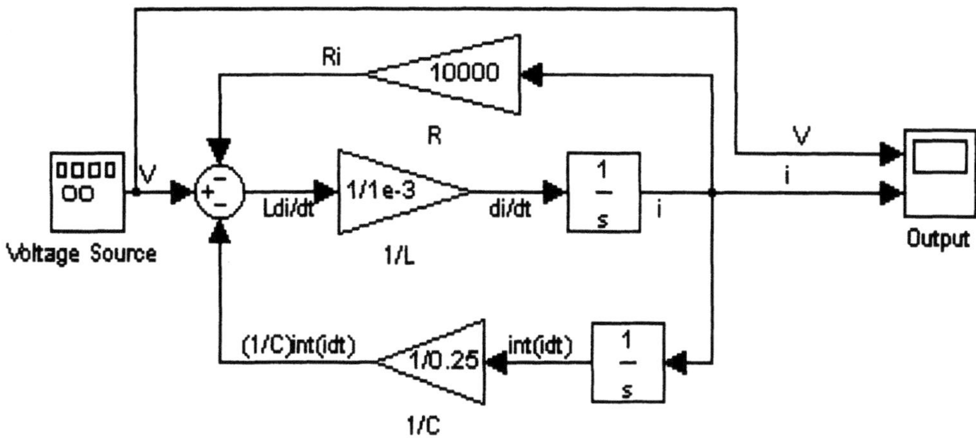

Figure 8.20. Simulation of RLC Series Circuit by state space variable approach

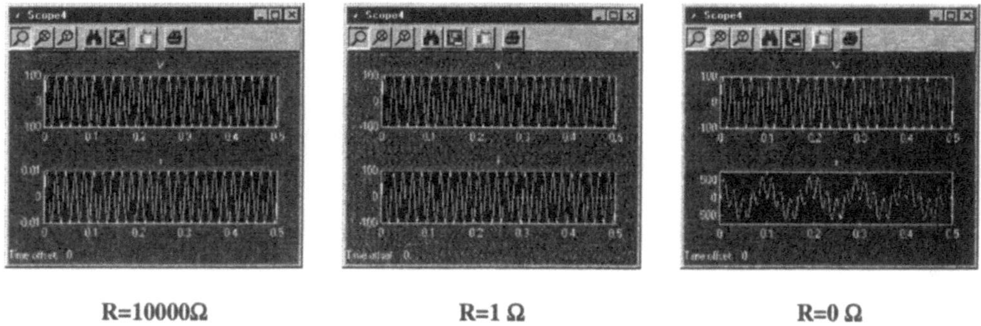

R=10000Ω   R=1 Ω   R=0 Ω

Figure 8.21. Response of RLC Series Circuit to sinusoidal input for different values of R

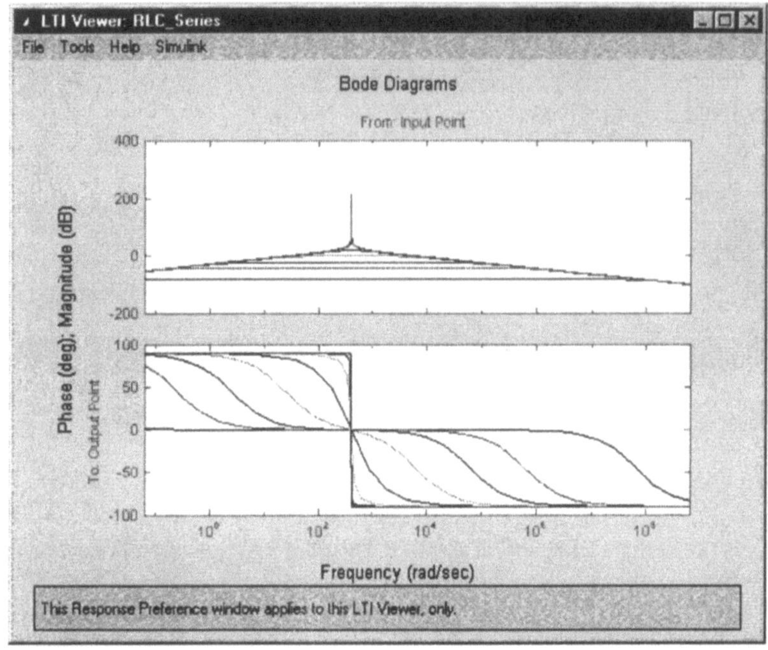

Figure 8.22. Frequency response plot of Series RLC Circuit for different values of R (R=10000 Ω (blue at the bottom), 100 Ω (green), 10 Ω (red), 1 Ω (cyan), 0.1 Ω(magenta), 0.01 Ω (yellow), 0.0001 Ω (black), 0.00001 Ω (blue), 0 Ω (green at the top))

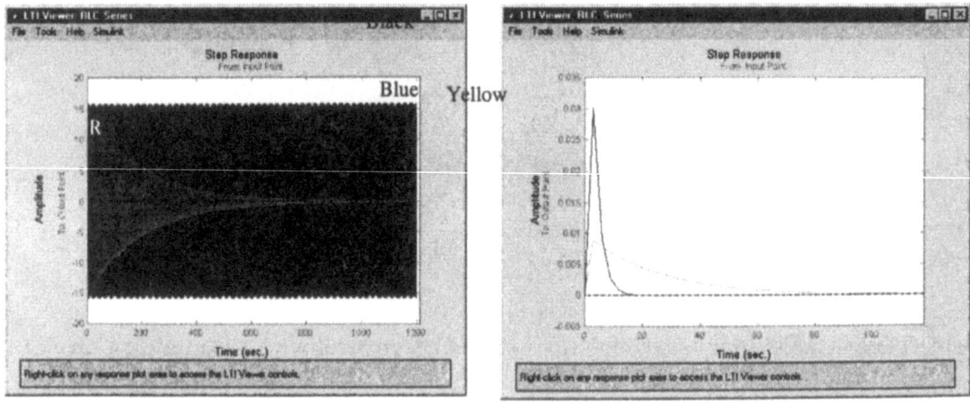

a. R=0 Ω (blue), 0.00001 Ω(green), 0.001 Ω(red), 0.01 Ω(cyan)

b. R=10000 Ω (black), 100 Ω(yellow), 10 Ω(black), 1 Ω(blue)

Figure 8.23. Step Response of the series RLC circuit for different values of R

8.2.7 Bandpass Filter

You are familiar with what filters and attenuators do in electronic circuits. We now take up for simulation, circuit of a typical bandpass filter. This circuit (Figure 8.24) should reject or attenuate signals of high and low frequency and allow signals in a particular range (called pass band) to pass through it.

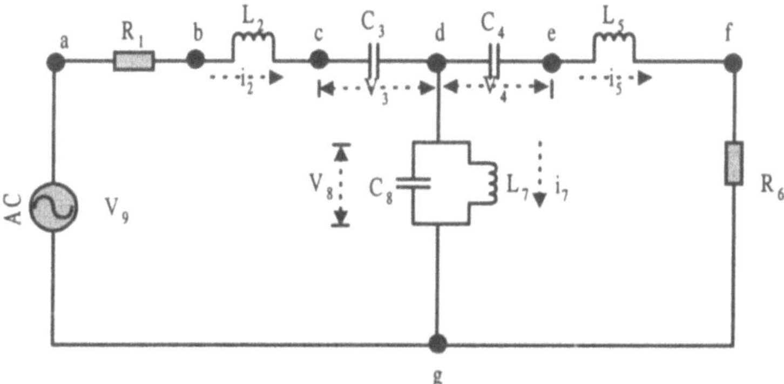

Figure 8.24. Bandpass Filter Circuit

Applying Kirchoff's Voltage and Current laws, the equations for the circuit of the filter can be written down as:

$$R_1 i_2 + L_2 \frac{di_2}{dt} + V_3 + V_8 - V_9 = 0$$

$$V_4 + L_5 \frac{di_5}{dt} + R_6 i_5 - V_8 = 0$$

$$L_7 \frac{di_7}{dt} = V_8$$

$$V_3 = \frac{1}{C_3} \int i_2 dt$$

$$V_4 = \frac{1}{C_4} \int i_5 dt$$

$$i_2 = C_8 \frac{dV_8}{dt} + i_7 + i_5$$

Taking the Laplace Transform of the above equations and eliminating i_7, V_3 and V_4, the transfer function of the circuit can be derived to be:

$$\frac{I_2(s)}{V_9(s)} = \frac{s^5 L_5 L_7 C_4 C_8 + s^4 R_6 L_7 C_4 C_8 + s^3 (L_5 C_4 + L_7 C_4 + L_7 C_8) + s^2 R_6 C_4 + s}{\Delta}$$

$$\frac{I_5(s)}{V_9(s)} = \frac{s^3 L_7 C_4}{\Delta}$$

where,

$$\Delta = s^6 L_2 L_5 L_7 C_4 C_8 + s^5 (R_1 L_5 L_7 C_4 C_8 + R_6 L_2 L_7 C_4 C_8) +$$

$$s^4 [(L_5 C_4 + L_7 C_4 + L_7 C_8) L_2 + R_1 R_6 L_7 C_4 C_8 + L_5 L_7 C_4 C_8 + \frac{L_5 L_7 C_4}{C_3}] +$$

$$s^3 [R_6 L_2 C_4 + R_1 (L_5 C_4 + L_7 C_4 + L_7 C_8) + R_6 L_7 C_8 C_4 + R_6 L_7 C_4] +$$

$$s^2 [L_2 + L_7 + R_1 R_6 C_4 + \frac{(L_5 C_4 + L_7 C_4 + L_7 C_8)}{C_3}] + s(R_1 + R_6 C_4) + \frac{1}{C_3}$$

The state space representation of the system can also be derived to be

$$\frac{d}{dt}\begin{bmatrix} i_2 \\ i_5 \\ i_7 \\ v_3 \\ v_4 \\ v_8 \end{bmatrix} = \begin{bmatrix} \dfrac{-R_1}{L_2} & 0 & 0 & \dfrac{-1}{L_2} & 0 & \dfrac{-1}{L_2} \\ 0 & \dfrac{-R_6}{L_5} & 0 & 0 & \dfrac{-1}{L_5} & \dfrac{1}{L_5} \\ 0 & 0 & 0 & 0 & 0 & \dfrac{1}{L_7} \\ \dfrac{1}{C_3} & 0 & 0 & 0 & 0 & 0 \\ 0 & \dfrac{1}{C_4} & 0 & 0 & 0 & 0 \\ \dfrac{1}{C_8} & \dfrac{-1}{C_8} & \dfrac{-1}{C_8} & 0 & 0 & 0 \end{bmatrix} \begin{bmatrix} i_2 \\ i_5 \\ i_7 \\ v_3 \\ v_4 \\ v_8 \end{bmatrix} + \begin{bmatrix} \dfrac{1}{L_2} \\ 0 \\ 0 \\ 0 \\ 0 \\ 0 \end{bmatrix} [v_9]$$

$$\begin{bmatrix} i_2 \\ i_5 \end{bmatrix} = \begin{bmatrix} 1 & 0 & 0 & 0 & 0 & 0 \\ 0 & 1 & 0 & 0 & 0 & 0 \end{bmatrix} \begin{bmatrix} i_2 \\ i_5 \\ i_7 \\ v_3 \\ v_5 \\ v_8 \end{bmatrix} + \begin{bmatrix} 0 \\ 0 \end{bmatrix} [v_9]$$

Choose the values of the circuit parameters as:

$R_1 = R_6 = 500\ \Omega$; $L_2 = L_5 = 39.9$ mH; $L_7 = 3.32$ mH; $C_3 = C_4 = 0.0264\ \mu$F; $C_8 = 0.319\ \mu$F;
$V_9 = 20\ \cos\omega t$

We will now vary the value of ω and see the response of the circuit. As mentioned earlier, we expect the circuit to attenuate low and high frequency signals. We will simulate the state space equations of the circuit using the block diagram approach as shown in Figure 8.25 and verify if it really does so.

Substitute the values of the parameters in the equation above and simulate using the block diagram approach (Figure 8.26). Vary the value of frequency from 50 Hz to 15 kHz and observe how the attenuation of signals takes place at low and high frequency (Figure 8.26). Observe the Bode plot (Figure 8.28). It is indeed the frequency response plot of a bandpass filter with stop bands at the extremes and pass band in the middle portion of the magnitude plot.

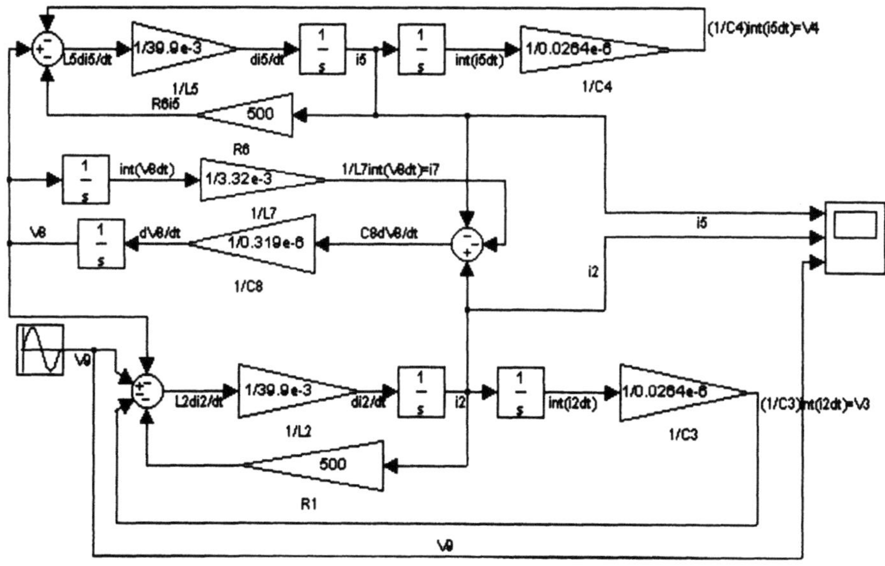

Figure 8.25. Simulation diagram of Bandpass Filter

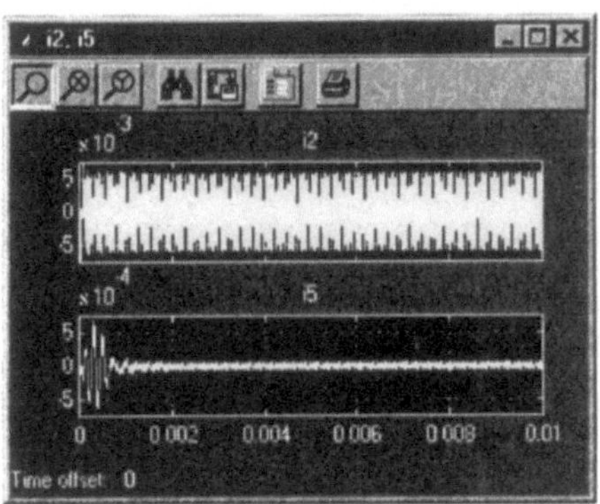

f= 15 kHz
$i_1 \rightarrow$ highly attenuated
$i_2 \rightarrow$ highly attenuated

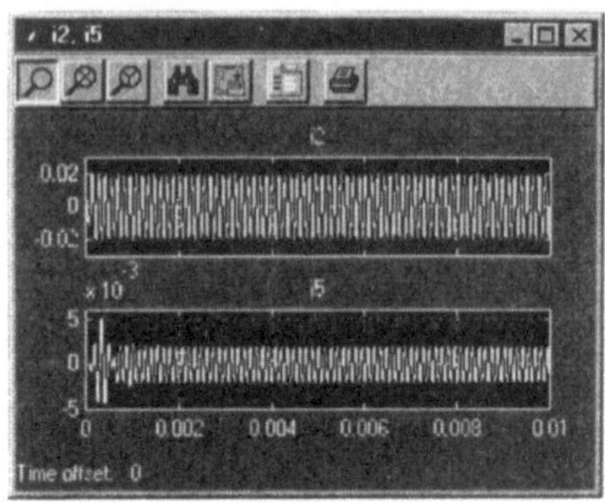

f= 7.5 kHz
$i_1 \rightarrow$ no attenuation
$i_2 \rightarrow$ highly attenuated

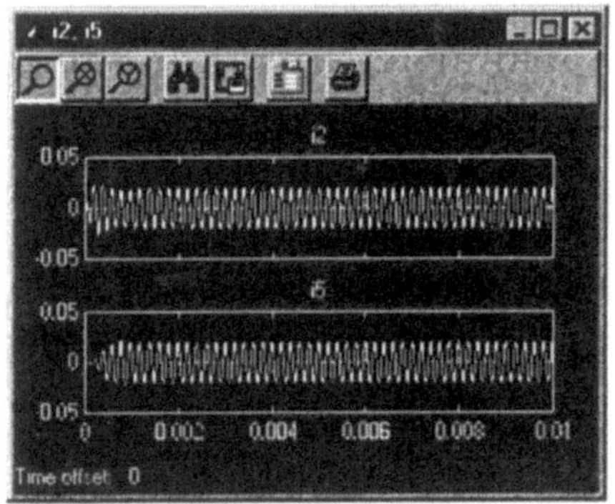

f= 5 kHz
$i_1 \rightarrow$ no attenuation
$i_2 \rightarrow$ no attenuation

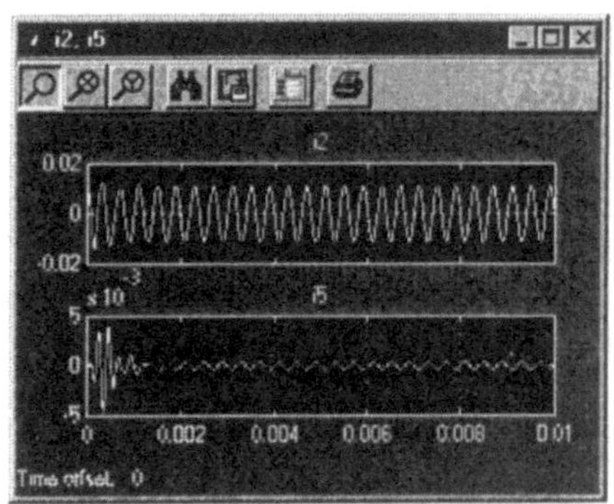

f= 2.5 kHz
$i_1 \rightarrow$ no attenuation
$i_2 \rightarrow$ highly attenuated

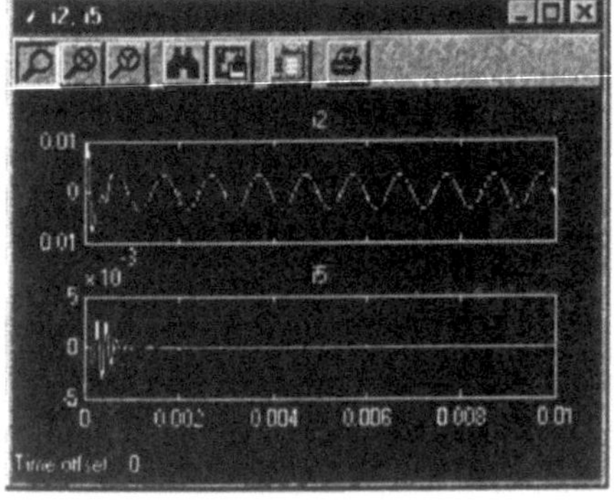

f= 1 kHz
$i_1 \rightarrow$ appreciably attenuated
$i_2 \rightarrow$ highly attenuated

f= 50 Hz
$i_1 \rightarrow$ highly attenuated
$i_2 \rightarrow$ highly attenuated

Figure 8.26. Response of Bandpass Filter at various frequencies

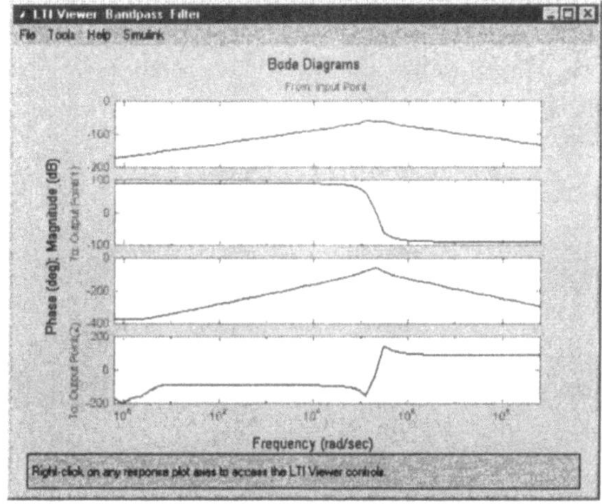

Figure 8.27. Frequency response (Bode) plot of Bandpass Filter

Practice Test 8.7.

1. Simulate the circuits shown in the Figure below by any method suitable to you and obtain their frequency response. Comment on the behaviour of the circuit and its use.

   Let the parameters of the circuits have the following values:
   $R_1 = 500\ \Omega$; $R_2 = 1\ M\Omega$; $C_1 = 0.1\mu F$ and $C_2 = 0.9\mu F$

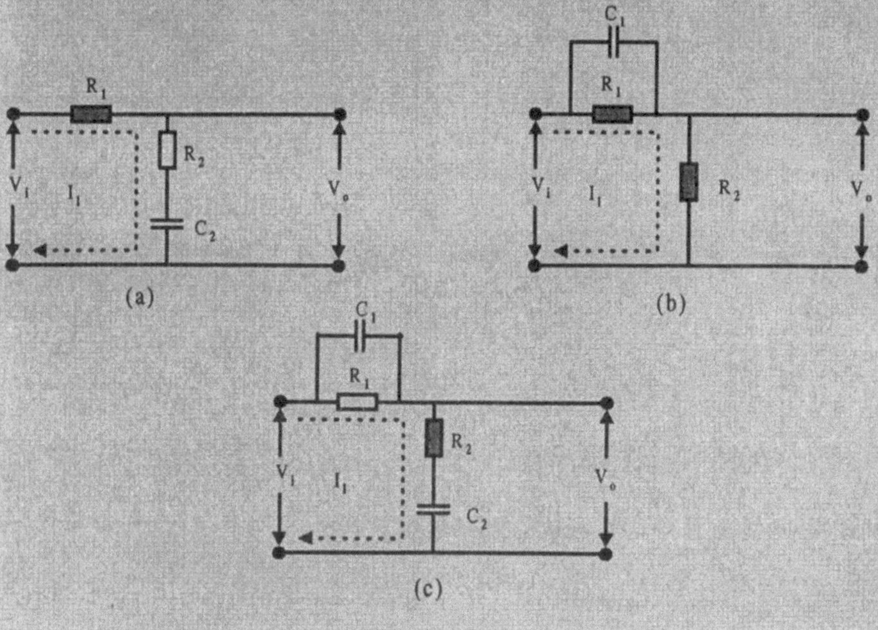

> 2. For the bandpass filter circuit simulated above, find out the cut-off frequencies for i_2 and i_5, keeping in mind that at cut-off frequency, the magnitude of i_2 and i_5 is $\dfrac{1}{\sqrt{2}}$ times the maximum magnitude.

8.3 Mechanical Systems

Most of the systems we come across in our day to day life are mechanical systems, resulting into translational or rotational motion. In the paragraphs to follow, we shall simulate some simple mechanical systems of both the categories.

8.3.1 Translational Mechanical Systems

You must be aware of the fact that the dynamics of an electrical system are determined by its resistance, inductance and capacitance components. In similar way, the dynamics of a translational mechanical system are determined by spring, mass and friction (or damper/dashpot) (see Table C.1 (b) of Appendix C for exact analogy). In the articles to follow, you will learn to simulate some simple physical systems involving these elements.

8.3.1.1 Force-displacement System
In most of the mechanical systems, application of force and calculating the resultant displacement forms an inevitable part. So, let us begin with simulation of such a problem.

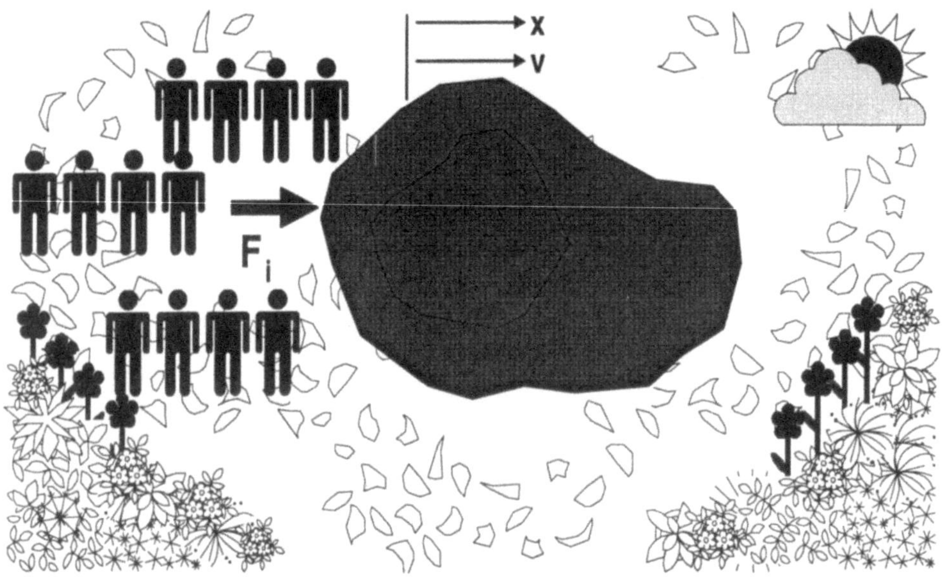

Figure 8.28. Force-displacement System

Consider a situation, in which a troop of army men is required to clear an area in a hilly terrain to dig trenches for making bunkers. A huge boulder of 3000 Kg is desired to be shifted for the purpose. The troop applies a constant force of 500 Newtons. We have to simulate the problem in order to find out the distance through which the boulder moves and the velocity and acceleration of this movement with respect to time.

The problem can be represented by the following equations:

$$\frac{M d^2 x}{dt^2} = F$$

$$\text{Acceleration} = acc. = \frac{F}{M}$$

$$\text{Velocity} = vel. = \int acc.dt$$

$$\text{Displacement} = \int vel.dt$$

These equations are simulated using block diagram approach (Figure 8.29) and the simulation results are shown in Figure 8.30.

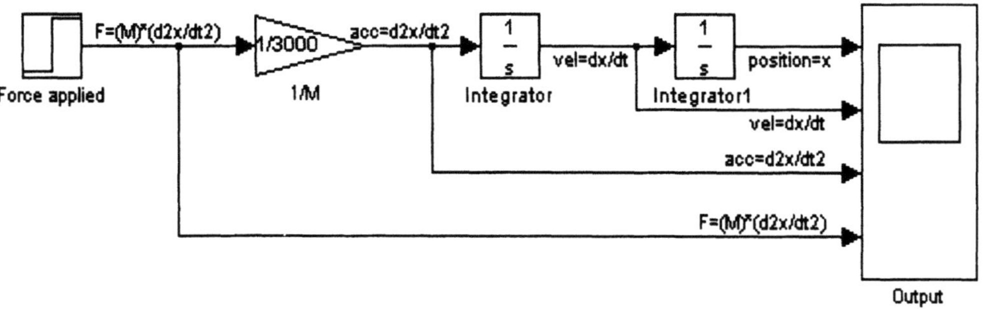

Figure 8.29. Simulation of Force-displacement Problem

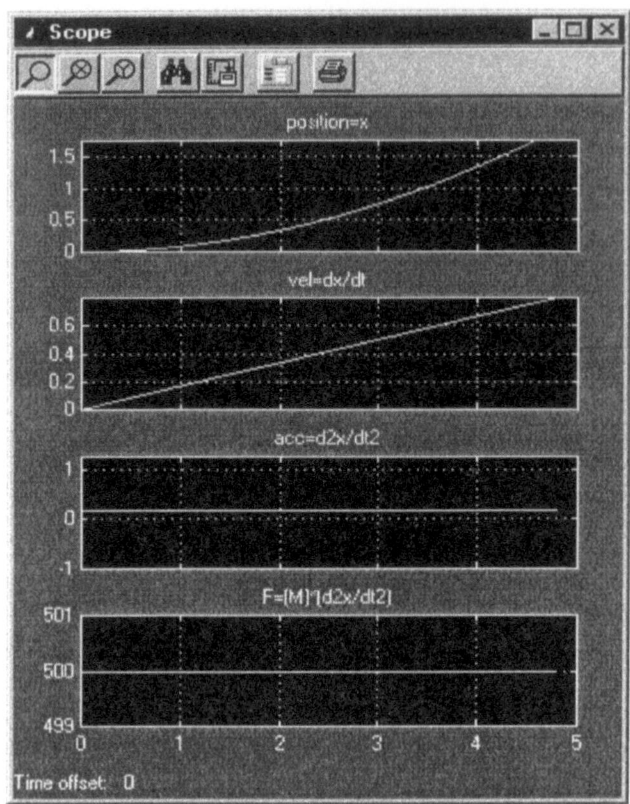

Figure 8.30. Simulation results of Force-displacement Problem

8.3.1.2 Spring-mass-damper-lever System

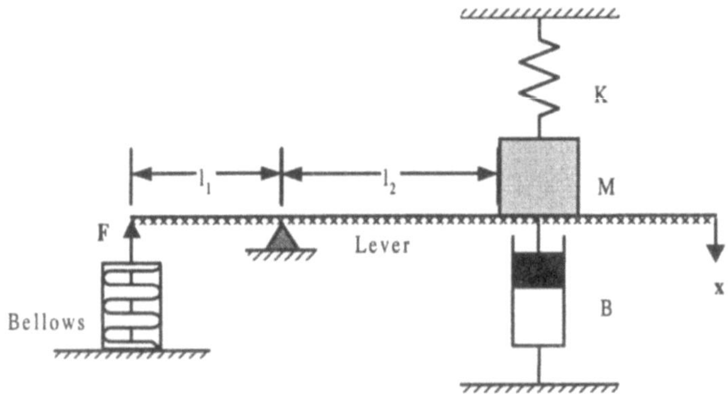

Figure 8.31. Spring-mass-damper-lever System

Let us next simulate a portion of a mechanical transducer, consisting of a spring-mass-damper-lever system arranged in a manner shown in Figure 8.31. The input to this system is force through a bellow and the output is displacement of the lever. The equations governing such a system can be written down as:

$$M\frac{d^2x}{dt^2} + B\frac{dx}{dt} + Kx = -\frac{l_1}{l_2}F$$

This equation is simulated using block diagram approach as shown in Figure 8.32 for values of parameters as:

$l_1= 0.5$ m, $l_2=1.5$ m, M=10 kg, B=5 Newton/(m/sec), K=4 Newton/m, F= 10 Newton

The transfer function for the model can be derived from the equation as follows:

$$\frac{X(s)}{F(s)} = \frac{-l_1/l_2}{Ms^2 + Bs + K}$$

The state space variable model of the system can also be obtained as follows:

let,

$$x_1 = x$$

$$x_2 = \dot{x}_1 = \frac{dx}{dt}$$

$$\dot{x}_2 = \ddot{x}_1 = \frac{d^2x}{dt^2} = -\frac{B}{M}\frac{dx}{dt} - \frac{K}{M}x - \frac{l_1/l_2}{M}F$$

putting in state variable matrix form,

$$\begin{bmatrix} \dot{x}_1 \\ \dot{x}_2 \end{bmatrix} = \begin{bmatrix} 0 & 1 \\ -\dfrac{K}{M} & -\dfrac{B}{M} \end{bmatrix} \begin{bmatrix} x_1 \\ x_2 \end{bmatrix} + \begin{bmatrix} 0 \\ \dfrac{-l_1/l_2}{M} \end{bmatrix} [F]$$

$$[y] = \begin{bmatrix} 1 & 0 \end{bmatrix} \begin{bmatrix} x_1 \\ x_2 \end{bmatrix} + [0][F]$$

You may simulate the system using transfer function model or state space model directly. However, in either case the response of the system will remain the same as shown in Figure 8.33.

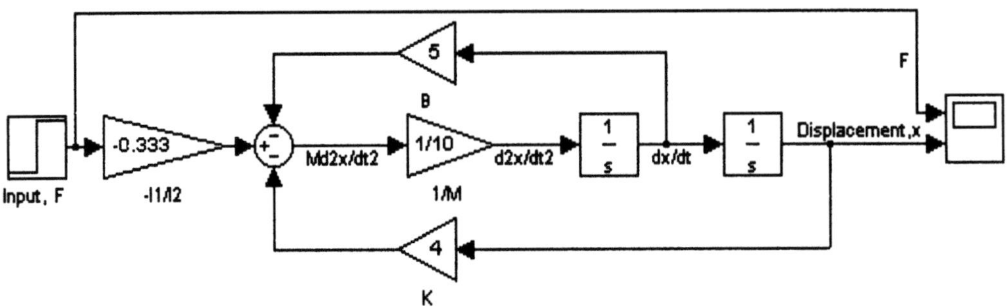

Figure 8.32. Simulation of Spring-mass-damper-lever System

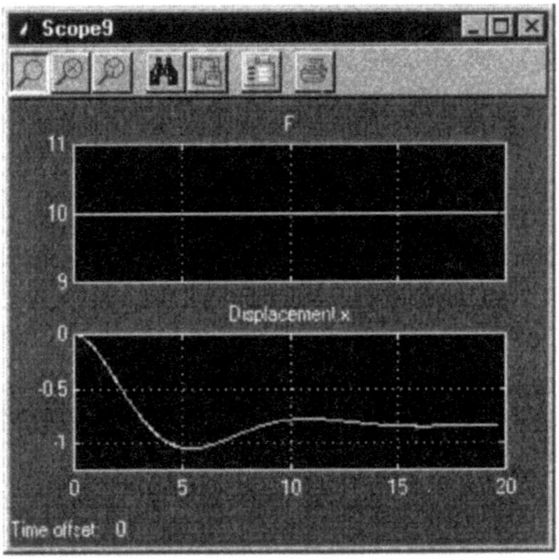

Figure 8.33. Response of Spring-mass-damper-lever System

8.3.1.3 Double Spring-mass-damper System

Next let us simulate a very common problem of mechanical system consisting of two springs, two masses and two dampers as arranged in Figure 8.34. Let the simulation parameters be:

$M_1 = 20$ kg. ; $M_2 = 10$ kg.; $K_1 = 2$ Newtons/m ; $K_2 = 3$ Newtons/m ; $B_1 = 5$ Newtons/(m/sec); $B_2 = 4$ Newtons/(m/sec); and F= ramp signal of magnitude 10 Newtons

Equations describing the system can be written down as:

$$M_1 \frac{d^2 y_1}{dt^2} = -(K_1 + K_2)y_1 + K_2 y_2 - (B_1 + B_2)\frac{dy_1}{dt} + B_2 \frac{dy_2}{dt}$$

$$M_2 \frac{d^2 y_2}{dt^2} = K_2 y_1 - K_2 y_2 + B_2 \frac{dy_1}{dt} - B_2 \frac{dy_2}{dt} + F$$

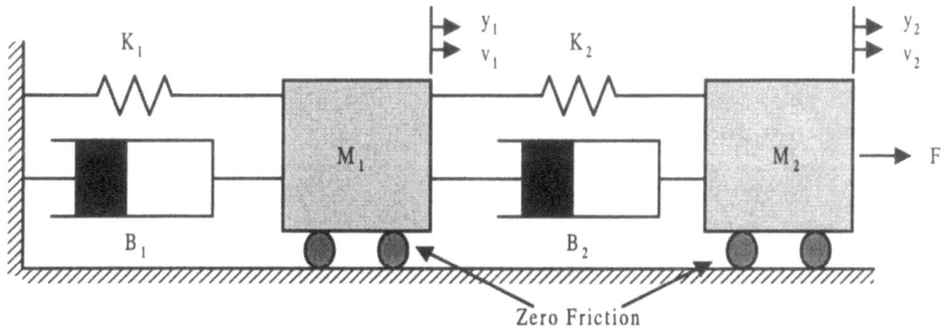

Figure 8.34. Double Spring-mass-damper System

The transfer function of the model can be derived to be:

$$\frac{y_1(s)}{F(s)} = \frac{sB_2 + K_2}{[M_1M_2s^4 + (B_1M_2 + B_2M_2 + M_1B_2)s^3 + (K_1M_2 + K_2M_2 + B_1B_2 + M_1K_2)s^2 + (K_1B_2 + B_1K_2)s + K_1K_2]}$$

$$\frac{y_2(s)}{F(s)} = \frac{M_1s^2 + (B_1 + B_2)s + (K_1 + K_2)}{[M_1M_2s^4 + (B_1M_2 + B_2M_2 + M_1B_2)s^3 + (K_1M_2 + K_2M_2 + B_1B_2 + M_1K_2)s^2 + (K_1B_2 + B_1K_2)s + K_1K_2]}$$

The state space variable model for the system can be derived as:

Choosing state variables as:

$$x_1 = y_1$$

$$x_2 = y_2$$

$$x_3 = \dot{x}_1 = \frac{dy_1}{dt}$$

$$x_4 = \dot{x}_2 = \frac{dy_2}{dt}$$

$$\dot{x}_3 = \ddot{x}_1 = \frac{d^2y_1}{dt^2} = -\frac{(K_1 + K_2)}{M_1}x_1 + \frac{K_2}{M_2}x_2 - \frac{1}{M_1}(B_1 + B_2)x_3 + \frac{B_2}{M_1}x_4$$

$$\dot{x}_4 = \ddot{x}_2 = \frac{d^2 y_2}{dt^2} = \frac{K_2}{M_2} x_1 - \frac{K_2}{K_1} x_2 + \frac{B_2}{M_2} x_3 - \frac{B_2}{M_2} x_4 + \frac{1}{M_2} F$$

Putting the above in state variable matrix form,

$$\begin{bmatrix} \dot{x}_1 \\ \dot{x}_2 \\ \dot{x}_3 \\ \dot{x}_4 \end{bmatrix} = \begin{bmatrix} 0 & 0 & 1 & 0 \\ 0 & 0 & 0 & 1 \\ -\dfrac{1}{M_1}(K_1 + K_2) & \dfrac{K_2}{M_1} & -\dfrac{1}{M_1}(B_1 + B_2) & \dfrac{B_2}{M_1} \\ \dfrac{K_2}{M_2} & -\dfrac{K_2}{M_2} & \dfrac{B_2}{M_2} & -\dfrac{B_2}{M_2} \end{bmatrix} \begin{bmatrix} x_1 \\ x_2 \\ x_3 \\ x_4 \end{bmatrix} + \begin{bmatrix} 0 \\ 0 \\ 0 \\ \dfrac{1}{M_2} \end{bmatrix} [F]$$

$$\begin{bmatrix} y_1 \\ y_2 \end{bmatrix} = \begin{bmatrix} 1 & 0 & 0 & 0 \\ 0 & 1 & 0 & 0 \end{bmatrix} \begin{bmatrix} x_1 \\ x_2 \\ x_3 \\ x_4 \end{bmatrix} + [0][F]$$

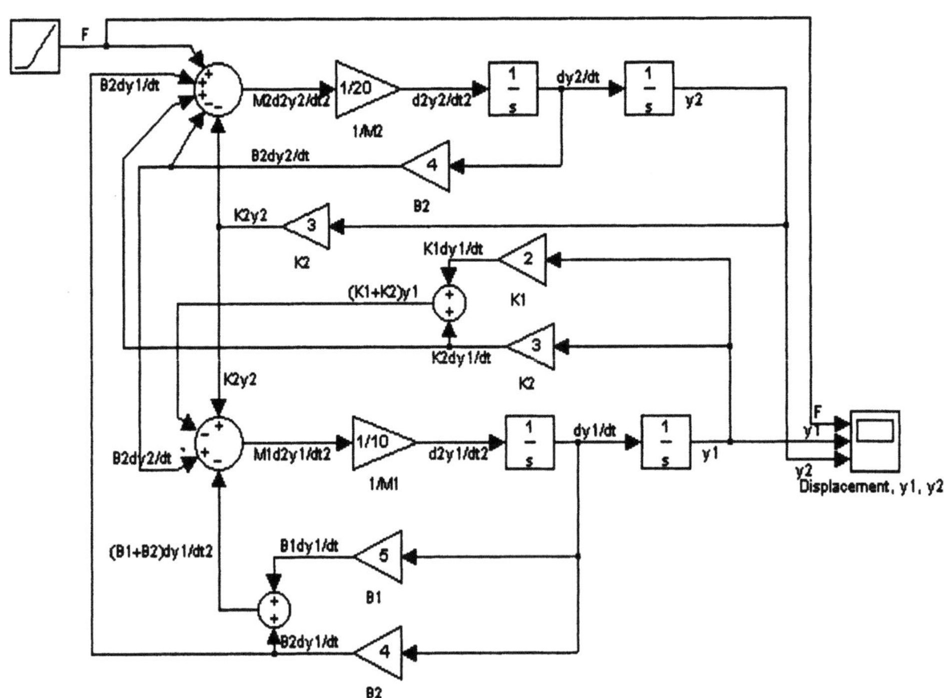

Figure 8.35. Simulation of Double Spring-mass-damper System

You can use transfer function model or the state space model directly for simulating the system. The response of the system will remain same irrespective of which model you choose for simulation.

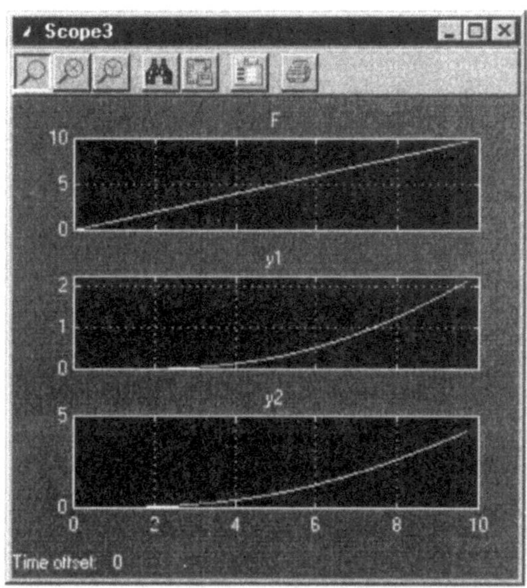

Figure 8.36. Response of Double Spring-mass-damper System to a ramp input signal

Practice Test 8.9.

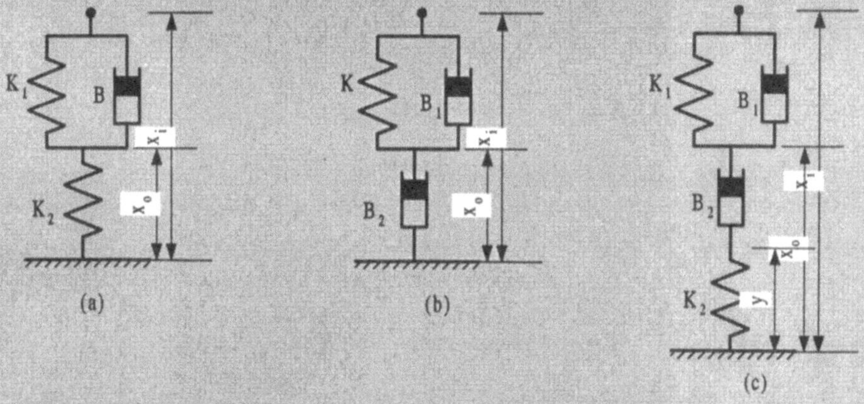

1. Simulate the translational mechanical systems shown above and obtain their frequency response. Comment on the behaviour of the systems and their use. Let the parameters be:
   $K= 15$; $K_1=12$; $K_2=13$; $B=16$; $B_2=14$; $B_3=18$; $x_1= 15\sin 314t$; 25

2. Simulate the following translational mechanical systems for parameters chosen as:
   $K= 5$; $K_1=2$; $K_2=3$; $B=6$; $B_1= 4$; $M =15$; $M_1=10$; $M_2=12$;
   $F= 7\cos 7t$; $5 \times 10^{-1.5t}$ and 13

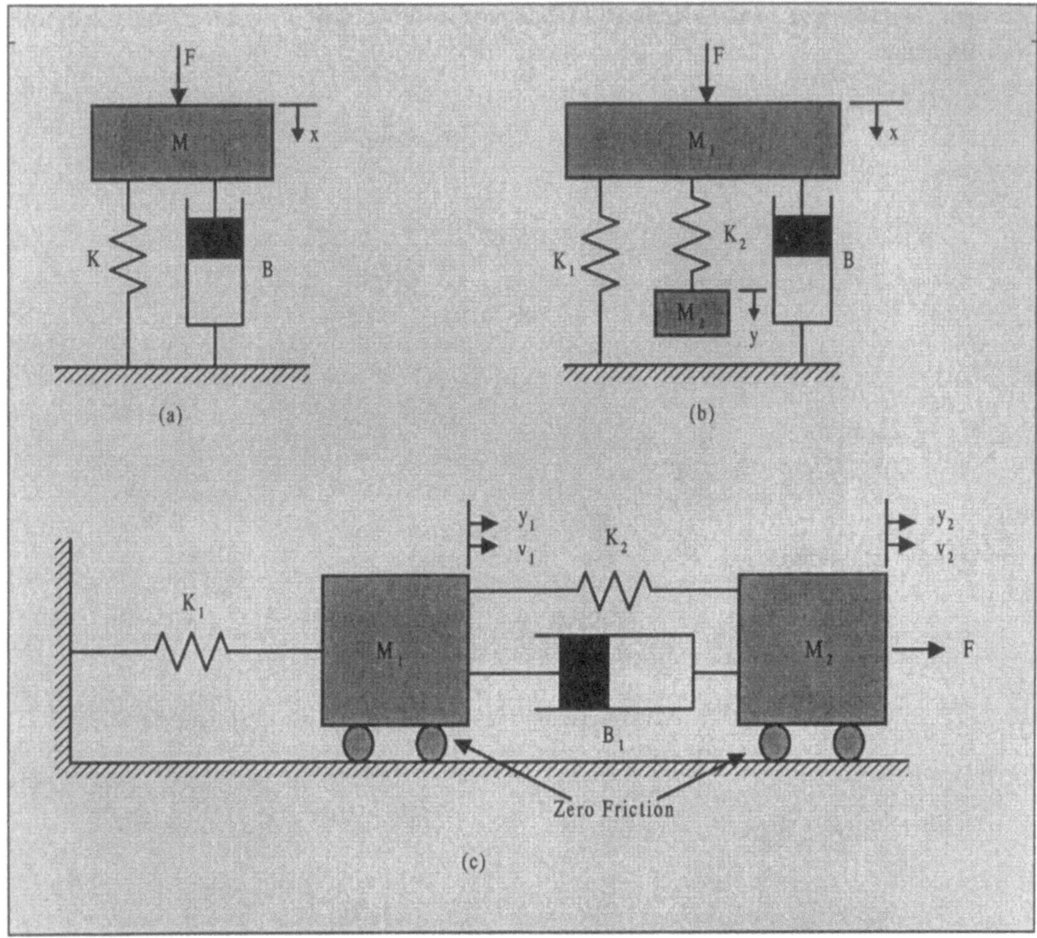

8.3.2 Rotational Mechanical Systems

Rotational mechanical systems, forms an integral part of many machines. Modeling of such a system is similar to the translational mechanical system, inertia, torsional spring constant, and viscous friction forming the respective elements of dynamics. In the paragraphs to follow, you will learn to simulate some simple physical rotational systems.

8.3.2.1 Spring-inertia-damper System

Consider a system consisting of a spring, an inertia element and a damper arranged as shown in Figure 8.37. The system is connected to a cam, which provides a square input torque signal to the system. Let the parameters of the system be:

K= 10 Newton-m/(rad/sec^2); J = 8 kg-m^2; B = 2 Newton-m/(rad/sec);
T= a square wave input of magnitude ±5 Newton-m and time period 0.2 second.

The equation governing the system can be written down as:

$$J\frac{d^2\theta}{dt^2} + B\frac{d\theta}{dt} + K\theta = T$$

The equation given above is simulated as shown in Figure 8.38 using block diagram approach.. The response of the system for square input is shown in Figure 8.39.

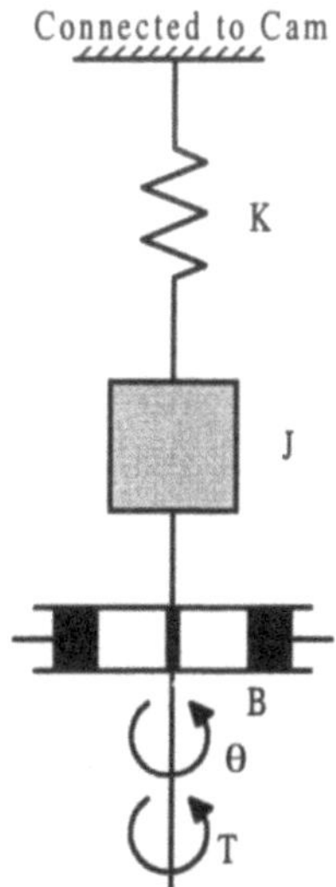

Figure 8.37. Spring-inertia-damper system

From the equations above the transfer function of the system can be derived to be:

$$\frac{\theta(s)}{T(s)} = \frac{1}{Js^2 + Bs + K}$$

The state space variable representation of the system can be derived to be:

$$x_1 = \theta$$

$$\dot{x}_1 = x_2 = \frac{d\theta}{dt}$$

$$\dot{x}_2 = \frac{d^2\theta}{dt^2} = -\frac{K}{J}\theta + \frac{T}{J}$$

Hence,

$$\begin{bmatrix} \dot{x}_1 \\ \dot{x}_2 \end{bmatrix} = \begin{bmatrix} 0 & 1 \\ -\frac{K}{J} & 0 \end{bmatrix}\begin{bmatrix} x_1 \\ x_2 \end{bmatrix} + \begin{bmatrix} 0 \\ \frac{1}{J} \end{bmatrix}[T]$$

$$[y] = [1 \quad 0]\begin{bmatrix} x_1 \\ x_2 \end{bmatrix}$$

You may use transfer function model or the state space model directly for simulation to obtain the same result.

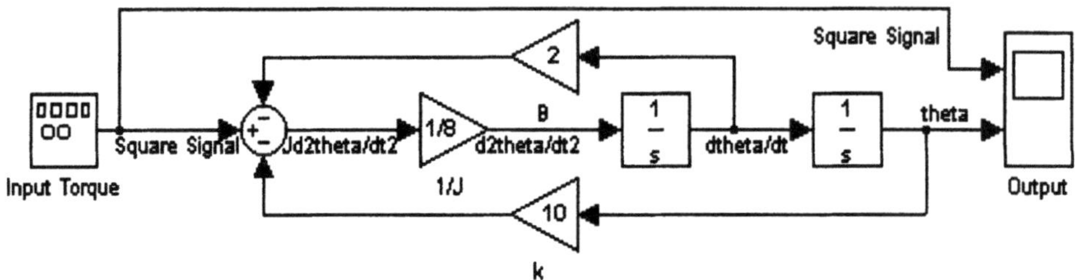

Figure 8.38. Simulation of Spring-inertia-damper System

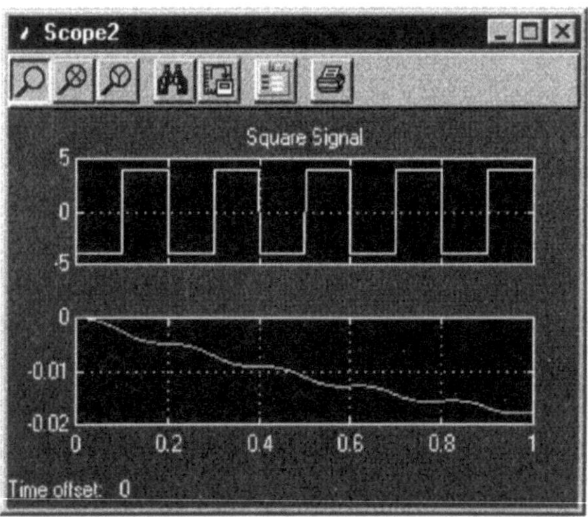

Figure 8.39. Response of Spring-inertia-damper System

8.3.2.2 Inertia-spring-inertia-damper System

This is again a very common example of rotational systems. It consists of two inertia elements and one each of torsional spring and damper. Let these be arranged as shown in Figure 8.40. Let the simulation parameters be:

K= 3 Newton-m/(rad/sec^2); J_1 = 6 kg-m^2; J_2 = 8 kg-m^2; B = 2 Newton-m/(rad/sec);
T= a square wave input of magnitude ±4 Newton-m with time period of 0.2 seconds

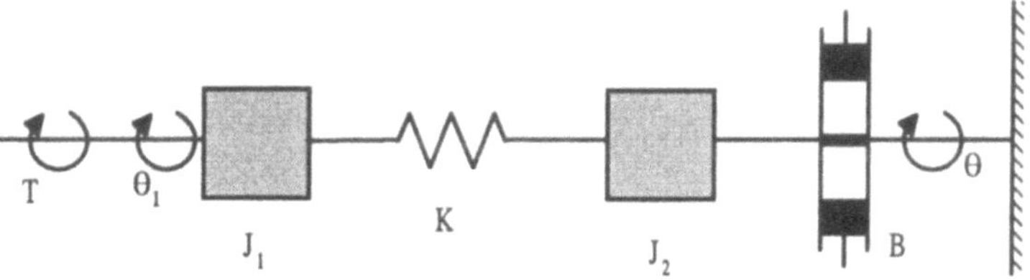

Figure 8.40. Inertia-spring-inertia-damper system

The equations describing the system can be written down as:

$$J_1 \frac{d^2\theta_1}{dt^2} + K(\theta_1 - \theta) = T$$

$$J_2 \frac{d^2\theta_2}{dt^2} + f \frac{d\theta}{dt} = K(\theta_1 - \theta)$$

These equations are simulated using the block diagram approach (Figure 8.41). And the response of the system is shown in Figure 8.42.

The transfer function model of the system can be derived for the system to be:

$$\frac{\theta(s)}{T(s)} = \frac{K}{J_1 J_2 + J_1 Bs^3 + K(J_1 + J_2)s^2 + KBs}$$

The state space variable representation of the system can be derived as:

Let,

$$x_1 = \theta_1$$

$$\dot{x}_1 = x_3 = \frac{d\theta_1}{dt}$$

$$\dot{x}_3 = \frac{d^2\theta_1}{dt^2} = -\frac{K}{J_1}\theta_1 + \frac{K}{J_1}\theta + \frac{T}{J_1}$$

$$x_2 = \theta$$

$$\dot{x}_2 = x_4 = \frac{d\theta}{dt}$$

$$\dot{x}_4 = \frac{d^2\theta}{dt^2} = -\frac{B}{J}\frac{d\theta}{dt} + K\theta_1 - K\theta$$

Hence,

$$\begin{bmatrix} \dot{x}_1 \\ \dot{x}_2 \\ \dot{x}_3 \\ \dot{x}_4 \end{bmatrix} = \begin{bmatrix} 0 & 0 & 1 & 0 \\ 0 & 0 & 0 & 1 \\ -\dfrac{K}{J_1} & \dfrac{K}{J_1} & 0 & 0 \\ \dfrac{K}{J_2} & -\dfrac{K}{J_2} & 0 & -\dfrac{B}{J_2} \end{bmatrix} \begin{bmatrix} x_1 \\ x_2 \\ x_3 \\ x_4 \end{bmatrix} + \begin{bmatrix} 0 \\ 0 \\ \dfrac{1}{J_1} \\ 0 \end{bmatrix} [T]$$

$$y = \begin{bmatrix} 0 & 0 & 1 & 0 \end{bmatrix} \begin{bmatrix} x_1 \\ x_2 \\ x_3 \\ x_4 \end{bmatrix}$$

You may use transfer function model or the state space model directly for simulation. In any case the response of the system would remain the same as shown in Figure 8.42.

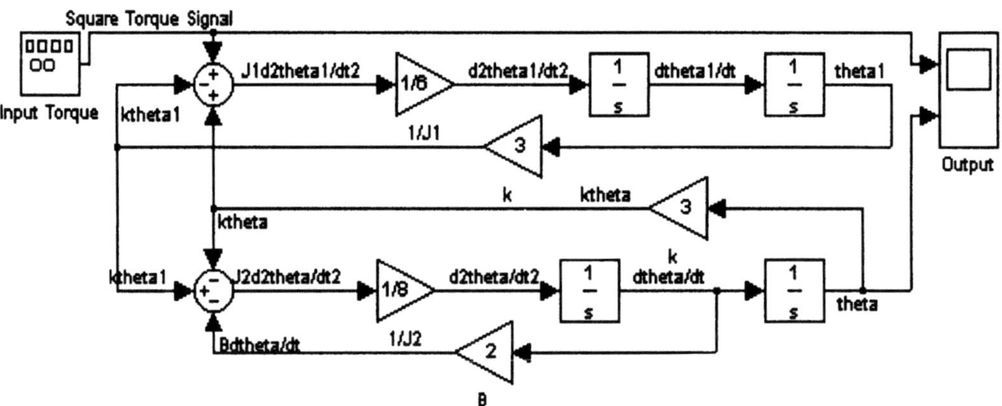

Figure 8.41. Simulation of Inertia-spring-inertia-damper System

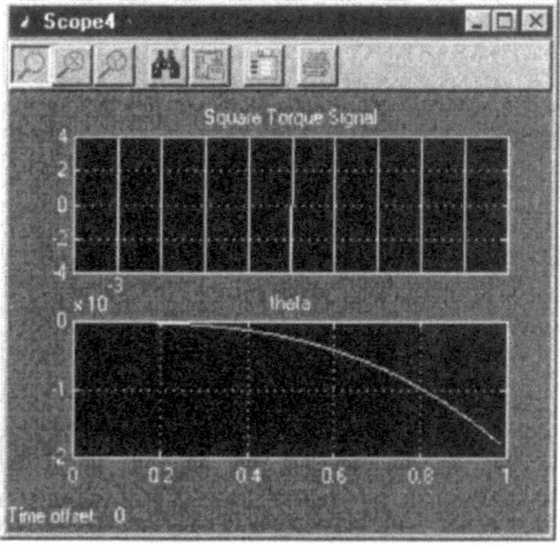

Figure 8.42. Response of Inertia-spring-inertia-damper System

Practice Test 8.10.

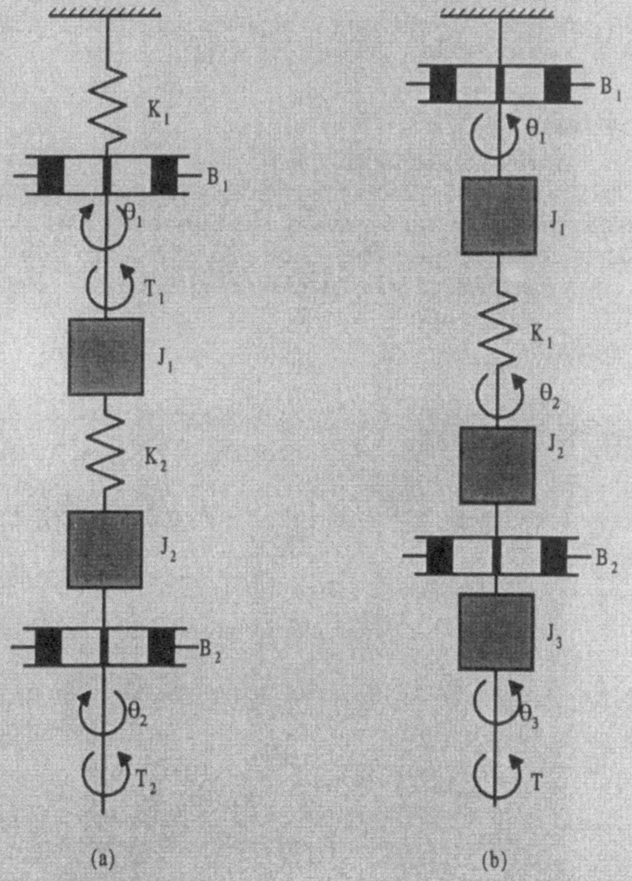

> 1. Simulate the given rotational mechanical systems for the following parameters:
>
> $K_1=2$; $K_2=3$; $B_1=6$; $B_2=4$; $J_1=15$; $J_2=10$; $J_3=15$; $T_1 = 8\cos5t$; $T_2 = 5e^{-1.5t}$

8.4. Fluid Systems

Large portion of our ambient is made of fluids of different types. We cannot imagine our life without existence of fluids be it water or air so important for our survival. So, to learn simulation of fluid systems have become almost inevitable. In addition to the natural fluid systems, man has built several systems using the properties of fluids. The world of fluid can be subdivided into two broad categories:

- Hydraulic system comprising of the incompressible fluids
  e.g., mainly liquids like water, oil *etc.*
- Pneumatic system comprising of the compressible fluids
  e.g., mainly gases like air, oxygen, hydrogen *etc.*

In the forthcoming paragraphs we will learn to simulate some fluid systems belonging to both the categories.

8.4.1 Hydraulic Systems

We discussed above that the hydraulic systems comprise of incompressible fluids. But we know that all liquids are compressible under high pressure. However, in the examples to follow, we will consider liquids to be incompressible. This results into what is known as an approximate model of the system, which is sufficiently accurate for most systems.

8.4.1.1 Isolated-tank System

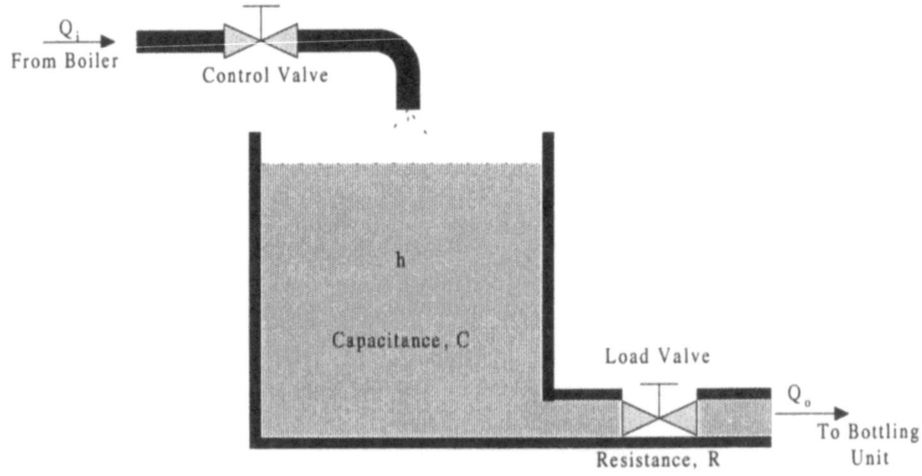

Figure 8.43. Isolated-tank System

Let us consider a jam and jelly factory that runs from morning 9 a.m. to evening 5.30 p.m. Every day during forenoon from 10 am to 11 am, 75 m^3/sec of cooked hot liquid jelly from boiler is strained into an isolated pre-bottling storage tank. Again, in the afternoon the process is repeated from 2 p.m. to 3 p.m. when 50 m^3/sec of jelly is strained into the same tank. Assuming the hydraulic resistance, R of the system to be 20 $\dfrac{\text{Newtons/m}^2}{m^3/\text{sec}}$ and hydraulic capacitance of the system to be 9 $\dfrac{m^3}{\text{Newtons/m}^2}$, the system is required to be simulated to obtain the height of hot liquid jelly in the storage tank and the discharge of the hot liquid jelly from the tank into the bottling unit over the working hours in a day.

The system can be represented by the following mathematical equation:

$$C\frac{dh}{dt} = (Q_i - Q_o)$$

But,

$$Q_o = \frac{h}{R}$$

Hence,

$$RC\frac{dh}{dt} + h = RQ_i$$

This equation is simulated using block diagram approach as shown in Figure 8.44. The response of the system from 8 am to 6 pm is shown in Figure 8.45.

The transfer function of the system may be derived to be:

$$\frac{H(s)}{Q_i(s)} = \frac{R}{sRC + 1}$$

and,

$$\frac{Q_o(s)}{Q_i(s)} = \frac{1}{sRC + 1}$$

And, the state space variable representation of the system can be written down as:

$$\frac{d}{dt}[H] = \left[\frac{-1}{RC}\right][H] + \left[\frac{1}{C}\right][Q_i]$$
$$[H] = [1][H] + [0][Q_i]$$

or,

$$\left[Q_o \right] = \left[\frac{1}{R} \right] \left[H \right] + \left[0 \right] \left[Q_i \right]$$

These models may be used directly to simulate the system.

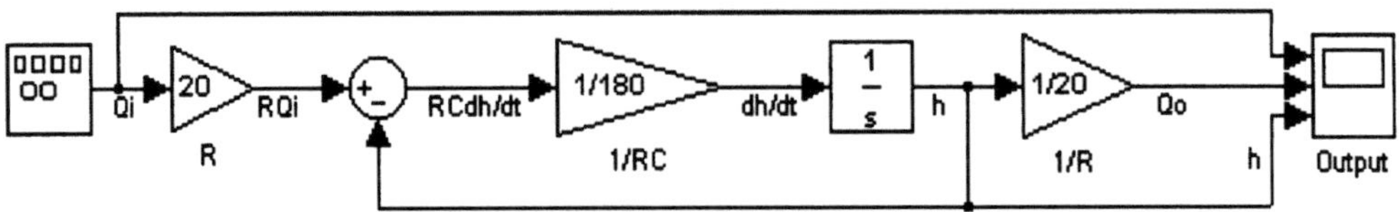

Figure 8.44. Simulation of an Isolated-tank System

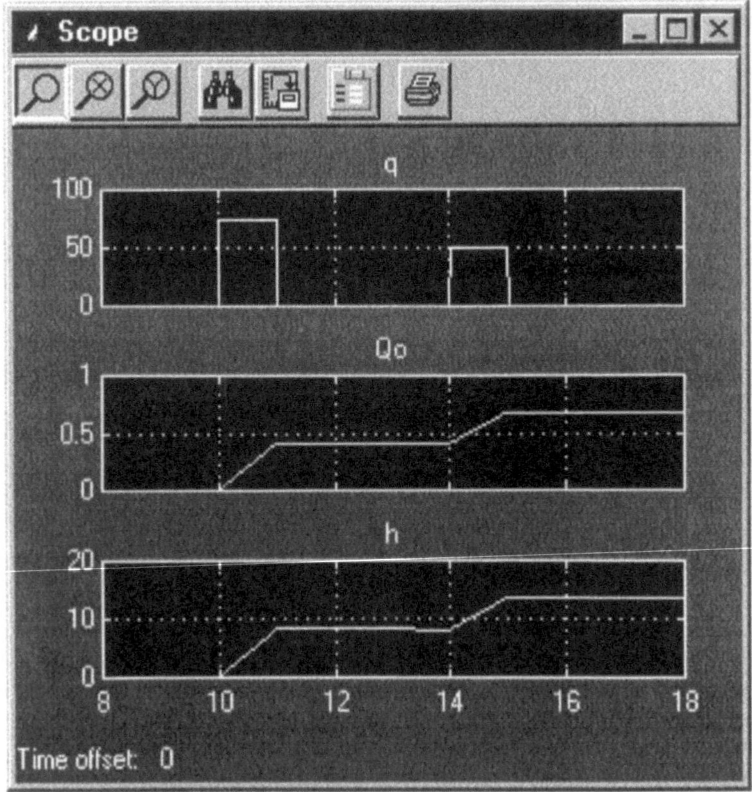

Figure 8.45. Response of an Isolated-tank System

8.4.1.2 Interacting-fluid System

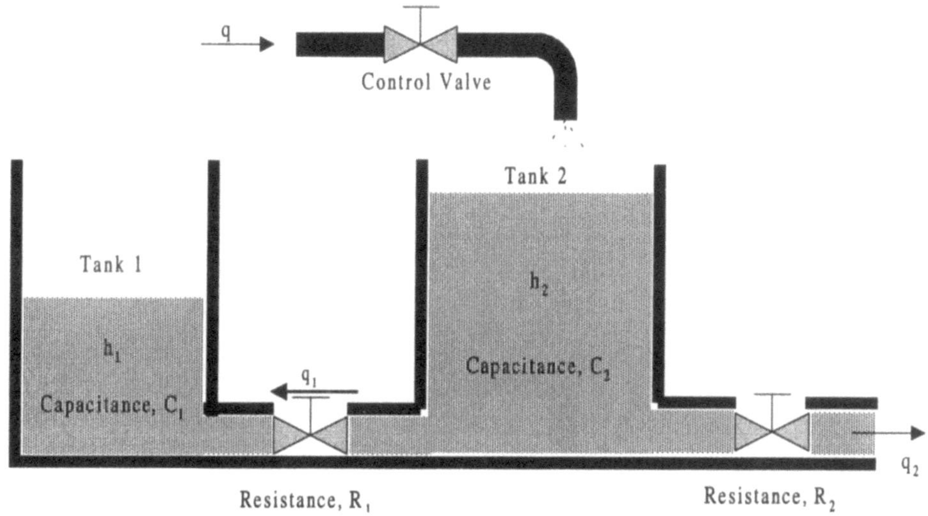

Figure 8.46. Interacting-fluid System

Let us consider yet another typical problem. A construction site received water from Public Health Engineering department twice a day -- from 6 a.m. to 7 a.m. in the morning and at the same time in the evening. As the building grew, curing area increased resulting into more water requirement. Hence, one water storage tank was found to be insufficient. The contractor arranged for a second storage tank, and the two were connected in a manner as shown in Figure 8.46. Since the head of water in tank 2 decides the head and flow of water into tank 1. Therefore, this system may be said to be an interacting-fluid system. Considering the tap supplying water to tank 2 has a discharge of q (=10m³/sec); the hydraulic resistance and capacitance of the tanks to

be respectively R_1 (=5 $\dfrac{\text{Newtons/m}^2}{\text{m}^3/\text{sec}}$), R_2 (=20 $\dfrac{\text{Newtons/m}^2}{\text{m}^3/\text{sec}}$), C_1 (=10 $\dfrac{\text{m}^3}{\text{Newtons/m}^2}$) and C_2

(=3 $\dfrac{\text{m}^3}{\text{Newtons/m}^2}$). We will simulate the system in order to observe its response over a period of seven days.

The equations governing the system can be written down as:

For Tank 1:

$$C_1 \frac{dh_1}{dt} = q_1$$

But,

$$q_1 = \frac{h_2 - h_1}{R_1}$$

Hence,

$$R_1 C_1 \frac{dh_1}{dt} + h_1 = h_2 \qquad\qquad \text{-------(1)}$$

For Tank 2:

$$C_2 \frac{dh_2}{dt} = q - q_1 - q_2$$

where,

$$q_2 = \frac{h_2}{R_2}$$

Hence,

$$R_2 C_2 \frac{dh_2}{dt} + \frac{R_2}{R_1} h_2 + h_2 = R_2 q + \frac{R_2}{R_1} h_1 \qquad\qquad \text{----------(2)}$$

Equation 1 and Equation 2 are simulated using block diagram approach (Figure 8.47). The result of the simulation is shown in Figure 8.48.

Transfer function of the system can be derived from Equation 1 and Equation 2 as:

$$\frac{h_1(s)}{q(s)} = \frac{R_2}{R_1 C_1 R_2 C_2 s^2 + (R_1 C_1 + R_2 C_2 + R_2 C_1)s + 1}$$

$$\frac{h_2(s)}{q(s)} = \frac{R_2(R_1 C_1 s + 1)}{R_1 C_1 R_2 C_2 s^2 + (R_1 C_1 + R_2 C_2 + R_2 C_1)s + 1}$$

$$\frac{q_2(s)}{q(s)} = \frac{R_1 C_1 s + 1}{R_1 C_1 R_2 C_2 s^2 + (R_1 C_1 + R_2 C_2 + R_2 C_1)s + 1}$$

The state space representation of the system too can be obtained as:

$$\frac{d}{dt}\begin{bmatrix} h_1 \\ h_2 \end{bmatrix} = \begin{bmatrix} -\dfrac{1}{R_1 C_1} & \dfrac{1}{R_1 C_1} \\ \dfrac{1}{R_1 C_2} & -\dfrac{1}{C_2}\left(\dfrac{1}{R_1} + \dfrac{1}{R_2}\right) \end{bmatrix} \begin{bmatrix} h_1 \\ h_2 \end{bmatrix} + \begin{bmatrix} 0 \\ \dfrac{1}{C_2} \end{bmatrix} [q]$$

$$\begin{bmatrix} h_1 \\ h_2 \end{bmatrix} = \begin{bmatrix} 1 & 0 \\ 0 & 1 \end{bmatrix} \begin{bmatrix} h_1 \\ h_2 \end{bmatrix} + [0][q]$$

You can simulate the system using transfer function model or the state space variable model directly.

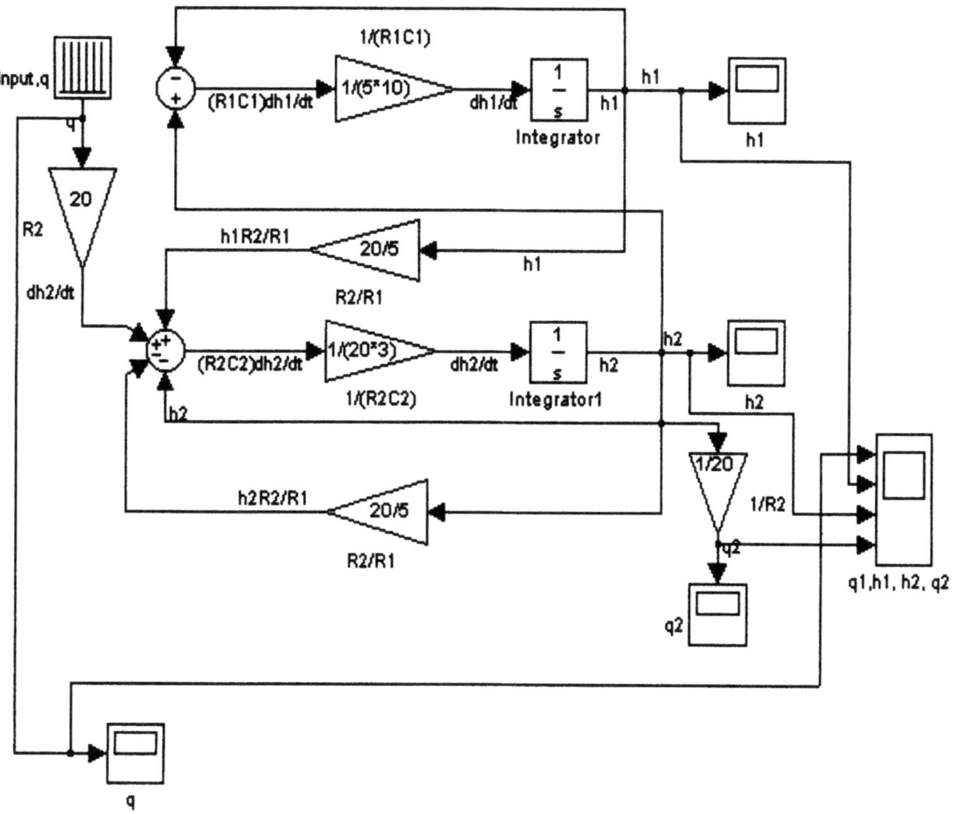

Figure 8.47. Simulation of Interacting-fluid System

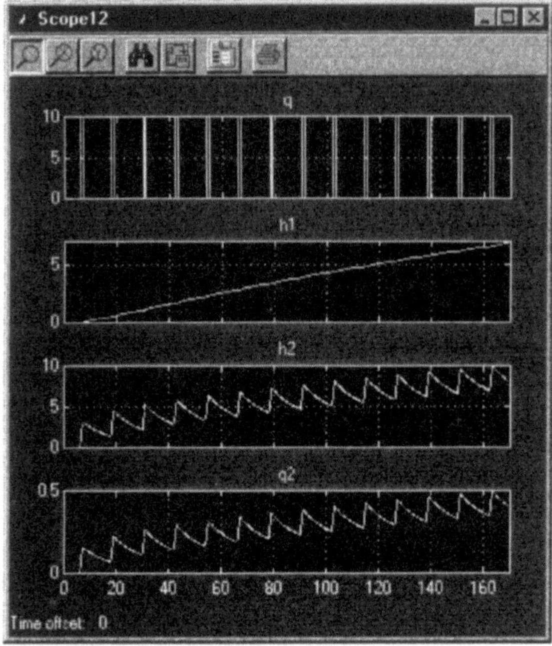

Figure 8.48. Response of Interacting-fluid System

8.4.2 Pneumatic Systems

As mentioned earlier, pneumatic systems form an important part of Control System in industries specially, where processes involving large power are to be controlled. The force exerted by gases especially air, provides the desired control. In the forthcoming paragraph, you will learn to simulate a simple pneumatic system.

8.4.2.1 Pneumatic-valve System

Simple schematic of a Control System involving pneumatic valve for control of pressure is shown in Figure 8.49. Let the input pressure, output pressure, pneumatic resistance and capacitance of the valve be P_i, P_o, R and C respectively.

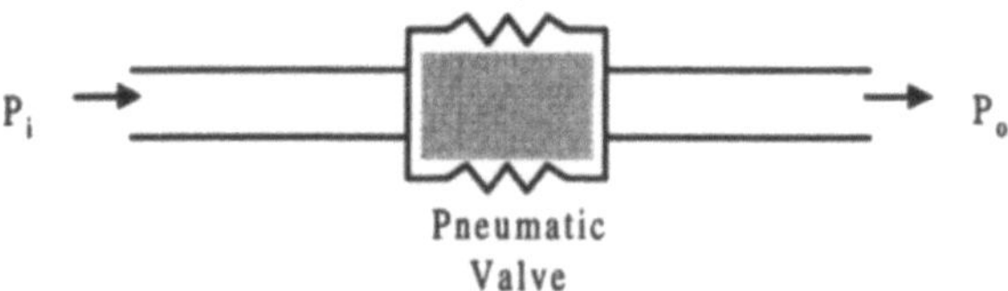

Figure 8.49. Pneumatic-valve System

The system can be described by the following equation:

$$P_i = RC\frac{dP_O}{dt} + P_O$$

This is clearly a first-order system. Let the time constant RC of the system be 1000 seconds. The input pressure signal is a step signal of 500 Newton magnitude. Simulate the system using block diagram approach (Figure 8.50). The response of the system is shown in Figure 8.51.

Transfer function for the system can be derived from the above equation to be:

$$\frac{P_O}{P_i} = \frac{1}{RCs + 1}$$

State space representation of the system can be derived to be:

$$x_1 = P_O$$

$$\dot{x}_1 = \frac{dP_O}{dt} = \frac{1}{RC}P_i - \frac{1}{RC}P_O$$

Hence,

$$\left[\dot{x}_1\right] = \left[-\frac{1}{RC}\right]\left[x_1\right] + \left[\frac{1}{RC}\right]\left[P_i\right]$$

$$\left[y\right] = \left[1\right]\left[x_1\right] + \left[0\right]\left[P_i\right]$$

You may use the transfer function model or the state space model directly for simulation to obtain the same response as shown in Figure 8.51.

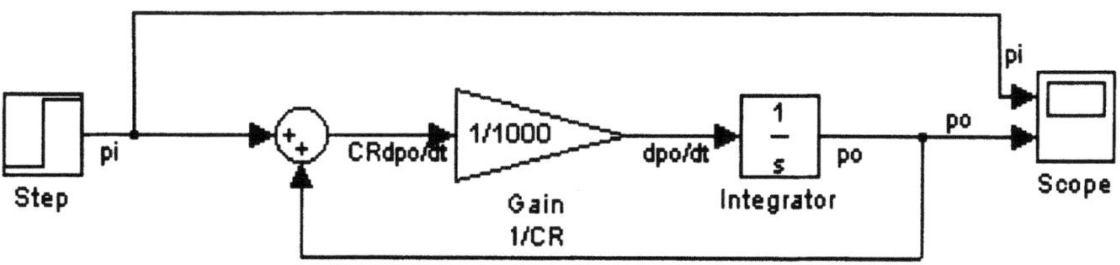

Figure 8.50. Simulation of Pneumatic-valve System

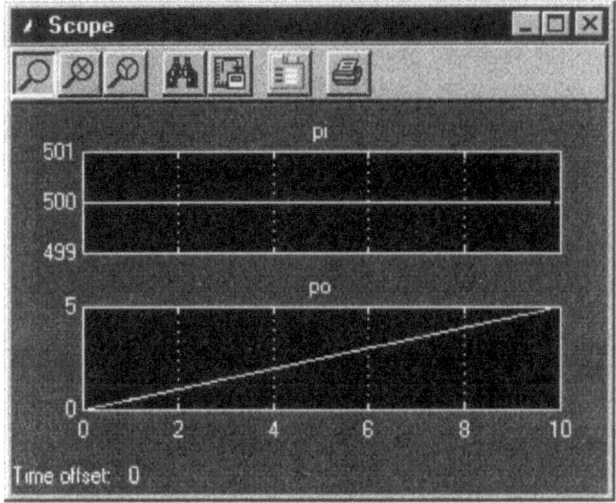

Figure 8.51. Response of Pneumatic-valve System

Practice Test 8.11.

1. Consider the same tank system of Figure 8.46. If the contractor at the building site made the arrangement as shown in Figure given below, find out the response of the system. Let the parameters for simulation be the same as before. Except for magnitude of water supply for the two taps which are of different sizes (receiving water from PHE department as before) $q_{i1}=10\text{m}^3/\text{sec}$ and $q_{i2}=20\text{m}^3/\text{sec}$.

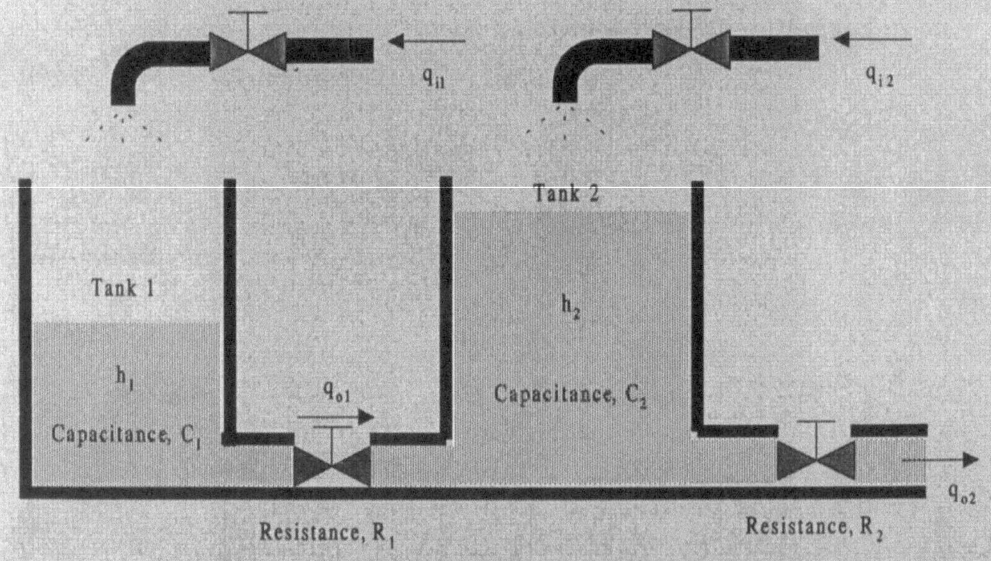

8.5 Thermal Systems

Thermal control systems form an integral part of many control systems. The temperature control of our body through perspiration is the most common example we are all familiar with. The heat loss phenomenon due to conduction, convection or radiation forms the basis of measurement and heat control in process industries. In the paragraphs to follow we shall learn to simulate different thermal systems using these effects.

8.5.1 A Mercury-thermometer System

To begin with, let us take a very simple system, consisting of a mercury thermometer in a liquid bath as shown in Figure 8.52 and simulate it.

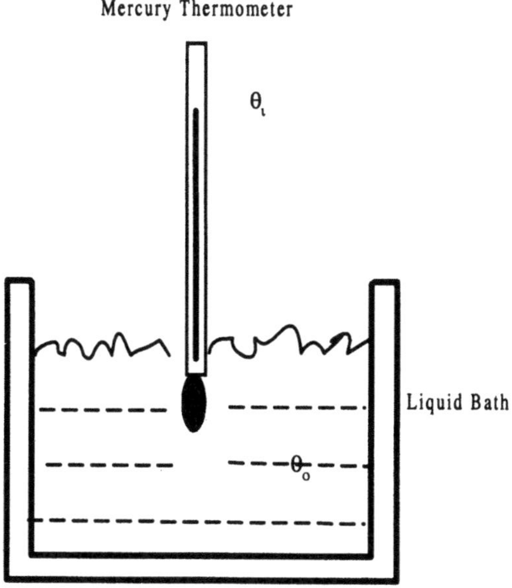

Figure 8.52. Thin-glass-wall-mercury-thermometer System

Assume that the thermometer is at a uniform temperature of θ_i °C and the temperature of the liquid is θ_o °C. The equation describing the system is:

$$C\frac{d\theta_o}{dt} = \frac{\theta_o - \theta_i}{R}$$

or,

$$RC\frac{d\theta_o}{dt} + \theta_o = \theta_i$$

This is obviously a first-order system.

Let us assume the simulation parameters of the system to be RC= 0.5 seconds, θ_2= 10°C and simulate the system using block diagram approach as shown in Figure 8.53. The mercury thermometer, which is initially at θ_i°C will attain a temperature of θ_o°C after some time which, is shown in response Figure 8.54.

From the same equations given above the transfer function of the model can be derived to be:

$$\frac{\theta o(s)}{\theta_i(s)} = \frac{1}{sRC+1}$$

The state space variable model of the system too can be found out to be:

$$\frac{d}{dt}[\theta_o] = \left[\frac{-1}{RC}\right][\theta_o] + \left[\frac{1}{RC}\right][\theta_i]$$

$$[\theta_o] = [1][\theta_o] + [0][\theta_i]$$

Either of these equations may be used directly for simulation but the response obtained would be the same in all the cases (Figure 8.55).

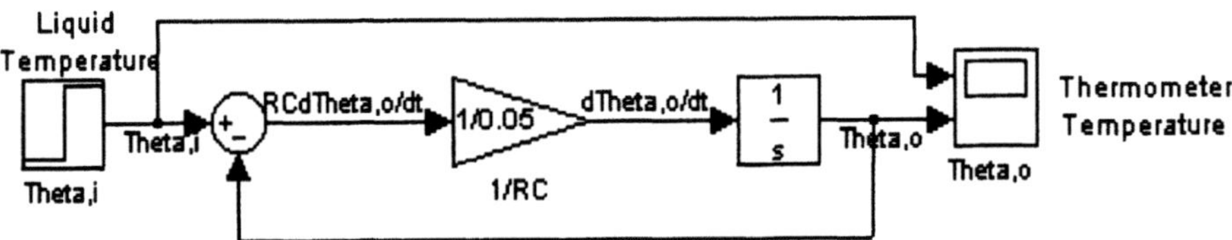

Figure 8.53. Block diagram for simulation of Mercury-thermometer System

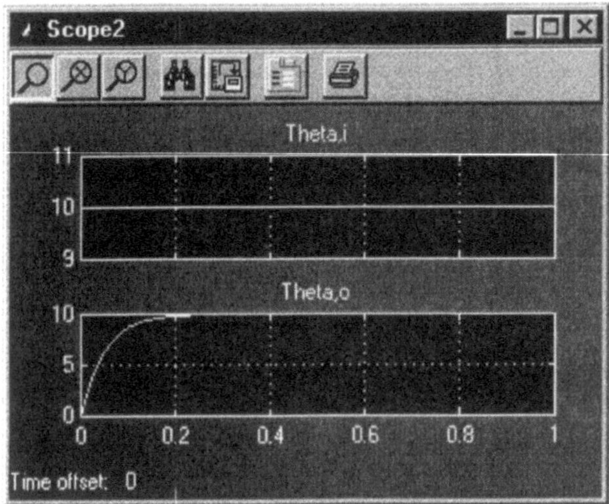

Figure 8.54. Response of Mercury-thermometer System to step input of magnitude 10

8.5.2 Oil-heating System

Next, we consider yet another thermal system that electrically heats mineral oils mainly used in transformers and circuit breakers, for removing moisture. The arrangement is as shown in Figure 8.55.

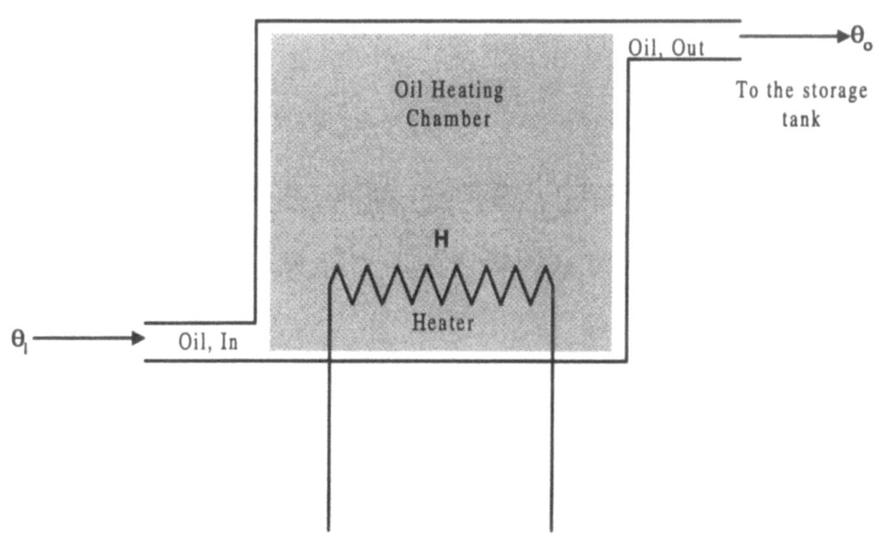

Figure 8.55. Oil-heating System

It consists of an oil-heating chamber into which oil flows in at a temperature of θ_i °C and leaves at θ_o °C. The chamber which has a thermal resistance of R °C sec/kcal and a thermal capacitance of kcal/°C is heated by an electric heater which provides a steady state heat input of H kcal/sec. The equations governing such a system can be written down as:

$$RC\frac{d\theta_o}{dt}+\theta_o = RH+\theta_i$$

This equation is simulated as shown in Figure 8.56. The value of the two inputs of the system *i.e.,* θ_i and H are taken as 5 °C and 20 kcal/sec (delayed step input of magnitude 5 and 20 with step time 0.3 and 5 respectively). The value of R and C are chosen to be 2 °C sec/kcal and 20 kcal/°C. The response of the system *i.e.,* variation of θ_o with respect to time is shown in Figure 8.57. With the value of the parameters chosen, it indicates a constant rise in the temperature of the oil leaving the chamber.

From the equations given above the transfer function of the model can be derived to be:

$$\theta_o(s)=\frac{R}{sRC+1}H(s)+\frac{1}{sRC+1}\theta_i(s)$$

The state space variable model of the system too can be found out as:

$$\frac{d}{dt}\left[\theta_{o}\right]=\left[\frac{-1}{RC}\right]\left[\theta_{o}\right]+\left[\frac{1}{RC}\quad\frac{1}{C}\right]\left[\begin{array}{c}\theta_{i}\\ H\end{array}\right]$$

$$\left[\theta_{o}\right]=[1]\left[\theta_{o}\right]+\left[0\quad 0\right]\left[\begin{array}{c}\theta_{i}\\ H\end{array}\right]$$

Either of these equations may be directly used for simulation. But, the response obtained would be the same in both the cases.

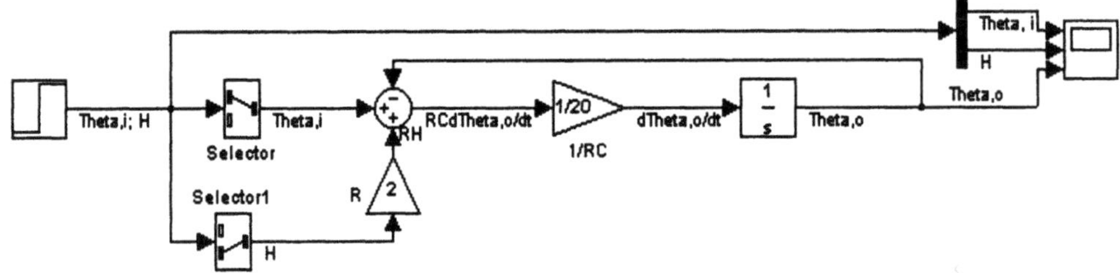

Figure 8.56. Simulation of Oil-heating System

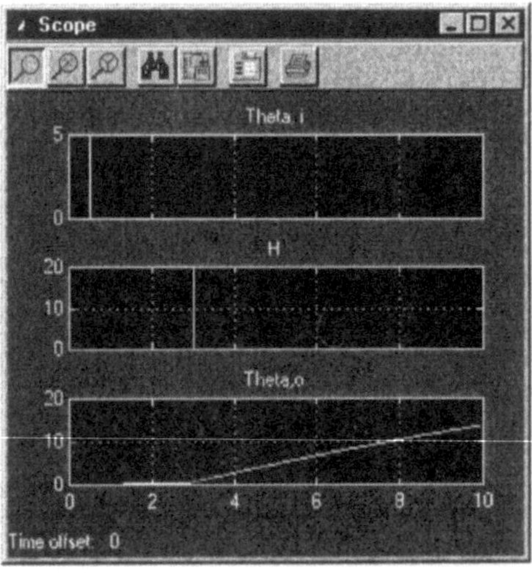

Figure 8.57. Response of Oil-heating System

Practice Test 8.12.

1. In the oil-heating system above, let the heater remain on for one hour, and off for another half an hour. Simulate the system for this situation.

2. What changes would you incorporate in the simulation diagram if you wish the temperature of the oil to go upto 500°C and then become constant?

Exercise for Chapter 8:

1. Simulate all the examples/problems considered in this chapter using transfer function and state space variable representation equations in SIMULINK® environment. Also simulate these examples in MATLAB® environment.

2. Consider the mechanical and hydraulic systems shown below. Simulate the two systems to prove that the they are equivalent to series RLC circuit and resonate at a resonant frequency given by:

$$f_o = \frac{1}{2\pi}\sqrt{\frac{K}{M}} \quad \text{and} \quad f_o = \frac{1}{2\pi\ \sqrt{M_h C_h}}$$

for all values of K, M, Mh and Ch.

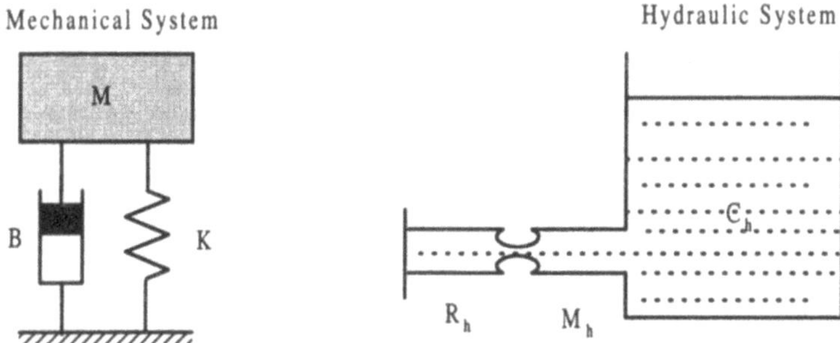

Figure 8.58. Mechanical and hydraulic system exhibiting resonant characteristics

Chapter 9

Some Complex Applications

After learning the tricks of the trade for simulations and proving your ability with some simple examples, apply your learning, to simulate some complex physical systems which you find scattered all around yourself.

In the previous chapter, we simulated some very simple pure physical systems of just one type *viz.*, electrical, mechanical, fluid and thermal. However, most of the real physical systems involve more than one of these effects. In addition, there are a few systems that have a number of components that involve several of these effects. In this chapter, we shall learn to simulate some of these complex systems.

9.1 DC Motors

As an Electrical Engineer, you must be quite familiar with the potentials of DC machines. Though, there is lot of hue and cry about rapid progress in the every field of technology, but so far, we have hardly been able to design an ac machine which can boast of providing the versatile and varied characteristics of a DC machine. No wonder, in the arena of ac transmission and distribution, DC machines still have their existance. In the paragraphs to follow, we shall simulate DC motors and study their characteristics for different inputs.

9.1.1 Separately-excited DC Motor

A separately-excited DC motor with shaft torque unimpressed and field energized, is suddenly connected to a 100 V DC supply. You have to simulate the system in order to observe the transients of current and speed as the armature picks up the speed. The parameters of the DC motor may be chosen as:

$R_a = 0.5\ \Omega$; $L_a = 0.0005H$; $J = 0.5kg\text{-}m^2$

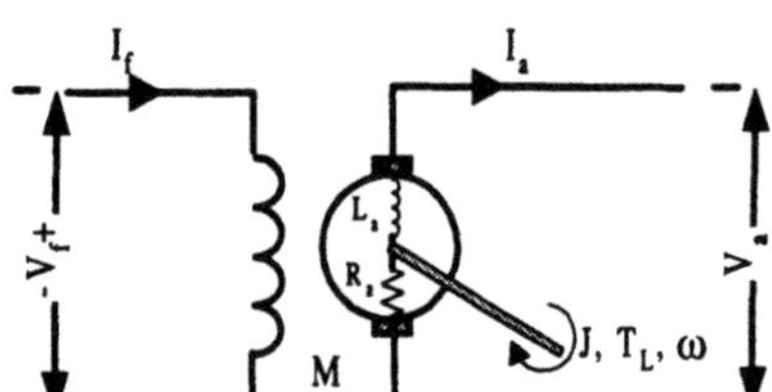

Figure 9.1. Separately-excited DC Motor

Equations describing the motion of the separately-excited DC motor can be written down as:

$$L_a \frac{dI_a}{dt} + R_a I_a + MI_f \omega = V_a$$

$$J \frac{d\omega}{dt} + T_L = MI_f I_a$$

The simulation is done as shown in Figure 9.2 for i_f= 2.5 A and 5 A, load torque T_L= 0 and 55 Newton-m. The response of the system is shown in Figure 9.3. Observe that initially the current is high, but as the rotor picks up the speed, it lowers down.

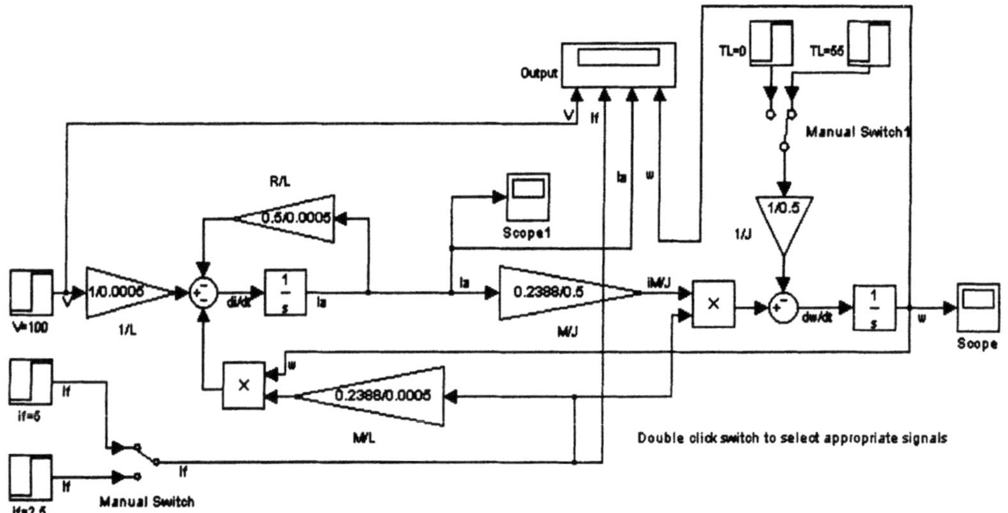

Figure 9.2. Simulation of Separately-excited DC Motor

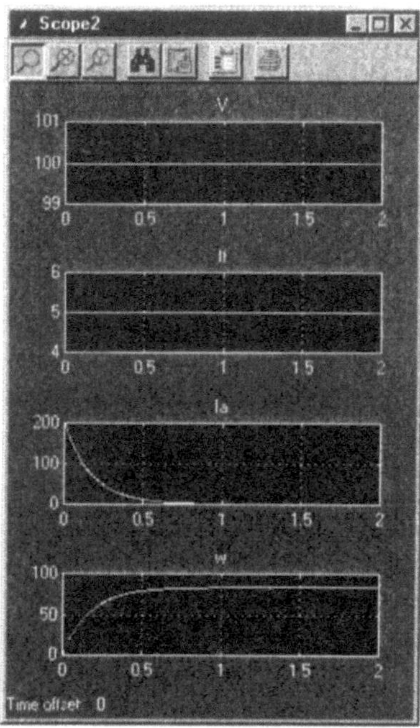

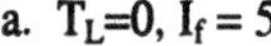

a. $T_L=0$, $I_f = 5$

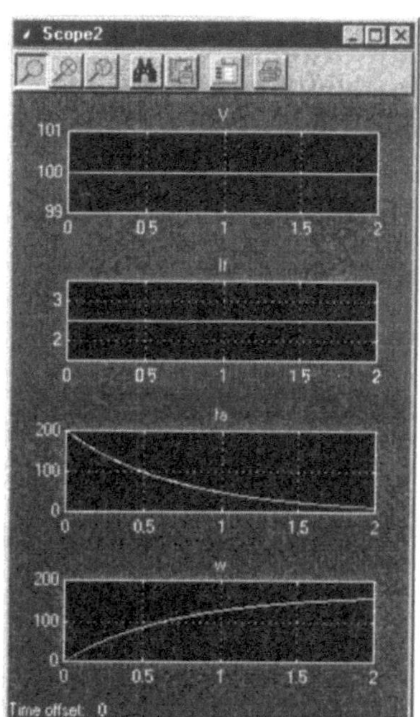

b. $T_L=0$, $I_f = 2.5$

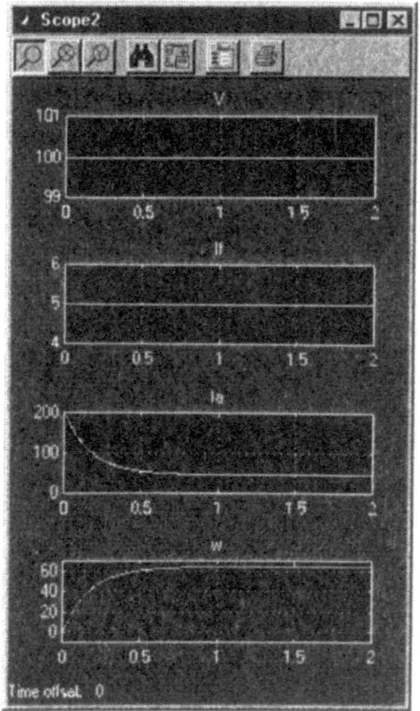

c. $T_L=55$, $I_f = 5$

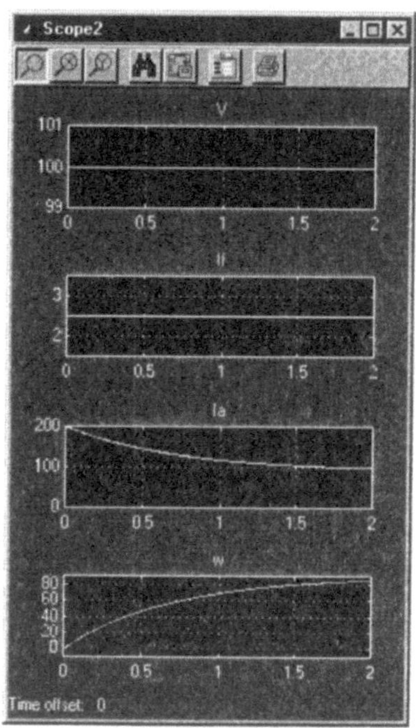

d. $T_L=55$, $I_f = 2.5$

Figure 9.3. Response of a Separately-excited DC Motor

9.1.2 DC Series Motor

The equations describing a separately-excited DC motor, mentioned above can also be used for studying the behavior of a series motor. The only consideration to be made is to treat $I_f=I_a$. Then, the equations become:

$$L_a \frac{dI_a}{dt} + (R_a + R_{se})I_a + MI_a\omega = V$$

$$J \frac{d\omega}{dt} + T_L = MI_a^2$$

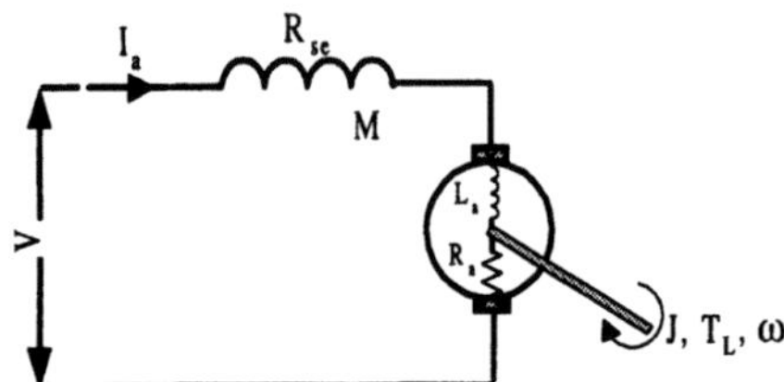

Figure 9.4. DC Series Motor

These equations are simulated (Figure 9.5) for the same parameters and the response of the system is shown in Figure 9.6. Observe the runaway speed at no load *i.e.,* $T_L=0$, and the change of current from high to low value as the motor picks up speed -- both with $T_L= 0$ or $T_L=100$ Newton-m.

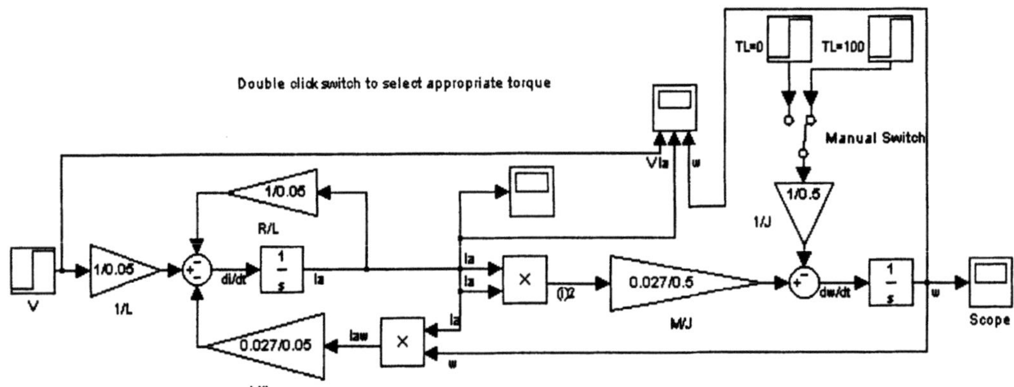

Figure 9.5. Simulation of DC Series Motor

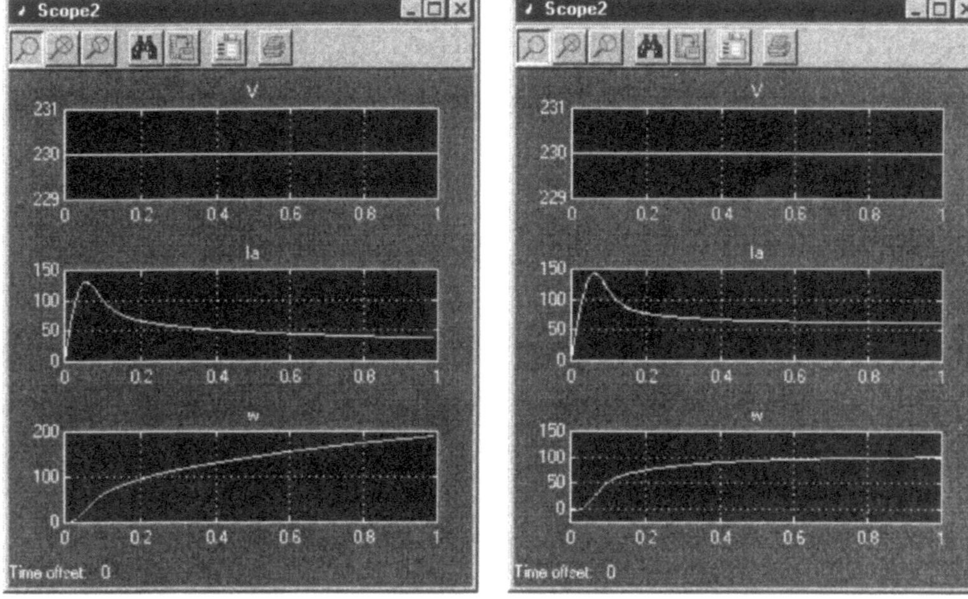

a. $T_L=0$ b. $T_L=100$

Figure 9.6. Response of DC Series Motor

9.2 Plunger System

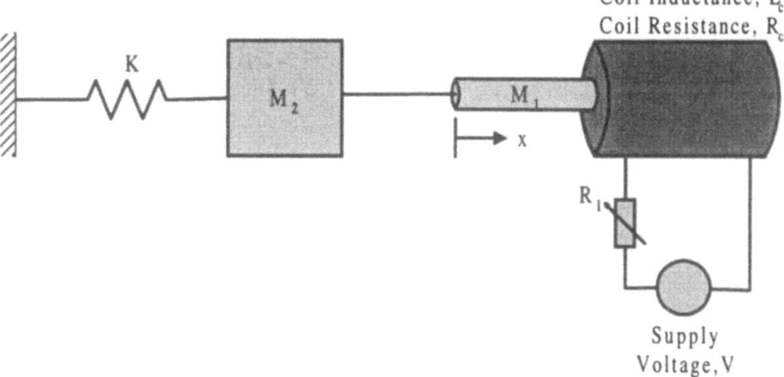

Figure 9.7. Plunger System

Next consider an electro-mechanical system as shown in Figure 9.7. For analysis, the system may be assumed to consist of two parts:

- electrical as shown in Figure 9.8a and described by the following equations:

$$(R_1 + R_c)i + L_c \frac{di}{dt} + k' \frac{dx}{dt} = V$$

- mechanical as shown in Figure 9.8b and described by the following equations:

$$f_p = (M_1 + M_2)\frac{d^2x}{dt^2} + (B_1 + B_2)\frac{dx}{dt} + (k'+K)x$$

where,

$$k'x = B_1 \frac{dx}{dt} = B_2 \frac{dx}{dt} = 0$$

The electrical and mechanical quantities are related as:

$$fp = k'i$$

substituting these in the equations for mechanical parts given above, we get:

$$k'i = (M_1 + M_2)\frac{d^2x}{dt^2} + Kx$$

Solving these equations, we get the transfer function of the system as:

$$\frac{X(s)}{V(s)} = \frac{k'}{s^3 L_c(M_1 + M_2) + s^2(R_1 + R_c)(M_1 + M_2) + s(L_c + k'^2) + K(R_1 + R_c)}$$

The system is simulated using the transfer function, as shown in Figure 9.9 for the following conditions:

Condition 1
The simulation parameters are chosen as given below:

M_1=0.25 kg; M_2= 5 kg; K= 3 Newtons/m; k'=0.6; R_1=50 Ω; R_c= 5 Ω; L_c= 100 mH

The system is connected to a 230V DC source. The result of simulation is shown in Figure 9.10a. The yellow wave predicting sustained oscillations indicates that the plunger will move in and out of the solenoid, about an equilibrium point never coming to rest.

Condition2
With the same parameters, we increase the DC voltage to 1000V. We now observe sustained oscillations again but of higher magnitude indicating that the plunger cannot be brought to rest by increasing the input voltage. Rather, the plunger starts vibrating with higher amplitude.

Condition 3

Reduce the value of resistance R_1 to zero; spring constant K to one-fourth its initial value *i.e.*, 0.75 Newton/m and let the supply voltage be 230V DC again. The response of the system shows decaying oscillations indicating that the plunger will come to rest after finite time and will rest inside the solenoid displaced by some 15 units from the original position.

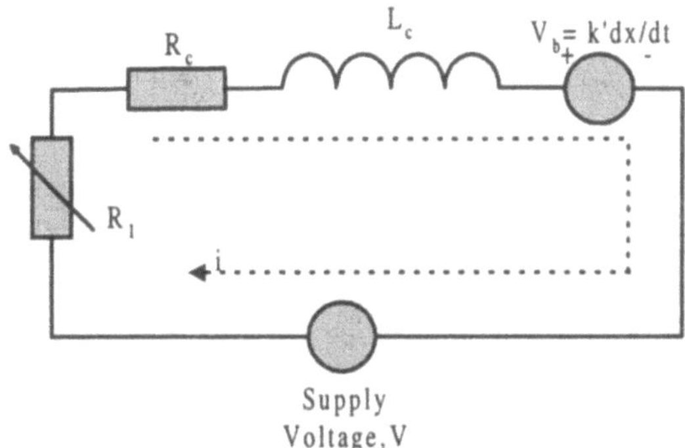

a. Equivalent electrical representation of Plunger System

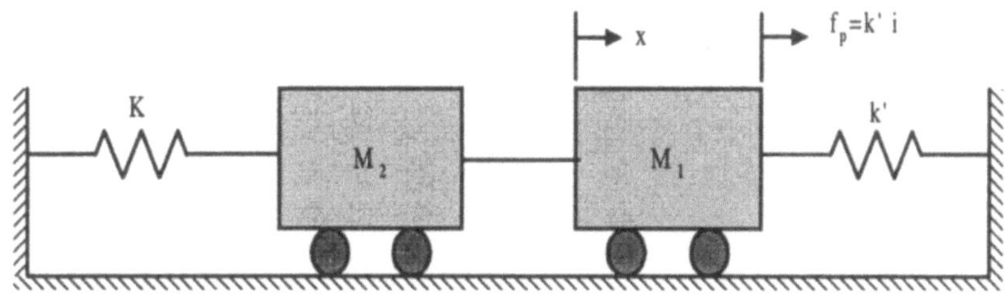

b. Equivalent mechanical representation of Plunger System

Figure 9.8. Electrical and mechanical equivalents of Plunger System

Condition 4

Let us next try out sinusoidal wave of 230V, 50 Hz as the input with parameters of Condition 1. The response of the system to sinusoidal input is shown in Figure 9.10b. The yellow wave of sustained oscillations of extremely low value indicates that due to inertia the plunger does not move from its initial position of equilibrium or rest.

Condition 5

We now change the parameters of the system to those of Condition 3 still having the same sinusoidal supply voltage as condition 4. In Figure 9.10b, we observe decaying oscillations indicating the plunger would come to rest in finite time around its equilibrium position without any final displacement.

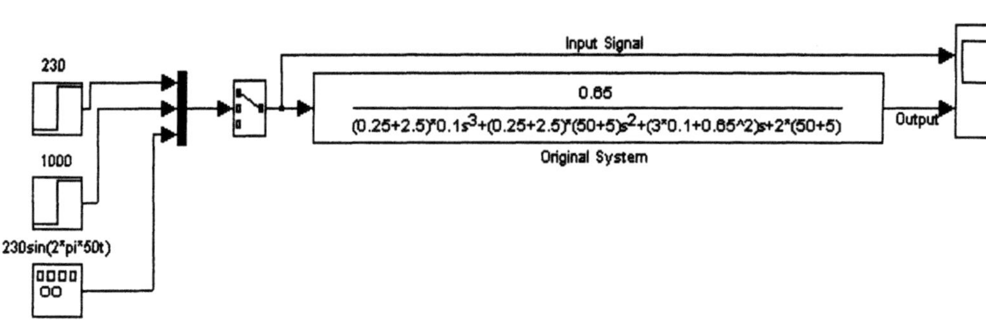

a.

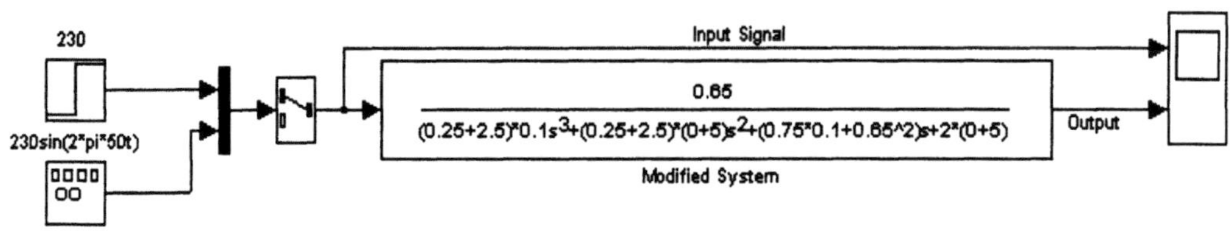

b.

Figure 9.9. Simulation of Plunger System using transfer function

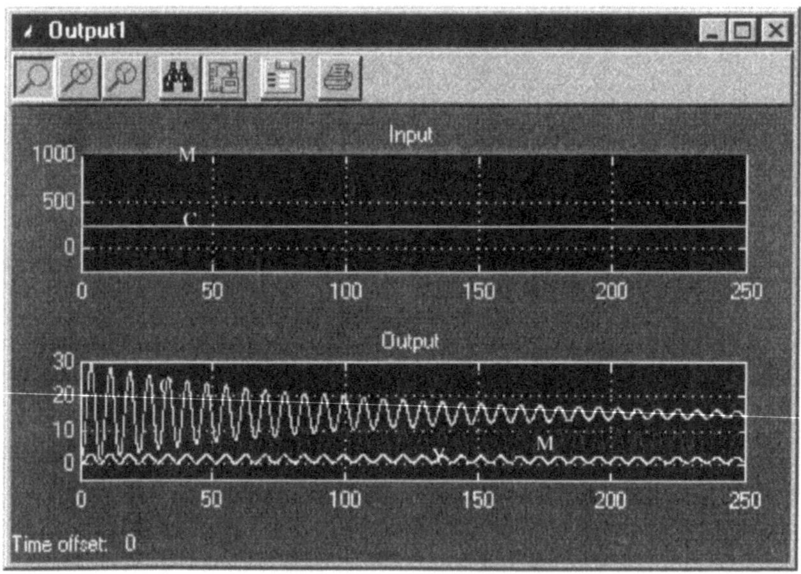

a. Response of the Plunger System to DC input

Figure 9.10. Response of the Plunger System to various input signals (Panel b. facing page)

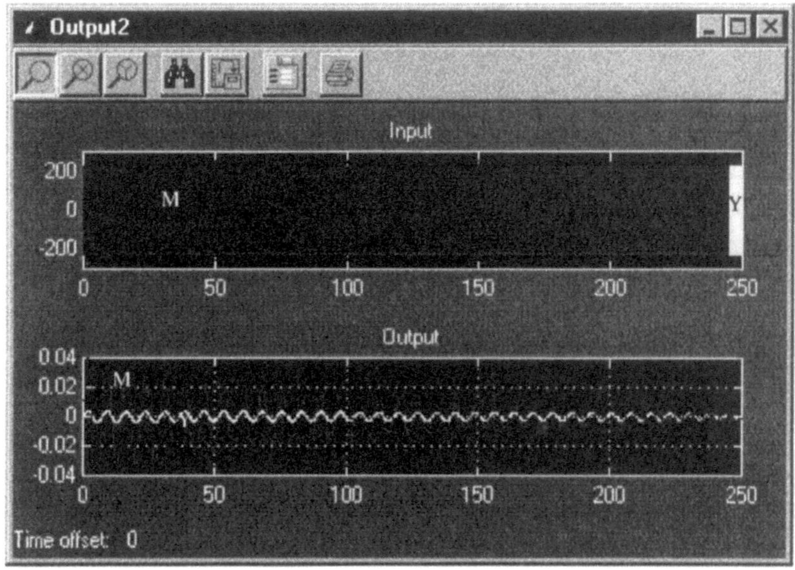

Figure 9.10b. Response of the Plunger System to sinusoidal input

9.3 Power-generating Systems

Simulated below are three power-generating systems with hydraulic turbine (Figure 9.11), steam turbine with reheat system (Figure 9.12), and steam turbine with non-reheat system (Figure 9.13). The systems are compared and the normalized values of the valve/gate opening, mechanical power and speed deviation are shown in Figure 9.14. without governor and with governor (shaded blocks in the simulation Figures). Clearly, there is an obvious improvement in the performance of the system.

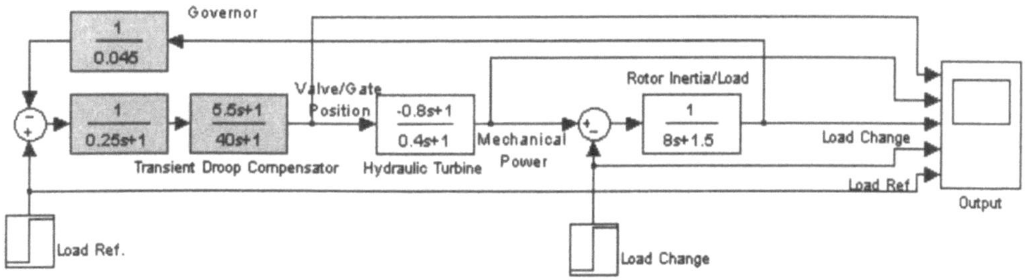

Figure 9.11. Generating System with Hydraulic Turbine

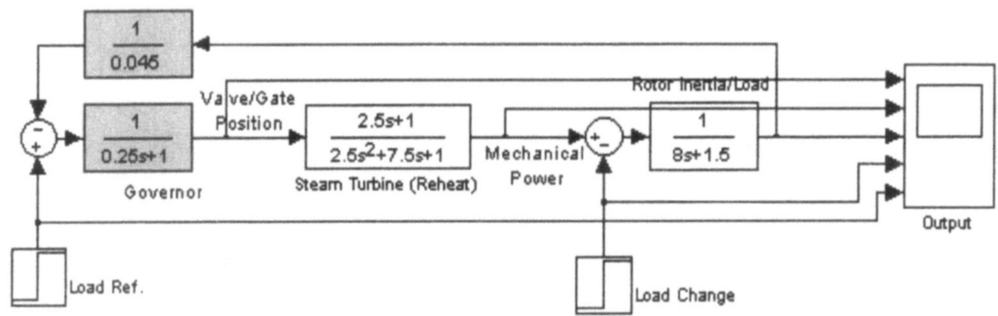

Figure 9.12. Generating System with Steam Turbine incorporating Reheat System

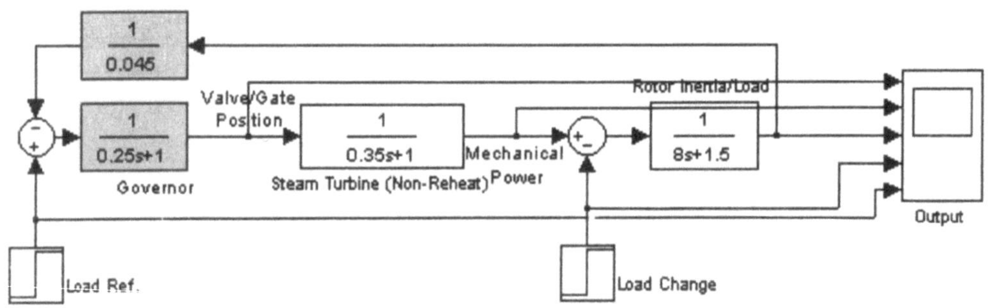

Figure 9.13. Generating System with Steam Turbine not incorporating Reheat System

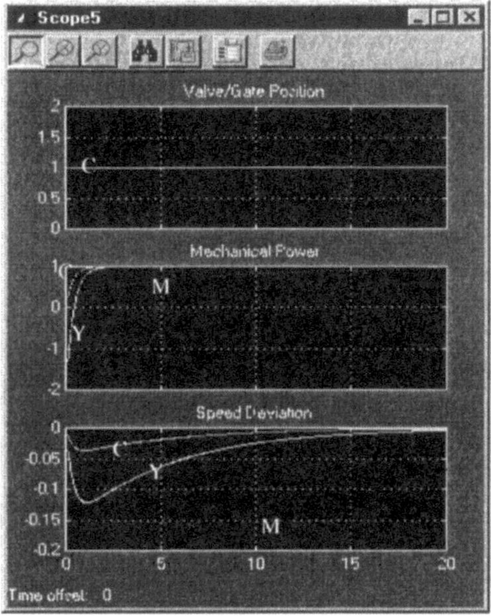

a. Without governor

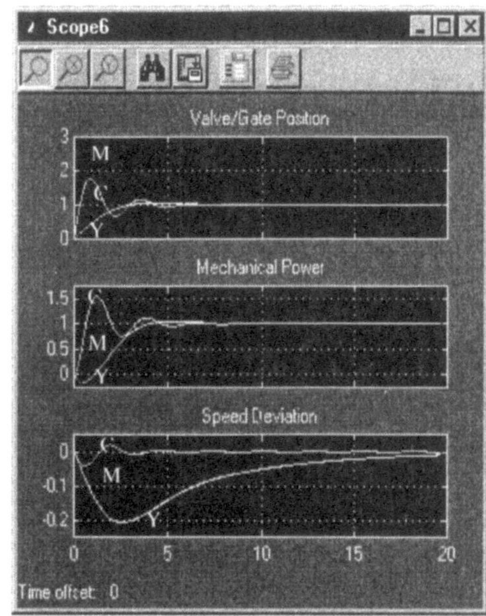

b. With governor

Figure 9.14. Comparison of valve/gate position; mechanical power and speed deviation of Hydraulic Turbine System (yellow color); Steam Turbine with Reheat System (magenta color) and Turbine with Non-reheat System (cyan color) respectively

9.4 Power-plant System

Let a certain power plant be represented by the block diagram shown in Figure 9.15a. Its response to unit positive step input is shown in Figure 9.16a. It clearly indicates an unstable system. The parameters of transient reduction element and Regulator 2 are then changed as indicated in Figure 9.15b. The response of the system after incorporating the change is shown in Figure 9.16b. Clearly, the system has become stable.

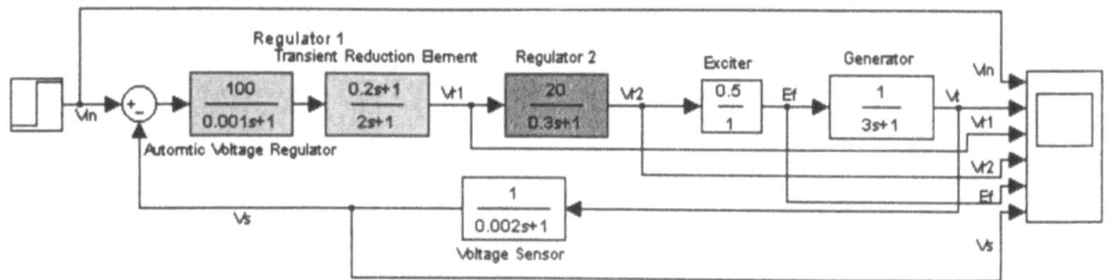

b. Original Power-plant System

Figure 9.15. Block diagram of Power-plant System before and after modification (Panel b. following page)

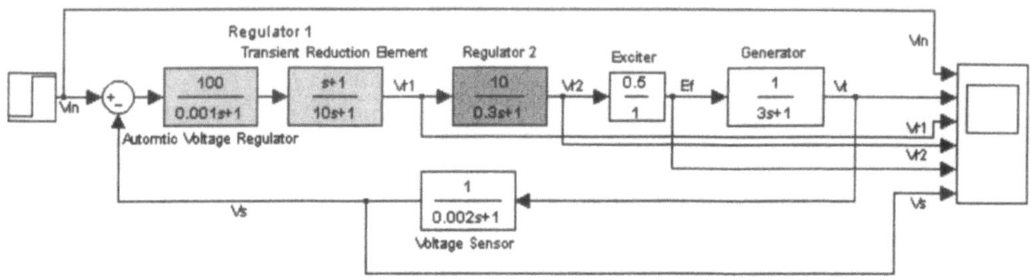

Figure 9.15b. Modified Power-plant System

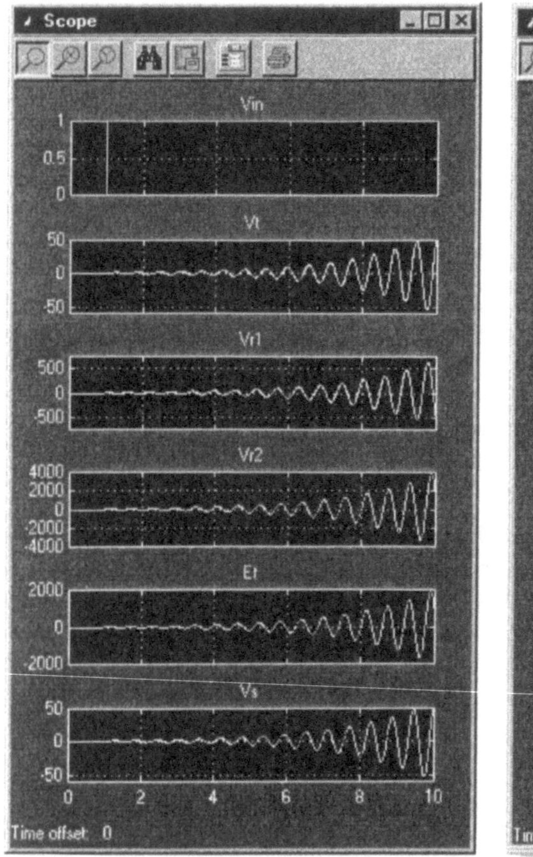

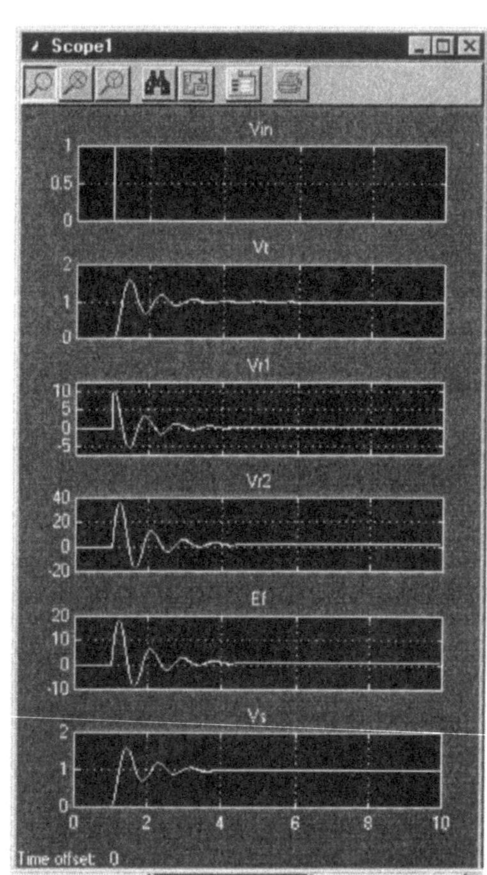

a. Response of original Power-plant System   b. Response of modified Power-plant System

Figure 9.16. Response of Power-plant System

9.5 Pacemaker System

The rate of heart-beat of a heart patient is controlled by an implanted pacemaker. The system block diagram is simulated as shown in Figure 9.17a. The input signal to the system is $(72+0.1t)u(t)$. From Figure 9.18, we observe that this system has a large error, which takes a long time to reduce to zero. However, if we use a gain compensator as shown in Figure 9.17b, the error is reduced to zero within one second. Response of both the systems is compared in Figure 9.18.

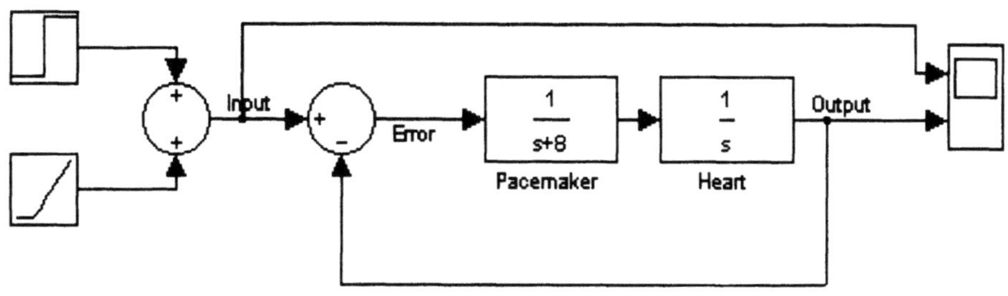

a. Uncompensated Pacemaker System

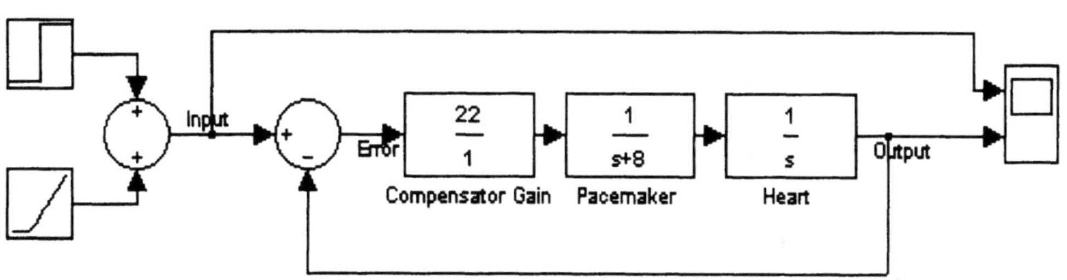

b. Compensated Pacemaker System

Figure 9.17. Pacemaker System

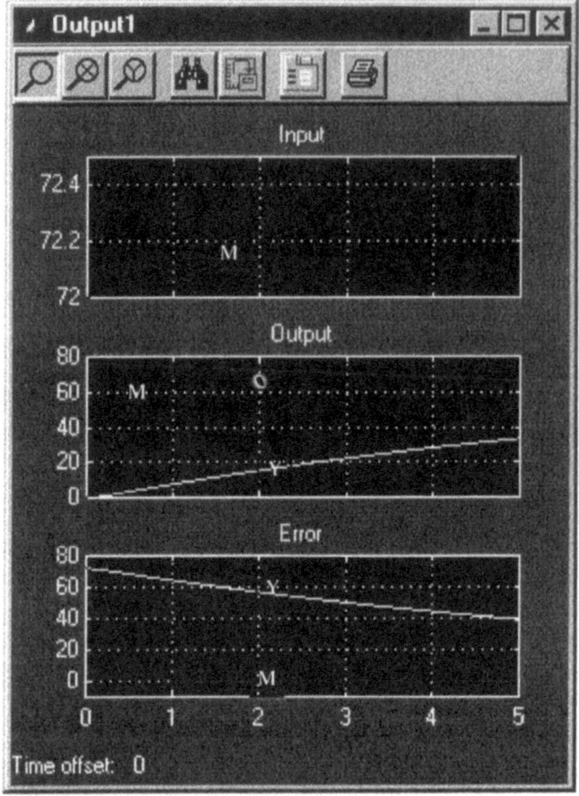

Figure 9.18. Response of Pacemaker System (compensator reduces error to zero within one second)

9.6 Inverted-pendulum-on-cart System

Quite a popular example of Control System Engineering consists of an inverted pendulum placed on a moving cart as shown in Figure 9.19. The mass of the cart is M and that of the pendulum bob is m. The pendulum of length l makes an angle of θ radians with the vertical. The force applied on the cart is F which produces a displacement of x from the initial position. The problem is to balance the pendulum by moving the cart according to the Newton's laws of motion.

Let the value of the system parameters be:

M=10kg, m=0.1 kg, l=1 m. Consider the initial value of θ to be 0.25 radians and that of x to be 0.15 m.

The state space equations representing the system can be found out as indicated below:

The equation of motion for displacement in the x direction is obtained by applying Newton's Laws of motion:

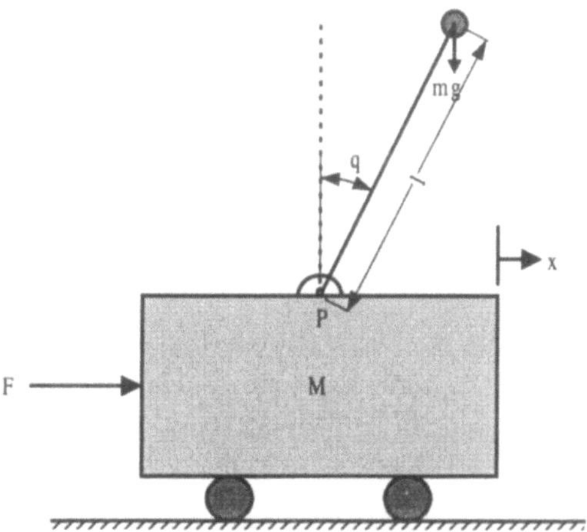

Figure 9.19. Inverted-pendulum-on-cart System

$$F = M\frac{d^2x}{dt^2} + m\frac{d^2(x + l\sin\theta)}{dt^2}$$

The conservation of momentum equation with respect to rotation about point P implies:

$$mgl\sin\theta = ml\cos\theta\,\frac{d^2(x + l\sin\theta)}{dt^2} - ml\sin\theta\,\frac{d^2(l\cos\theta)}{dt^2}$$

Considering θ being very small, these equations reduce to:

$$F = (M + m)\frac{d^2x}{dt^2} + ml\frac{d^2\theta}{dt^2}$$

$$g\theta = \frac{d^2x}{dt^2} + l\frac{d^2\theta}{dt^2}$$

Choosing the states, input and output as:

$$x_1 = \theta$$
$$x_2 = \dot{x}_1 = \dot{\theta}$$
$$x_3 = x$$
$$x_4 = \dot{x}_3 = \dot{x}$$
$$u = F$$
$$y_1 = \theta = x_1$$
$$y_2 = x = x_3$$

The state equations may then be written down as follows:

$$
\begin{bmatrix} \dot{x}_1 \\ \dot{x}_2 \\ \dot{x}_3 \\ \dot{x}_4 \end{bmatrix} = \begin{bmatrix} 0 & 1 & 0 & 0 \\ \dfrac{M+m}{Ml}g & 0 & 0 & 0 \\ 0 & 0 & 0 & 1 \\ -\dfrac{m}{M}g & 0 & 0 & 0 \end{bmatrix} \begin{bmatrix} x_1 \\ x_2 \\ x_3 \\ x_4 \end{bmatrix} + \begin{bmatrix} 0 \\ -\dfrac{1}{Ml} \\ 0 \\ \dfrac{1}{M} \end{bmatrix} [u]
$$

$$
\begin{bmatrix} y_1 \\ y_2 \end{bmatrix} = \begin{bmatrix} 1 & 0 & 0 & 0 \\ 0 & 0 & 1 & 0 \end{bmatrix} \begin{bmatrix} x_1 \\ x_2 \\ x_3 \\ x_4 \end{bmatrix} + [0][u]
$$

For the values of the parameters chosen, the system is simulated in Figure 9.20. Obviously, the system is highly unstable. Little disturbance will cause the pendulum to topple down. The respone of the system to unit step input is shown in Figure 9.21.

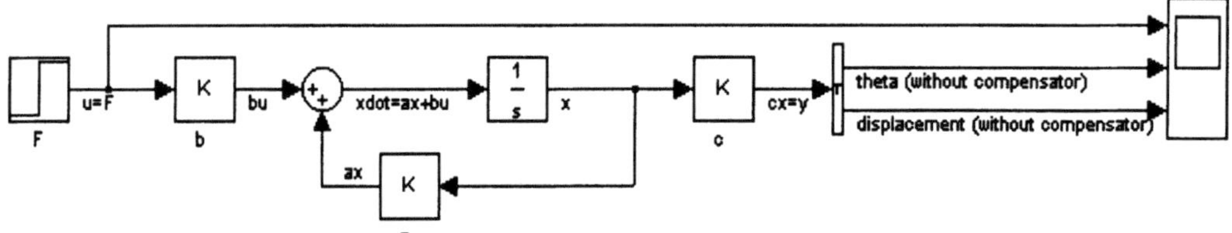

Figure 9.20. Inverted-pendulum-on-cart System

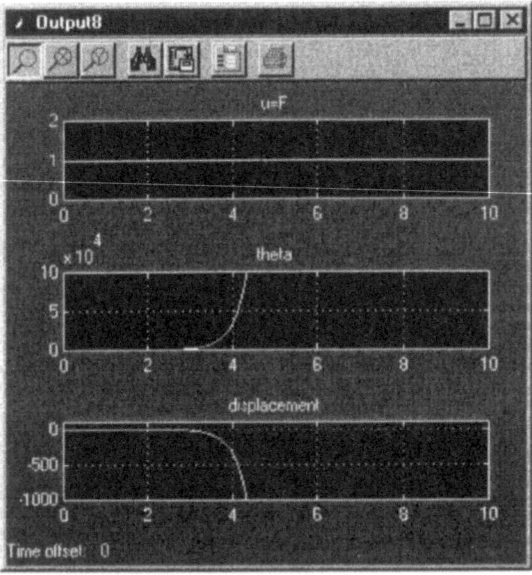

Figure 9.21. Response of Inverted-pendulum-on-cart System to unit step input

Let us design a regulator for the system to make it stable. Obviously, lead-lag cascade compensation is not going to make the system stable. Let us design the regulator using the pole placement technique about which you learnt in Chapter 7. Let us find the values of k for closed loope poles p given by:

p=[-1+j -1-j -1.5 -2]

place(a,b,p)
returns
k=[-225.1024 -68.2653   -6.1224 -13.2653]

Using these values of k, the system is again simulated as shown in Figure 9.22 with other parameters remaining unchanged. The response of the system is then obtained for unit step input as shown in Figure 9.23. The system has definitely become stable. You may try experimenting with some other inputs like a sine or a square signal. The system still remains stable.

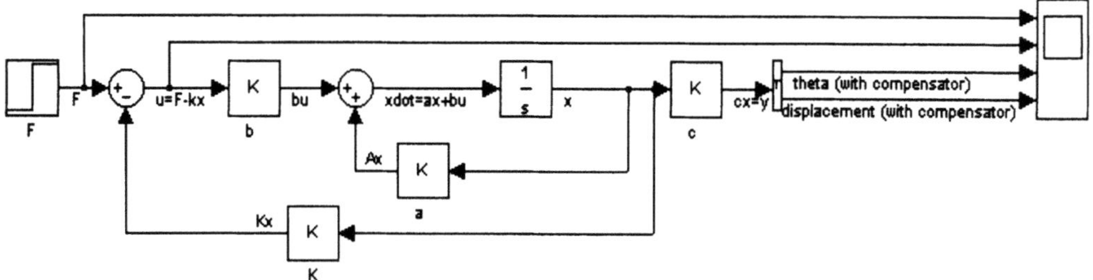

Figure 9.22. Inverted-pendulum-on-cart System with feedback regulator

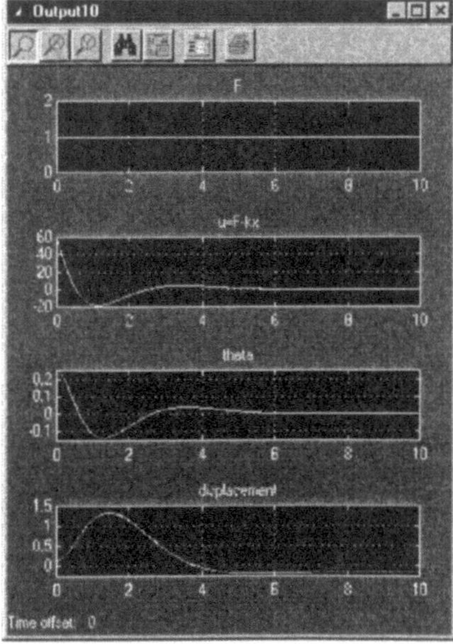

Figure 9.23. Response of Inverted-pendulum-on-cart System with feedback regulator to unit step input

9.7 Essential-oil-extraction Plant

Consider a steam-heated essential-oil-extraction plant. The aroma of the oil spoils if the temperature of the plant exceeds 200°C. A temperature sensing element which takes 0.2 seconds to sense the change in temperature is used to sense the temperature inside the oil extractor. The steam input to the oil-extraction plant is controlled by a valve operated by solenoid that receives its input signal through an amplifier connected to the temperature-sensing element. The arrangement in the Figure 9.24. Choose the parameters as following:

gain of steam inlet valve = 0.5,
gain of temperature sensing element=$0.5e^{-0.2s}$
gain of feedback amplifier = 4
approximate transfer function describing the process = 12/(s-2)
approximate transfer function of describing the solenoid = 2.5/(s+10)

The system has to be simulated assuming the steam input to be $25m^3$ constant.

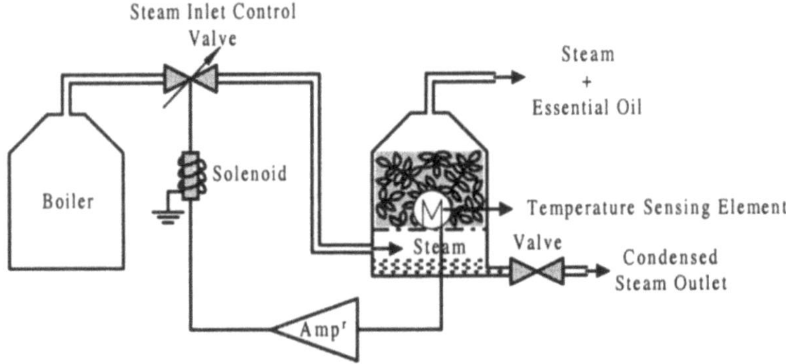

Figure 9.24. Essential-oil-extraction Plant

The simulation of the system is simple except for the time delay provided by the temperature sensing element to sense the temperature and then send signals to the solenoid (through amplifier) which operates the valve controlling the steam input to the system. The time delay may be tackled by any of the following two approaches:

- using transport delay block provided with the Non-Linear sub-library of SIMULINK® (Figure 9.25b).
- using **Pade** approximation of the time delay (Figure 9.25c and Figure 9.25d use 1st and 2nd order of pade approximation for the time delay).

Equivalent transfer functions for the expression $e^{-0.2s}$ using **pade** function is obtained as follows:

First-order approximation:

```
[num,den]=pade(0.2,1)
returns
num =
    -1   10
den =
     1   10
```

Second-order approximation:

```
[num,den]=pade(0.2,2)
returns
num =
     1   -30   300
den =
     1    30   300
```

The results of the simulation is shown in Figure 9.26 with and without feedback using all of the above mentioned approaches.

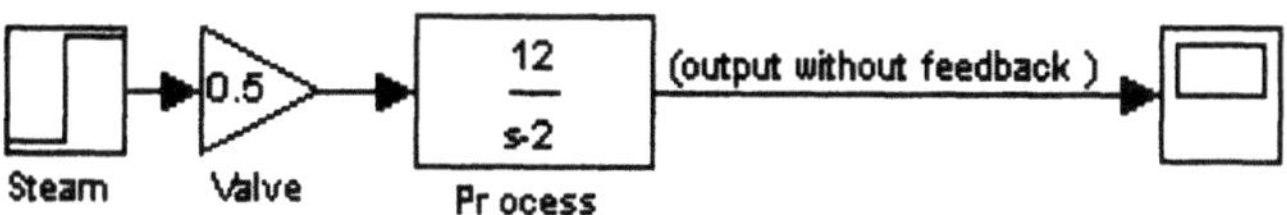

a. System without any feedback

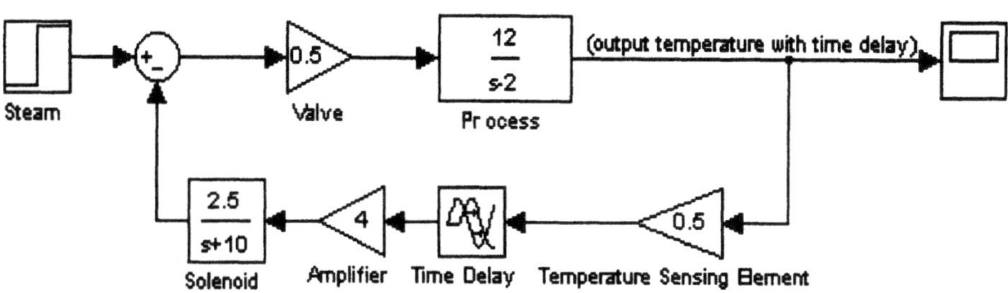

b. Time delay simulated using Transport Delay block of SIMULINK®

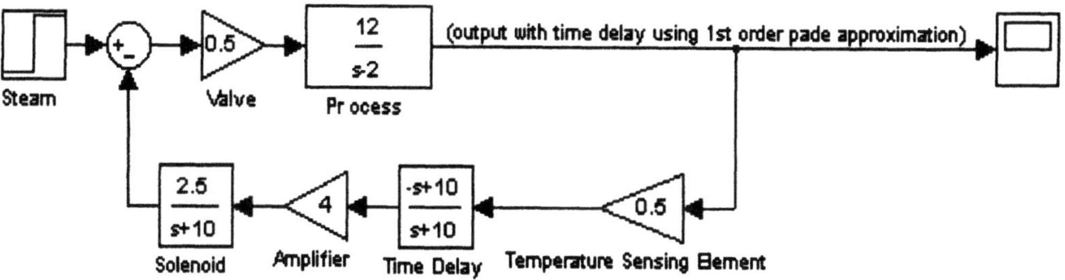

c. Time delay simulated using Pade approximation of 1st order

Figure 9.25. Simulation of Oil-extraction Plant using block diagram approach (Panel d. following page)

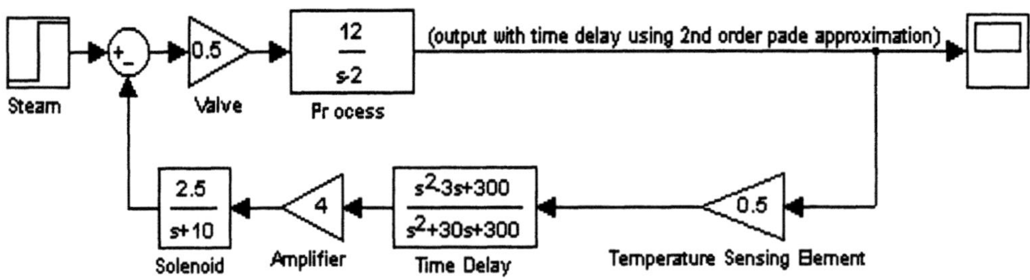

Figure 9.25d. Time delay simulated using Pade approximation of 2nd order

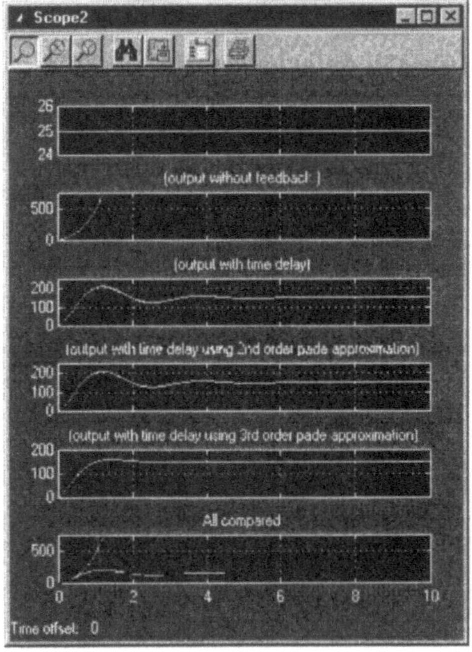

Figure 9.26. Response of Oil-extraction Plant with and without feedback using different methods for simulating time delay

9.8 Musical-octave System

As we have almost reached the end of this book, let us celebrate with music. Let us design a musical-octave system and enjoy the seven musical notes known as **sa, re, ga, ma, pa, dha, ni, sa** in Indian musical system, and as **doe, re, mi, fa, so, la, ti, doe** in the western musical system.

The system is simulated as shown in Figure 9.27 using plenty of subsystems as indicated in Figure 9.28. It also uses the subsystem available with MATLAB® demo of merge blocks as it is incorporating the use of accompanying s-function named **mergefcn,** thus teaching you how to use the already available resources around you to achieve your objective.

Figure 9.29 shows the waveform of the musical notes individually and collectively.

The fundamental frequency for the system was chosen as 256 Hz. Sinusoidal waveform of amplitude 100 was synthesized using this frequency for the note **sa**. Rest of the notes were generated multiplying the fundamental with the well-established successive ratios as shown in Figure 9.28c.

You may keep the same fundamental frequency or change it to and simulate the system. You can enjoy the music of the individual notes by entering the following at the command prompt of the MATLAB® window:

sound(sa)
sound(re)
sound(ga)......and so on.

or, you may listen to all the notes together by entering the following at the command prompt:

sound(oct)

which sounds as if all the octave keys of some musical instrument have been pressed sequentially for a short duration.

Since, the simulation takes a comparatively longer time to run you may save on your time by just loading the workspace you downloaded from the web-site of Springer-Verlag at `ftp://ftp.springer.co.uk` as octave and enter the commands as before to enjoy the notes.

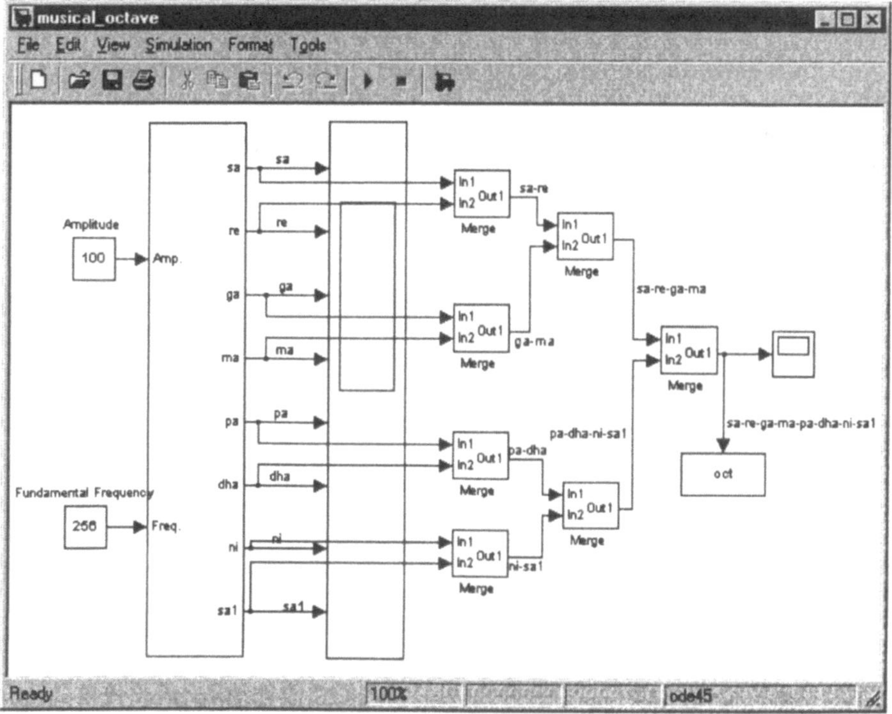

Figure 9.27. Simulation of Musical-octave System

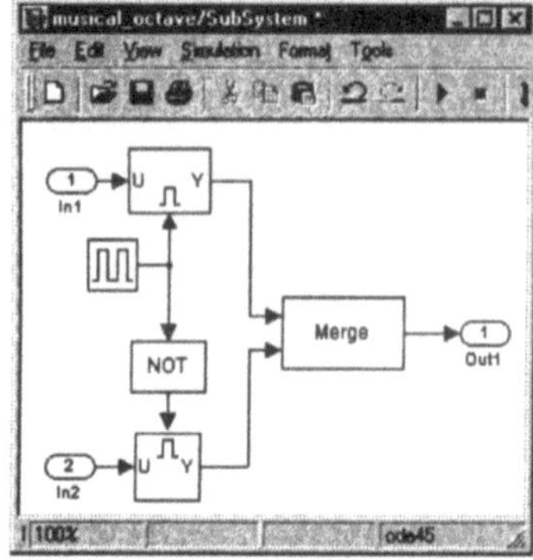

a.

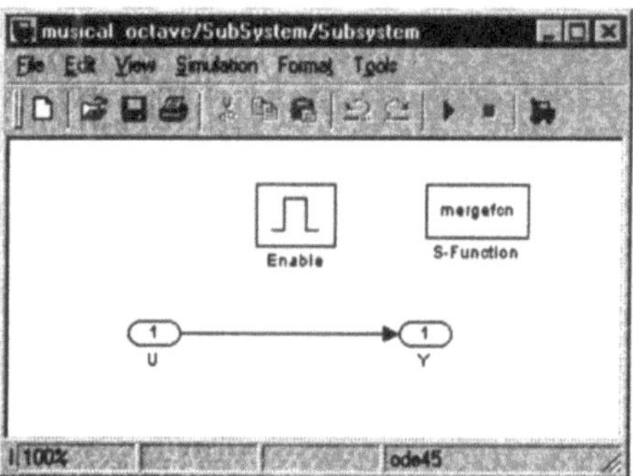

b.

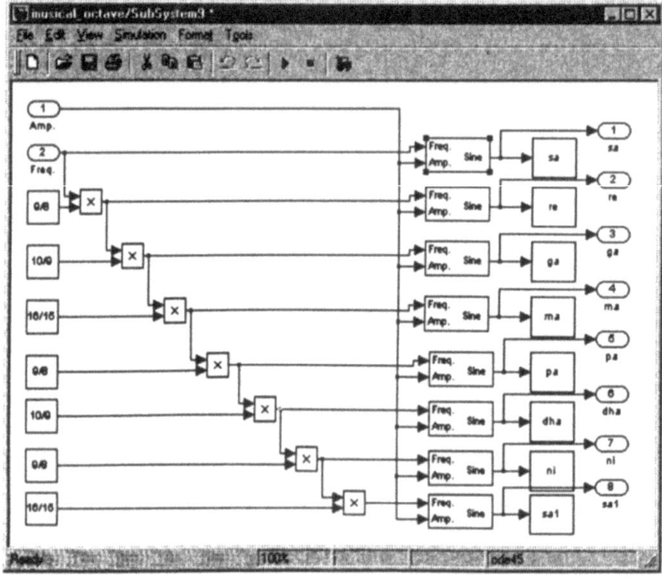

c.

Figure 9.28. Subsystems of Musical-octave System (Panel d. following page)

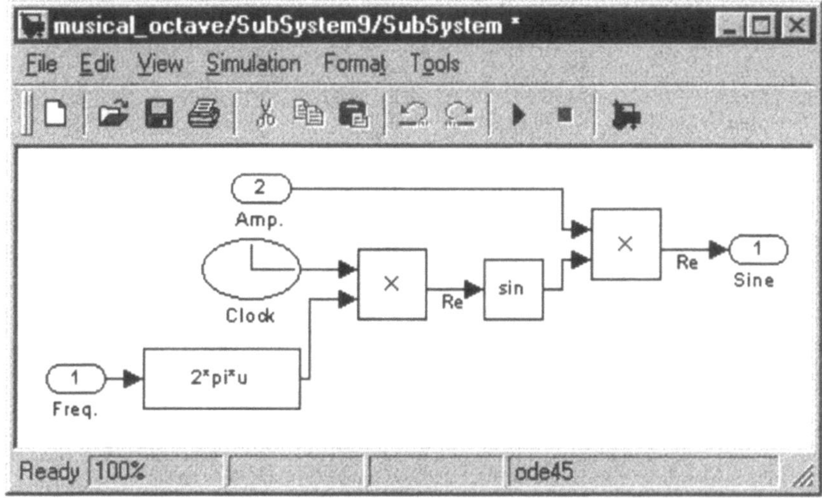

Figure 9.28d.

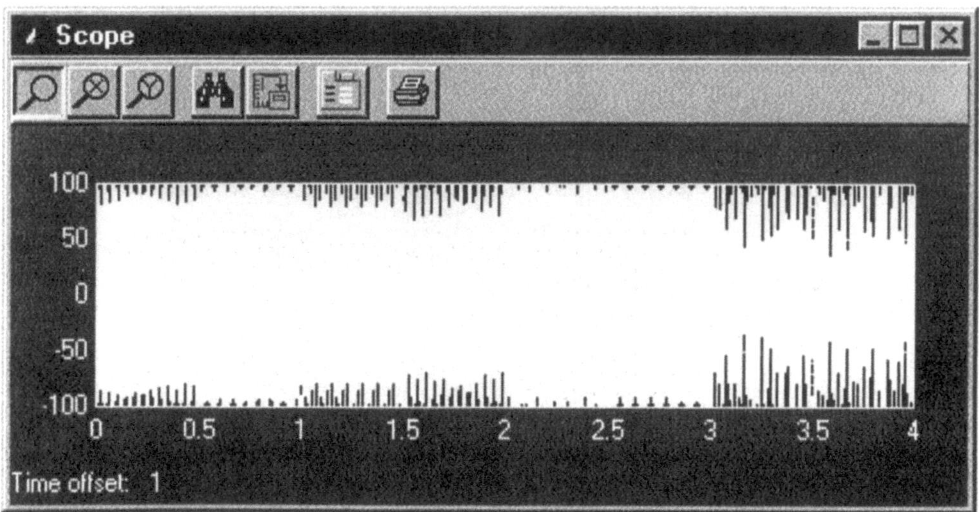

a. All notes together for 0.5 units of simulation time

Figure 9.29. Waveform of Musical-octave System (Panel b. facing page)

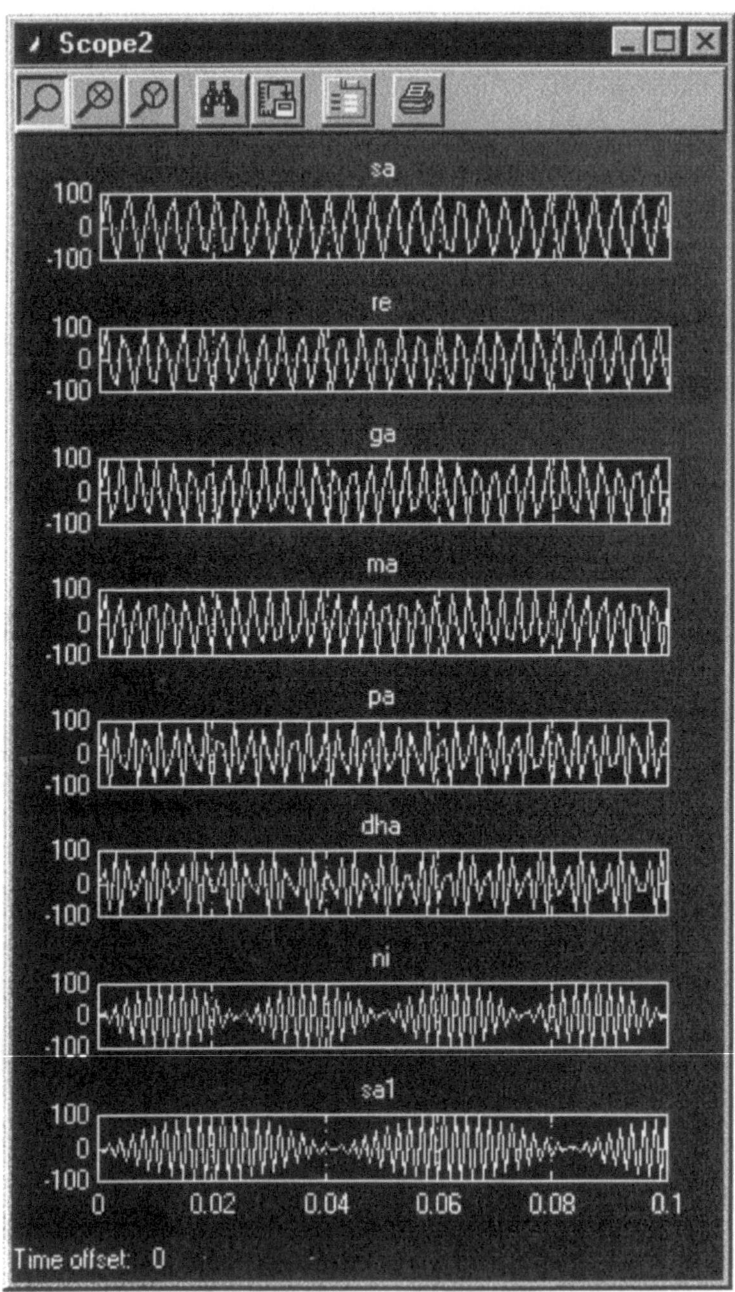

Figure 9.29b. All notes separate

Try experimenting to produce another musical sequence and enjoy a fiesta at the end of the book.

Exercise for Chapter 9:

1. Consider the robot arm shown in Figure 9.30. Prove that the transfer function of the system is:

$$\frac{\theta_{L(s)}}{i_{a(s)}} = \frac{K_i(Bs+K)}{s^4 J_L J_m + s^3[J_L(B_m+B)+J_m(B_L+B)]+s^2[B_L B_m +(B_L+B_m)B+(J_m+J_L)K]+s(K(B_L+B_m))}$$

Using this transfer function simulate the system for parameters indicated using SIMULINK® or otherwise. Design a series compensator for the system with unity feedback (negative) and compare the system response with the original system.

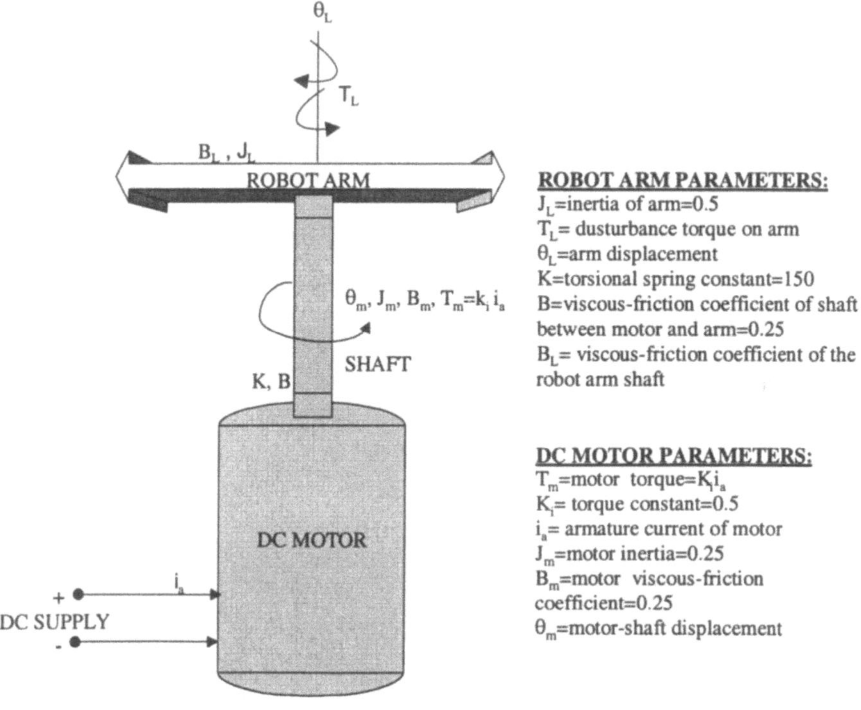

Figure 9.30.

2. For the Plunger system discussed in Article 9.2. Design a suitable compensator such that the plunger exhibits negligible vibrations even for simulation parameters chosen for Condition1.

3. Consider a system whose block diagram is shown in the Figure given below. The saturation block parameters are set as upper limit at 0.25 and lower limit at -0.25. Obtain the response of the system for a unit step signal having step time 1. How does the response change if the saturation block is removed. Comment on the response of the system if the feedback loop is also removed.

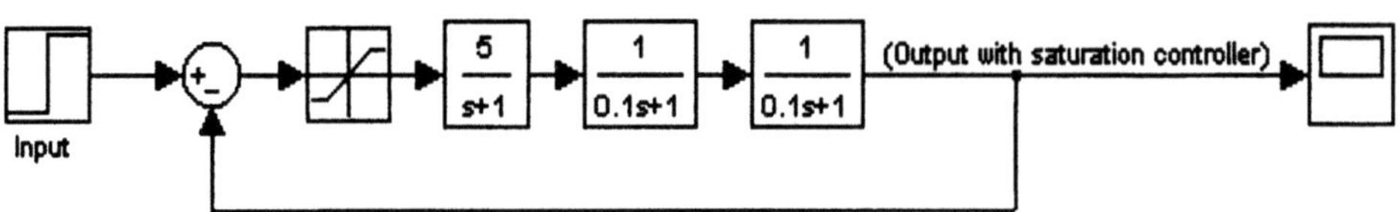

4. Consider a non-linear position servo system is shown in the Figure given below. The output of the system is obtained through a gear-box having backlash described by deadband width 1 and initial output zero. Vary the gain of the proportional controller and observe the response of the system for a unit step input signal having unit step time. Comment on the nature of the response.

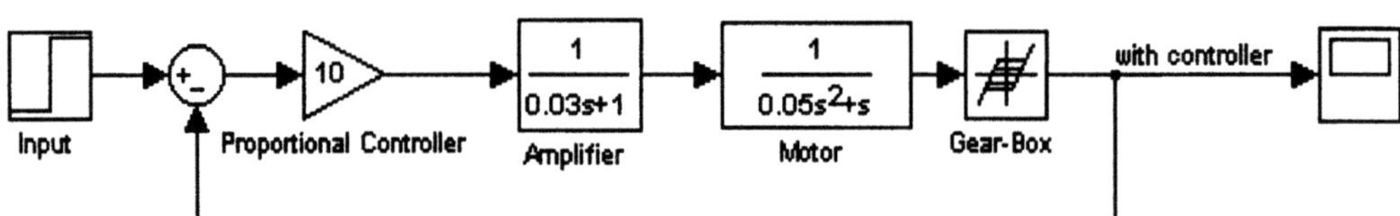

5. The table below summarises the test data obtained during harmonic analysis of a gadget. The transducer used to measure the output had a transfer function given by $\dfrac{10}{s^2 + 2s + 2}$. Is the system stable? If not design a suitable compensator for the same.

Frequency (rad/sec)	0.25	0.5	0.75	1	2	3	4	5
Amplitude	0.0504	0.1035	0.1642	0.2425	0.4851	0.1403	0.0701	0.0428
Phase Angle	84.39	79.56	76.34	75.96	165.96	190.78	191.11	189.86

Frequency (rad/sec)	10	20	30	40	50	60	70
Amplitude	0.0102	0.0025	0.0011	6.25×10^{-4}	4.00×10^{-4}	2.77×10^{-4}	2.04×10^{-4}
Phase Angle	185.53	182.84	181.90	181.42	181.14	180.95	180.81

Frequency (rad/sec)	90	100	1000	2000	2500	5000
Amplitude	1.56×10^{-4}	1.23×10^{-4}	1.00×10^{-4}	4.44×10^{-7}	1.60×10^{-7}	4.00×10^{-8}
Phase Angle	180.71	180.63	180.57	180.05	180.03	180.02

6. An aircraft-attitude-control System is shown in Figure 9.31. Simulate the system for the parameters indicated in each block assuming unit step input for different values of the constants of the PID controller. Find the best possible combination for the system to have critically damped response.

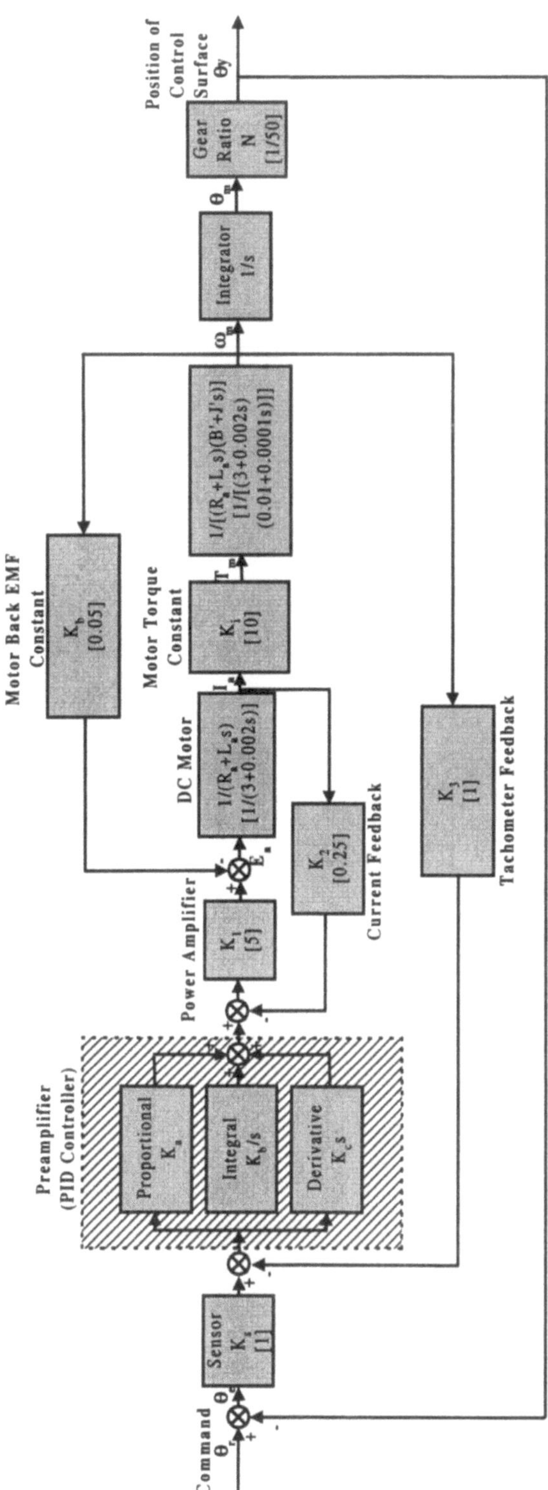

Figure 9.31. Block diagram representation of an Attitude-control System for an aircraft

Appendix A

This Appendix provides information about features of MATLAB®.

Given below is an alphabetical list of built-in functions with a short explanation, as provided with MATLAB®. You can obtain more information on these functions by entering the following at the command prompt:

Help *FunctionName*

where, *FunctionName* is the name of the function given below.

:	-	Regularly spaced vector and index into matrix
all	-	True if all elements of vector are nonzero
and	-	Logical AND (&)
ans	-	Most recent answer
any	-	True if any element of vector is nonzero
assignin	-	Assign variable in workspace
autumn	-	Shades of red and yellow color map
axes	-	Create axes in arbitrary positions
axis	-	Control axis scaling and appearance
bitand	-	Bit-wise AND
bitcmp	-	Complement bits
bitget	-	Get bit
bitmax	-	Maximum floating point integer
bitor	-	Bit-wise OR
bitset	-	Set bit
bitshift	-	Bit-wise shift
bitxor	-	Bit-wise XOR
blkdiag	-	Block diagonal concatenation
bone	-	Gray-scale with tinge of blue color map
box	-	Axis box
break	-	Terminate execution of WHILE or FOR loop
brighten	-	Brighten or darken color map
builtin	-	Execute built-in function from overloaded method

calendar	-	Calendar
camdolly	-	Dolly camera
cameramenu	-	Interactively manipulate camera
camlight	-	Creates or sets position of a light
camlookat	-	Move camera and target to view specified object
camorbit	-	Orbit camera
campan	-	Pan camera
campos	-	Camera position
camproj	-	Camera projection
camroll	-	Roll camera
camtarget	-	Camera target
camup	-	Camera up vector
camva	-	Camera view angle
camzoom	-	Zoom camera
case	-	SWITCH statement case
catch	-	Begin CATCH block
caxis	-	Pseudocolor axis scaling
clear	-	Clear variables and functions from memory
clock	-	Current date and time as date vector
colon	-	Colon (:)
colorbar	-	Display color bar (color scale)
colorcube	-	Enhanced color-cube color map
colordef	-	Set color defaults
colormap	-	Color look-up table
compan	-	Companion matrix
cool	-	Shades of cyan and magenta color map
copper	-	Linear copper-tone color map
cputime	-	CPU time in seconds
ctranspose	-	Complex conjugate transpose (')
daspect	-	Data aspect ratio
date	-	Current date as date string
datenum	-	Serial date number
datestr	-	String representation of date
datetick	-	Date-formatted tick labels
datevec	-	Date components
demo	-	Run demonstrations
diag	-	Diagonal matrices and diagonals of matrix
diffuse	-	Diffuse reflectance
disp	-	Display an array
disp	-	Display matrix or text
display	-	Overloaded function to display an array
edit	-	Edit M-file
else	-	IF statement condition
elseif	-	IF statement condition

end	-	Terminate scope of FOR, WHILE, SWITCH, TRY and IF statements
end	-	Last index
eomday	-	End of month
eps	-	Floating point relative accuracy
eq	-	Equal (==)
error	-	Display error message and abort function
errortrap	-	Skip error during testing
etime	-	Elapsed time
eval	-	Execute string with MATLAB expression
evalc	-	Evaluate MATLAB expression with capture
evalin	-	Evaluate expression in workspace
exist	-	Check if variables or functions are defined
eye	-	Identity matrix
feval	-	Execute function specified by string
fill3	-	Filled 3-D polygons
find	-	Find indices of nonzero elements
flag	-	Alternating red, white, blue, and black color map
flipdim	-	Flip matrix along specified dimension
fliplr	-	Flip matrix in left/right direction
flipud	-	Flip matrix in up/down direction
flops	-	Floating point operation count
for	-	Repeat statements a specific number of times
fprintf	-	Display formatted message
function	-	Add new function
gallery	-	Higham test matrices
ge	-	Greater than or equal (>=)
global	-	Define global variable
gray	-	Linear gray-scale color map
graymon	-	Set graphics defaults for gray-scale monitors
grid	-	Grid lines
gt	-	Greater than (>)
gtext	-	Mouse placement of text
gtext	-	Place text with mouse
hadamard	-	Hadamard matrix
hankel	-	Hankel matrix
help	-	On-line help, display text at command line
helpdesk	-	Comprehensive hypertext documentation and troubleshooting
helpwin	-	On-line help, separate window for navigation
hidden	-	Mesh hidden line removal mode
hilb	-	Hilbert matrix
hold	-	Hold current graph
horzcat	-	Horizontal concatenation ([,])
hot	-	Black-red-yellow-white color map
hsv	-	Hue-saturation-value color map

i, j	-	Imaginary unit
if	-	Conditionally execute statements
ind2sub	-	Multiple subscripts from linear index
inf	-	Infinity
inmem	-	List functions in memory
input	-	Prompt for user input
inputname	-	Input argument name
intersect	-	Set intersection
invhilb	-	Inverse Hilbert matrix
isempty	-	True for empty matrix
isequal	-	True if arrays are identical
isfinite	-	True for finite elements
isglobal	-	True for global variables
isinf	-	True for infinite elements
islogical	-	True for logical array
ismember	-	True for set member
isnan	-	True for Not-a-Number
isnumeric	-	True for numeric arrays
jet	-	Variant of HSV
keyboard	-	Invoke keyboard from M-file
kron	-	Kronecker tensor product (kron)
lasterr	-	Last error message
lastwarn	-	Last warning message
ldivide	-	Left array divide (\)
le	-	Less than or equal (<=)
legend	-	Graph legend
length	-	Length of vector
lightangle	-	Spherical position of a light
lighting	-	Lighting mode
lines	-	Color map with the line colors
linspace	-	Linearly spaced vector
lists	-	Comma separated lists
load	-	Load workspace variables from disk
logical	-	Convert numeric values to logical
loglog	-	Log-log scale plot
logspace	-	Logarithmically spaced vector
lookfor	-	Search all M-files for keyword
lt	-	Less than (<)
magic	-	Magic square
material	-	Material reflectance mode
mesh	-	3-D mesh surface
meshgrid	-	X and Y arrays for 3-D plots
mex	-	Compile MEX-function
mfilename	-	Name of currently executing M-file

minus	-	Minus (-)
mislocked	-	True if M-file cannot be cleared
mldivide	-	Backslash or left matrix divide (\)
mlock	-	Prevent M-file from being cleared
mpower	-	Matrix power (^)
mrdivide	-	Slash or right matrix divide (/)
mtimes	-	Matrix multiply (*)
munlock	-	Allow M-file to be cleared
NaN	-	Not-a-Number
nargchk	-	Validate number of input argument
nargin	-	Number of function input arguments
nargout	-	Number of function output arguments
ndims	-	Number of dimensions
ne	-	Not equal (~=)
not	-	Logical NOT (~)
now	-	Current date and time as date number
ones	-	Ones array
open	-	Edit M-file
or	-	Logical OR (|)
orient	-	Set paper orientation
otherwise	-	Default SWITCH statement case
pack	-	Consolidate workspace memory
paren	-	Parentheses and subscripting (( ))
paren	-	Brackets ([ ])
paren	-	Braces and subscripting ({ })
pascal	-	Pascal matrix
pause	-	Wait in seconds
pause	-	Wait for user response
pbaspect	-	Plot box aspect ratio
pcode	-	Create pre-parsed pseudo-code file (P-file)
persistent	-	Define persistent variable
pi	-	3.1415926535897....
pink	-	Pastel shades of pink color map
plot	-	Linear plot
plot3	-	Plot lines and points in 3-D space
plotedit	-	Tools for editing and annotating plots
plotedit	-	Experimental graph editing and annotation tools
plotyy	-	Graphs with y tick labels on the left and right
plus	-	Plus (+)
polar	-	Polar coordinate plot
power	-	Array power (.^)
precedence	-	Operator Precedence in MATLAB
print	-	Print graph or SIMULINK system; or save graph to M-file
printopt	-	Printer defaults

prism	-	Prism color map
punct	-	Decimal point (.)
punct	-	Structure field access (.)
punct	-	Parent directory (..)
punct	-	Continuation (...)
punct	-	Separator (,)
punct	-	Semicolon (;)
punct	-	Comment (%)
punct	-	Invoke operating system command (!)
punct	-	Assignment (=)
punct	-	Quote (')
quit	-	Quit MATLAB session
rand	-	Uniformly distributed random numbers
randn	-	Normally distributed random numbers
rdivide	-	Right array divide (./)
Readme	-	What's new in MATLAB
realmax	-	Largest positive floating point number
realmin	-	Smallest positive floating point number
repmat	-	Replicate and tile array
reshape	-	Change size
return	-	Return to invoking function
rosser	-	Classic symmetric eigenvalue test problem
rot90	-	Rotate matrix 90 degrees
rotate3d	-	Interactively rotate view of 3-D plot
run	-	Run script
save	-	Save workspace variables to disk
script	-	About MATLAB scripts and M-files
semilogx	-	Semi-log scale plot
semilogy	-	Semi-log scale plot
setdiff	-	Set difference
setxor	-	Set exclusive-or
shading	-	Color shading mode
size	-	Size of matrix
specular	-	Specular reflectance
spring	-	Shades of magenta and yellow color map
sprintf	-	Write formatted data to a string
sub2ind	-	Linear index from multiple subscripts
subplot	-	Create axes in tiled positions
subsasgn	-	Subscripted assignment (( ),{ },.)
subsindex	-	Subscript index
subsref	-	Subscripted reference (( ),{ },.)
summer	-	Shades of green and yellow color map
surf	-	3-D colored surface
surfl	-	3-D shaded surface with lighting

surfnorm	-	Surface normals
switch	-	Switch among several cases based on expression
texlabel	-	Produces TeX format from a character string
text	-	Text annotation
tic	-	Start stopwatch timer
times	-	Array multiply (.*)
title	-	Graph title
toc	-	Stop stopwatch timer
toeplitz	-	Toeplitz matrix
transpose	-	Transpose (.')
tril	-	Extract lower triangular part
triu	-	Extract upper triangular part
try	-	Begin TRY block
type	-	List M-file
uicontrol	-	Create user interface control
uimenu	-	Create user interface menu
uminus	-	Unary minus (-)
union	-	Set union
unique	-	Set unique
uplus	-	Unary plus (+)
vander	-	Vandermonde matrix
varargin	-	Variable length input argument list
varargout	-	Variable length output argument list
ver	-	MATLAB, SIMULINK, and toolbox version information
vertcat	-	Vertical concatenation ([;])
vga	-	Windows colormap for 16 colors
view	-	3-D graph viewpoint specification
viewmtx	-	View transformation matrix
vrml	-	Save graphics to VRML 2.0 file
warning	-	Display warning message
weekday	-	Day of week
what	-	List MATLAB-specific files in directory
whatsnew	-	Display Readme files
which	-	Locate functions and files
while	-	Repeat statements an indefinite number of times
white	-	All white color map
who	-	List current variables
whos	-	List current variables, long form
why	-	Succinct answer
wilkinson	-	Wilkinson's eigenvalue test matrix
winter	-	Shades of blue and green color map
workspace	-	Display Workspace Browser, a GUI for managing the workspace
xlabel	-	X-axis label
xlim	-	X limits

xor	-	Logical EXCLUSIVE OR
ylabel	-	Y-axis label
ylim	-	Y limits
zeros	-	Zeros array
zlabel	-	Z-axis label
zlim	-	Z limits
zoom	-	Zoom in and out on a 2-D plot

Appendix B

This Appendix provides information about features of SIMULINK®.

Below is the SIMULINK® Library showing the block symbols and giving a brief description of the blocks (as actually found written down below in the SIMULINK® Library browser) for reference:

Table B1. Continuous Library

Blocks	Symbol	Description
Derivative	du/dt Derivative	Numerical derivative: du/dt.
Integrator	$\frac{1}{s}$ Integrator	Continuous-time integration of the input signal.
Memory	Memory	Apply a one integration step delay. The output is the previous input value.
State-Space	$\dot{x} = Ax+Bu$ $y = Cx+Du$ State-Space	State-space model: $dx/dt = Ax + Bu$ $y = Cx + Du$
Transfer Fcn	$\frac{1}{s+1}$ Transfer Fcn	Matrix expression for numerator, vector expression for denominator. Output width equals the number of rows in the numerator. Coefficients are for descending powers of s.
Transport Delay	Transport Delay	Apply specified delay to the input signal. Best accuracy is achieved when the delay is larger than the simulation step size.
Variable Transport Delay	Variable Transport Delay	Apply a delay to the first input signal. The second input specifies the delay time. Best accuracy is achieved when the delay is larger than the simulation step size.
Zero-Pole	$\frac{(s-1)}{s(s+1)}$ Zero-Pole	Matrix expression for zeros. Vector expression for poles and gain. Output width equals the number of columns in zeros matrix, or one if zeros is a vector.

Table B2. Discrete Library

Blocks	Symbol	Description
Discrete Transfer Fcn	$\dfrac{1}{z+0.5}$ Discrete Transfer Fcn	Matrix expression for numerator, vector expression for denominator. Output width equals the number of rows in the numerator. Coefficients are for descending powers of z.
Discrete Zero-Pole	$\dfrac{(z-1)}{z(z-0.5)}$ Discrete Zero-Pole	Matrix expression for zeros. Vector expression for poles and gain. Output width equals the number of columns in zeros matrix, or one if zeros is a vector.
Discrete Filter	$\dfrac{1}{1+0.5z^{-1}}$ Discrete Filter	Vector expression for numerator and denominator. Coefficients are for ascending powers of 1/z.
Discrete State-Space	y(n)=Cx(n)+Du(n) x(n+1)=Ax(n)+Bu(n) Discrete State-Space	Discrete state-space model: $x(n+1) = Ax(n) + Bu(n)$ $y(n) \; = Cx(n) + Du(n)$
Discrete-Time Integrator	$\dfrac{T}{z-1}$ Discrete-Time Integrator	Discrete-time integration of the input signal.
First-Order Hold	First-Order Hold	First-order hold.
Unit Delay	$\dfrac{1}{z}$ Unit Delay	Sample and hold with one sample period delay.
Zero-Order Hold	Zero-Order Hold	Zero-order hold.

Table B3. Functions & Tables Library

Blocks	Symbol	Description
Fcn	f(u) Fcn	General expression block. Use "u" as the input variable name. Example: sin(u[1] * exp(2.3 * -u[2]))
Look-Up Table	Look-Up Table	Perform 1-D linear interpolation of input values using the specified table. Extrapolation is performed outside the table boundaries.
Look-Up Table (2-D)	Look-Up Table (2-D)	Performs 2-D linear interpolation of input values using the specified input/output table. Extrapolation is performed outside the table boundaries.
MATLAB Fcn		Pass the input values to a MATLAB function for evaluation. The function must return a single vector argument of length 'Output width'. Examples: sin, sin(u), foo(u(1), u(2))
S-Function	system S-Function	User-definable block. Blocks may be written in M, C or Fortran and must conform to S-function standards. t,x,u and flag are automatically passed to the S-function by Simulink. "Extra" parameters may be specified in the 'S-function parameters' field.

Table B4. Math Library

Blocks	Symbol	Description
Abs	$|u|$ Abs	$y = |u|$
Algebraic Constraint	Solve $f(z)$ $f(z) = 0$ z Algebraic Constraint	Constrains input signal f(z) to zero and outputs an algebraic state z. This block outputs the value necessary to produce a zero at the input. The output must affect the input through some feedback path. Provide an initial guess of the output to improve algebraic loop solver efficiency.
Complex to Magnitude-Angle	$|u|$ $\angle u$ Complex to Magnitude-Angle	Compute magnitude and/or radian phase angle of the input.
Complex to Real-Imag	Re(u) Im(u) Complex to Real-Imag	Output the real and/or imaginary components of the input.
Dot Product	• Dot Product	Inner (dot) product. y=sum(u1.*u2)
Gain	1 Gain	Scalar or vector gain. $y = k.*u$
Logical Operator	AND Logical Operator	Logical operators. For a single input, operators are applied across the input vector. For multiple inputs, operators are applied across the inputs.
Magnitude-Angle to Complex	$|\cdot|$ $\angle$ Magnitude-Angle to Complex	Construct a complex output from magnitude and/or radian phase angle input.
Math Function	e^u Math Function	Mathematical functions including logarithmic, exponential, power, and modulus functions.
Matrix Gain	K Matrix Gain	Matrix Gain.
Min Max	min MinMax	Output min or max of input. For a single input, operators are applied across the input vector. For multiple inputs, operators are applied across the inputs.

Table B4. (continued)

Blocks	Symbol	Description
Product	× Product	Multiply or divide inputs. Specify one of the following: a) * or / for each input port (e.g., **/*) b) scalar greater than 1 multiplies all input ports c) '1' multiplies all elements of single input vector
Real-Imag to Complex	Re Im Real-Imag to Complex	Construct a complex output from real and/or imaginary input.
Relational Operator	<= Relational Operator	Relational operators.
Rounding Function	floor Rounding Function	Rounding operations.
Sign	Sign	Output 1 for positive input, -1 for negative input, and 0 for 0 input. $y = signum(u)$
Slider Gain	1 Slider Gain	This is the 'simulink3/Math/Slider Gain' block
Sum	+ + 	Add or subtract inputs. Specify one of the following: a) string containing + or - for each input port, | for spacer between ports (e.g. ++|-|++) b) scalar >= 1. A value > 1 sums all inputs; 1 sums elements of a single input vector
Trigonometric Function	sin Trigonometric Function	Trigonometric and hyperbolic functions.

Table B5. Non-linear Library

Blocks	Symbol	Description
Backlash	Backlash	Model backlash where the deadband width specifies the amount of play in the system.
Coulomb & Viscous Friction	Coulomb & Viscous Friction	A discontinuity offset at zero models coulomb friction. Linear gain models viscous friction. $y = \text{sign}(x) * (\text{Gain} * \text{abs}(x) + \text{Offset})$
Dead Zone	Dead Zone	End value when outside of the deadzone.
Manual Switch	Manual Switch	Output toggles between two inputs by double-clicking on the block.
Multiport Switch	Multiport Switch	Pass through the input signals corresponding to the rounded value of the first input.
Quantizer	Quantizer	Discretize input at given interval.
Rate Limiter	Rate Limiter	Limit rising and falling rates of signal.
Relay	Relay	Output the specified 'on' or 'off' value by comparing the input to the specified thresholds. The on/off state of the relay is not affected by input between the upper and lower limits.
Saturation	Saturation	Limit input signal to the upper and lower saturation values.
Switch	Switch	Pass through input 1 when input 2 is greater than or equal to threshold; otherwise, pass through input 3.

Table B6. Signals and Systems Library

Blocks	Symbol	Description
Sub System		This is the 'simulink3/Signals & Systems/SubSystem' block
In 1		Provide an input port for a subsystem or model. The 'Sample time' parameter may be used to specify the rate at which a signal enters the system.
Enable		Place this block in a subsystem to create an enabled subsystem.
Trigger		Place this block in a subsystem to create a triggered subsystem.
Bus Selector		This block accepts input from a Mux or Bus Selector block. The left listbox shows the signals in the input bus. Use the Select button to select the output signals. The right listbox shows the selections. Use the Up, Down, or Remove button to reorder the selections. Check 'Muxed output' to multiplex the output.
Configurable Subsystem		This block may be configured to represent any of the top-level blocks and subsystems in a user-specified Simulink Library.
Data Store Memory		Define a memory region for use by the Data Store Read and Data Store Write blocks.
Data Store Read		Read values from specified data store.
Data Store Write		Write values to specified data store.
Data Type Conversion		Convert input signal to specified data type.
Demux		Split vector signal into scalars or smaller vectors.

Table B6. (continued)

Blocks	Symbol	Description
From	[A] From	Receive signals from the Goto block with the specified tag. If the tag is defined as 'scoped' in the Goto block, then a Goto Tag Visibility block must be used to define the visibility of the tag. After 'Update Diagram', the block icon displays the selected tag name (local tags are enclosed in brackets, [], and scoped tag names are enclosed in braces, { }).
Function-Call Generator	f0 Function-Call Generator	Execute a function-call subsystem at a specified rate. Demux the block's output to execute multiple function-call subsystems in a prescribed order. The system connected to first demux port is executed first, the system connected to second demux port is executed second, and so on.
Goto	[A] Goto	Send signals to From blocks that have the specified tag. If tag visibility is 'scoped', then a Goto Tag Visibility block must be used to define the visibility of the tag. The block icon displays the selected tag name (local tags are enclosed in brackets, [], and scoped tag names are enclosed in braces, { }).
Goto Tag Visibility	{A} Goto Tag Visibility	Used in conjunction with Goto and From blocks to define the visibility of scoped tags. For example, if this block resides in a subsystem (or root system) called MYSYS, then the tag is visible to From blocks that reside in MYSYS or in subsystems of MYSYS.
Ground	Ground	Used to "ground" input signals. (Prevents warnings about unconnected input ports.) Outputs zero.
Hit Crossing	Hit Crossing	Force simulation to locate ("hit") zero crossing of the input signal. Outputs 1 when hit crossing is detected, otherwise outputs 0.
IC	[1] IC	Initial condition for signal.
Merge	Merge Merge	Merge the input signals into a single output signal whose initial value is specified by the 'Initial output' parameter. If 'Initial output' is empty, the Merge block outputs the initial output of one of its driving blocks.
Model Info	Model Info	This block allows revision control information to be displayed within the model.
Mux		Combine scalar or vector signals into larger vectors.

Table B6. (continued)

Blocks	Symbol	Description
Probe	W:0, Ts:[0 0], C:0 Probe	Probe a line for its width, sample time, or complex signal flag.
Selector	Selector	Select or re-order the specified elements of the input vector. y=u(Elements).
Terminator	Terminator	Used to "terminate" output signals. (Prevents warnings about unconnected output ports.)
Width	0 Width	Output the width of the input signal.
Out 1	1 Out1	Provide an output port for a subsystem or model. The 'Output when disabled' and 'Initial output' parameters only apply to conditionally executed subsystems. When a conditionally executed subsystem is disabled, the output is either held at its last value or set to the 'Initial output'. The 'Initial output' parameter can be specified as the empty matrix, [], in which case the initial output is equal to the output of the block feeding the outport.

Table B7. Sinks Library

Blocks	Symbol	Description
Display	0 Display	Numeric display of input values.
Scope	Scope	This is the 'simulink3/Sinks/Scope' block
Stop simulation	STOP Stop Simulation	Stop simulation when input is non-zero.
To File	untitled.mat To File	Write time and input to specified MAT file in row format. Time is in row 1.
To Workspace	simout To Workspace	Write input to specified matrix in MATLAB's main workspace. The matrix has one column per input element and one row per simulation step. Data is not available until the simulation is stopped or paused.
XY Graph	XY Graph	XY scope using MATLAB graph window. First input is used as time base. Enter plotting ranges.

Table B8. Sources Library

Blocks	Symbol	Description
Band Limited White Noise	Band-Limited White Noise	White noise for continuous (s-domain) systems. Band-limited using zero-order-hold.
Chirp Signal	Chirp Signal	Chirp Signal. (Sine wave with increasing frequency)
Clock	Clock	Output the current simulation time.
Constant	1 Constant	Output a constant.
Digital Clock	12:34 Digital Clock	Output current simulation time at the specified rate.
Discrete Pulse Generator	Discrete Pulse Generator	Generate pulses at regular intervals. Specify parameters as integer multiples of the sample time.
From Workspace	[T,U] From Workspace	Read data values specified in matrix or structure format from MATLAB's workspace. Matrix format: var=[TimeValues DataValues] Structure format: var.time=[TimeValues] var.signals.values=[DataValues] Select interpolation to interpolate or extrapolate at time steps for which data does not exist.
From File	untitled.mat From File	Read time and output values from the first matrix in the specified MAT file. The matrix must contain time values in row one. Additional rows correspond to output elements. Interpolates between columns.
Pulse Generator	Pulse Generator	Pulse Generator
Ramp	Ramp	ramp
Random Number	Random Number	Output a normally (Gaussian) distributed random signal. Output is repeatable for a given seed.

Table B8. (continued)

Blocks	Symbol	Description
Repeating Sequence	Repeating Sequence	Repeating table.
Signal Generator	Signal Generator	Output various wave forms.
Sine Wave	Sine Wave	Output a sine wave.
Step	Step	Output a step.
Uniform Random Number	Uniform Random Number	Output a uniformly distributed random signal. Output is repeatable for a given seed.

Alphabetical List of Built-in Functions for Model Analysis and Construction, with a Short Explanation as Provided with MATLAB® and SIMULINK®:

add_block	-	Add new block
add_line	-	Add new line
addterms	-	Add terminators to unconnected ports
bdclose	-	Close a Simulink window
bdroot	-	Root level model name
bool	-	Convert numeric array to boolean
close_system	-	Close open model or block
delete_block	-	Remove block
delete_line	-	Remove line
dlinmod	-	Extract linear model from discrete-time system
find_system	-	Search a model
frameedit	-	Edit print frames for annotated model printouts
gcb	-	Get the name of the current block
gcbh	-	Get the handle of the current block
gcs	-	Get the name of the current system
get_param	-	Get simulation parameter values from model
getfullname	-	Get the full path name of a block
hasmask	-	Check for mask
hasmaskdlg	-	Check for mask dialog
hasmaskicon	-	Check for mask icon
iconedit	-	Design block icons using g input function
libinfo	-	Get library information for a system
linmod	-	Extract linear model from continuous-time system
linmod2	-	Extract linear model, advanced method
load_system	-	Load existing model without making model visible
maskpopups	-	Return and change masked block's popup menu items
movemask	-	Restructure masked built-in blocks as masked subsystems
new_system	-	Create new empty model window
open_system	-	Open existing model or block
orient	-	Set paper orientation
print	-	Print graph or Simulink system; or save graph to M-file
printopt	-	Printer defaults
replace_block	-	Replace existing blocks with a new block
save_system	-	Save an open model
set_param	-	Set parameter values for model or block
simulink	-	Open main block library
simulink_extras	-	Simulink Extras block library
sim	-	Simulate a Simulink model
simget	-	Get SIM Options structure
simset	-	Define options to SIM Options structure
sldebug	-	Debug a Simulink model
slupdate	-	Update older 1.x models to 3.x
trim	-	Find steady-state operating point

Appendix C

This Appendix provides information about features of Control System as a whole and about Control System Toolbox information about features of SIMULINK®.

Some important information related to Control System Engineering:

Table C.1. Control System Quantities

a. Electrical System

Quantity	Symbol	Metric Units	Component Schematic	Component Equation
Charge	q	Coloumb	-	$i = \dfrac{dq}{dt}$
Flux Linkage	ψ	Ampere Turns	-	$e = \dfrac{d\psi}{dt}$
Current	i	Ampere		i=I (specified value)
Voltage	e or V	Volt	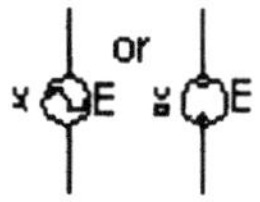	e=E or V (specified value)
Resistance	R	Ohm	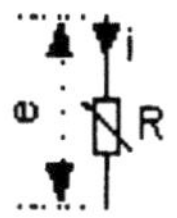	$e = Ri$ or $i = \dfrac{e}{R}$
Inductance	L	Henry		$e = L\dfrac{di}{dt}$ or $i = L\dfrac{d\psi}{dt}$
Capacitance	C	Farad	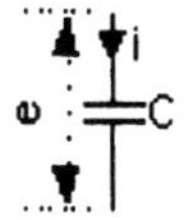	$e = \dfrac{1}{C}\int i\,dt$ or $i = C\dfrac{de}{dt}$

b. Mechanical System (Tranlational)

Quantity	Symbol	Metric Units	Component Schematic	Component Equation	Analogy w.r.t. Electrical Circuit	
					Force-Current Analogy	Force-Voltage Analogy
Displacement	x or y	meter (m)	-	$v = \dfrac{dx}{dt}$	Flux Linkage	Charge
Force	F	Newton or $\dfrac{kg \text{-} m}{sec^2}$		F (specified value)	Current	Voltage
Velocity	v	$\dfrac{m}{sec}$		v (specified value)	Voltage	Current
Acceleration	$\dot{v}$ or a	$\dfrac{m}{sec^2}$	-	F = Ma	-	$\dfrac{d}{dt}(\text{Current})$ or $\dfrac{d^2}{dt^2}(\text{Charge})$
Mass	M	Kg		$F = M\dfrac{dv}{dt}$	Capacitance	Inductance
Spring Constant or Mechanical Stiffness	K	$\dfrac{Newton}{m}$		$F = K\int v\,dt$	Inductance	$\dfrac{1}{Capacitance}$
Damping Coefficient or Coefficient of Friction	B or f	$\dfrac{Newton}{m/sec}$		F = Bv	$\dfrac{1}{Resistance}$	Resistance

c. Mechanical System (Rotational)

Quantity	Symbol	Metric Units	Component Schematic	Component Equation	Analogy w.r.t. Electrical Circuit	
					Torque-Current Analogy	Torque-Voltage Analogy
Angular Displacement	θ	Radians (rad)	-	$\omega = \dfrac{d\theta}{dt}$	Flux Linkage	Charge
Angular Velocity	ω	$\dfrac{radian}{sec}$		ω (specified)	Current	Current
Torque	T	Newton-m		T (specified)	Voltage	Voltage
Angular Acceleration	$\dot{\omega}$	$\dfrac{radian}{sec^2}$	-	$\dot{\omega} = \dfrac{d\omega}{dt}$	-	$\dfrac{d}{dt}(\text{Current})$ or $\dfrac{d^2}{dt^2}(\text{Charge})$
Moment of Inertia	J	$\dfrac{Newton-m}{rad/sec^2}$ or $Kg\text{-}m^2$		$T = J\dfrac{d\omega}{dt}$	Capacitance	Inductance
Tortional Spring Constant or Mechanical Stiffness	K	$\dfrac{Newton-m}{rad}$		$T = K\int\omega\,dt$	Inductance	$\dfrac{1}{Capacitance}$
Damping Coefficient	B or f	$\dfrac{Newton-m}{rad/sec}$		$T = B\omega$	$\dfrac{1}{Resistance}$	Resistance

d. Fluid System (Hydraulic)

Quantity	Symbol	Metric Units	Component Schematic	Component Equation	Analogy w.r.t. Electrical Circuit
Volume Flow rate	q	$\dfrac{m^3}{sec}$	-	q (specified)	Current
Pressure	p	$\dfrac{Newton}{m^2}$	-	p (specified)	Voltage
Hydraulic Capacitance	C	$\dfrac{m^3}{Newton/m^2}$		$q = C\dfrac{dp}{dt}$	Capacitance
Hydraulic Resistance	R	$\dfrac{Newton/m^2}{m^3/sec}$		$q = \dfrac{p}{R}$	Resistance
Inertance	L	$\dfrac{Newton\text{ - }sec^2}{m^5}$		$q = \dfrac{1}{L}\int p\,dt$	Inductance

e. Fluid System (Pneumatic)

Quantity	Symbol	Metric Units	Component Schematic	Component Equation	Analogy w.r.t. Electrical Circuit
Mass Flow rate	q	$\dfrac{kg}{sec}$	-	q (specified)	Current
Pressure	p	$\dfrac{Newton}{m^2}$	-	p (specified)	Voltage
Pneumatic Capacitance	C	$\dfrac{kg}{Newton/m^2}$	-	$q = C\dfrac{dp}{dt}$	Capacitance
Pneumatic Resistance	R	$\dfrac{Newton/m^2}{kg/sec}$	-	$q = \dfrac{p}{R}$	Resistance
Pneumatic Inertance	L	$\dfrac{Newton\text{ - }sec^2}{m^2 kg}$	-	$q = \dfrac{1}{L}\int p\,dt$	Inductance

f. Thermal System

Quantity	Symbol	Metric Units	Component Schematic	Component Equation	Analogy w.r.t. Electrical Circuit
Rate of heat flow	h	$\dfrac{Joule}{sec}$	-	h (specifed)	Current
Temperature	θ	$^\circ C$	-	θ (specified)	Voltage
Thermal Capacitance	C	$\dfrac{Joule}{^\circ C}$	-	$h = \dfrac{1}{C}\int \theta\,dt$	Capacitance
Thermal Resistance	R	$\dfrac{^\circ C}{Joule/sec}$	-	$h = \dfrac{\theta}{R}$	Resistance

Apart from basic elements of various types of systems discussed above, there are some other elements, which are quite frequently used in control systems. Details of some of these systems are given in table below:

Table C.2. Some typical systems

System	System Schematic	Mathematical Model of the System
Vacuum Tube	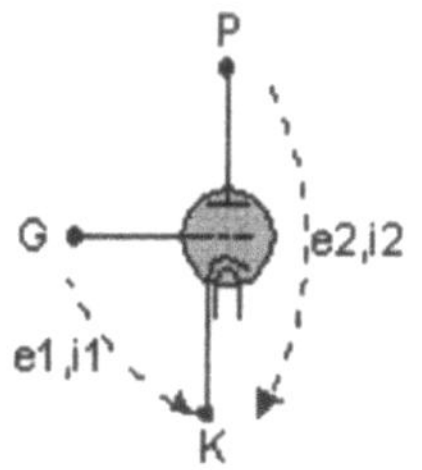	$$\begin{bmatrix} i_1 \\ i_2 \end{bmatrix} = \begin{bmatrix} 0 & 0 \\ g_m & \dfrac{1}{r_p} \end{bmatrix} \begin{bmatrix} e_1 \\ e_2 \end{bmatrix}$$
Transistor	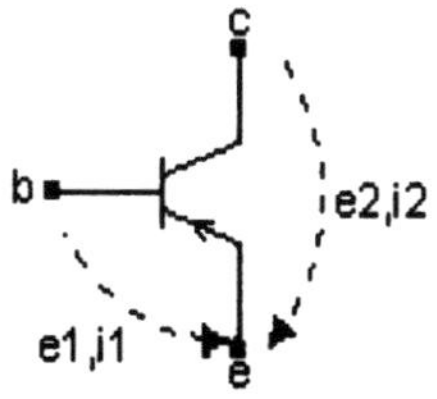	$$\begin{bmatrix} e_1 \\ e_2 \end{bmatrix} = \begin{bmatrix} h_{11} & h_{12} \\ h_{21} & h_{22} \end{bmatrix} \begin{bmatrix} i_1 \\ i_2 \end{bmatrix}$$
Magnetically Coupled Inductor Pair		$$e_1 = L_1 \frac{di_1}{dt} \pm M \frac{di_2}{dt}$$ $$e_2 = L_2 \frac{di_2}{dt} \pm M \frac{di_1}{dt}$$
Transformer	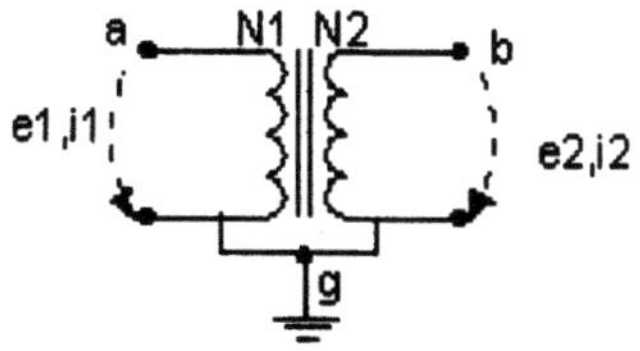	$$\begin{bmatrix} e_1 \\ i_2 \end{bmatrix} = \begin{bmatrix} 0 & \dfrac{N_1}{N_2} \\ -\dfrac{N_1}{N_2} & 0 \end{bmatrix} \begin{bmatrix} i_1 \\ i_2 \end{bmatrix}$$

Table C.2. (continued)

System	System Schematic	Mathematical Model of the System
Operational Amplifier in Inverting Mode	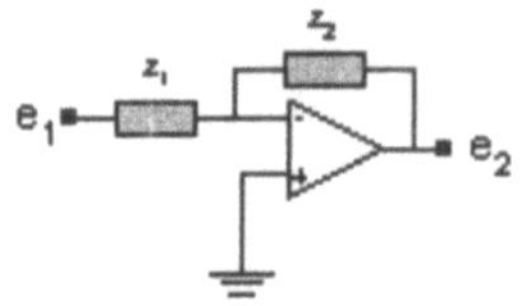	$$\frac{e_2}{e_1} = -\frac{z_2}{z_1}$$
Fixed Lever	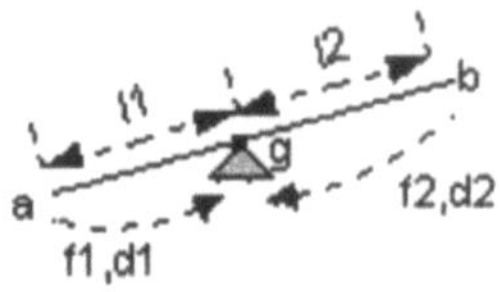	$$\begin{bmatrix} f_1 \\ d_2 \end{bmatrix} = \begin{bmatrix} 0 & \dfrac{l_2}{l_1} \\ -\dfrac{l_2}{l1} & 0 \end{bmatrix} \begin{bmatrix} d_1 \\ f_2 \end{bmatrix}$$
Gear Train	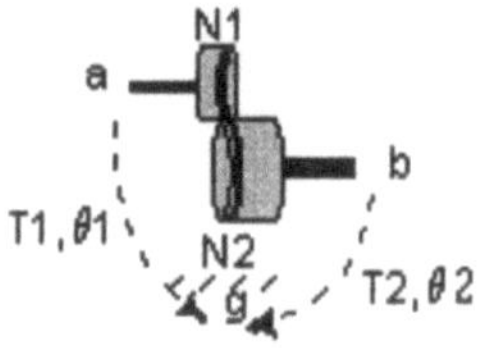	$$\begin{bmatrix} T_1 \\ \theta_2 \end{bmatrix} = \begin{bmatrix} 0 & \dfrac{N_1}{N_2} \\ -\dfrac{N_1}{N_2} & 0 \end{bmatrix} \begin{bmatrix} \theta_1 \\ T_2 \end{bmatrix}$$
Reservoir	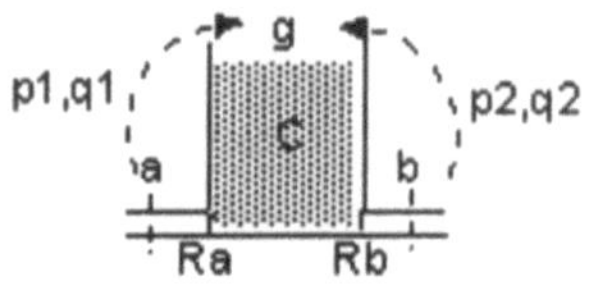	$$\begin{bmatrix} p_1 \\ p_2 \end{bmatrix} = \begin{bmatrix} R_a + \dfrac{1}{C} & \dfrac{1}{C} \\ \dfrac{1}{C} & R_b + \dfrac{1}{C} \end{bmatrix} \begin{bmatrix} q_1 \\ q_2 \end{bmatrix}$$
Hydraulic Piston	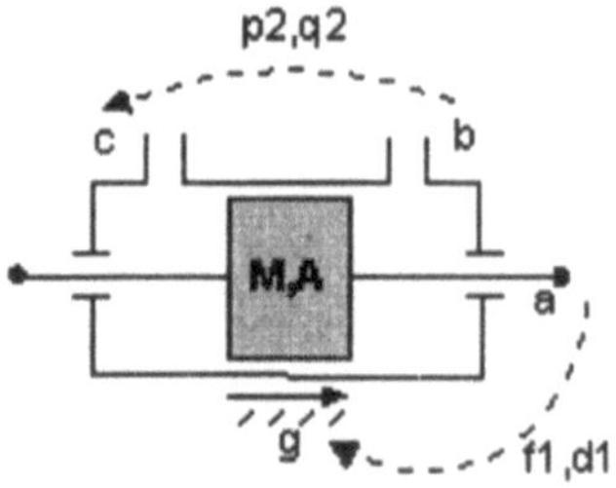	$$\begin{bmatrix} f_1 \\ q_2 \end{bmatrix} = \begin{bmatrix} Mp^2 & A \\ -Ap & 0 \end{bmatrix} \begin{bmatrix} d_1 \\ p_2 \end{bmatrix}$$
Pulley	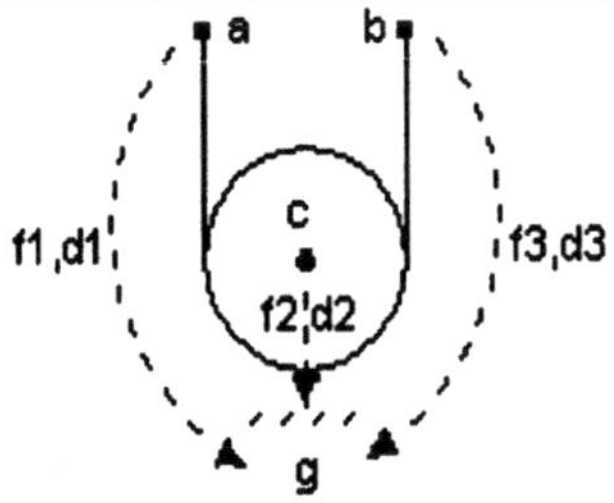	$$\begin{bmatrix} f_1 \\ d_2 \\ f_3 \end{bmatrix} = \begin{bmatrix} 0 & 1 & 0 \\ -1 & 0 & 2 \\ 0 & -2 & 0 \end{bmatrix} \begin{bmatrix} d_1 \\ f_2 \\ d_3 \end{bmatrix}$$

C.1 Plotting some Typical Signals

Often, it is required to plot a particular signal. In the examples to follow, you will learn to achieve this through the MATLAB® command window as well as through SIMULINK®. As you already know, some signals like impulse are available only with MATLAB®, while some are available only with SIMULINK®. So you have to be judicious while choosing an environment for plotting of signals. Experience alone will tell you which environment you should choose for obtaining and plotting a particular signal.

C.1.1 Generating Signals through the MATLAB® Window

The **fplot** function or the **ezplot** function can be used at the command prompt of the MATLAB® window to obtain plots of different functions in the Figure window. Some typical signals are plotted below using the **fplot** function.

Table C.3. Signals generated through the MATLAB® Window

Name	Command	Plot
Step Signal (Function: K *i.e.,* Constant)	fplot('5',linspace(0,500))	
Ramp Signal (Function: Kt)	t=linspace(0,500); fplot('5*t',t)	

Table C.3. (continued)

Name	Command	Plot
Parabolic Signal (Function: $Kt^2/2$)	t=linspace(0,500); fplot('(5*t^2)/2',t)	
Sinusoidal Signal (Function: $K\sin(x)$)	fplot('5*sin(x)',[0 10*pi])	
Exponentially Rising Sinusoidal Signal (Function: $x\sin(x)$)	fplot('x*sin(x)',[0 10*pi])	

Table C.3. (continued)

Name	Command	Plot
Exponentially Decaying Sinusoidal Signal (Function: $(1/x)\sin(x)$)	fplot('(1/x)*sin(x)', [0 10*pi])	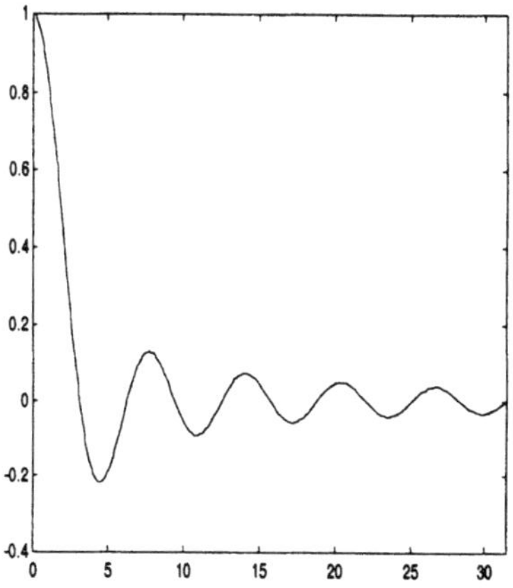
Exponentially Rising Signal (Function: e^x)	t=linspace(0,500); fplot('exp(x)',t)	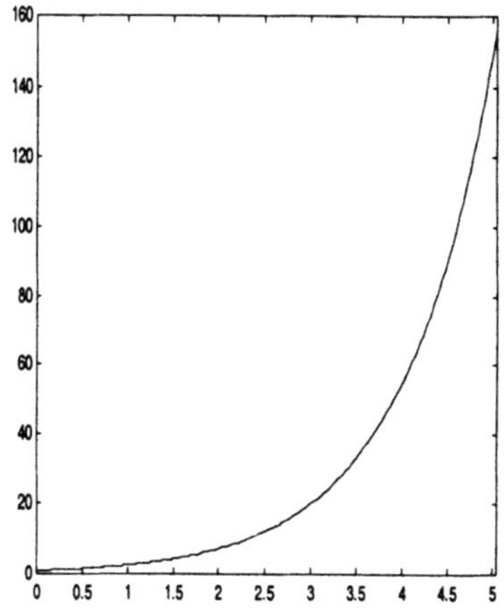
Exponentially Decaying Signal (Function: e^{-x})	t=linspace(0,500); fplot('exp(-x)',t)	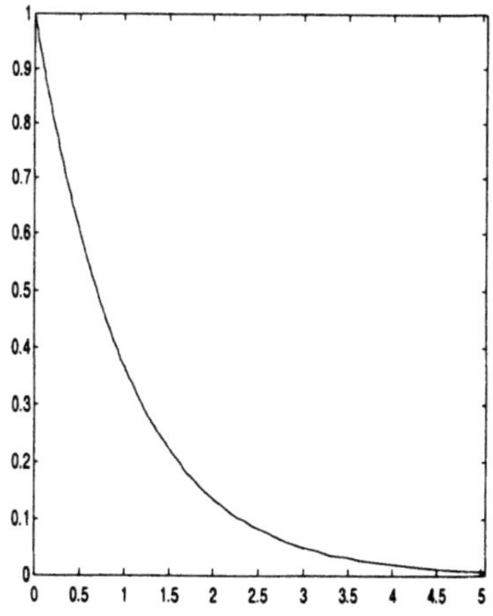

Table C.3. (continued)

Name	Command	Plot
Exponential Distribution Curve/ Signal (Function: te^{-xt})	t=linspace(0,500); fplot('x*exp(-x)',t)	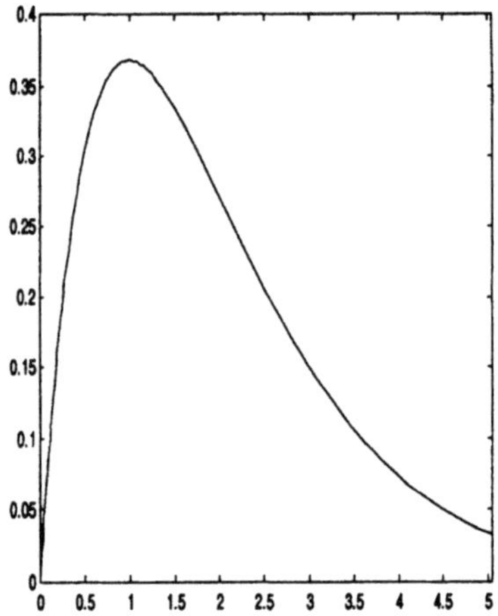
Exponential Distribution Curve/ Signal (Function: t^2e^{-xt})	t=linspace(0,500); fplot('x^2*exp(-x)',t)	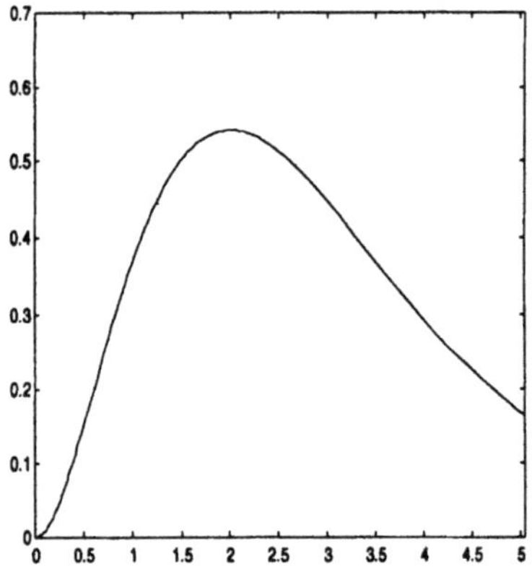
Exponential Distribution Curve/ Signal (Function: $t^{-1}e^{-xt}$)	t=linspace(0,500); fplot('x^-1*exp(-x)',t)	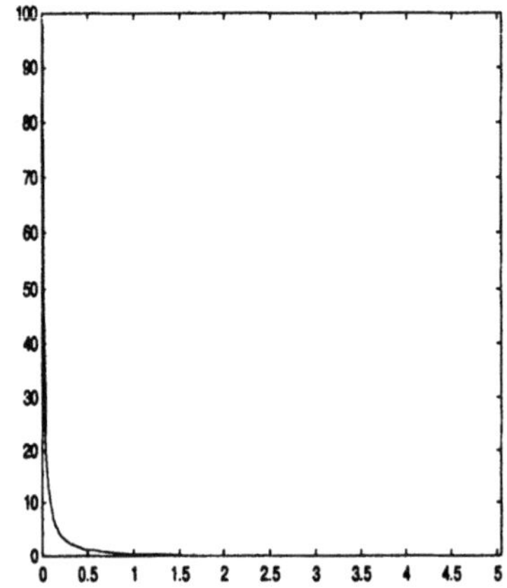

Table C.3. (continued)

Name	Command	Plot
Step Signal With Initial Exponential Rise (Function: $1-e^{-x}$)	t=linspace(0,500); fplot('1-exp(-x)',t)	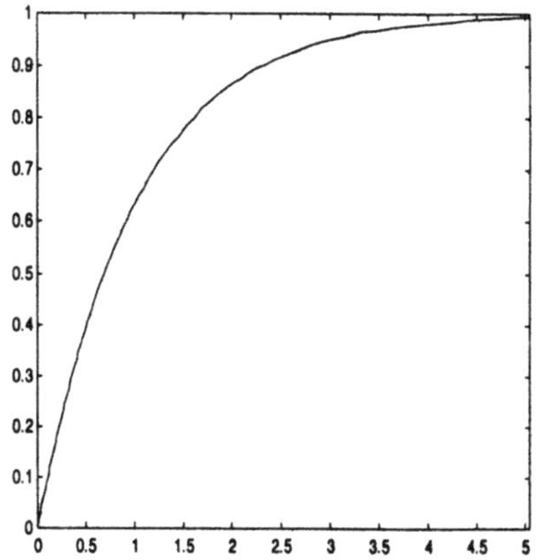
Sine Wave Signal With A Phase Difference Of 30° (Function: $K\sin(x-\pi/6)$)	fplot('5*sin(x-pi/6)',[0 10*pi])	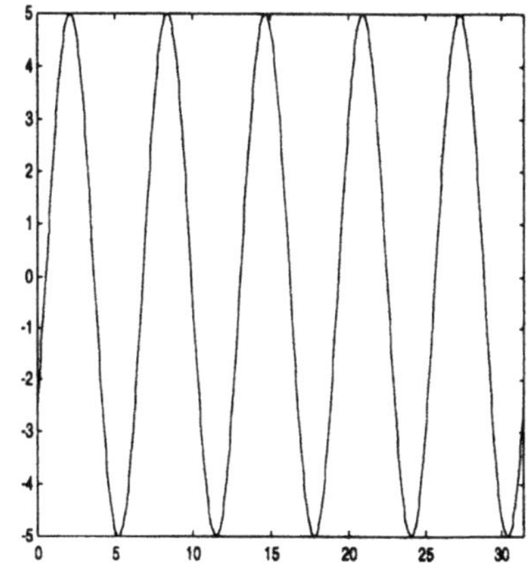
Rectified Sine wave Signal (Function: abs(sinx))	fplot('abs(5*sin(x))',[0 10*pi])	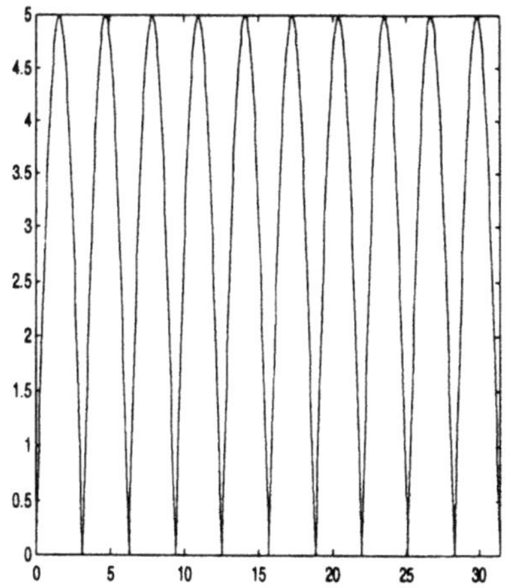

Table C.3. (continued)

Name	Command	Plot
Hyperbolic Sinusoidal Signal (Function: Ksinh(x))	fplot('5*sinh(x)',[0 10*pi])	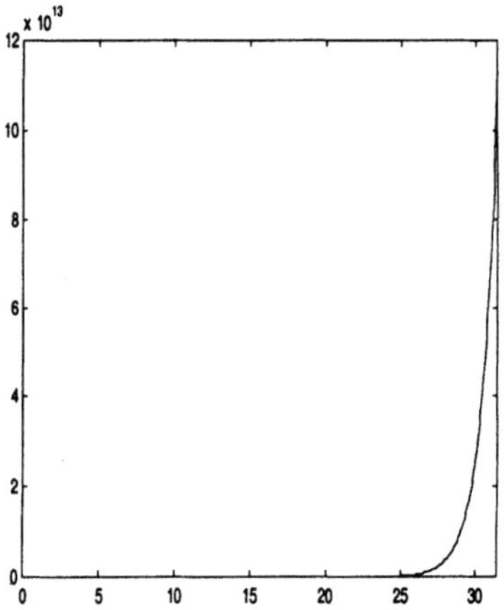
Hyperbolic Cosinusoidal Signal (Function: Kcosh(x))	fplot('5*cosh(x)',[0 10*pi])	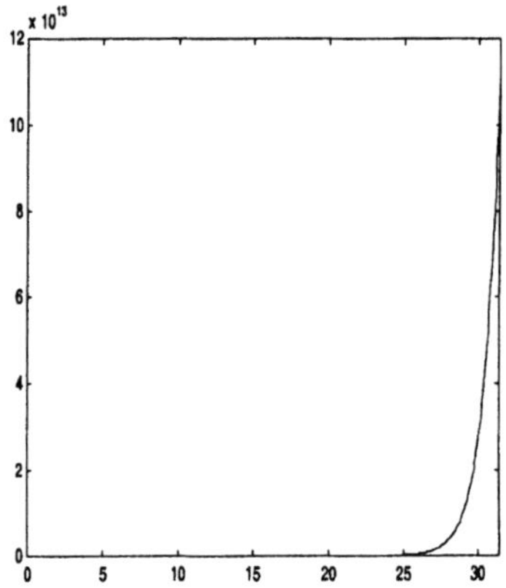

Three Phase Signal	Plotting of this signal has a slightly different approach. Make a function file called ThreePhase and enter the following lines into it. Save it as m file.
	
	`function Y = ThreePhase(x)`
	`Y(:,1) = 5*sin(x(:));`
	`Y(:,2) = 5*sin(x(:)+2*pi/3);`
	`Y(:,3) = 5*sin(x(:)+4*pi/3);`
	
	Enter the following at the command prompt.
	
	`fplot('ThreePhase', [0 10])`
		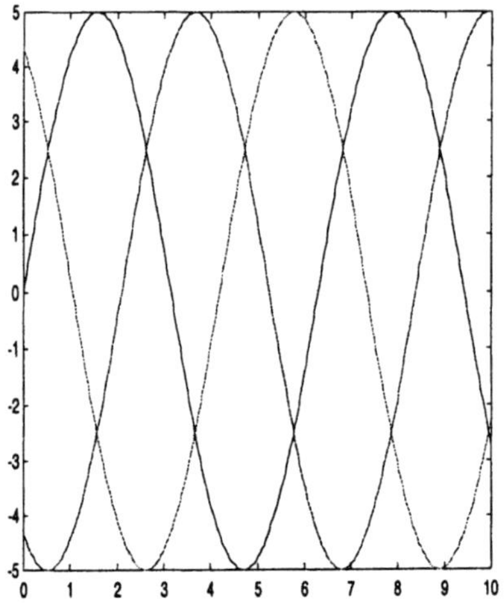

C.1.2 Generating Signals through the SIMULINK® Window

The same signals and many others can be obtained using SIMULINK®. Some typical signals are plotted below. To see the parameters set for the blocks check Signals, AM_Modulation and PWM file in the accompanying floppy.

Table C.4. Signals Generated through the SIMULINK® window

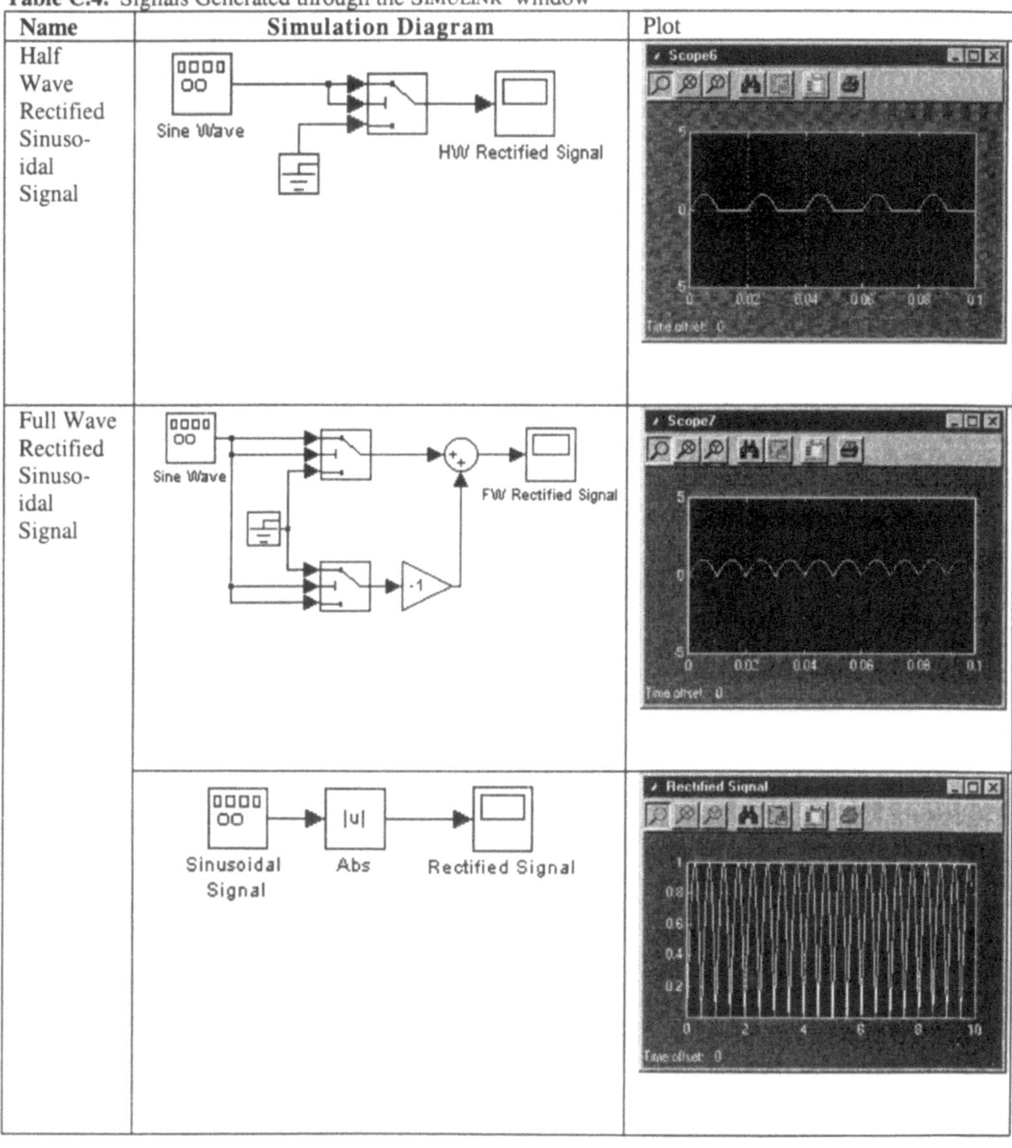

Name	Simulation Diagram	Plot
Half Wave Rectified Sinuso-idal Signal		
Full Wave Rectified Sinuso-idal Signal		

Table C.4. (continued)

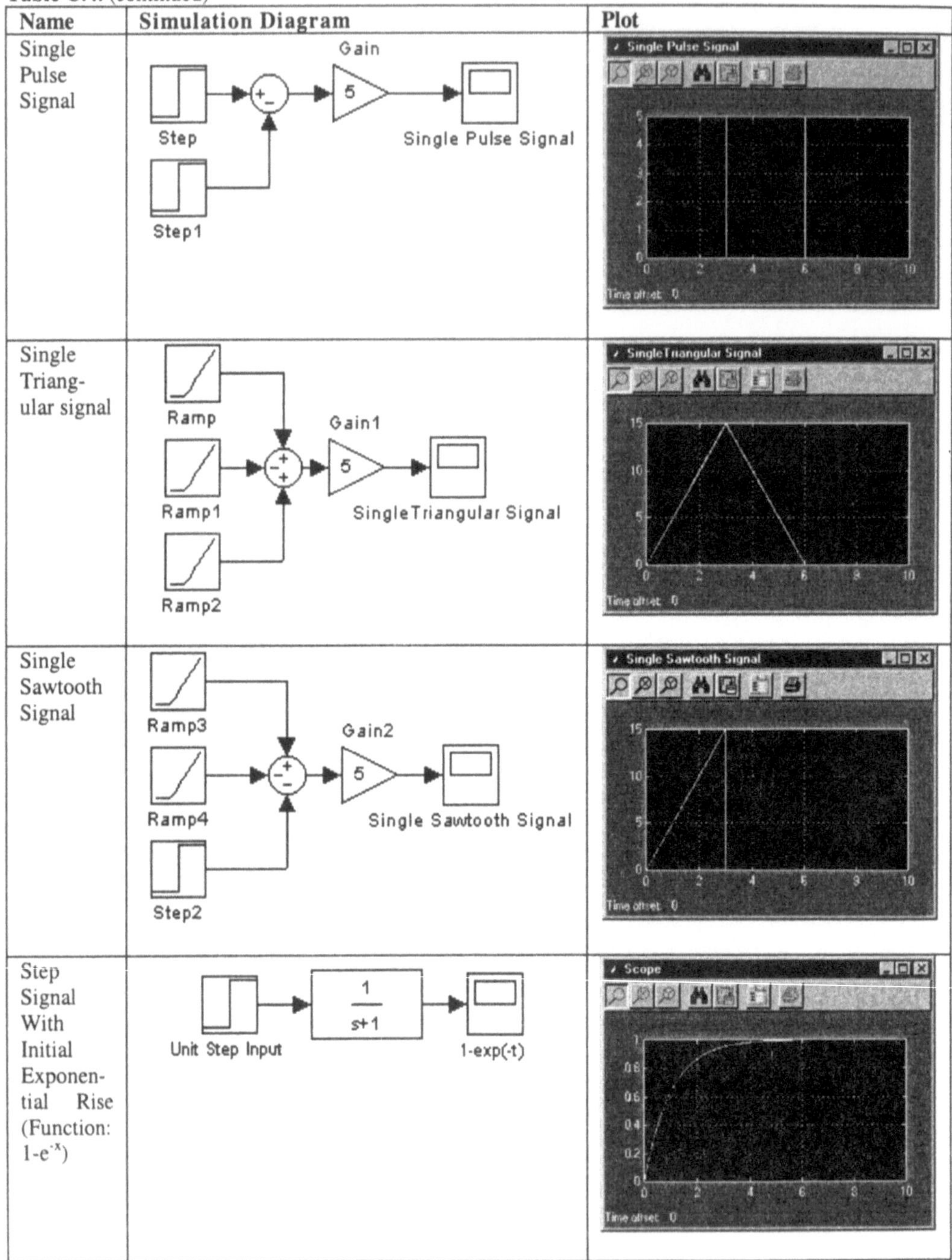

Name	Simulation Diagram	Plot
Single Pulse Signal		
Single Triang-ular signal		
Single Sawtooth Signal		
Step Signal With Initial Exponen-tial Rise (Function: $1-e^{-x}$)		

Table C.4. (continued)

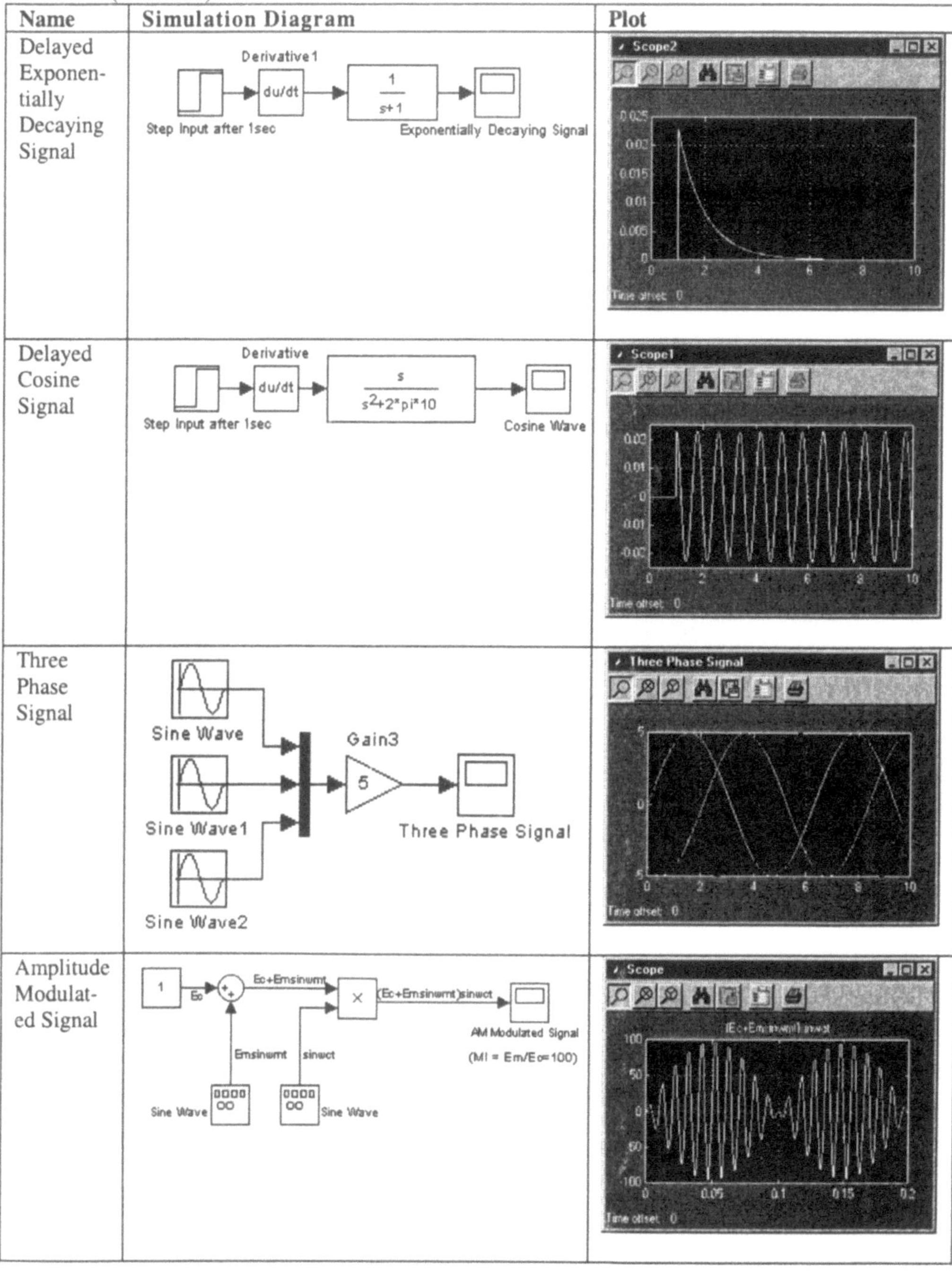

Table C.4. (continued)

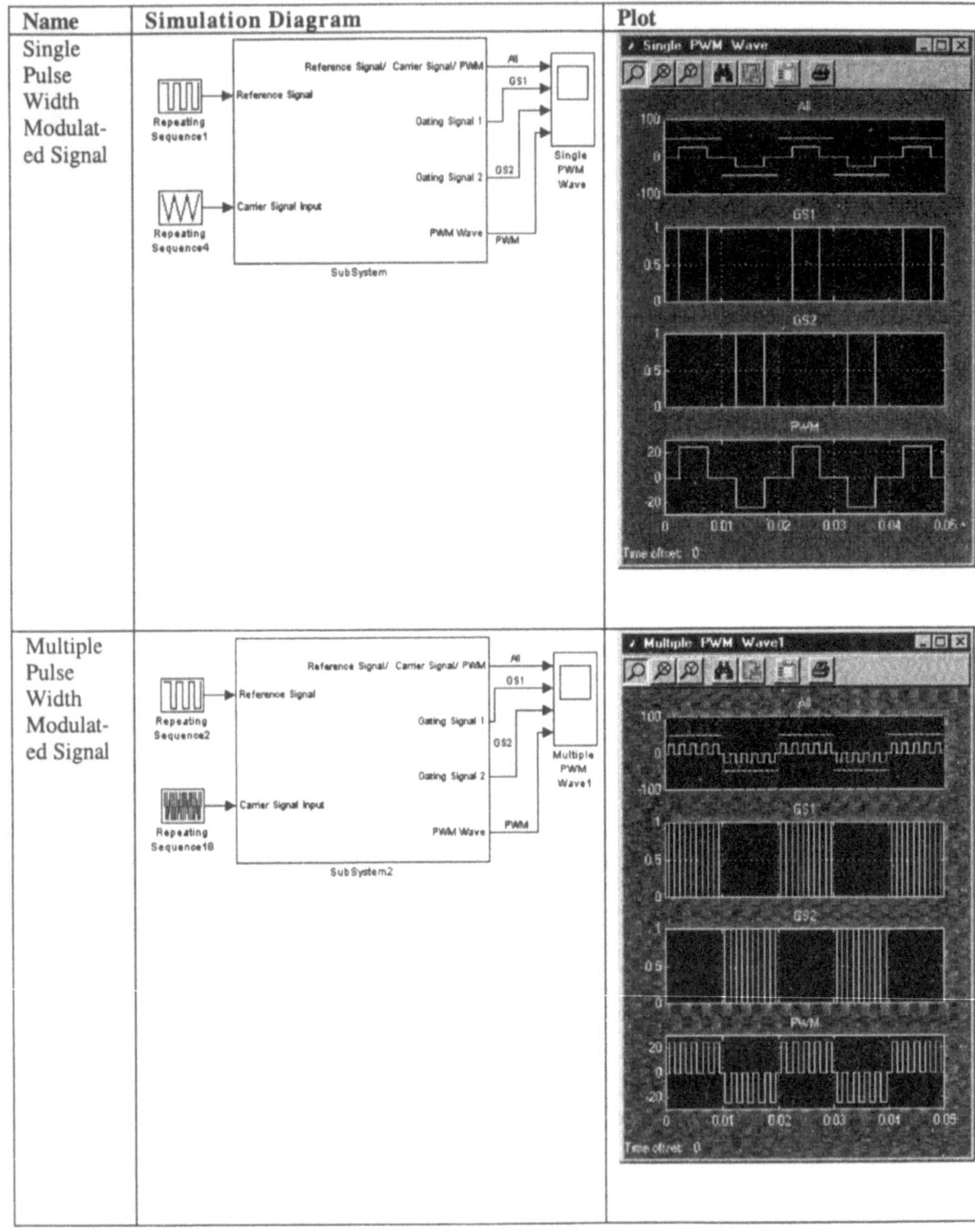

Table C.4. (continued)

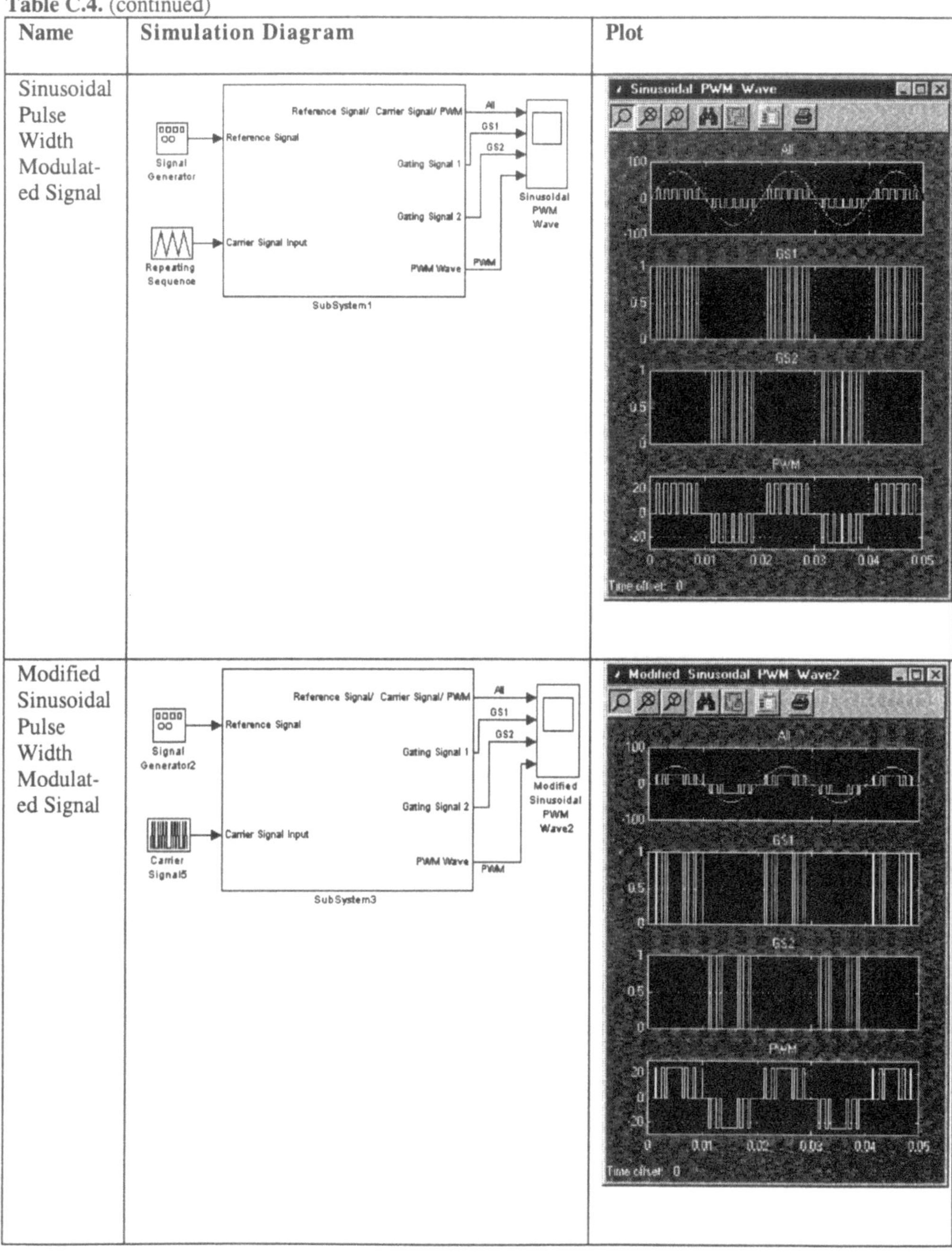

Name	Simulation Diagram	Plot
Sinusoidal Pulse Width Modulated Signal		
Modified Sinusoidal Pulse Width Modulated Signal		

The Subsystems of the PWM Signals are shown in Figure C.1:

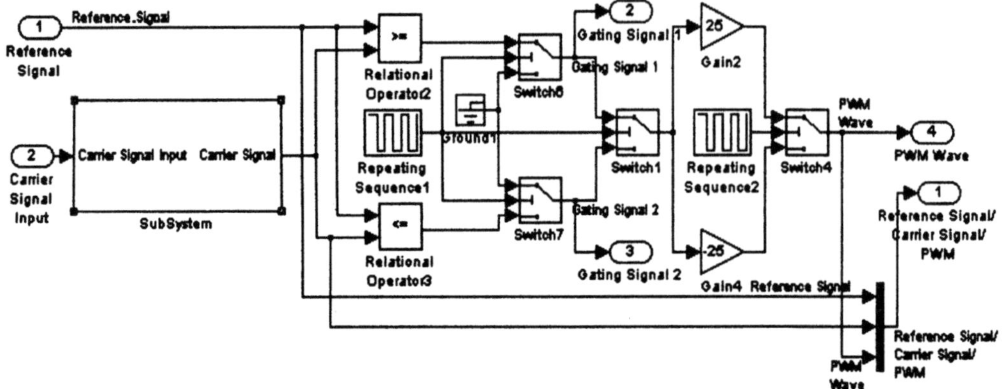

a. Details of subsystem for obtaining PWM signals

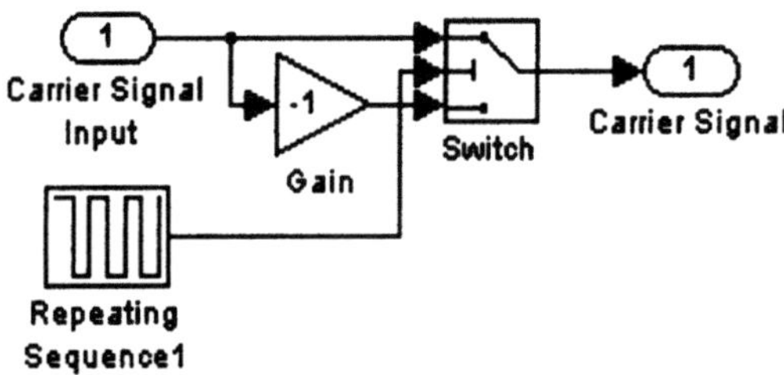

b. Details of subsystem shown in Panel a.

Figure C.1. Subsystems for PWM signal

Alphabetical List of Built-in Functions as Provided with Control System Toolbox of MATLAB®:

'	-	Pertransposition
*	-	Multiply LTI systems (series connection)
.'	-	Transposition of input/output map
/	-	Right divide -- sys1/sys2 means sys1*inv(sys2)
[..]	-	Concatenate LTI models along inputs or outputs
\	-	Left divide -- sys1\sys2 means inv(sys1)*sys2
^	-	LTI model powers
+ and -	-	Add and subtract LTI systems (parallel connection)
acker	-	SISO pole placement
append	-	Group LTI systems by appending inputs and outputs
augstate	-	Augment output by appending states
balreal	-	Gramian-based input/output balancing
bode	-	Bode plot of the frequency response
c2d	-	Continuous to discrete conversion
canon	-	State-space canonical forms
care	-	Solve continuous algebraic Riccati equations
chgunits	-	Change units of FRD model frequency points
class	-	Model type ('tf', 'zpk', 'ss', or 'frd')
connect	-	Derive state-space model from block diagram description
covar	-	Covariance of response to white noise
ctrb, obsv	-	Controllability and Observability matrices
ctrldemo	-	Introduction to the Control System Toolbox
d2c	-	Discrete to continuous conversion
d2d	-	Resample discrete-time model
damp	-	Natural frequency and damping of system poles
dare	-	Solve discrete algebraic Riccati equations
dcgain	-	D.C. (low frequency) gain
delay2z	-	Replace delays by poles at z=0 or FRD phase shift
diskdemo	-	Digital design of hard-disk-drive controller
dlyap	-	Solve discrete Lyapunov equations
dsort	-	Sort discrete poles by magnitude
dss	-	Create a descriptor state-space model
dssdata	-	Descriptor version of SSDATA
esort	-	Sort continuous poles by real part
estim	-	Form estimator given estimator gain
evalfr	-	Evaluate frequency response at given frequency
feedback	-	Feedback connection of two systems
filt	-	Specify a digital filter
frd	-	Create a frequency response data model
frd	-	Conversion to frequency data
frdata	-	Extract frequency response data

freqresp	-	Frequency response over a frequency grid
gensig	-	Generate input signal for LSIM
get	-	Access values of LTI model properties
gram	-	Controllability and Observability Gramians
hasdelay	-	True for models with time delays
impulse	-	Impulse response
initial	-	Response of state-space system with given initial state
inv	-	Inverse of an LTI system
isa	-	Test if LTI model is of given type
isct	-	True for continuous-time models
isdt	-	True for discrete-time models
isempty	-	True for empty LTI models
isproper	-	True for proper LTI models
issiso	-	True for single-input/single-output models
jetdemo	-	Classical design of jet transport yaw damper
kalman	-	Kalman estimator
kalmd	-	Discrete Kalman estimator for continuous plant
kalmdemo	-	Kalman filter design and simulation
lft	-	Generalized feedback interconnection (Redheffer star product)
lqgreg	-	Form LQG regulator given LQ gain and Kalman estimator
lqr,dlqr	-	Linear-quadratic (LQ) state-feedback regulator
lqrd	-	Discrete LQ regulator for continuous plant
lqry	-	LQ regulator with output weighting
lsim	-	Response to arbitrary inputs
ltimodels	-	Detailed help on various types of LTI models
ltiprops	-	Detailed help on available LTI properties
ltiview	-	Response analysis GUI (LTI Viewer)
ltiview	-	Response analysis GUI (LTI Viewer)
lyap	-	Solve continuous Lyapunov equations
margin	-	Gain and phase margins
milldemo	-	SISO and MIMO LQG control of steel rolling mill
minreal	-	Minimal realization and pole/zero cancellation
modred	-	Model state reduction
ndims	-	Number of dimensions
nichols	-	Nichols chart
norm	-	Norms of LTI systems
nyquist	-	Nyquist plot
pade	-	Pade approximation of time delays
parallel	-	Generalized parallel connection (see also overloaded +)
place	-	MIMO pole placement
pole, eig	-	System poles
pzmap	-	Pole-zero map
reg	-	Form regulator given state-feedback and estimator gains
reshape	-	Reshape array of LTI models

rlocfind	-	Interactive root locus gain determination
rlocus	-	Evans root locus
rltool	-	Root locus design GUI
rss,drss	-	Random stable state-space models
series	-	Generalized series connection (see also overloaded *)
set	-	Set/modify properties of LTI models
sigma	-	Singular value frequency plot
size	-	Model sizes and order
sminreal	-	Structurally minimal realization
ss	-	Create a state-space model
ss	-	Conversion to state space
ss2ss	-	State coordinate transformation
ssbal	-	Diagonal balancing of state-space realizations
ssdata	-	Extract state-space matrices
stack	-	Stack LTI models/arrays along some array dimension
step	-	Step response
stepfun	-	Generate unit-step input
tf	-	Create a transfer function model
tf	-	Conversion to transfer function
tfdata	-	Extract numerator(s) and denominator(s)
totaldelay	-	Total delay between each input/output pair
zero	-	System (transmission) zeros
zpk		Create a zero/pole/gain model
zpk		Conversion to zero/pole/gain
zpkdata		Extract zero/pole/gain data

Appendix D

This Appendix provides information about features of material accompanying this book available on the web-site.

In this Appendix, you will find details of the contents and use of the material accompanying this book that has been placed on the web-site of Springer-Verlag at `ftp://ftp.springer.co.uk`. As you must have already read in the Introduction, this material has been especially prepared for you keeping in mind that you are in the process of learning. It is designed to save you precious time and energy that you might otherwise be spending on typing and troubleshooting the commands at the command prompt in the MATLAB® command window or designing and troubleshooting the models in the SIMULINK® window. You will find all the examples given in this book in the grey area of 12.5% intensity or otherwise here. You will also find some of the problems solved in more than one way here, which might not have been mentioned in the book or might be featuring in the form of unsolved examples. Through this Appendix you will learn to derive maximum benefit out of the material provided.

D.1 What to Do First

First and foremost thing that we would suggest you to do is to open the web-site and download the material. If you are well conversant with Windows operating system and have little knowledge of Internet accessing and downloading the material, which is an automatic process, it it should not be much of a problem for you.

D.2 What the Downloaded Material contains

As already mentioned above, the downloaded material mainly contains the examples provided in this book. Examples of Chapter 1 to Chapter 9 along with examples from Appendix C and Appendix E are what the files you have downloaded contain. Figure D.1 and Figure D.2 give details of the structure and content of the downloaded material.

D.3 How to use the Material

Once you are done with downloading the material into hard disk, transfer the material to the

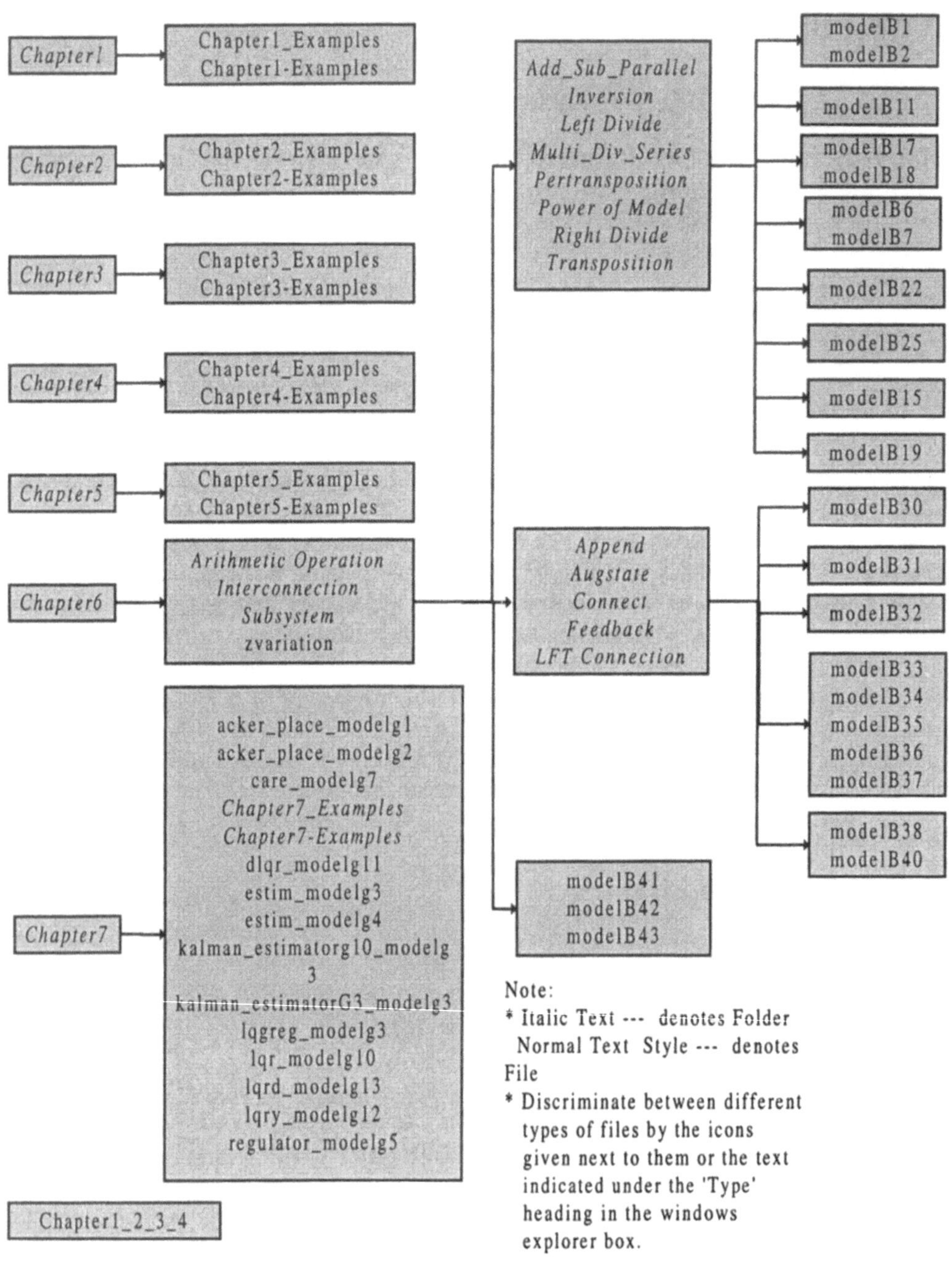

Figure D.1. Details of material downloaded

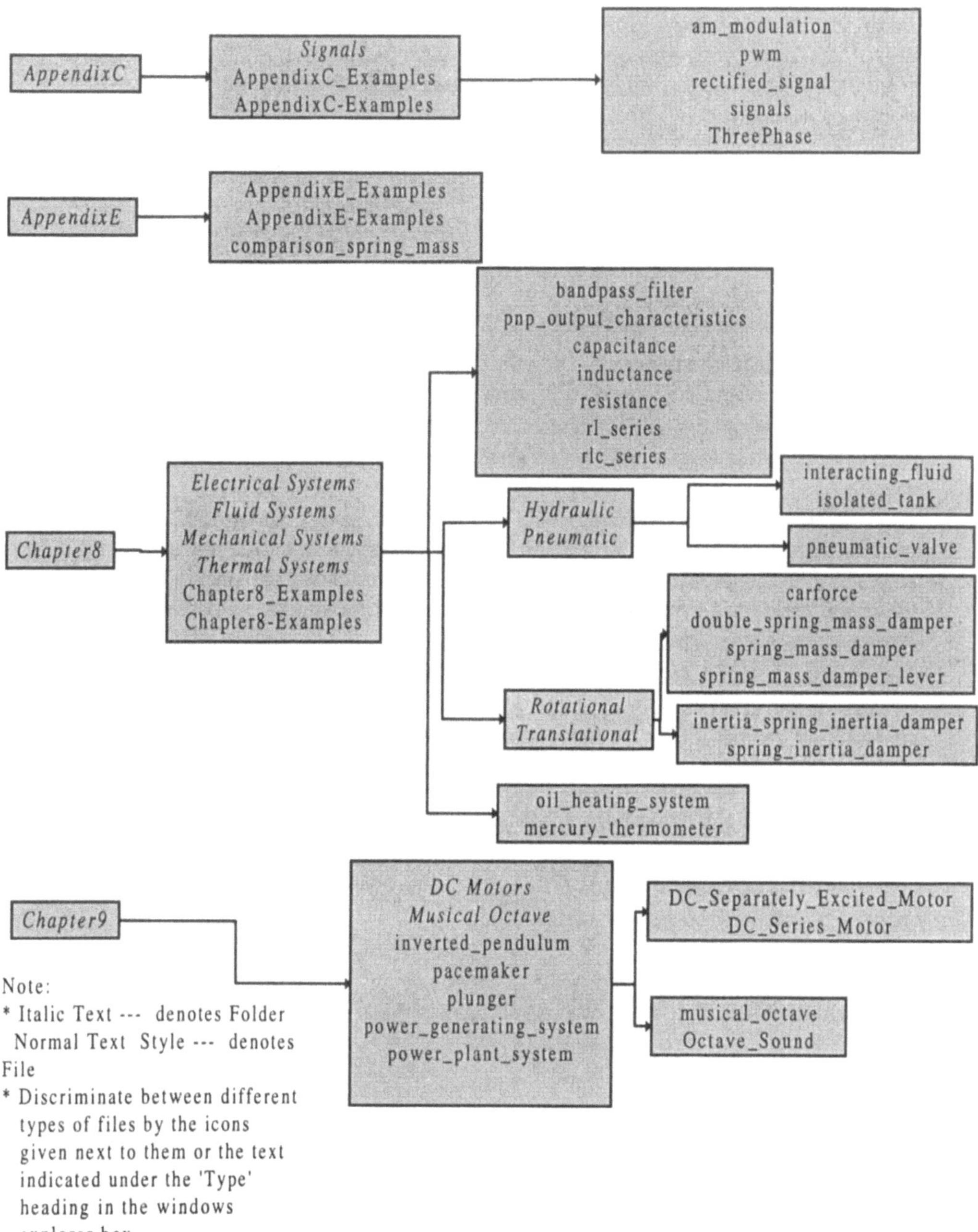

Figure D.2. Details of material downloaded

work directory of MATLAB® . Now, you can straight away start using them with and extract the maximum benefit out of them. We would recommend you to sit in front of the computer with the chapter of the book you wish to learn open in front of you. As you read, you should try out the examples given in the grey (12.5% intensity) area of the book first on your own. Only if you fail or do not have sufficient time, you should turn towards the examples you downloaded and copied to your hard disk. After all, failures are supposed to be the pillar stones of success. However in the lines to follow, more details are provided about the material.

For Chapter 1 to Chapter 4

The statement of the examples given in the grey (12.5% intensity) area of each of these chapters is provided in two types of files:

i. Word File (with **.doc** extension) -- If you have **Notebook** feature of MATLAB® installed in your computer, you can open the word file Chapter1-Examples, Chapter2-Examples, Chapter3-Examples or Chapter4-Examples, select the example you wish to work with and press Control+Enter keys simultaneously. You will get the desired results.

ii. MATLAB® File (with **.m** extension) -- If you intend to work only in MATLAB® environment without taking the trouble of opening another program like MS Word having m-book template, then this is just for you! The same contents as above are saved as m-file. You can open the m-files Chapter1_Examples, Chapter2_Examples, Chapter3_Examples or Chapter4_Examples and using cut and paste feature of windows (or MATLAB®) you can execute the examples from the MATLAB® command prompt. Or, you can simply write filename at the command prompt in which case the results of the examples are displayed in the MATLAB® window. Using **more** function of MATLAB® you can control the scrolling display on the screen.

However, if you have just switched on your computer and have invoked MATLAB® with a desire to work with examples of Chapter 2, you will get unexpected results with either file. For example you write modelA1 or modelB1 expecting the computer to give you the desired result all you get is a statement of the following nature:

```
>> modelA1
??? Undefined function or variable 'modelA1'.
>> modelB1
??? Undefined function or variable 'modelB1'.
```

This has happened since the data involved in making of these models is missing. You can avoid this by either of the following two ways:

i. you work in a systematic way.
i.e., first generate models of Chapter1 then of Chapter 2 and so on. But, this is not very convenient especially if you desire to work with examples of Chapter 4.

ii. you can simply load the workspace Chapter1_2_3_4 (with **.mat** extension) and then proceed!

Loading workspace called Chapter1_2_3_4.mat actually loads all the models generated in Chapter 1 to Chapter 4.

For Chapter 5
The nature of files in this directory is same as for Chapter 1 to Chapter 4 explained above.

For Chapter 6
Examples of this chapter are infact models created in SIMULINK® environment. You will notice that this is the smallest chapter of the book. This is because the examples of this chapter have been covered in the folder/directory named Chapter 6. How to perform arithmetic operation on the models, interconnect them or work with subsystems in SIMULINK® environment is illustrated in several ways in these model files having **.mdl** extension. You will observe that the same result can be obtained by several ways to give the same result. Scope block with foreground color as black and background color as grey compares all the results. You can now stop bothering for setting the simulation and/or block parameteres to obtain the correct simulation results and concentrate on the real problem.

For Chapter 7
The examples covered in this chapter are a cocktail of both the types explained above. You will find MATLAB® file (named Chapter7_Examples) and word file (named Chapter7-Examples) having same contents. Rest of the examples are SIMULINK® models. These can be used in a similar way as explained above.

For Chapter 8 and Chapter 9
The examples of these chapters are stored in the folder/directory named Chapter 8 and Chapter 9. Applications covered in Chapter 8 are explained only using block diagram approach in the book. However, in the examples, you will find the same applications simulated using transfer function and state space variable approach too. By observing the response of scope block with black foreground color and grey background color, you will find that the result obtained by any of these methods is the same. Similar treatment has been extended to the examples of Chapter 9.

For Appendix C and Appendix E
Examples covered in these two appendices are stored in the floppy under the names Appendix C and Appendix E respectively using the same approach as explained above.

Appendix E

This Appendix provides information in question and answer form on several aspects relevant to the topics covered in this book.

Down here, you will find solution to some of the problems you might have come across while going through this book and trying out many things given in this book or on your own.

Q. I get confused with the use the three brackets have, specific to matrices in MATLAB®. Please help!

A. Let us understand this by using a matrix **a** as given below:

$$a = \begin{bmatrix} 1 & 2 & 3 \\ 4 & 5 & 6 \\ 7 & 8 & 9 \\ 10 & 11 & 12 \end{bmatrix}$$

( ) - Small brackets or parentheses are used for indexing subscripts of vectors and matrices. A comma (,) sign separates the row and column subscripts. If elements of a complete row (or column) are desired, a colon (:) sign is used in place of numeral identifying row (or column).

Example:
```
a(1,1)
returns

ans =
   1

a(4,3)
returns

        ans =
        12

a(:,2)
returns

        ans =
        2
        5
```

```
     8
     11
```

[ ] - *Square brackets are used for entering matrices*. The row elements are separated by spaces where as one row is separated from the other by a semicolon (;) sign.

Example:
a=[1 2 3; 4 5 6; 7 8 9; 10 11 12]
returns

```
a =
1    2    3
4    5    6
7    8    9
10   11   12
```

{ } - *Curly brackets or braces are used for entering cell arrays.*

Example:

b={[1 2],[3 4]; [5 6], [7 8]; [9 10], [11 12]}
returns

```
b =
[1x2 double]   [1x2 double]
[1x2 double]   [1x2 double]
[1x2 double]   [1x2 double]
```

The values of b can be obtained by cell referencing method as follows:
b{1,:,:}
returns

```
ans =
1    2
ans =
3         4
4
```

and so on...

Q. How can a matrix have magnitude and direction both? And how are such vectors written? Does a cell array denote a vector because that is a large form of matrix and a matrix having magnitude and direction both should also be large?

A. Oh no! You have all wrong concepts. A vector in MATLAB® language is different from vector defined in your physics text. *It is rather a special case of matrix having a single row or column.* It has nothing to do with magnitude or direction of a matrix or cell array. Similarly, a scalar in MATLAB® is an element, which are not placed within brackets.

Q. While using commands like [u,t]=**gensig**('sin',2*pi,60) or [m,p,w]=**bode**(model), MATLAB® produces a long list of values which goes on scrolling on the screen. I do not want this display. What should I do?

A. Probably you have forgotten the basic things MATLAB® taught you. *Put a semicolon sign (;) at the end of the command. The display of result on the screen will be suppressed.* The result would however, remain in the memory of the workplace and you can always retrieve and use it by as usual using the variable name.

Q. While working with figures, the tick labels and the tick marks on the axis do not appear quite clear due to grey background in that area of Figure window. Can I change this color?

A. Yes, why not! If you want to change the background color of the Figure window for the whole MATLAB® session, *enter the following command at the start of the session*:

set(0,'defaultfigurecolor','w')

Your Figure window will now have a white background. Its appearance will change from the undesired Figure window shown in Figure E.1 to desired Figure window as shown in Figure E.2. Instead of using letters for desired color that was white in this example, you can use the equivalent vector of rgb (for example, [1 1 1] for black, [0 0 0] for white, [1 1 0] for yellow and so on) and get not only white but any other color or shade desire.

set(0, 'defaultfigurecolor', [1 1 0])
or alternatively,
set(0,'defaultfigurecolor','w')
returns Figure windows with yellow background.

In case you want to change color of only a particular Figure window say 5th, in the command above instead of 0, use the 5 i.e. the handle of the Figure window.

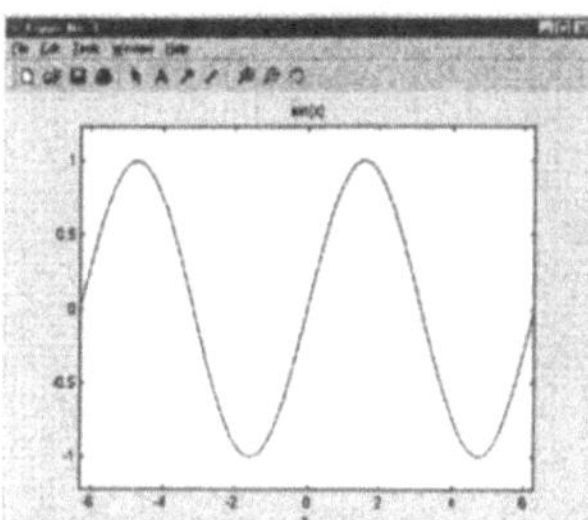 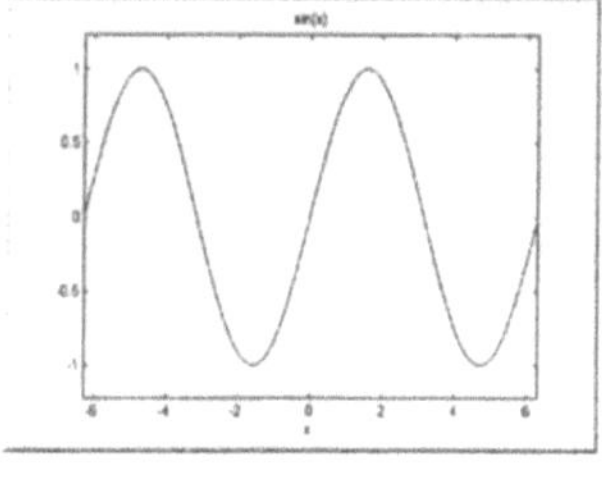 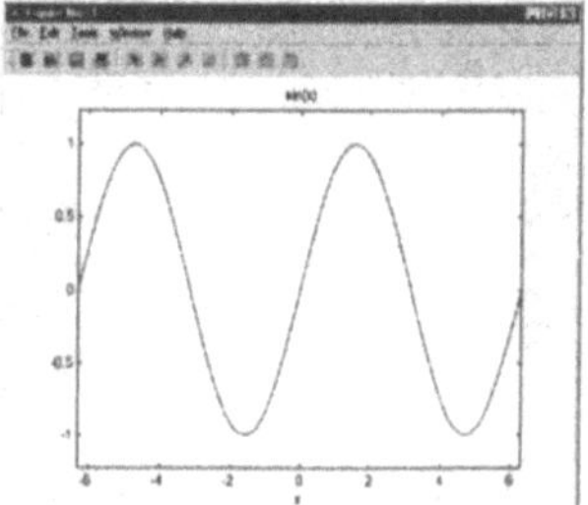

Figure E.1. Default Figure window

Figure E.2. Figure window with white background

Figure E.3. Figure window with yellow background

Q. For the two matrices a and b, I wish to multiply the corresponding elements of a to those of b. When I issue the product command (a*b), the two matrices get multiplied instead of the corresponding element. I get the desired answer when I perform multiplication using subscripts (*i.e.,* c[1 1]= a[1 1]*b[1 1] and so on) though, but this is a very long process for such small operation I believe. The same problem exists for other arithmetic operations too. Is there a shorter way out?

A. Yes, for sure. What you desire to do is known as *bit wise operation* in the language of MATLAB®. And to achieve this is very simple. *You only have to put a dot sign (.) before the arithmetic operator* and you get the result!

Q. I created transfer functions using following command:

model555=**tf**([1 2], [5 6 7])
model666=**tf**([1], [2 3])

Later on while trying out addition of the two models, I entered the following command:
Model777=Model555+model666

MATLAB® returned the following error message:

??? Undefined function or variable 'Model555'.

What happened? Where did I go wrong?
A. Dear me! How can you forget that MATLAB® is case sensitive? The transfer function model555 you created is stored as model555 and not Model555. And so the error message! Remember, *for MATLAB® model555, Model555 MODEL555, mODEL555 and mOdEl555 are five different models stored in the workspace.*

Q. While assigning properties to model555 above, I wanted to enter the following in 'User Data' property:

'This is Krishna's example'
I entered the following at the MATLAB® command prompt:

model555=**tf**([1 2], [5 6 7], 'UserData', 'This is Krishna's example')

However MATLAB® returned the following error message:

??? ,'This is Krishna's

       |

Improper function reference. A "," or ")" is expected.

What should I do to fix up the problem?
A. Write the command as follows :

model555=**tf**([1 2], [5 6 7], 'UserData', 'This is Krishna"s example')

Did you notice the difference?
You have to use a double quote in place of single one for apostrophe s.

Q. I tried **augstate** function on model555 and model666 only to receive an error message. Can you tell me why?
A. Sure, what are we here for? *Augstate is only meaningful only for state space models and hence it cannot be used for tf, zpk or frd models.* MATLAB® gave you this message only. Is'nt it? However, If you still insist upon augstating a tf, zpk or frd model, first convert them to ss model and then proceed.

Q. When trying to simulate physical systems with large orders or with too many elements, deriving transfer function expression is a real cumbersome job that puts me off. Can you suggest some simple method, which is not so taxing?
A. When you have a computer with such a powerful package as MATLAB® loaded onto it, there should be little reasons for you to worry for these trivial things. *MATLAB® has a Symbolic Math Toolbox, which comes to your rescue at such trying moments.* In order to illustrate how

this Toolbox can be used to achieve our purpose, let us take up the example of Double Spring Mass Damper system. Using the basic equations we will derive the transfer function expression. This is how you achieve it:

```
% Define symbols
syms m1 m2 k1 k2 b1 b2 y1 y2 f s p q;
%Solve the two equations
p=solve('s^2*m1*y1+s*(b1+b2)*y1+(k1+k2)*y1=s*b2*y2+k2*y2','s^2*m2*y2+s*b2*y2+k
2*y2=s*b2*y1+k2*y1+f');
%obtain the values of the solution
p.y1;
p.y2;
%Arrange the equation in descending order of s
Y1=collect(p.y1)
Y2=collect(p.y2)
%Obtain transfer functions
tf1=Y1/f
tf2=Y2/f
%Subtitute the values of the parameters of simulation
m1=10; m2=20; b1=5; b2=4; k1=2; k2=3;
TF1=subs(tf1)
TF2=subs(tf2)
returns


Y1 =
f*(k2+s*b2)/(s^4*m2*m1+(m2*b1+b2*m1+m2*b2)*s^3+(m2*k2+k2*m1+b2*b1+m2*k1)*
s^2+(k2*b1+b2*k1)*s+k2*k1)
Y2 =
f/(s^4*m2*m1+(m2*b1+b2*m1+m2*b2)*s^3+(m2*k2+k2*m1+b2*b1+m2*k1)*s^2+(k2*b
1+b2*k1)*s+k2*k1)*(s^2*m1+(b1+b2)*s+k1+k2)
tf1 =
(k2+s*b2)/(s^4*m2*m1+(m2*b1+b2*m1+m2*b2)*s^3+(m2*k2+k2*m1+b2*b1+m2*k1)*s
^2+(k2*b1+b2*k1)*s+k2*k1)
tf2 =
1/(s^4*m2*m1+(m2*b1+b2*m1+m2*b2)*s^3+(m2*k2+k2*m1+b2*b1+m2*k1)*s^2+(k2*b
1+b2*k1)*s+k2*k1)*(s^2*m1+(b1+b2)*s+k1+k2)
TF1 =
(3+4*s)/(200*s^4+220*s^3+150*s^2+23*s+6)
TF2 =
1/(200*s^4+220*s^3+150*s^2+23*s+6)*(10*s^2+9*s+5)
```

Is this not quite easy!

Q. If the computer can really do so much, then I am sure it must be having some solution for yet another problem of mine. Entering mathematical expressions and checking them for errors has always been a phobia to me. Things appear to be so much jumbled up with signs like / for division, * for multiplication and ^ raising the power to and so on .... that I can never really interpret the jargon with ease. Can MATLAB® not give me expression in the form as we have been taught to write down on paper since childhood?

A. MATLAB® has been developed by experts who have kept all these little difficulty of yours into mind. *They incorporated **pretty** function in MATLAB®'s built-in functions library for just this purpose.* Let us take up the same expression from the previous question and see how it is done.

```
pretty(tf1)
```

```
pretty(TF1)
returns
```

```
(k2 + s b2)/(s4m2 m1 + (m2 b1 + b2 m1 + m2 b2) s3+ (m2 k2 + k2 m1 + b2 b1 + m2 k1)
s2 + (k2 b1 + b2 k1) s + k2 k1)
                        3 + 4 s
            -------------------------------------
            200 s4  + 220 s3  + 150 s2  + 23 s + 6
```

This is what you had always desired for. Is it not!

Q. Oh! That was really very exciting. Hey, don't laugh if I sound silly, but I really crave to know if MATLAB® can give me Laplace Transform, Z Transform and their inverses too for mathematical expressions?

A. Sure, MATLAB® can do these little things for you. *Again the Symbolic Math Toolbox is where you will find refuge.* We will see how things are required to be carried out for achieving this through some examples:

```
Syms s z a b c k t x w
laplace(t^2)
returns


ans =
2/s^3

ilaplace(2/s^3 )
returns


ans =
t^2

laplace (cos(w*t)+exp(-a*t)-sin(w*t))
returns


ans =
s/(s^2+w^2)+1/(s+a)-w/(s^2+w^2)
ilaplace(s/(s^2+w^2)+1/(s+a)-w/(s^2+w^2))
returns


ans =
cos((w^2)^(1/2)*t)+exp(-a*t)-w/(w^2)^(1/2)*sin((w^2)^(1/2)*t)

ztrans(t^2)
returns


ans =
z*(z+1)/(z-1)^3

iztrans(z*(z+1)/(z-1)^3)
returns


ans =
n^2
```

```
ztrans(cos(w*t)+exp(-a*t)-sin(w*t))
returns
```

```
ans =
(-z+cos(t))*z/(-z^2+2*z*cos(t)-1)+exp(-a*t)*z/(z-1)+z*sin(t)/(-z^2+2*z*cos(t)-1)
iztrans(-z+cos(t))*z/(-z^2+2*z*cos(t)-1)+exp(-a*t)*z/(z-1)+z*sin(t)/(-z^2+2*z*cos(t)-1)
returns
```

```
ans =
(iztrans(-z,z,n)+charfcn[0](n)*cos(t))*z/(-z^2+2*z*cos(t)-1)+exp(-a*t)*z/(z-1)+z*sin(t)/(-
z^2+2*z*cos(t)-1)
```

So, Howzzaaattt? Pretty easy! Then let us see you finding out Fourier transform and its inverse similarly!

Q. When all models can be transformed to another type, which model should I choose to work as?

A. *MATLAB® converts all models to state space model internally and then carries out the desired operations on it.* Sometimes this conversion and re-conversion may result into loosing some data. Hence, *it is the best policy to work with state space model for highest accuracy in the result.*

Q. While trying simulation of Series RL circuit, through SIMULINK®, I observed that when I use Figure 8.12 (a) and (c) for simulation, the transient part in the response is missing and the scope shows a constant value of 10. Why is it so?

A. Oh! That shows you are taking things seriously and are quite observant too. That was really a question we had expected from a serious reader like you. By the way did you go back to the MATLAB® window and read the error message there? The answer lies in this statement. *The algebraic loop is actually responsible for eating up the exponentially rising part of the circuits response.* For more details about the algebraic loop refer the SIMULINK® manual. However, we will just drop you a hint for avoiding it. *Use integrating blocks instead of differentiating blocks for simulation.* As you go through other examples incorporated in this book, you will understand what we mean.

Q. Can you give me some more tips that would prove helpful to me while working?

A. Sure, why not! Here are a few of them:
   - *when you issue **rlocfind** function, make sure that you have a root locus plot in the Figure window. This function operates even if there is no root locus plot in the Figure window, but that may cause confusion at times.*
   - *while working with SIMULINK®, if the response on the scope does not appear, try reducing simulation time from the Parameter window or set the reduced time in the setting window of the scope.*
   - *if the response signal is not visible in the scope window of SIMULINK®, click the binocular icon.*
   - *for a MIMO system, keep the sample time same for all inputs and outputs. As MATLAB® is not multirate (but SIMULINK® is) hence for other operations also see that your models have same sample time.*
   - *while computing covariance of systems, MATLAB® gives error message if you have specified the delay of the system.*

- *in the MATLAB® command window you must already be using up and down arrow keys to save yourself from typing previous commands again and again at the prompt. However, pressing the arrows key say 20 times to obtain the command you had entered 20 steps earlier too can be very irritating. You can save some of this labour by just typing the first letter of the command in conjunction with the arrow keys when the number of steps involved reduces as MATLAB® writes only those commands at the prompt that start with the letter you typed.*

Q. While trying out simulation examples given in this book, I felt there are better ways of doing the same thing. Then why have you chosen a particular method of simulation only?

A. The Series RL circuit design example of Chapter 8 has been solved by many methods demonstrating the fact that a problem can be solved in several ways. As for the other examples of Chapter 8, in the book, only block diagram of the system has been given but the floppy contains simulation of the system using transfer function model as well as state variable model. You will observe that all the models produce the same result. *As far as the block diagram representation of models is concerned, even that can be done in several ways again to get the same result.* To emphasise this point, let us consider the Mass-spring System whose demo is presented in the MATLAB® Demo window under Examples and Demos sub-menu of help main menu. The block diagram representation of the system as given in MATLAB® demos is shown in Figure E.4. The same system can be represented by block diagram shown in Figure E.5. However, the result of the two representations is the same as shown in Figure E.6.

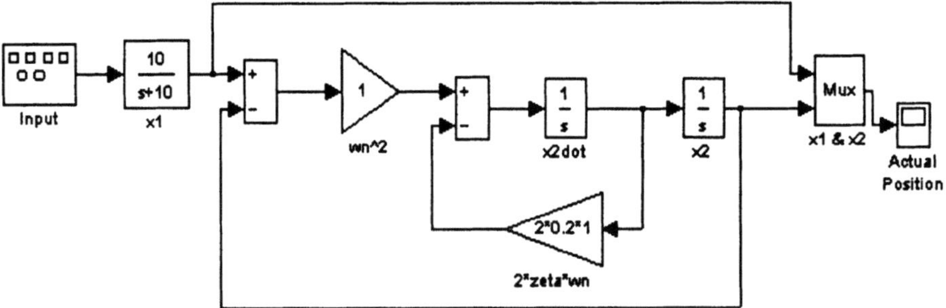

Figure E.4. Representation of Mass-spring System as given in MATLAB® demo

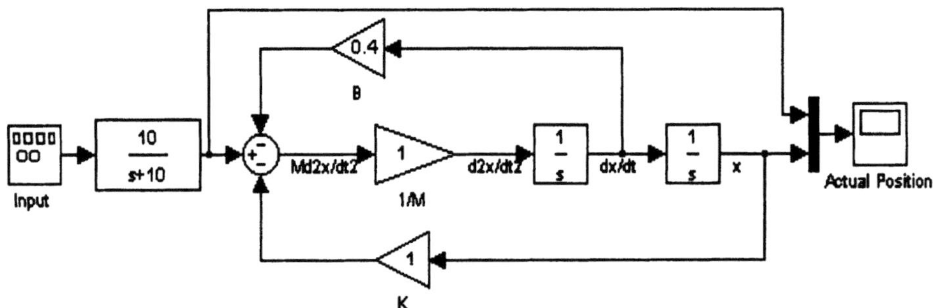

Figure E.5. Different way of representation of Mass-spring System

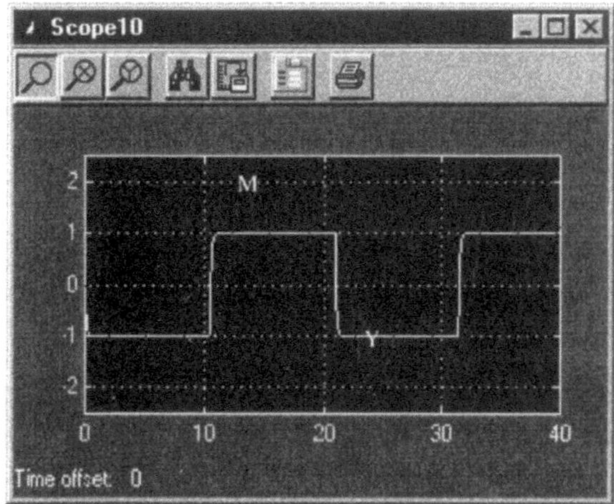

Figure E.6. Response of Mass-spring System by any method of representation to square input

Q. Sometimes the response of a system displayed on the MATLAB® window screen is too large. The screen goes on scrolling, and when it stops, I find that the initial data is not visible on the screen. Is there a way out by which I can view the screen page wise as I used to do in DOS by attaching **/p** to the command?

A. MATLAB® has a built-in function **more** for you to solve your problem by directing MATLAB® to show one screen at a time. All you have to do is to type **more on** at the command prompt to achieve this. When you are done and wish to return back to normal type **more off** at the command prompt and that's it. *If you want only a particular number of lines to be displayed on the screen, type **more(N)** at the prompt where N is the numeral specifying the number of lines you wish to be displayed on the screen at one time.*

Q. I found out transfer functions of many systems as you had explained earlier. But, the system on which I am working right now, has equations containing numeric values only as given below, and I wish to find out the roots of these equations. Do I have to substitute a symbol for each numeric and then proceed as directed above using functions provided in the Symbolic Toolbox?

5x+4y+6z=9
x-y-z=4
16x+14y-8z=2

A. No. You are getting confused and forgetting the basics of MATLAB®. Okay, let us revise the procedure for solving simultaneous equations:

```
%arrange the Right Hand Side and Left Hand Side of equations in matrix form
a=[5 4 6 ; 1 -1 -1 ; 16 14 -8 ];
b=[9; 4; 2];
%obtain the roots
x=a\b
%check the roots a*x should be equal to b
a*x
returns

x =
```

```
    2.5814
   -2.3023
    0.8837
ans =
    9.0000
    4.0000
    2.0000
```

Remember that Symbolic Math Toolbox has to be used when you are dealing with arithmetic expressions involving symbols instead of numeric values. While for expressions involving numeric values, you use the same old tricks that you learnt while doing MATLAB®.

Q. When I try to open the examples given in the floppy under Ch7-SIMULINK, I get blocks having question marks like shown below and two-error message block also appears. Why is this happening?

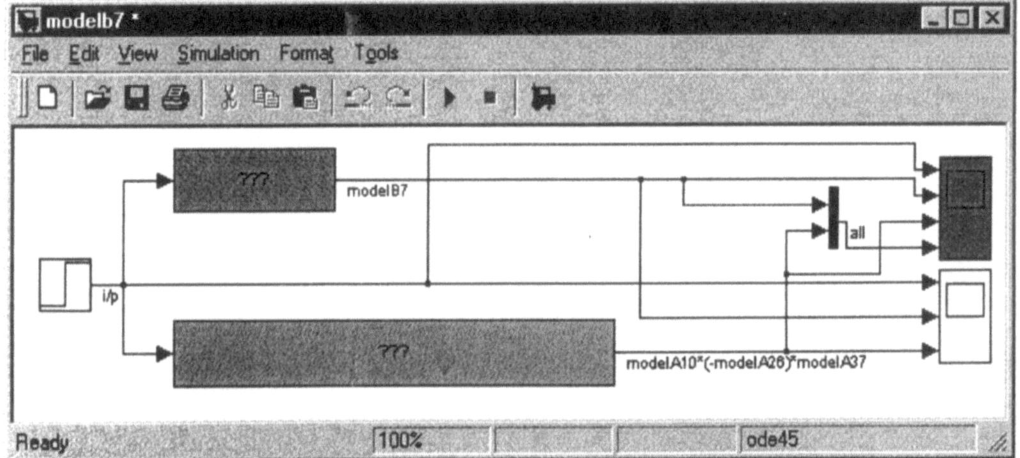

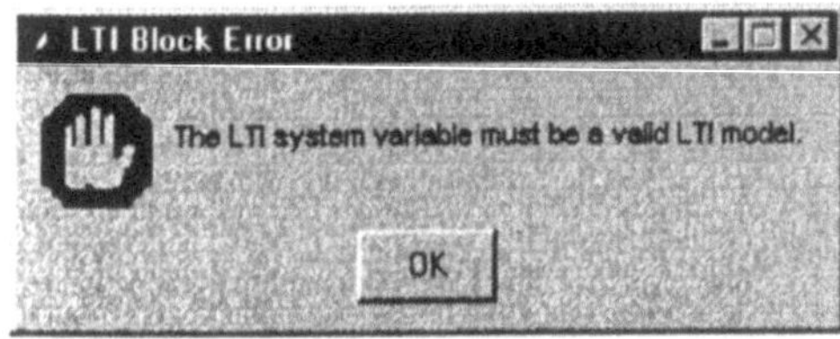

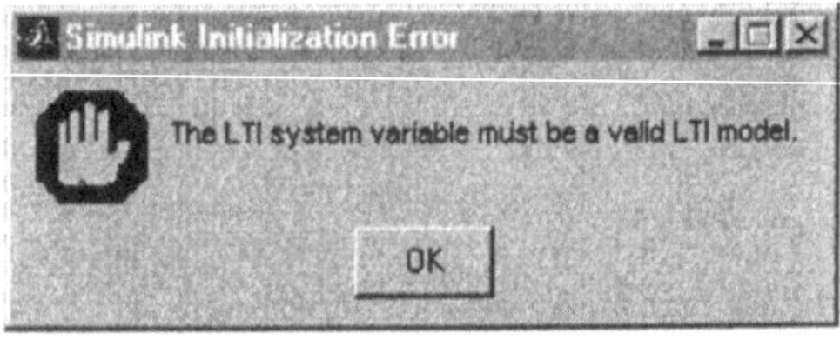

A. *Whenever you are using LTI system block in SIMULINK® window for modelling, and entering a model name existing in the workspace, be sure to first load the workspace containing the model before you open the SIMULINK® file containing that model name.* In the example you have quoted, you should first load Chapter1_2 workspace before opening modelB7 or any other present in Ch6-SIMULINK directory. For loading the workspace go to the file menu of MATLAB® window and choose the load workspace sub-menu. In the Window Explorer click Chapter1_2.MAT file. Now, when you open SIMULINK® file modelB7, your window will appear like given below without any error message window popping up.

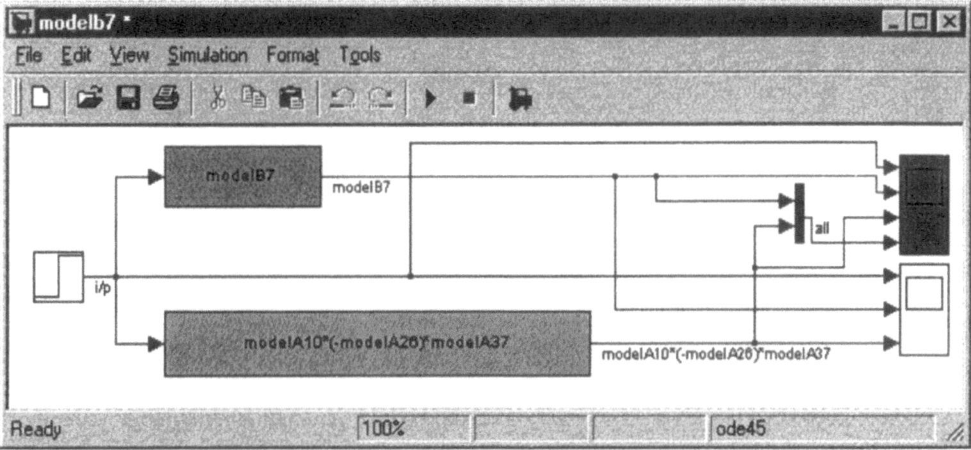

Q. While trying out examples of model manipulation using SIMULINK®, I tried to simulate product of modelA1 and modelA17 in three ways to obtain response same as modelB6 as shown in Figure given below. I obtained correct result in case (a) and (b) but model (c) gives me a wrong result (see the results in the scope window). I cannot understand the reason for this. Please explain.

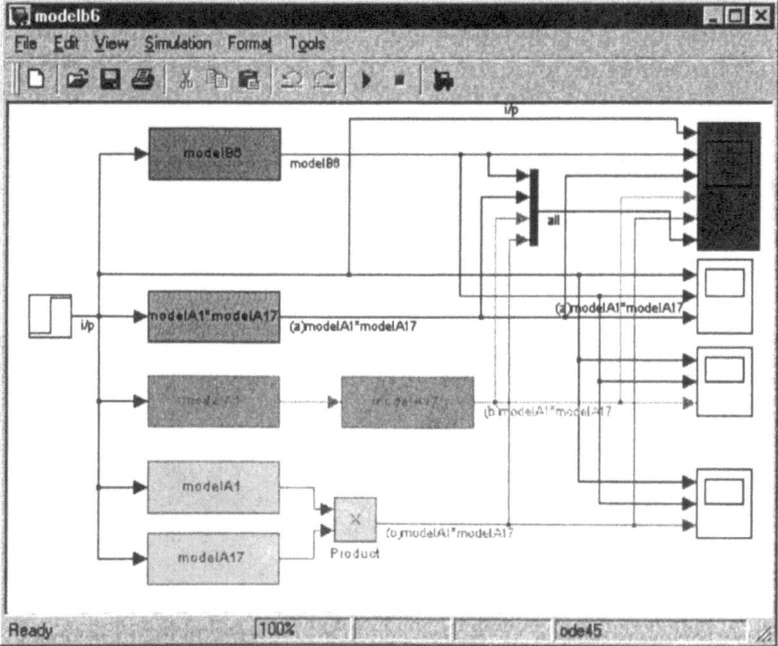

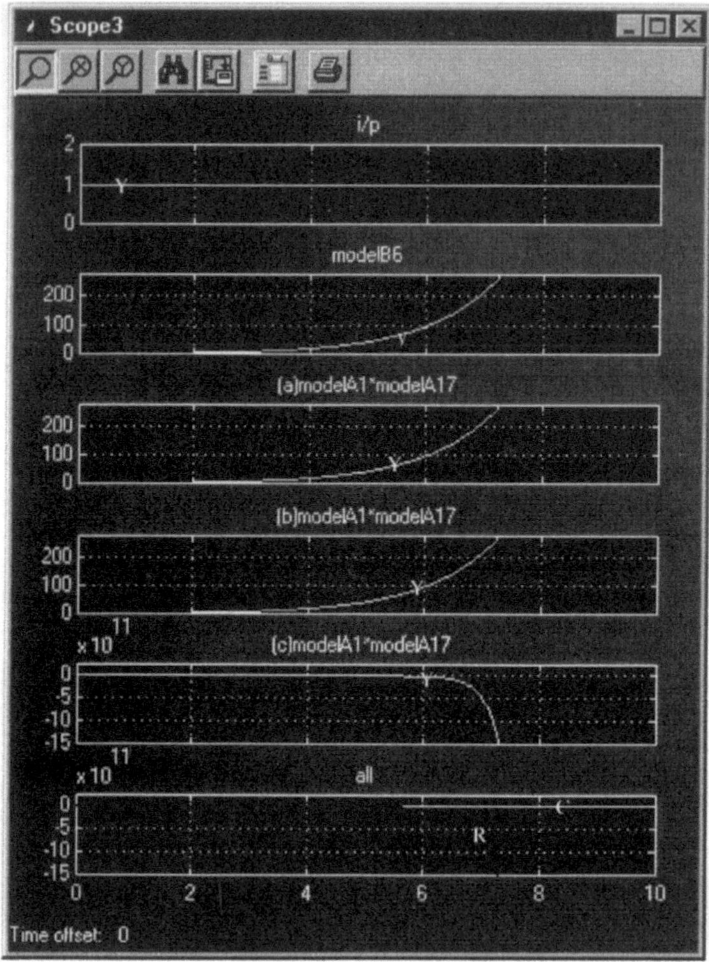

A. *Actually, in case (c) you are multiplying the signal outputs from the blocks of modelA1 and modelA17. In other cases the models were being multiplied and the signal obtained was the output signal of the overall product block instead of those of individual signals.* And hence the difference.

Q. While working with **rlocfind** function, I entered the following at the command prompt:

[k,Poles]=**rlocfind**(model1)

Contrary to what is written in this book, regarding this function, all I found was a blank Figure window as shown below:

Though a message appeared in the MATLAB®'s command window saying

'Select a point in the graphics window'

but I could not even find the cross-hair in the Figure window. When I tried moving the mouse over the window area, the cursor disappeared. However, when I clicked the mouse accidentally, I did get a result as follows:

selected_point =
  0.4816 + 0.6257i
k =
  0.3227
poles =
  0.5305

Can you explain me what happened?

A. Sure. We are here for just that! That is to solve your problems regarding Control System Toolbox of MATLAB®. *First of all change the color of the Figure window to white.* Hope you remember how to do that. *Now, again enter the command at the prompt.* I think now the crosshair is visible to you. *Click at any point now, and you get the associated values of gain and pole at that point.*

Even though you got an answer, but here is a word of caution. The function **rlocus** assumes that there is a root locus plot already in the Figure window. *So, before entering the rlocus function at the prompt, first draw a root locus using rlocus function. Then, a root locus plot will be visible in your Figure window and you can then choose a point placing the cross-hair to obtain the associated values of gain and poles with the selected point.*

Q. I downloaded all the material provided on the web-site in a separate directory in the hard-disk on the desktop. When I accessed files from this directory, some files open fine but others created problem. What is going wrong?

A. If you have trouble with opening any file (workspace files like Chapter1_2_3_4 or Octave_Sound *etc.,* sometimes seem to create problem) try copying them in **work** folder/directory and then open the contents of these files directly from there.

References

MATLAB® On-line Help Manuals Provided by The MathWorks, Inc.

MATLAB® : The Language of Technical Computing (1998). Getting Started with MATLAB® Version 5.

MATLAB® : The Language of Technical Computing (1999). Using MATLAB® Version 5.3.

Symbolic Math Toolbox: For Use with MATLAB® (1998). User's Guide Version 2.

MATLAB® : The Language of Technical Computing (1999). Using MATLAB® Graphics Version 5.3.

MATLAB® : The Language of Technical Computing (1999). Installation Guide for PC Release 11 (MATLAB® 5.3 Product Family)

MATLAB® : The Language of Technical Computing (1998). MATLAB® Notebook User's Guide Version 5.2

MATLAB® : The Language of Technical Computing (1999). MATLAB® Function Reference (Volume 1: Language)Version 5.

MATLAB® : The Language of Technical Computing (1999). MATLAB® Function Reference (Volume 1: Language) Version 5.

MATLAB® : The Language of Technical Computing (1999). MATLAB® Function Reference (Volume 2: Graphics) Version 5.

Control System Toolbox: For Use with MATLAB® (1999). User's Guide Version 4.2.

SIMULINK®: Dynamic System Simulation for MATLAB® (1999). Using SIMULINK® Version 3.

Books

Rudra Pratap (1998) Getting Started with MATLAB® 5. Oxford University Press, Inc. New York.

William S. Levine (1996) The Control Handbook (The electrical engineering handbook series). CRC Press, Inc. Florida.

J. Lowen Shearer, Bohdan T. Kulakowski, John F. Gardner (1997) Dynamic Modeling and Control of Engineering Systems (ed. 2). Prentice-Hall, Inc. New Jersey.

Stanley M. Shinners (1992) Modern Control System Theory and Design. John Wiley & Sons, Inc. New York Chichester Brisbane Toronto Singapore.

Raymond T. Stefani, Clement J. Savant, Jr., Bahram Shahian, Gene H. Hostetter (1994) Saunders College Publishing, Harcourt Brace College Publishers Boston New York.

Philip Thomas (1999) Simulation of Industrial Processes for Control Engineer. Butterworth Heinmann Oxford Auckland Boston Johannesburg Melbourne New Delhi

John Van De Vegte (1990) Feedback Control Systems (ed. 2). Prentice-Hall, Inc. New Jersey.

Benjamin C. Kuo (1995) Automatic Control Systems (ed. 7) Prentice-Hall, Inc. New Jersey.

Katsuhiko Ogata (1987) Discrete-Time Control Systems. Prentice-Hall, Inc. New Jersey.

Katsuhiko Ogata (1995) Modern Control Engineering (ed. 2) Prentice-Hall, Inc. New Jersey.

Rolf Isermann (1989) Digital Control Systems Volume I and Volume II. Springer-Verlag Berlin Heidelberg.

William J. Palm III (1986) Control Systems Engineering. John Wiley & Sons, Inc. New York Chichester Brisbane Toronto Singapore.

Peter Harriott (1997) Process Control. Tata McGraw-Hill Publishing Company Limited. New Delhi.

Ernest O. Doebelin (1985) Control System Principles and Design. John Wiley & Sons, Inc. New York Chichester Brisbane Toronto Singapore.

M. Gopal (1997) Digital Control and State Variable Methods Tata McGraw-Hill Publishing Company Limited. New Delhi.

M. Gopal (1989) Modern Control System Theory. Wiley Eastern Limited. New Delhi.

I.J. Nagrath, M. Gopal (1987) Control Systems Engineering (ed. 2). Wiley Eastern Limited New Delhi.

I.J. Nagrath, M. Gopal (1982) Systems Modelling and Analysis. Tata McGraw-Hill Publishing Company Limited. New Delhi.

Prabha Kundur (1994) Power System Stability and Control McGraw-Hill, Inc. New York.

D.P. Sen Gupta, J.W. Lynn (1980) The Macmillan Press Limited. London and Basingstoke.

Paul C. Krause (1987) Analysis of Electric Machinery. McGraw-Hill, Inc. New York.

Herman E. Koenig, Yilmaz Tokad, Hiremaglur K. Kesavan, Harry G. Hedges (1967) McGraw-Hill, Inc. New York

Arthur B. Williams, Fred J. Taylor (1995) Electronic Filter Design Handbook (ed. 3) McGraw-Hill, Inc. New York

Journals

Chi-Jui Wu, Shih-Shong Yen, Wei-Nan Chang, Chui-Nan Chang, Chao-Hui Li, Tzong-Yih Guo, Enhancement of static excitation system performance for generators near electric arc furnace loads. IEEE Transactions on Energy Conversion, Vol. 14 No.2, June1999.

Index